Vector Space Projections

WILEY SERIES IN TELECOMMUNICATIONS AND SIGNAL PROCESSING

John G. Proakis, Editor
Northeastern University

Introduction to Digital Mobil Communications
Yoshihiko Akaiwa
Digital Telephony, 2nd Edition
John Bellamy
Elements of Information Theory
Thomas M. Cover and Joy A. Thomas
Practical Data Communications
Roger L. Freeman
Radio System Design for Telecommunications, 2nd Edition
Roger L. Freeman
Telecommunication System Engineering, 3rd Edition
Roger L. Freeman
Telecommunications Transmission Handbook, 4th Edition
Roger L. Freeman
Introduction to Communications Engineering, 2nd Edition
Robert M. Gagliardi
Optical Communications, 2nd Edition
Robert M. Gagliardi and Sherman Karp
Active Noise Control Systems: Algorithms and DSP Implementations
Sen M. Kuo and Dennis R. Morgan
Mobile Communications Design Fundamentals, 2nd Edition
William C. Y. Lee
Expert System Applications for Telecommunications
Jay Liebowitz
Digital Signal Estimation
Robert J. Mammone, Editor
Digital Communication Receivers: Synchronization, Channel Estimation, and Signal Processing
Heinrich Meyr, Marc Moeneclaey, and Stefan A. Fechtel
Synchronization in Digital Communications, Volume I
Heinrich Meyr and Gerd Ascheid
Business Earth Stations for Telecommunications
Walter L. Morgan and Denis Rouffet
Wireless Information Networks
Kaveh Pahlavan and Allen H. Levesque
Satellite Communications: The First Quarter Century of Service
David W. E. Rees
Fundamentals of Telecommunication Networks
Tarek N. Saadawi, Mostafa Ammar, with Ahmed El Hakeem
Meteor Burst Communications: Theory and Practice
Donald L. Schilling, Editor
Vector Space Projections: A Numerical Approach to Signal and Image Processing, Neural Nets, and Optics
Henry Stark and Yongyi Yang
Signaling in Telecommunication Networks
John G. van Bosse
Telecommunication Circuit Design
Patrick D. van der Puije
Worldwide Telecommunications Guide for the Business Manager
Walter H. Vignault

Vector Space Projections

A Numerical Approach to Signal and Image Processing, Neural Nets, and Optics

Henry Stark

Yongyi Yang

Illinois Institute of Technology

A Wiley-Interscience Publication

JOHN WILEY & SONS, INC.

New York / Chichester / Weinheim / Brisbane / Singapore / Toronto

Copyright © 1998 by John Wiley & Sons, Inc.

Library of Congress Cataloging in Publication Data:

Stark, Henry, 1938–
     Vector space projections : a numerical approach to signal and
image processing, neural nets, and optics / Henry Stark and Yongyi
Yang.
         p.   cm. — (Wiley series in telecommunications and signal
processing)
     "A Wiley-Interscience publication."
     Includes bibliographical references and index.
     ISBN 0-471-24140-7 (cloth : alk. paper)
     1. Image processing—Digital techniques.   2. Signal processing—
Digital techniques.   3. Neural networks (Computer science)
I. Yang, Yongyi.   II. Title.   III. Series.
TA1637.S727   1998
621.36'7—dc21
                                            98-10921
                                            CIP

Printed in the United States of America

10 9 8 7 6 5 4 3 2 1

To
Dan Youla
and
Heywood Webb (in memory)
Primus inter pares
and
Shujun, Xianglun,
and my parents, Dingshi and Jinrong

Contents

Preface   xi

1 Vector Space Concepts   **1**
   1.1   Introduction   1
   1.2   Vector Spaces   1
   1.3   Normed Vector Spaces   6
   1.4   Inner Product Spaces   13
   1.5   Hilbert Spaces   21
   1.6   Summary   29
       References   29

2 Projections Onto Convex Sets   **33**
   2.1   Introduction   33
   2.2   Convex Sets   33
   2.3   Projection   35
   2.4   Properties of Projections   42
   2.5   Fundamental Theory of POCS   49
   2.6   The POCS Algorithm and Its Numerical Aspects 57
   2.7   Application of POCS Method   67
   2.8   Proof of Opial's Theorem   72
   2.9   POCS Algorithm Formulated in a Product Space 78

2.10 *Projection Algorithms in the Presence of Non-intersecting Sets*   82
2.11 *Summary*   86
      References   87

3 Elementary Projectors   **91**
3.1 *Introduction*   91
3.2 *Linear Type Constraint*   91
3.3 *Soft Linear Type Constraint*   97
3.4 *Error Type Constraint*   102
3.5 *Similarity Constraint*   109
3.6 *Time-Domain Constraints*   114
3.7 *Frequency-Domain Constraints*   119
3.8 *Summary*   125
     Appendix   126
     References   130

4 Solutions of Linear Equations   **133**
4.1 *Introduction*   133
4.2 *Convex-Set Formulation*   136
4.3 *Convergence Analysis*   142
4.4 *Convergence Acceleration*   148
4.5 *Minimum Distance Property*   154
4.6 *A Hybrid Algorithm*   157
4.7 *Application Example: Computed Tomography*   162
4.8 *Application Example: Resolution Enhancement*   170
4.9 *Summary*   176
     References   177

5 Generalized Projections   **181**
5.1 *Introduction*   181
5.2 *Convex versus Non-Convex Sets*   182
5.3 *Examples of Non-convex Sets*   184
5.4 *Theory of Generalized Projection*   185
5.5 *Traps and Tunnels*   192
5.6 *Proof of Theorem 5.4-2*   193
5.7 *Generalized Projection Algorithm in a Product Vector Space*   196
5.8 *Summary*   202

References   202

6 Applications to Communications   **205**
  6.1 Introduction   205
  6.2 Non-uniform Sampling   205
  6.3 Digital Communications   211
  6.4 Digital Filters   221
  6.5 Estimation of Probability Density Functions   239
  6.6 Spread Spectrum   249
  6.7 Image Compression   262
  6.8 Summary   272
       References   275

7 Applications to Optics   **281**
  7.1 Introduction   281
  7.2 Image Sharpening   282
  7.3 The Phase Retrieval Problem   285
  7.4 Beam Forming   288
  7.5 Color Matching   291
  7.6 Blind Deconvolution   295
  7.7 Design of Diffractive Optical Elements   302
  7.8 Summary   312
       References   312

8 Applications to Neural Nets   **319**
  8.1 Introduction   319
  8.2 The Brain as a Neural Network   319
  8.3 The Hopfield Net   321
  8.4 POCS and the Continuous Hopfield ACAM   328
  8.5 A Hopfield-Net Based Classifier   331
  8.6 Learning in Multi-Layer Nets   337
  8.7 Back-Propagation Rule for the Two-Layer Net   339
  8.8 Projection Method Learning Rule   343
  8.9 Experiments and Results   346
  8.10 Summary   357
       References   358

9   Applications to Image Processing   **363**

   9.1   Introduction   363

   9.2   Noise Smoothing   364

   9.3   Image Synthesis   369

   9.4   Restoration of Blurred and Noisy Images   375

   9.5   High-Resolution Image Restoration   382

   9.6   Restoration of Quantum-Limited Images   389

   9.7   Summary   395

       References   397

Index   401

Preface

This book was written in response to the growing demand for a source that explains and illustrates the method of vector-space projections, especially the method of *projections onto convex sets* (POCS), a Hilbert space approach for problem solving.

We believe that vector-space projections is an exceptionally useful problem-solving technique that finds a spectrum of applications in basic problems in science and engineering. Indeed as the method is applied to an ever larger collection of problems, it is discovered that many convergent iterative algorithms already in existence are vector-space projection algorithms.

Vector-space projection algorithms are usually iterative and involve a great deal of number crunching. Without computers, vector-space projection methods would be primarily of theoretical interest. Fortunately, as modern computers have become more powerful, it is rare indeed when a personal computer (PC) or modest work-station is insufficient to solve a projection problem in a reasonable time. All of the examples furnished in this book were solved on PCs or low-end workstations.

As a rule vector-space projection methods do not furnish "optimum" solutions such as those associated with minimum-mean square error, maximum entropy, maximum likelihood, maximum *a posteriori* estimation, maximum signal-to-noise, etc. Vector-space projection methods, specifically convex projection methods, always yield a solution consistent with a set of constraints furnished by the user. Of course the constraints themselves must be consistent. For example, one cannot say "I seek to find a continuous, non-trivial, function that is both absolutely time-limited and band-limited." A function with such properties will not be found, since it is easy to demonstrate that only the trivial zero function can have both properties simultaneously.

To the best of the authors' knowledge the fundamental theory of vector-space projection methods was developed in the 1960s by Gubin et al. [1], Bregman [2], Halperin [3] and Opial [4]. In turn, these researchers built on von Neumann's work on alternating orthogonal projections in 1950 [5]. Major additional contributions to the theory were furnished by Youla [6] in the late 1970s and Youla [7] and Levi [8] in the early 1980s. Since then the theory has been extended by a large number of workers from all over the world. Indeed much research on refining, extending, and applying projection methods is going on at present. It would be foolish and arrogant on our part to try to list all of the important contributions made in this area in recent years. This book is very far from being a survey of the work going on in vector-space projection methods. Our goal in writing this book was to produce a self-teaching text with a large number of meaningful examples. Each example highlights some interesting aspect of the method or suggests a new algorithm.

A self-teaching text must begin with basic background material before going on to more advanced topics. For this reason we include some basic material from vector space theory early on. It is assumed that the reader has, at the least, a first degree in a physics, mathematics, engineering, or allied discipline. Knowledge of calculus is essential, and some exposure to linear algebra helps also.

In Chapter 1 we review vector space concepts, bases, dimensionality, inner products and inner-product spaces, and Hilbert spaces. Chapter 2 introduces the notion of convexity, projectors and projections, properties of projections, the fundamental theorem of vector-space projections, some elementary examples, and a proof of the fundamental theorem that can easily be omitted on a first reading without affecting one's understanding of subsequent material. In Chapter 3 we introduce a number of useful constraint sets in vector spaces. We show how to determine set properties such as convexity and closedness and demonstrate how to compute the associated projectors. We derive a significant number of projectors of use in signal processing and subdivide these into time-domain and frequency-domain projections. A number of elementary, but non-trivial examples are furnished.

When solved on a computer, a large number of engineering and scientific problems involve the solution of linear equations. In Chapter 4 we show why the method of vector-space projections is ideally suited to solve such equations, especially when additional constraints exist. An example of how this method reconstructs an image in computerized tomography from detector readings is furnished. In Chapter 5 we discuss an important result of vector-space projections called generalized projections (GP). In many problems the appropriate constraints sets are not convex and in such cases we cannot apply a POCS algorithm and expect convergence. The theory of generalized projections demonstrates that a restricted type of convergence, called *summed distance error convergence*, is possible under certain conditions. Chapter 6 is the first of four applications chapters in which we illustrate how the method of vector-space projections can solve problems of importance in various areas of science and engineering. In particular, in Chapter 6 we apply the method to signal reconstruction from non-uniform samples; digital communications corrupted by quantizing noise; digital filter design; artifact reduction in image compression; and spread-spectrum signal recovery in the presence of natural and jammer noise. Applications of the method of vector-space projections to optics are illustrated in Chapter 7. There we apply projection methods to the superresolution problem; (resolution beyond the diffraction-limit); the phase retrieval problem; beam forming; color matching; blind-deconvolution; and design of diffractive optics. In Chapter 8, projection methods are applied to neural nets and pattern recognition systems. It is shown that the operation of the synchronous binary Hopfield net is a generalized projection algorithm, which explains why some stable states may not be legitimate solutions. The classic perceptron algorithm for linearly separable pattern classes is shown to be an elementary convex projection algorithm, which can be improved by using additional prior knowledge. Lastly we use projection methods to derive a learning rule for feed-forward nets with hidden layers. As in all the application chapters, enough background material is furnished so that the reader need not have prior knowledge of the particular area in which the application is illustrated.

Finally, in Chapter 9 we describe applications of projection methods to the popular field of image processing. We apply these methods in connection with noise-smoothing, image synthesis, restoration of blurred and noisy images, high-resolution imaging, and restoration of quantum-limited images.

No attempt is made to be complete regarding the many interesting and novel applications of vector-space methods to the various topics discussed in this book and beyond. The book was not intended as a survey, and we sincerely hope that any omission to published work is not interpreted as a value judgement on our part.

Problems have been added at the end of some chapters to enable this book to be used as a text in a course on projection methods.

The authors are grateful to the many people who provided the time and facilities needed for creating this book. Bonnie Dow and Jane Wurster ably assisted in the difficult typing of the manuscript. Thanks are due to the National Science Foundation and the Department of Defense, for supporting much of the research reported in the text. Our greatest debt, however, is reserved for our many present and former students, whose work and insight have contributed to the education of their teacher. D. Cahana, S. Cruze, M.I. Sezan, A. Levi, E. Yudilevich, H. Peng, P. Oskoui-Fard, Shu-jen Yeh, W. Catino, Jia Pang, J.L. Wurster and K. Haddad deserve special mention.

REFERENCES

1. L. G. Gubin, B. T. Polyak, and E. V. Raik, The method of projections for finding the common point in convex sets, *USSR Comput. Math. Phys.*, **7**(6):1-24, 1967.

2. L. M. Bregman, Finding the common point of convex sets by the method of successive projections, *Dokl. Akad. Nauk. USSR*, **162**(3):487-490, 1965.

3. I. Halperin, The product of projection operators, *Acta Sci. Math.,* **23**:96-99, 1960.

4. Z. Opial, Weak convergence of the sequence of successive approximation for nonexpansive mappings, *Bull. Am. Math. Soc.,* **73**:591-597, 1967.

5. J. von Neumann, *Functional Operators,* vol. II of *Ann. Math. Stud.,* **22**(55), Princeton University Press, Princeton, N. J., 1950.

6. D.C. Youla, Generalized image restoration by the method of alternating orthogonal projections, *IEEE Trans. Circ. and Syst.,* **25**:694-702, Sept. 1978.

7. D.C. Youla and H. Webb, Image restoration by the method of convex projections: Part 1–theory," *IEEE Trans. Med. Imaging,* **MI-1**:81-94, Oct. 1982.

8. A. Levi and H. Stark, Signal restoration from phase by projections onto convex sets, *J. Opt. Soc. Am.,* **73**:810-822, 1983.

About the Authors

Henry Stark received the B.S. degree from the City College of New York and the M.S. and Dr. Eng. Sc. degrees from Columbia University, all in electrical engineering. He is the Bodine Professor of Electrical and Computer Engineering at the Illinois Institute of Technology and, formerly, the department chairman. His interests are in image recovery, signal processing, pattern recognition, medical imaging, and optical information processing. He has written some 130 papers in these areas and received an outstanding paper award from the IEEE Engineering in Medicine and Biology Society in 1980. In addition to co-authoring this text, Dr. Stark is a co-author of *Modern Electrical Communications* (Prentice-Hall, 1988), *Probability, Random Processes and Estimation Theory for Engineers* (Prentice-Hall, 1994) and the editor of two books on *Optical Fourier Transforms* and *Image Recovery*, both published by Academic Press. More recently, he co-edited (with P. M. Clarkson) the book *Signal Processing Methods for Audio, Images and Telecommunications* (Academic Press, 1995). Together with his students, he has written numerous book chapters on topics ranging from neural nets to sampling theory.

Dr. Stark is a former *Associate Editor of the IEEE Transaction on Medical Imaging* and a former Topical Editor on Imaging Processing for the *Journal of the Optical Society of America*. He is a consulting editor for the Oxford University Series on Optics and the Academic Press Series on Signal Processing. He is a Fellow of the IEEE and the Optical Society of America.

Yongyi Yang received his B.S. and M.S. degrees in electrical engineering for Northern Jiaotong University, Beijing, China. He received a M.S. degree in mathematics from the Illinois Institute of Technology (IIT) in 1992, and a Ph.D. in electrical engineering from the same institution in 1994. His Ph.D. thesis,"Projection-

Based Signal Processing for Visual Communication Problems," discusses in detail some of the problems discussed in this book.

In 1995, he joined the electrical engineering department at IIT, where he is currently an assistant professor. He is a principal contributor to the edited book *Signal Processing Methods for Audio, Images, and Telecommunications*, by P. Clarkson and H. Stark, published by Academic Press.

Dr. Yang has strong interests and publications in image processing, video compression, computer vision, applied mathematical and statistical methods, and undergraduate education.

1

Vector Space Concepts

1.1 INTRODUCTION

In this chapter we provide some basic concepts and results pertaining to linear vector spaces. We shall need this material to better understand the topics in subsequent chapters. Our treatment will be necessarily brief, yet broad enough to establish a base for the development of the theory of vector-space projections in upcoming chapters. While the symbols and theorems may look daunting, the material is readily available to anyone with a first degree in engineering or science. Further discussion of linear vector spaces can be found in standard texts on linear algebra and linear functional analysis. To name a few, see [1] to [5], for example.

1.2 VECTOR SPACES

Definitions

A *vector space* $\mathbf{V}$ is a non-empty collection of elements on which two operations (called *addition* and *scalar multiplication*, respectively) are defined so that for each pair of elements $\mathbf{x}, \mathbf{y}$ in $\mathbf{V}$ there is a unique element $\mathbf{x} + \mathbf{y}$ (called the *sum* of $\mathbf{x}$ and $\mathbf{y}$) in $\mathbf{V}$, and for each scalar (number) α and each element $\mathbf{x}$ in $\mathbf{V}$ there is a unique element $\alpha\mathbf{x}$ (called the *scalar multiple* of $\mathbf{x}$ by α) in $\mathbf{V}$, such that the following properties hold:

(i) For each pair of elements $\mathbf{x}$, $\mathbf{y}$ in $\mathbf{V}$, $\mathbf{x} + \mathbf{y} = \mathbf{y} + \mathbf{x}$.

(ii) For all $\mathbf{x}$, $\mathbf{y}$, $\mathbf{z}$ in $\mathbf{V}$, $(\mathbf{x} + \mathbf{y}) + \mathbf{z} = \mathbf{x} + (\mathbf{y} + \mathbf{z})$.

(iii) There exists an element $\mathbf{0}$ in $\mathbf{V}$, called the *zero vector* for $\mathbf{V}$, such that $\mathbf{x} + \mathbf{0} = \mathbf{x}$ for each $\mathbf{x}$ in $\mathbf{V}$.

(iv) For each element $\mathbf{x}$ in $\mathbf{V}$ there exists an element $\mathbf{y}$ in $\mathbf{V}$ such that $\mathbf{x} + \mathbf{y} = \mathbf{0}$.

(v) For each pair of scalars α, β and each $\mathbf{x}$ in $\mathbf{V}$, $(\alpha\beta)\mathbf{x} = \alpha(\beta\mathbf{x})$, and $\alpha(\mathbf{x} + \mathbf{y}) = \alpha\mathbf{x} + \alpha\mathbf{y}$.

(vi) For each pair of scalars α, β and each $\mathbf{x}$ in $\mathbf{V}$, $(\alpha + \beta)\mathbf{x} = \alpha\mathbf{x} + \beta\mathbf{x}$.

(vii) For each element $\mathbf{x}$ in $\mathbf{V}$, $1\mathbf{x} = \mathbf{x}$ for the scalar 1.

In the literature, vector spaces are also called *linear spaces* or *linear vector spaces*. The elements $\mathbf{x}$, $\mathbf{y}$, $\mathbf{z}$, etc. of a vector space are called *vectors*. Depending on the application, scalars used in a vector space may be real numbers or complex numbers. Vector spaces in which the scalars are complex numbers are called *complex vector spaces*, and those in which the scalars must be real are called *real vector spaces*.

In a vector space $\mathbf{V}$, a *set* (or to be exact, a *subset* of $\mathbf{V}$) is defined to be a collection of vectors in $\mathbf{V}$. The vectors that make up a set are called the *elements* of the set. For a set S in $\mathbf{V}$, a vector $\mathbf{x}$ is said *to belong to S* if $\mathbf{x}$ is an element of S, which is denoted by $\mathbf{x} \in S$; otherwise, it is said *not to* belong to S, which is accordingly denoted by $\mathbf{x} \notin S$. The *complement* of set S, denoted by S^c, is defined to be the set of all the vectors in the space $\mathbf{V}$ but not in S. Also, a set A is said to be a subset of a set B, which is denoted by $A \subset B$ or, equivalently, $B \supset A$, if every element of set A is also an element of set B. The *intersection* of two sets A and B, denoted by $A \cap B$, is the set of elements that belong to both A and B. Sometimes $A \cap B$ is written as AB. The *union* of two sets A and B, denoted by $A \cup B$, is the set of elements that belong to either A or B or both.

In a vector space $\mathbf{V}$, a subset $\mathbf{W}$ is called a *subspace* of $\mathbf{V}$ if $\mathbf{W}$ itself is a vector space under the same operations of addition and scalar multiplication defined on $\mathbf{V}$. Note that in order to prove that a subset $\mathbf{W}$ is a subspace it is not necessary to verify all the defining properties for a vector space. This is true because some of the properties that hold for elements of $\mathbf{V}$ automatically hold for that of $\mathbf{W}$. As a matter of fact, the following result holds:

Theorem 1.2-1 *If $\mathbf{W}$ is a set of one or more vectors from a vector space $\mathbf{V}$, then $\mathbf{W}$ is a subspace of $\mathbf{V}$ if and only if the following conditions hold:*

(a) *If $\mathbf{x}$ and $\mathbf{y}$ are in $\mathbf{W}$, then $\mathbf{x} + \mathbf{y}$ is also in $\mathbf{W}$.*

(b) *If $\mathbf{x}$ is in $\mathbf{W}$, then $\alpha\mathbf{x}$ is also in $\mathbf{W}$ for every scalar α.*

The proof of this result is left as an exercise to the interested reader.

Discussion

At first glance, a reader might find the defining properties for a vector space rather overwhelming. In fact, all these properties can be naturally abstracted from physical

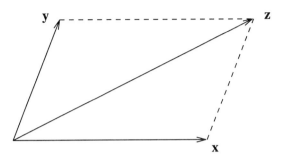

Fig. 1.2-1   A swimmer, swimming at velocity **y** relative to water flowing at velocity **x**, appears to be moving at velocity **z** = **x** + **y** to an observer on the shore.

observations in science and engineering. This is, perhaps, the major reason that linear algebra has become indispensable to modern scientists and engineers. If we examine closely the defining properties for a vector space, we find that the first four properties define how the vectors in a vector space interact among themselves, while the remaining ones define how they interact with scalars. In the following we present two practical examples where these properties agree with the laws of physics.

For the first example, let's consider the situation of a swimmer swimming in a river. Assume the water is running at a velocity **x** and the swimmer is swimming at a velocity **y** *relative* to the water. From physics we know that the velocities **x** and **y** can be represented by arrows (vectors), of which the length represents the speed and the direction represents the direction of the motion. In such a case, to an observer sitting on the river bank, the swimmer actually appears to be moving at the velocity of **z** = **x** + **y**, which is obtained via the parallelogram law (which obeys the defining properties (i) and (ii) of a vector space), as illustrated in Fig. 1.2-1. In the instance that the river is still, i.e., **x** = **0**, we have **z** = **y**. In words, the swimmer will appear to be moving at his real velocity to the observer (which obeys property (iii)). In the instance that the swimmer is swimming upstream, against the current, at the same speed as that of the running water, the swimmer will appear stationary to the observer (which obeys property (iv)).

For the second example we examine the behavior of electrical signals in electronic circuits and systems. It is known that an electric signal such as voltage can be viewed as a variable which is a function of time, say $v(t)$. As will be clear from examples of vector spaces to be furnished later in this section, such a signal can be treated as a vector in a vector space. From basic circuit theory it is well known that electrical signals can be superimposed together to yield new signals, of which the underlying principle completely agrees with the first four defining properties of a vector space. Also, an electric signal, say $v(t)$, can be amplified by an amplifier of gain α to yield a new signal $\alpha v(t)$ (which corresponds to scalar multiplication in a vector space). Clearly, if $\alpha = 1$, we will have property (vii). Furthermore, the following results hold: (1) If the signal $v(t)$ is first amplified by an amplifier of gain β and then by an amplifier of gain α, the amplified signal will be $\alpha(\beta v(t))$, which is identical to the resulting signal $(\alpha\beta)v(t)$, which would result if $v(t)$ is amplified directly by an amplifier of gain $\alpha\beta$ (which obeys property (v)); (2) If the signal $v(t)$ is first amplified by an amplifier of gain α and another amplifier of gain β separately, and then added together, the resulting signal $\alpha v(t) + \beta v(t)$ will be the same as that of $v(t)$ being directly amplified by an amplifier of gain $\alpha + \beta$ (which obeys property (v)); and lastly, (3) if two signals are added together and then amplified by an amplifier, the resulting signal will be the same as that of amplifying them individually by the same amplifier, and then adding the results together (which obeys property (vi)).

Examples of Vector Spaces

Having examined the defining properties of a vector space, we introduce in what follows several examples of vector spaces to illustrate their many forms. Note that in describing a vector space it is necessary to specify not only the vectors but also the operations of addition and scalar multiplication. It is also necessary to verify that each of the defining properties for a vector space is satisfied. In the following examples, however, we leave out the verification details for brevity.

Example 1.2-1 *The vector space* R^n. The set of all *n-tuples* of the form $(a_1, a_2, \cdots, a_n)$ with *real entries* or *components* a_i forms a vector space, denoted by R^n, under the operations of componentwise addition and scalar multiplication. Thus, if $\mathbf{x} = (x_1, x_2, \cdots, x_n) \in R^n$, $\mathbf{y} = (y_1, y_2, \cdots, y_n) \in R^n$, and α is any real number, then

$$\mathbf{x} + \mathbf{y} = (x_1 + y_1, x_2 + y_2, \cdots, x_n + y_n) \tag{1.2-1}$$

and

$$\alpha\mathbf{x} = (\alpha x_1, \alpha x_2, \cdots, \alpha x_n) \tag{1.2-2}$$

are elements of R^n.

As an example of a subspace, the set of vectors of the form $(0, a_2, \cdots, a_n)$ is a subspace in R^n (Exercise 1-1). ∎

Example 1.2-2 *The vector space* $\mathbf{P}(x)$ *of real polynomials*. A *real polynomial* is an expression of the form $a_n x^n + a_{n-1} x^{n-1} + \cdots + a_0$, where n is a nonnegative integer and the a_i's are real. The set of all such polynomials forms a vector space under the following operations: if $\mathbf{a} = a_n x^n + a_{n-1} x^{n-1} + \cdots + a_0$ and $\mathbf{b} = b_m x^m + b_{m-1} x^{m-1} + \cdots + b_0$ are elements of $\mathbf{P}(x)$ (without loss of generality, assume that $m < n$), and α is any real number, then

$$\mathbf{a} + \mathbf{b} = a_n x^n + \cdots + (a_m + b_m)x^m + \cdots + (a_0 + b_0) \tag{1.2-3}$$

and

$$\alpha\mathbf{a} = \alpha a_n x^n + \alpha a_{n-1} x^{n-1} + \cdots + \alpha a_0 \tag{1.2-4}$$

are elements of $\mathbf{P}(x)$.

Note that in this vector space each polynomial is treated as a single vector. As an example of subspace, the set of polynomials of the form $a_n x^n + a_{n-1} x^{n-1} + \cdots + a_2 x^2 + a_0$ forms a subspace in this vector space. All the elements of this subspace share the same property that $a_1 = 0$. ∎

Example 1.2-3 *The vector space* $\mathbf{F}(R)$ *of real-valued functions.* The set of all real-valued functions defined on the entire real line (i.e., the set of real numbers R) forms a vector space under the operations defined as follows: if $\mathbf{f} = f(x)$ and $\mathbf{g} = g(x)$ are two such functions, and α is any real number, then the sum function (vector) $\mathbf{f} + \mathbf{g}$ and scalar multiple $\alpha\mathbf{f}$ are defined respectively by

$$(\mathbf{f} + \mathbf{g})(x) = f(x) + g(x) \tag{1.2-5}$$

and

$$(\alpha\mathbf{f})(x) = \alpha f(x) \tag{1.2-6}$$

for any $x \in R$. Note that in order to emphasize the fact that an entire function rather than its value at any particular point is treated as a vector, we have used a bold letter to represent a function vector.

As an example of subspace, the set of all real *continuous* functions defined on R forms a subspace in this vector space. ∎

Bases and Dimensionality

Let $\mathbf{V}$ be a vector space, and S be a set of vectors in $\mathbf{V}$. Then S is said to be *linearly dependent* if there exist a finite number of distinct vectors $\mathbf{x}_1, \mathbf{x}_2, \cdots, \mathbf{x}_n$ in S such that $\alpha_1 \mathbf{x}_1 + \alpha_2 \mathbf{x}_2 + \cdots + \alpha_n \mathbf{x}_n = \mathbf{0}$ for some combination of scalars $\alpha_1, \alpha_2, \cdots, \alpha_n$, *not all zero*; Otherwise, it is said to be *linearly independent*.

In the vector space $\mathbf{V}$, a vector $\mathbf{y}$ is called a *linear combination* of the vectors $\mathbf{x}_1, \mathbf{x}_2, \cdots, \mathbf{x}_n$ if it can be expressed in the form $\mathbf{y} = \alpha_1 \mathbf{x}_1 + \alpha_2 \mathbf{x}_2 + \cdots + \alpha_n \mathbf{x}_n$ for some combination of scalars $\alpha_1, \alpha_2, \cdots, \alpha_n$.

If the set S is non-empty, then the set of all the possible linear combinations of the vectors in S is called the *span* of S and is denoted by $span(S)$. Interestingly, the span of S forms a subspace in $\mathbf{V}$ (see Exercise 1-3). Also, the set S is said to *span* the space $\mathbf{V}$ if $span(S) = \mathbf{V}$. Clearly, the set S spans the space $\mathbf{V}$ if and only if every vector in $\mathbf{V}$ can be expressed as a linear combination of the vectors in S.

In the vector space $\mathbf{V}$, the set of vectors S is called a *basis* of $\mathbf{V}$ if: (i) it is linearly independent; and (ii) it spans $\mathbf{V}$. A vector space may have many bases.[†]

[†]A fundamental question is whether every linear space has a basis. The answer is affirmative which is made possible by Zorn's lemma [4].

However, if a basis for a vector space contains a finite number of vectors, then any basis for the same vector space will contain exactly the same number of vectors (see Exercise 1-4). A vector space is said to be *finite-dimensional* if it has a basis that consists of only a finite number of vectors; otherwise it is said to be *infinite-dimensional*. For a finite-dimensional space **V**, the unique number of vectors in each basis for **V** is called the *dimension* of **V**, and is denoted $dim(\mathbf{V})$.

The most significant property of a basis for a vector space is that every vector in the vector space can be expressed as a unique linear combination of vectors in the basis. As a result, in a vector space **V** of dimension n, every set with more than n vectors is linearly dependent (Exercise 1-7). Also in the same vector space **V**, any linearly independent set of n vectors is a basis for **V**. Furthermore, if **W** is a subspace of **V**, then $dim(\mathbf{W}) \leq dim(\mathbf{V})$.

To help clarify these concepts, we furnish two examples below.

Example 1.2-4 In the space R^3, let $\mathbf{e}_1 = (1,0,0)$, $\mathbf{e}_2 = (0,1,0)$, and $\mathbf{e}_3 = (0,0,1)$. It is readily seen that the set of vectors $\{\mathbf{e}_1, \mathbf{e}_2, \mathbf{e}_3\}$ is a basis for R^3. Hence we have $dim(R^3) = 3$. In other words, R^3 is a 3-dimensional space. Similarly, if we let $\mathbf{u}_1 = (1,0,0)$, $\mathbf{u}_2 = (1,2,0)$, and $\mathbf{u}_3 = (1,1,1)$. Then $\{\mathbf{u}_1, \mathbf{u}_2, \mathbf{u}_3\}$ is also a basis for R^3. Note that the set of vectors $\{\mathbf{e}_1, \mathbf{e}_2, \mathbf{e}_3, \mathbf{u}_2\}$ is not linearly independent. This is true because $\mathbf{u}_2$ can be expressed as a linear combination of $\mathbf{e}_1$ and $\mathbf{e}_2$, i.e., $\mathbf{u}_2 = \mathbf{e}_1 + 2\mathbf{e}_2$. ∎

In general, in the space R^n the set of vector $\{\mathbf{e}_1, \mathbf{e}_2, \cdots, \mathbf{e}_n\}$ with $\mathbf{e}_1 = (1,0,0,\cdots,0,0)$, $\mathbf{e}_2 = (0,1,0,\cdots,0,0)$, $\cdots$, and $\mathbf{e}_n = (0,0,0,\cdots,0,1)$ is a basis and is called the *standard basis* for R^n. As a result, we have $\dim(R^n) = n$. For an obvious reason, the space R^2 is often referred as the 2-D (i.e., the two-dimensional) space, while the space R^3 is referred as the 3-D space.

Example 1.2-5 In the space $\mathbf{P}(x)$ (given in Example 1.2-2), the set of vectors $\{1, x, x^2, \cdots, x^n, \cdots\}$ is a basis. Since this basis does not have a finite number of elements, the space $\mathbf{P}(x)$ is infinite-dimensional. ∎

Finally, we mention in passing that even though theoretically it is true that every vector space has a basis, it is by no means a trivial matter to find a basis for some infinite-dimensional vector spaces. For example, it is rather challenging to find a basis for the space $\mathbf{F}(R)$ given in Example 1.2-3.

1.3 NORMED VECTOR SPACES

The concept of vector space provides us with a way of describing how the elements (vectors) of a vector space interact with each other under algebraic operations. In practical applications, physical quantities of interest in science and engineering can often be represented by vectors in vector spaces. In such applications, it is often the case that we are interested in extracting a *metric* related to the *size* of the vector. For example, in signal processing, we may want to know how much energy a signal has. In order to have a tool to compare such metrics in a vector-space setting we introduce the concept of *norm*.

Definitions

Let V be a vector space. A *norm* on V is a function that assigns a real number, denoted by $\|x\|$, to each vector x in V such that

   (i) $\|x\| \geq 0$, and $\|x\| = 0$ if and only if x=0.
  (ii) $\|\alpha x\| = |\alpha| \, \|x\|$ for any scalar α and any x in V.
 (iii) $\|x + y\| \leq \|x\| + \|y\|$ for all x, y in V.

The vector space V together with its norm $\|\cdot\|$ is called a *normed vector space*.

Some discussion is needed. In a normed vector space, the norm serves as measure of the size (length) of its vectors, which is made possible by the defining properties of the norm. Indeed, the first defining property states that the size of a vector should always be positive unless the vector is the zero vector. This agrees with the common sense that the length of a physical object can never assume a negative value. The second defining property states that if a vector is scaled by some amount then its size (length) should also be changed by the same amount, which again agrees perfectly with common sense. The third defining property, which is somewhat less straightforward, is called the *triangle inequality* of the norm. As explained in the following, the triangle property of the norm has important applications in science and engineering.

Even though the norm is useful as a measure of the size of individual vectors in a normed vector space, it can also be used as a measure for comparing vectors in a vector space. Indeed, for any two vectors x and y in a normed space V, the length of their difference vector $x - y$ naturally defines the *distance* between them. More specifically, if we define $d(x,y) = \|x - y\|$, then $d(x,y)$ gives the distance between x and y, which satisfies the following properties:

   (a) $d(x,y) \geq 0$, and $d(x,y) = 0$ if and only if $x = y$.
   (b) $d(x,y) = d(y,x)$.
   (c) $d(x,y) \leq d(x,z) + d(z,y)$.

Property (a) states that the distance between two *distinct* points (vectors) is always positive, which is a direct result of the defining property (i) for the norm. Property (b) states that the distance from x to y is the same as that from y to x, which is a direct result of the defining property (ii) for the norm. Property (c), also called the *triangle inequality* of the distance, states that the shortest route from point x to point y is the direct route, and not via any other third point. This property is made possible by the triangle inequality of the norm, i.e., $\|\mathbf{x} + \mathbf{y}\| \leq \|\mathbf{x}\| + \|\mathbf{y}\|$. Indeed, we have

$$
\begin{aligned}
d(\mathbf{x}, \mathbf{y}) &= \|\mathbf{x} - \mathbf{y}\| = \|(\mathbf{x} - \mathbf{z}) + (\mathbf{z} - \mathbf{y})\| \\
&\leq \|\mathbf{x} - \mathbf{z}\| + \|\mathbf{z} - \mathbf{y}\| = d(\mathbf{x}, \mathbf{z}) + d(\mathbf{z}, \mathbf{y}).
\end{aligned} \tag{1.3-1}
$$

From the above discussion it is clear that the norm in a normed vector space can be used as a measure for the distance between vectors. This allows us to introduce the concept of *convergence*, which is one of the most important ideas in normed vector spaces—both theoretically and numerically. Let $\{\mathbf{f}_n\}$ denote the sequence of vectors $\mathbf{f}_1, \mathbf{f}_2, \cdots, \mathbf{f}_n, \cdots$, in the vector space $\mathbf{V}$. The sequence $\{\mathbf{f}_n\}$ is said to *converge* to $\mathbf{f}$ if there is a vector $\mathbf{f}$ such that

$$
\lim_{n \to \infty} \|\mathbf{f}_n - \mathbf{f}\| = 0. \tag{1.3-2}
$$

We need to point out that there exists at most one such $\mathbf{f}$ (which is due to the triangle inequality of the norm), and when it exists it is called the *limit* of $\{\mathbf{f}_n\}$. In shorthand notation, we write $\mathbf{f}_n \longrightarrow \mathbf{f}$. A sequence that converges is also called a *convergent sequence* (see Exercise 1-8).

In a vector space, a sequence $\{\mathbf{f}_n\}$ is called a *Cauchy sequence* if it satisfies

$$
\lim_{m,n \to \infty} \|\mathbf{f}_m - \mathbf{f}_n\| = 0. \tag{1.3-3}
$$

In words, the distance between any two vectors in a Cauchy sequence *eventually* becomes zero. It is clear that a convergent sequence is automatically a Cauchy sequence, but the converse is not always true. A normed vector space $\mathbf{V}$ is said to be *complete* if every Cauchy sequence in $\mathbf{V}$ converges (see Exercise 1-9).

Let S be a set in a normed vector space $\mathbf{V}$, a point $\mathbf{x}_0 \in S$ is called an *interior point* of S if there exists $r > 0$ such that all the points satisfying $\|\mathbf{x} - \mathbf{x}_0\| < r$ belong to S. In other words, if $\mathbf{x}_0$ is an interior point of the set S then all the points in a small neighborhood of $\mathbf{x}_0$ also belong to S. A point $\mathbf{x}_0 \in S$ is called an *boundary point* of S if every neighborhood of $\mathbf{x}_0$ contains at least one point in S^c—the complement set of S. Clearly, a point in the set S can be only either an interior point of S or a boundary point of S but not both. The set S is said to be *open* if every point of S is an interior point of S. The set S is said to be *closed* if its complement set S^c is open. A point $\mathbf{x}$ is called the *limit point* of the set S if there exists a sequence in S that converges to $\mathbf{x}$. The *closure* of the set S, denoted by $\bar{S}$, is defined as the union of the set with all of its limit points. It can be shown that a set is closed if and only if it contains all of its limit points (see Exercise 1-10). In other words, the closure of a closed set is the set itself. This property is often used to test whether a set is closed or not. Namely, in order to prove that a set is closed, we can simply pick an arbitrary convergent sequence in the set and show that its limit also lies in the set.

Examples of Normed Vector Spaces

In the following, we present a number of examples of normed vector spaces which are often used in science and engineering. For the interest of brevity, we omit the details necessary to verify that the defining properties of the norm are satisfied by each of the given normed vector spaces.

Example 1.3-1 *The vector space* R^n. On R^n, there exist a variety of norms of which perhaps the most popular is the *Euclidean* norm. Consider a row vector $\mathbf{x} = (x_1, x_2, \cdots, x_n) \in R^n$; its Euclidean norm, denoted by $\|\mathbf{x}\|_2$, is given by

$$\|\mathbf{x}\|_2 = \left(\sum_{i=1}^{n} |x_i|^2 \right)^{1/2}. \tag{1.3-4}$$

One significant property of the Euclidean norm is that it completely agrees with our measurement of geometric quantities such as length and distance in the physical world, i.e., R, R^2 or R^3 space.

A variant of the Euclidean norm is the so-called *weighted Euclidean norm*, which is also quite useful in applications such as signal processing. Let $w_1, w_2, \cdots, w_n$ be n real numbers, which are called *weighting factors*; then for each vector $\mathbf{x} = (x_1, x_2, \cdots, x_n) \in R^n$, its weighted Euclidean norm $\|\mathbf{x}\|_{\mathrm{w}}$ is defined as

$$\|\mathbf{x}\|_{\mathrm{w}} = \left(\sum_{i=1}^{n} |w_i x_i|^2 \right)^{1/2}. \tag{1.3-5}$$

A generalization of the Euclidean norm is the so-called p-norm, which is defined for all the numbers $1 \le p \le \infty$. For $1 \le p < \infty$, the p-norm of a vector $\mathbf{x} = (x_1, x_2, \cdots, x_n) \in R^n$ is

$$\|\mathbf{x}\|_p = \left(\sum_{i=1}^{n} |x_i|^p \right)^{1/p}, \tag{1.3-6}$$

and for $p = \infty$, the p-norm of a vector $\mathbf{x} = (x_1, x_2, \cdots, x_n) \in R^n$ is given by

$$\|\mathbf{x}\|_{\infty} = \max_{1 \le i \le n} \{|x_i|\}. \tag{1.3-7}$$

The triangle inequality of the p-norm for $1 \le p < \infty$, i.e., $\|\mathbf{x} + \mathbf{y}\|_p \le \|\mathbf{x}\|_p + \|\mathbf{y}\|_p$ for all $\mathbf{x}, \mathbf{y} \in R^n$, follows directly from the famous Minkowski's inequality: If $1 \le p < \infty$ and a_i, b_i are complex numbers for $i = 1, 2 \cdots, n$, then

$$\left(\sum_{i=1}^{n} |a_i + b_i|^p \right)^{1/p} \le \left(\sum_{i=1}^{n} |a_i|^p \right)^{1/p} + \left(\sum_{i=1}^{n} |b_i|^p \right)^{1/p}. \tag{1.3-8}$$

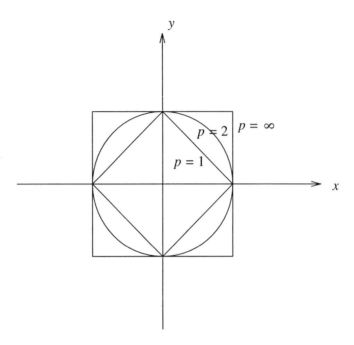

Fig. 1.3-1   The geometric shape of the unit-sphere $\left\{\mathbf{x} : \|\mathbf{x}\|_p \leq 1\right\}$ in the space R^2 for the cases of $p = 1, 2$, and ∞.

It is interesting to note that the p-norm for $p = 1$ is simply the absolute sum of all the components of the vector. That is, for $\mathbf{x} \in R^n$, we have $\|\mathbf{x}\|_1 = \sum_{i=1}^{n} |x_i|$. Also, for $p = 2$ the p-norm is simply the Euclidean norm. In the practical world, the p-norm for the space R^n has been proven to be very useful especially for the cases of $p = 1$, $p = 2$, and $p = \infty$. It is rather interesting to examine the geometrical shape of the "unit-sphere", i.e., the region $\left\{\mathbf{x} : \|\mathbf{x}\|_p \leq 1\right\}$ in the space R^n. In Fig. 1.3-1 are shown the unit-spheres for the cases of $p = 1, 2$, and ∞ in the space R^2. Only for $p = 2$ and R^3 is the unit-sphere truly a unit sphere as we visualize it.

To further clarify these concepts we examine the distance function corresponding to the p-norm in the space R^n. Consider two points $\mathbf{x}_1 = (x_1, y_1)$ and $\mathbf{x}_2 = (x_2, y_2)$ in the 2-D plane (or space) R^2. Let $d_p(\mathbf{x}_1, \mathbf{x}_2)$ denote the distance between them when the p-norm is used. Clearly, if the 1-norm is used then we have the sum of *absolute displacements*

$$d_1(\mathbf{x}_1, \mathbf{x}_2) = |x_1 - x_2| + |y_1 - y_2|. \tag{1.3-9}$$

If the 2-norm is used then we have the familiar *straight-line* distance.

$$d_2(\mathbf{x}_1, \mathbf{x}_2) = \sqrt{(x_1 - x_2)^2 + (y_1 - y_2)^2}. \tag{1.3-10}$$

If the ∞-norm is used then

$$d_\infty(\mathbf{x}_1, \mathbf{x}_2) = \max\left\{|x_1 - x_2|, |y_1 - y_2|\right\}. \tag{1.3-11}$$

In the literature the $d_1(\mathbf{x}_1, \mathbf{x}_2)$ is sometimes referred as *city-block* distance and $d_\infty(\mathbf{x}_1, \mathbf{x}_2)$ as *chessboard* distance.   ■

Example 1.3-2 *Sequence spaces* l^p. For $1 \leq p < \infty$, let l^p denote the set of all the sequences of the form $(x_1, x_2, \cdots, x_n, \cdots)$ which satisfy

$$\left(\sum_{i=1}^{\infty} |x_i|^p\right)^{1/p} < \infty. \tag{1.3-12}$$

For $p = \infty$, l^p denotes the set of all the *bounded sequences*. A sequence $(x_1, x_2, \cdots, x_n, \cdots)$ is said to be *bounded* if there exists a finite number M such that $|x_i| < M$ for all $i = 1, 2, \cdots$. Then it is straightforward to verify that the set l^p for each $1 \leq p \leq \infty$ indeed forms a vector space under the componentwise addition and scalar multiplication.

For $\mathbf{x} = (x_1, x_2, \cdots, x_n, \cdots) \in l^p$, we define

$$\|\mathbf{x}\|_p = \left(\sum_{i=1}^{\infty} |x_i|^p\right)^{1/p}, \tag{1.3-13}$$

when $1 \leq p < \infty$; and

$$\|\mathbf{x}\|_p = \sup_{i=1,2,\cdots} \left\{|x_i|\right\}, \tag{1.3-14}$$

when $p = \infty$.[†]

It can be shown that $\|\cdot\|_p$ defined above satisfies all the defining property of the norm for a vector space, and thus is a legitimate norm for the vector space l^p for each $1 \leq p \leq \infty$. In the literature, this norm is simply referred as the l^p *norm*. Of all these norms, the l^1, l^2, and l^∞ norms are among the most interesting to us.   ■

Example 1.3-3 *Function spaces* L^p. Let L^2 denote the set of all *square-integrable* functions defined on R. That is, a function $f(t), -\infty < t < \infty$, is in L^2 if and only if it satisfies

$$\int_{-\infty}^{\infty} |f(t)|^2 \, dt < \infty. \tag{1.3-15}$$

[†]Note that "sup" is used to denote the *least upper bound* of a set of real numbers. That is, it is the smallest number that is no less than any number in the set. For example, $\sup\{1-1/1, 1-1/2, \cdots, 1-1/n, \cdots\} = 1$.

Similarly, we will use "inf" to denote the *largest lower bound* of a set of real numbers. That is, it is the largest number that is no larger than any number in the set. For example, $\inf\{1 + 1/1, 1 + 1/2, \cdots, 1 + 1/n, \cdots\} = 1$.

Then it is readily seen that the set L^2 forms a vector space under the pointwise addition and scalar multiplication. For each vector $\mathbf{f} = f(t) \in L^2$, we define

$$\|\mathbf{f}\|_2 = \left(\int_{-\infty}^{\infty} |f(t)|^2 \, dt \right)^{1/2}. \tag{1.3-16}$$

Then $\|\mathbf{f}\|_2$ defines a norm for the space L^2. In the literature, it is frequently referred as the L^2 norm.

The L^2 norm is possibly the most widely used norm in science and engineering. One of its many important attributes is that it serves as a measure of the *energy* of the physical quantity that the function represents. For example, if $f(t)$ represents an electric current signal passing through a unit-valued resistor, where t denote the time, then its total energy E is measured by $E = \int_{-\infty}^{\infty} |f(t)|^2 \, dt$, which is simply the square of the L^2 norm of $f(t)$. In the real world it seems that all the known electric current signals are described by elements in L^2, otherwise, the looming energy crisis would no longer be an issue for our future generations.

A generalization of the L^2 space is the L^p space. A function $f(t)$ defined on R is said to be *p-integrable* if

$$\int_{-\infty}^{\infty} |f(t)|^p \, dt < \infty, \tag{1.3-17}$$

for $1 \le p < \infty$. Let L^p denote the set of all p-integrable functions defined on R. Then L^p is a vector space on which we can define the L^p norm

$$\|\mathbf{f}\|_p = \left(\int_{-\infty}^{\infty} |f(t)|^p \, dt \right)^{1/p}. \tag{1.3-18}$$

The triangle inequality of the L^p norm follows directly from the Minkowski's inequality for functions: If $1 \le p < \infty$ and $f(t), g(t)$ are p-integrable functions on R, then

$$\left(\int_{-\infty}^{\infty} |f(t) + g(t)|^p \, dt \right)^{1/p} \le \left(\int_{-\infty}^{\infty} |f(t)|^p \, dt \right)^{1/p}$$
$$+ \left(\int_{-\infty}^{\infty} |g(t)|^p \, dt \right)^{1/p}. \tag{1.3-19}$$

∎

Finally, we mention without proof that all the examples of normed vector spaces furnished above, i.e., the space R^n with p-norm for $1 \le p \le \infty$, the l^p space with l^p norm for $1 \le p \le \infty$, and the L^p space with L^p norm for $1 \le p < \infty$, are all complete normed vector spaces.

1.4 INNER PRODUCT SPACES

The notion of norm provides us with a means for measuring both the length of a vector and the distance between any two vectors in a vector space. From elementary algebra, we know that an equally important geometric notion in the Euclidean space R^n is the angle between two vectors, which is related to the *dot product* of two vectors. Consider two vectors $\mathbf{x} = (x_1, x_2)$ and $\mathbf{y} = (y_1, y_2)$ in R^2; their dot product is defined as

$$\mathbf{x} \cdot \mathbf{y} = x_1 y_1 + x_2 y_2, \tag{1.4-1}$$

and the angle θ between them is determined through the relation

$$\mathbf{x} \cdot \mathbf{y} = \|\mathbf{x}\|_2 \, \|\mathbf{y}\|_2 \cos \theta, \tag{1.4-2}$$

where $\|\mathbf{x}\|_2$ and $\|\mathbf{y}\|_2$ denote the Euclidean norm of $\mathbf{x}$ and $\mathbf{y}$, respectively. Clearly, when the angle $\theta = 90°$, we have $\mathbf{x} \cdot \mathbf{y} = 0$, and the vectors $\mathbf{x}$ and $\mathbf{y}$ are said to be *perpendicular*, or *orthogonal*, to each other.

It turns out that all these geometrical notions such as dot-product and orthogonality can be extended to an arbitrary vector space through the concept of *inner product*. The concept of orthogonality has tremendous application in engineering. For example, two different messages occupying the same frequency band and transmitted simultaneously are uniquely recoverable if their carriers are orthogonal.

Definitions

Let $\mathbf{V}$ be a vector space. An inner product on $\mathbf{V}$ is a function that assigns to every ordered pair of vectors $\mathbf{x}$ and $\mathbf{y}$ in $\mathbf{V}$ a scalar value, denoted by $\langle \mathbf{x}, \mathbf{y} \rangle$, such that for all $\mathbf{x}, \mathbf{y}$ and $\mathbf{z}$ in $\mathbf{V}$ and any scalar α we have

(a) $\langle \mathbf{x}, \mathbf{x} \rangle \geq 0$, and $\langle \mathbf{x}, \mathbf{x} \rangle = 0$ if and only if $\mathbf{x} = \mathbf{0}$.

(b) $\langle \mathbf{x} + \mathbf{y}, \mathbf{z} \rangle = \langle \mathbf{x}, \mathbf{z} \rangle + \langle \mathbf{y}, \mathbf{z} \rangle$.

(c) $\langle \alpha \mathbf{x}, \mathbf{y} \rangle = \alpha \langle \mathbf{x}, \mathbf{y} \rangle$.

(d) $\langle \mathbf{x}, \mathbf{y} \rangle = \overline{\langle \mathbf{y}, \mathbf{x} \rangle}$, where $\overline{\langle \mathbf{y}, \mathbf{x} \rangle}$ is the complex conjugate of $\langle \mathbf{y}, \mathbf{x} \rangle$.

A vector space together with its inner product is called an *inner product space*.

Note that properties (b) and (c) state that the inner product is linear in the first component. Property (d) implies that the inner product is, except for conjugacy, commutative over its two components. Indeed, in a real inner product space where only real scalars are used, property (d) simply reduces to $\langle \mathbf{x}, \mathbf{y} \rangle = \langle \mathbf{y}, \mathbf{x} \rangle$.

It is easily shown that except for conjugacy the inner product is also linear in its second component. Indeed, applying properties (b) and (d), we have

$$\langle \mathbf{x}, \mathbf{y} + \mathbf{z} \rangle = \overline{\langle \mathbf{y} + \mathbf{z}, \mathbf{x} \rangle} = \overline{\langle \mathbf{y}, \mathbf{x} \rangle + \langle \mathbf{z}, \mathbf{x} \rangle} = \langle \mathbf{x}, \mathbf{y} \rangle + \langle \mathbf{x}, \mathbf{z} \rangle, \tag{1.4-3}$$

and applying properties (c) and (d), we have

$$\langle \mathbf{x}, \, \alpha \mathbf{y} \rangle = \overline{\langle \alpha \mathbf{y}, \, \mathbf{x} \rangle} = \overline{\alpha \langle \mathbf{y}, \, \mathbf{x} \rangle} = \overline{\alpha} \langle \mathbf{x}, \, \mathbf{y} \rangle. \qquad (1.4\text{-}4)$$

As a result, the inner product in a real inner product space is a linear function in either of its two components, and is also commutative over its two components.

Now let's examine the dot product in the space R^n. Consider two vectors $\mathbf{x} = (x_1, x_2, \cdots, x_n)$ and $\mathbf{y} = (y_1, y_2, \cdots, y_n)$ in R^n; their dot product is given by

$$\mathbf{x} \cdot \mathbf{y} = \sum_{i=1}^{n} x_i \, y_i. \qquad (1.4\text{-}5)$$

It is readily seen that the dot product, as a function of two vector variables, satisfies all the properties of an inner product. Hence, if we define $\langle \mathbf{x}, \, \mathbf{y} \rangle = \mathbf{x} \cdot \mathbf{y}$, then it defines an inner product on the space R^n. In the literature, this inner product is called the *Euclidean inner product*.

An interesting observation on the Euclidean inner product is the following. Consider an arbitrary vector $\mathbf{x} = (x_1, x_2, \cdots, x_n) \in R^n$; if we compute the inner product of $\mathbf{x}$ with itself, we have

$$\langle \mathbf{x}, \, \mathbf{x} \rangle = \sum_{i=1}^{n} x_i x_i = \sum_{i=1}^{n} |x_i|^2 = \|\mathbf{x}\|_2^2, \qquad (1.4\text{-}6)$$

where $\|\cdot\|_2$ is the Euclidean norm on R^n. Equation (1.4-6) reveals a rather spectacular property of the Euclidean inner product: For every $\mathbf{x} \in R^n$, we have $\|\mathbf{x}\|_2 = \sqrt{\langle \mathbf{x}, \, \mathbf{x} \rangle}$. In words, the Euclidean norm is associated with the Euclidean inner product.

As mentioned at the beginning of this section, the inner product is a direct extension of the dot product, i.e., the Euclidean inner product, to general vector spaces. Then a natural question is: Does this association of the Euclidean inner product with the Euclidean norm also carry over to an arbitrary inner product? Interestingly, the answer is yes and we have:

Theorem 1.4-1 *Let* $\mathbf{V}$ *be an inner product space. Then the inner product* $\langle \cdot, \, \cdot \rangle$ *on* $\mathbf{V}$ *induces a norm* $\|\cdot\|$ *for* $\mathbf{V}$ *via the relation* $\|\mathbf{x}\| = \sqrt{\langle \mathbf{x}, \, \mathbf{x} \rangle}$ *for all* $\mathbf{x}$ *in* $\mathbf{V}$.

For a proof of this result, we need to verify that the induced norm $\|\cdot\|$ satisfies all the defining properties of a norm. It is clear that $\|\mathbf{x}\| = 0$ if and only if $\mathbf{x} = \mathbf{0}$. Also, for any $\mathbf{x} \in \mathbf{V}$ and for any scalar α, we have

$$\|\alpha \mathbf{x}\| = \sqrt{\langle \alpha \mathbf{x}, \, \alpha \mathbf{x} \rangle} = \sqrt{\alpha \langle \mathbf{x}, \, \alpha \mathbf{x} \rangle} = \sqrt{\alpha \overline{\alpha} \langle \mathbf{x}, \, \mathbf{x} \rangle} = |\alpha| \cdot \|\mathbf{x}\|. \qquad (1.4\text{-}7)$$

What remains to be shown is the triangle inequality expressed as $\|\mathbf{x} + \mathbf{y}\| \leq \|\mathbf{x}\| + \|\mathbf{y}\|$, which follows directly from the *Schwarz inequality* for an inner product:

$$|\langle \mathbf{x}, \, \mathbf{y} \rangle| \leq \|\mathbf{x}\| \cdot \|\mathbf{y}\|. \qquad (1.4\text{-}8)$$

Equation (1.4-8) is also referred to, sometimes, as Cauchy-Schwarz or Cauchy-Bunyakovski inequality [5]. Based on Eq. (1.4-8), we have

$$
\begin{aligned}
\|\mathbf{x} + \mathbf{y}\|^2 &= \langle \mathbf{x} + \mathbf{y}, \ \mathbf{x} + \mathbf{y} \rangle \\
&= \langle \mathbf{x}, \ \mathbf{x} \rangle + \langle \mathbf{x}, \ \mathbf{y} \rangle + \langle \mathbf{y}, \ \mathbf{x} \rangle + \langle \mathbf{y}, \ \mathbf{y} \rangle \\
&= \|\mathbf{x}\|^2 + 2\Re(\langle \mathbf{x}, \ \mathbf{y} \rangle) + \|\mathbf{y}\|^2 \\
&\leq \|\mathbf{x}\|^2 + 2|\langle \mathbf{x}, \ \mathbf{y} \rangle| + \|\mathbf{y}\|^2 \\
&\leq \|\mathbf{x}\|^2 + 2\|\mathbf{x}\| \cdot \|\mathbf{y}\| + \|\mathbf{y}\|^2 \\
&= (\|\mathbf{x}\| + \|\mathbf{y}\|)^2 .
\end{aligned}
\tag{1.4-9}
$$

Thus the triangle inequality for the norm follows immediately by taking the square root of both sides in Eq. (1.4-9). Note that in the third row from the top of the above equation $\Re(\cdot)$ is used to denote the real part of a complex number.

The Schwarz inequality in Eq. (1.4-8) is an interesting result by itself. Clearly, it holds if $\mathbf{y} = \mathbf{0}$. Assume that $\mathbf{y} \neq \mathbf{0}$ and let $c = \langle \mathbf{x}, \ \mathbf{y} \rangle / \|\mathbf{y}\|^2$. Then we have

$$
\begin{aligned}
0 \leq \|\mathbf{x} - c\mathbf{y}\|^2 &= \langle \mathbf{x} - c\mathbf{y}, \ \mathbf{x} - c\mathbf{y} \rangle = \langle \mathbf{x}, \ \mathbf{x} - c\mathbf{y} \rangle - c\langle \mathbf{y}, \ \mathbf{x} - c\mathbf{y} \rangle \\
&= \langle \mathbf{x}, \ \mathbf{x} \rangle - \bar{c}\langle \mathbf{x}, \ \mathbf{y} \rangle - c\langle \mathbf{y}, \ \mathbf{x} \rangle + |c|^2 \langle \mathbf{y}, \ \mathbf{y} \rangle \\
&= \|\mathbf{x}\|^2 - |\langle \mathbf{x}, \ \mathbf{y} \rangle|^2 / \|\mathbf{y}\|^2 ,
\end{aligned}
\tag{1.4-10}
$$

from which Eq. (1.4-8) follows. Also, from Eq. (1.4-10) it is clear that

$$
\|\mathbf{x}\|^2 - |\langle \mathbf{x}, \ \mathbf{y} \rangle|^2 / \|\mathbf{y}\|^2 = 0
\tag{1.4-11}
$$

if and only if $\mathbf{x} = c\mathbf{y}$. In other words, the Schwarz inequality $|\langle \mathbf{x}, \ \mathbf{y} \rangle| \leq \|\mathbf{x}\| \cdot \|\mathbf{y}\|$ assumes equality if and only if $\mathbf{x}$ and $\mathbf{y}$ are *linearly dependent*.

To summarize, from the above development it is clear that the inner product operation automatically induces a norm for the vector space, which is called the *induced norm* for the inner product space. Also, from the discussion in the last section, we see that there may well exist many different norms for a same vector space. One may wonder if each of these different norms can be induced from an inner product. The answer to this question leads to the following result, of which the proof is omitted for the sake of brevity.

Theorem 1.4-2 (*Jordan-von Neumann*) *Let $\|\cdot\|$ be a norm on a linear vector space* **V**. *Then there exists an inner product $\langle \cdot, \ \cdot \rangle$ on* **V** *such that $\langle \mathbf{x}, \ \mathbf{x} \rangle = \|\mathbf{x}\|^2$ for all $\mathbf{x} \in$* **V** *if and only if the norm satisfies the parallelogram law*

$$
\|\mathbf{x} + \mathbf{y}\|^2 + \|\mathbf{x} - \mathbf{y}\|^2 = 2\left(\|\mathbf{x}\|^2 + \|\mathbf{y}\|^2 \right)
\tag{1.4-12}
$$

for all $\mathbf{x}, \mathbf{y} \in$ **V**. *Furthermore, if this is the case, such an inner product is unique and is given by*

$$\langle \mathbf{x}, \mathbf{y} \rangle = \frac{1}{4} \left(\|\mathbf{x} + \mathbf{y}\|^2 - \|\mathbf{x} - \mathbf{y}\|^2 + j\|\mathbf{x} + j\mathbf{y}\|^2 - j\|\mathbf{x} - j\mathbf{y}\|^2 \right), \quad (1.4\text{-}13)$$

where $j \triangleq \sqrt{-1}$.

As pointed out at the beginning of this section, two vectors in R^2 are perpendicular, or orthogonal, to each other if their dot product is zero. The concept of orthogonality can be extended to an inner product space directly. Let $\mathbf{V}$ be an inner product space, then two vectors $\mathbf{x}, \mathbf{y} \in \mathbf{V}$ are said to be *orthogonal* to each other is their inner product $\langle \mathbf{x}, \mathbf{y} \rangle$ is zero. A set S in $\mathbf{V}$ is said to be an *orthogonal set* if any two *distinct* vectors of S are orthogonal; if in addition, the norm of every vector in S is 1, then the set S is said to be *orthonormal* in $\mathbf{V}$.

For example, in the Euclidean space R^n the set of standard basis vectors $\{\mathbf{e}_1, \mathbf{e}_2, \cdots, \mathbf{e}_n\}$ with $\mathbf{e}_1 = (1, 0, 0, \cdots, 0, 0)$, $\mathbf{e}_2 = (0, 1, 0, \cdots, 0, 0)$, $\cdots$, and $\mathbf{e}_n = (0, 0, 0, \cdots, 0, 1)$ is an orthonormal set.

Orthonormal sets in an inner product space are discussed in the next section and are of special interest to us because of their useful properties in practical problems. It is straightforward to show that an orthonormal set is automatically an independent set (Exercise 1-14). On the other hand, we can construct an orthonormal set from an arbitrary linearly independent set. A formal statement of this construction is given below.

Theorem 1.4-3 (*Gram-Schmidt orthonormalization*) *Let* $\{\mathbf{x}_n : n = 1, 2, \cdots\}$ *be a linearly independent set in an inner product space* $\mathbf{V}$. *Define*

$$\mathbf{y}_1 = \mathbf{x}_1, \quad \mathbf{u}_1 = \frac{1}{\|\mathbf{y}_1\|}\mathbf{y}_1, \quad (1.4\text{-}14)$$

and

$$\mathbf{y}_n = \mathbf{x}_n - \sum_{i=1}^{n-1} \langle \mathbf{x}_n, \mathbf{u}_i \rangle \mathbf{u}_i, \quad \mathbf{u}_n = \frac{1}{\|\mathbf{y}_n\|}\mathbf{y}_n. \quad (1.4\text{-}15)$$

Then the set $\{\mathbf{u}_n : n = 1, 2, \cdots\}$ *so constructed is an orthonormal set in* $\mathbf{V}$ *and*

$$span\{\mathbf{u}_1, \mathbf{u}_1, \cdots, \mathbf{u}_n\} = span\{\mathbf{x}_1, \mathbf{x}_1, \cdots, \mathbf{x}_n\} \quad (1.4\text{-}16)$$

for every $n = 1, 2, \cdots$.

This result can be proved by using mathematical induction. Clearly, the set $\{\mathbf{u}_1\}$ is an orthonormal set in $\mathbf{V}$, and we have $span\{\mathbf{u}_1\} = span\{\mathbf{x}_1\}$. Assume that the set $\{\mathbf{u}_1, \mathbf{u}_1, \cdots, \mathbf{u}_n\}$ obtained from Eqs. (1.4-14) and (1.4-15) is an orthonormal set and $span\{\mathbf{u}_1, \mathbf{u}_1, \cdots, \mathbf{u}_n\} = span\{\mathbf{x}_1, \mathbf{x}_1, \cdots, \mathbf{x}_n\}$. Let

$$\mathbf{y}_{n+1} = \mathbf{x}_{n+1} - \sum_{i=1}^{n} \langle \mathbf{x}_{n+1}, \mathbf{u}_i \rangle \mathbf{u}_i. \quad (1.4\text{-}17)$$

Then for $m < n$ we have

$$
\begin{aligned}
\langle \mathbf{y}_{n+1},\ \mathbf{u}_m \rangle &= \langle \mathbf{x}_{n+1} - \sum_{i=1}^{n} \langle \mathbf{x}_{n+1},\ \mathbf{u}_i \rangle \mathbf{u}_i,\ \mathbf{u}_m \rangle \\
&= \langle \mathbf{x}_{n+1},\ \mathbf{u}_m \rangle - \langle \mathbf{x}_{n+1},\ \mathbf{u}_m \rangle \langle \mathbf{u}_m,\ \mathbf{u}_m \rangle \\
&= 0.
\end{aligned}
\tag{1.4-18}
$$

Furthermore, $\mathbf{y}_{n+1} \neq \mathbf{0}$, since otherwise we would have $\mathbf{x}_{n+1} \in span\{\mathbf{u}_1, \mathbf{u}_2,$ $\cdots, \mathbf{u}_n\} = span\{\mathbf{x}_1, \mathbf{x}_2, \cdots, \mathbf{x}_n\}$, which contradicts the fact that $\{\mathbf{x}_n : n = 1,$ $2, \cdots\}$ is a linearly independent set. Let $\mathbf{u}_{n+1} = \mathbf{y}_n / \|\mathbf{y}_n\|$. Then $\|\mathbf{u}_{n+1}\| = 1$, and for $m < n$ we have $\langle \mathbf{u}_{n+1},\ \mathbf{u}_m \rangle = 0$. Therefore, $\{\mathbf{u}_1, \mathbf{u}_1, \cdots, \mathbf{u}_n, \mathbf{u}_{n+1}\}$ is an orthonormal set. Also, $span\{\mathbf{u}_1, \mathbf{u}_1, \cdots, \mathbf{u}_n, \mathbf{u}_{n+1}\} = span\{\mathbf{x}_1, \mathbf{x}_1, \cdots,$ $\mathbf{x}_n, \mathbf{u}_{n+1}\} = span\{\mathbf{x}_1, \mathbf{x}_1, \cdots, \mathbf{x}_n, \mathbf{x}_{n+1}\}$. The theorem follows by induction.

Discussion

The inner product has proven to be quite useful in practical problems. Perhaps the most convincing example is its application in signal detection for pattern recognition and communications.[†] In signal processing, where signals are represented by vectors in a vector space, the inner product is alternatively termed *correlation* which is used as a measure of resemblance or similarity between two signals. Let two signals be represented by two vectors $\mathbf{f}$ and $\mathbf{g}$, respectively, in an inner product space $\mathbf{V}$, then their correlation $r(\mathbf{f}, \mathbf{g})$ is defined as

$$
r(\mathbf{f}, \mathbf{g}) = \langle \mathbf{f},\ \mathbf{g} \rangle.
\tag{1.4-19}
$$

In addition, their *normalized correlation* $\Upsilon(\mathbf{f}, \mathbf{g})$ is defined as

$$
\Upsilon(\mathbf{f}, \mathbf{g}) = \frac{\langle \mathbf{f},\ \mathbf{g} \rangle}{\|\mathbf{f}\| \cdot \|\mathbf{g}\|},
\tag{1.4-20}
$$

where both $\mathbf{f}$ and $\mathbf{g}$ are assumed to be non-zero. From the Schwarz inequality, we have

$$
0 \leq |\Upsilon(\mathbf{f}, \mathbf{g})| \leq 1,
\tag{1.4-21}
$$

for all non-zero vectors $\mathbf{f}, \mathbf{g} \in \mathbf{V}$. Clearly, the normalized correlation $r(\mathbf{f},\ \mathbf{g})$ assumes its largest value 1 in magnitude when $\mathbf{f}$ and $\mathbf{g}$ are scalar multiples of each other (in which case they are most similar because one is an amplified or attenuated replica of the other), and assumes its smallest value 0 in magnitude when they are orthogonal (in which case they are most dissimilar because they have no component in common).

In certain applications, it is advantageous to use the similarity measure rather than the distance measure to compare the closeness between two vectors. One

---

[†]For example, modulation by an in-phase and quadrature carrier as in *quadrature amplitude modulation*, or modulation by orthogonal code sequences as in code-division multiple access (CDMA).

such an example is its application in digital communications. For simplicity, let us consider the case of binary communications where a transmitter sends a series of binary symbols '0' and '1' to a receiver through a communication channel. At the transmitter, the binary symbols '0' and '1' are represented by electrical waveforms, say $f_0(t)$ and $f_1(t)$, respectively. Due to degradation caused by the channel, the signal observed at the receiver, say $f(t)$, is not exactly the same as the one initially sent by the transmitter. Let's assume that the channel degradation can be modeled by pure scalar attenuation and additive interference, then we have

$$f(t) = \begin{cases} \alpha f_0(t) + n(t) & \text{if '0' is sent} \\ \alpha f_1(t) + n(t) & \text{if '1' is sent,} \end{cases} \qquad (1.4\text{-}22)$$

where α is the channel attenuation factor such that $0 < \alpha < 1$, and $n(t)$ is independent, additive interference called *noise*. In such a case, the receiver faces the task of determining (or detecting) whether a '0' or '1' was sent based on the received noisy signal $f(t)$. Note that the channel attenuation factor α is not known; and even if it were, it varies from channel to channel.

A seemingly straightforward but ineffective strategy might be simply to compare the distances $\|f(t) - f_0(t)\|$ with $\|f(t) - f_1(t)\|$ and decide whether '0' or '1' was sent based on which correlation yields the smaller distance. Conceivably, the performance of such a detection system will be severely affected by the channel attenuation and interference noise.

A better strategy is to compare the correlations $r(f(t), f_0(t))$ with $r(f(t), f_1(t))$ and decide whether '0' or '1' was sent based on which one yields the larger value. In such a case, we have

$$r(f(t), f_0(t)) = \begin{cases} \alpha\|f_0(t)\|^2 + \langle n(t), f_0(t) \rangle & \text{if '0' is sent} \\ \alpha\langle f_1(t), f_0(t) \rangle + \langle n(t), f_0(t) \rangle & \text{if '1' is sent,} \end{cases} \qquad (1.4\text{-}23)$$

and

$$r(f(t), f_1(t)) = \begin{cases} \alpha\langle f_0(t), f_1(t) \rangle + \langle n(t), f_1(t) \rangle & \text{if '0' is sent} \\ \alpha\|f_1(t)\|^2 + \langle n(t), f_1(t) \rangle & \text{if '1' is sent.} \end{cases} \qquad (1.4\text{-}24)$$

If we choose the waveforms $f_0(t)$ and $f_1(t)$ to be orthogonal to each other, then $\langle f_0(t), f_1(t) \rangle = \langle f_1(t), f_0(t) \rangle = 0$. Also, the inner products $\langle n(t), f_0(t) \rangle$ and $\langle n(t), f_1(t) \rangle$ have the tendency of averaging out the noise $n(t)$ so that they are often negligible. Then we have

$$r(f(t), f_0(t)) \approx \begin{cases} \alpha\|f_0(t)\|^2 & \text{if '0' is sent} \\ 0 & \text{if '1' is sent,} \end{cases} \qquad (1.4\text{-}25)$$

and

$$r(f(t), f_1(t)) \approx \begin{cases} 0 & \text{if '0' is sent} \\ \alpha\|f_1(t)\|^2 & \text{if '1' is sent.} \end{cases} \qquad (1.4\text{-}26)$$

Clearly, we have $r(f(t), f_0(t)) > r(f(t), f_1(t))$ if '0' was sent; and $r(f(t), f_1(t))$

$> r(f(t), f_0(t))$ if '1' was sent. Therefore, it is reasonable to expect this strategy to work well against channel degradation. As a matter of fact, it is known that this strategy achieves the optimal performance, i.e., minimizes the probability of error, under certain conditions. As a result, it has been successfully applied to practical communication systems.

Examples of Inner Product Spaces

Earlier in this section, it was pointed out that the dot product in the space R^n is the Euclidean inner product on R^n. It turns out that this inner product induces the Euclidean norm. In the following, we present a few more examples of inner product spaces. The details of the demonstration that each of them satisfies the properties of the inner product are omitted in favor of brevity.

Example 1.4-1 *The l^2 space.* In the l^2 space, that is, the space of all the sequences of the form $(x_1, x_2, \cdots, x_n, \cdots)$ which satisfy

$$\|\mathbf{x}\|_2 = \left(\sum_{i=1}^{\infty} |x_i|^2 \right)^{1/2} < \infty, \tag{1.4-27}$$

we define

$$\langle \mathbf{x}, \mathbf{y} \rangle = \sum_{i=1}^{\infty} x_i \overline{y_i} \tag{1.4-28}$$

for all $\mathbf{x}, \mathbf{y} \in l^2$. Then Eq. (1.4-28) defines an inner product on l^2. Moreover, we have

$$\langle \mathbf{x}, \mathbf{x} \rangle = \sum_{i=1}^{\infty} x_i \overline{x_i} = \sum_{i=1}^{\infty} |x_i|^2 = \|\mathbf{x}\|_2^2, \tag{1.4-29}$$

for all $\mathbf{x} \in l^2$. Hence the inner product so-defined in Eq. (1.4-28) induces the l^2 norm. ∎

Unless otherwise exclusively specified, the notion of the l^2 space in what follows is meant to be the l^2 space together with the inner product in Eq. (1.4-28) as well as the induced l^2 norm built on it.

Example 1.4-2 *The L^2 space.* In the L^2 space, that is, the space of all the square-integrable functions defined on R, the L^2 norm of $\mathbf{f} = f(t), -\infty < t < \infty$, is given by

$$\|\mathbf{f}\|_2 = \left(\int_{-\infty}^{\infty} |f(t)|^2 \, dt \right)^{1/2} < \infty. \tag{1.4-30}$$

For all $\mathbf{f} = f(t), \mathbf{g} = g(t) \in L^2$, we define

$$\langle \mathbf{f}, \ \mathbf{g} \rangle = \langle f(t), \ g(t) \rangle = \int_{-\infty}^{\infty} f(t)\overline{g(t)} \ dt. \tag{1.4-31}$$

Then Eq. (1.4-31) defines an inner product on the L^2 space. Furthermore,

$$\langle \mathbf{f}, \ \mathbf{f} \rangle = \langle f(t), \ f(t) \rangle = \int_{-\infty}^{\infty} f(t)\overline{f(t)} \ dt = \|\mathbf{f}\|_2^2, \tag{1.4-32}$$

for all $\mathbf{f} \in L^2$. Hence the inner product defined in Eq. (1.4-31) induces the L^2 norm. ∎

Unless otherwise exclusively specified, the notion of L^2 space in what follows is meant to be the L^2 space together with the inner product in Eq. (1.4-31) as well as the induced L^2 norm built on it.

Example 1.4-3 *The space of pairs of L^2 functions.* Even though most practical problems can be formulated as a problem in either the l^2 or L^2 space, sometimes it is simpler if we treat these problems in a variant of these spaces or even some problem-dependent inner product spaces. As an example we extend the L^2-space inner product to the space of L^2 *function pairs*, which is ideal for dealing with problems such as the classical blind deconvolution problem (Section 7.6) and neural net training (Section 8.8).

Let $\mathbf{V}$ be the set of all the function pairs of the form $(f(t), g(t))$, where both $f(t)$ and $g(t)$ are in L^2. For all such function pairs $(f(t), g(t)), (u(t), v(t)) \in \mathbf{V}$, and α a scalar, we define,

$$(f(t), g(t)) + (u(t), v(t)) = (f(t) + u(t), g(t) + v(t)), \tag{1.4-33}$$

and

$$\alpha\,(f(t), g(t)) = (\alpha f(t), \alpha g(t)). \tag{1.4-34}$$

Then it is readily seen that the set $\mathbf{V}$ forms a vector space. To avoid notational complexity, in the following we simply use the function pair such as $(f(t), g(t))$ to denote a vector in $\mathbf{V}$ instead of a designated bold letter.

On the space $\mathbf{V}$, we define the inner product (Exercise 1-16) as

$$\langle (f(t), g(t)), \ (u(t), v(t)) \rangle \stackrel{\triangle}{=} \int_{-\infty}^{\infty} f(t)\overline{u(t)} \ dt + \int_{-\infty}^{\infty} g(t)\overline{v(t)} \ dt, \tag{1.4-35}$$

for all $(f(t), g(t)), (u(t), v(t)) \in \mathbf{V}$. Furthermore, this inner product also induces a norm $\|\cdot\|$ for $\mathbf{V}$. More specifically, for $(f(t), g(t)) \in \mathbf{V}$, we have

$$\|(f(t), g(t))\| \ = \ \langle (f(t), g(t)), \ (f(t), g(t)) \rangle^{1/2}$$

$$= \left(\int_{-\infty}^{\infty} |f(t)|^2 \, dt + \int_{-\infty}^{\infty} |g(t)|^2 \, dt \right)^{1/2}. \quad (1.4\text{-}36)$$

Note that if we use the notations $\langle \cdot, \ \cdot \rangle_2$ and $\|\cdot\|_2$ to denote the inner product and norm on the L^2 space, respectively, then Eq. (1.4-35) can be rewritten as

$$\langle (f(t), g(t)), \ (u(t), v(t)) \rangle = \langle f(t), \ u(t) \rangle_2 + \langle g(t), \ v(t) \rangle_2, \quad (1.4\text{-}37)$$

and Eq. (1.4-36) can be rewritten as

$$\|(f(t), g(t))\| = \left(\|f(t)\|_2^2 + \|g(t)\|_2^2 \right)^{1/2}. \quad (1.4\text{-}38)$$

■

1.5 HILBERT SPACES

From Secs. 1.3 and 1.4 it should be clear that the concept of norm allows us to talk about length, distance, as well as convergence in a vector space, while the concept of inner product enables us to talk about geometrical concepts such as angles, directions, and orthogonality. Furthermore, the inner product in an inner product space also induces a legitimate norm for the vector space. In this section, we briefly discuss an important class of inner product spaces called *Hilbert spaces*.

Definition

An inner product space which is *complete* with respect to the norm induced by the inner product is called a *Hilbert space*. In other words, an inner product space is a Hilbert space if and only if every Cauchy sequence converges with respect to the norm induced by the inner product.[†] To help clarify this concept, let's examine the examples of inner product spaces furnished in the last section.

Example 1.5-1 *The Euclidean space* R^n. On the Euclidean space R^n, the Euclidean inner product is defined as

$$\langle \mathbf{x}, \ \mathbf{y} \rangle = \sum_{i=1}^{n} x_i \, y_i, \quad (1.5\text{-}1)$$

[†] In the literature an inner product space is also called a *pre-Hilbert space*, the reason being that any inner product space can be modified to form a Hilbert space.

for all $\mathbf{x} = (x_1, x_2, \cdots, x_n), \mathbf{y} = (y_1, y_2, \cdots, y_n)$ in R^n. The induced Euclidean norm is

$$\|\mathbf{x}\|_2 = \left(\sum_{i=1}^{n} |x_i|^2 \right)^{1/2} , \tag{1.5-2}$$

for all $\mathbf{x} \in R^n$. Also, the space R^n is complete with respect to the Euclidean norm. Hence, the Euclidean space R^n is a Hilbert space.

From Section 1.3, it is known that a generalization of the Euclidean norm on the space R^n is the p-norm which is defined as

$$\|\mathbf{x}\|_p = \left(\sum_{i=1}^{n} |x_i|^p \right)^{1/p} , \tag{1.5-3}$$

for all $\mathbf{x} \in R^n$. For each $1 \le p \le \infty$, the space R^n associated with the p-norm defines a normed vector space which is also complete. An interesting question is whether all these p-normed spaces are Hilbert spaces. The answer is no except for the case of $p = 2$. The reason is that the p-norm for $p \ne 2$ cannot be induced by an inner product so that no inner product space will have the p-norm for $p \ne 2$ as its induced norm. Indeed, consider $\mathbf{x} = (1, 0, 0, \cdots, 0), \mathbf{y} = (0, 1, 0, \cdots, 0) \in R^n$, if the p-norm were induced by an inner product, then the parallelogram law (Eq. (1.4-12))

$$\|\mathbf{x} + \mathbf{y}\|_p^2 + \|\mathbf{x} - \mathbf{y}\|_p^2 = 2 \left(\|\mathbf{x}\|_p^2 + \|\mathbf{y}\|_p^2 \right) , \tag{1.5-4}$$

would need to be satisfied. For $1 \le p < \infty$, Eq. (1.5-4) yields

$$\left(2^{1/p} \right)^2 + \left(2^{1/p} \right)^2 = 2(1 + 1), \tag{1.5-5}$$

which is true if and only if $p = 2$. For the case $p = \infty$, Eq. (1.5-4) yields

$$1 + 1 = 2(1 + 1), \tag{1.5-6}$$

which, of course, is a contradiction. ∎

Example 1.5-2 *The l^2 space*. On the l^2 space, the inner product is defined as

$$\langle \mathbf{x}, \mathbf{y} \rangle = \sum_{i=1}^{\infty} x_i \overline{y_i} \tag{1.5-7}$$

for all $\mathbf{x}, \mathbf{y} \in l^2$. The induced l^2 norm is

$$\|\mathbf{x}\|_2 = \left(\sum_{i=1}^{\infty} |x_i|^2 \right)^{1/2} , \tag{1.5-8}$$

for all $\mathbf{x} \in l^2$. Also, the l^2 space is complete with respect to the l^2 norm. Hence, the l^2 space is a Hilbert space.

From Section 1.3, a generalization of the l^2 space is the l^p space on which the p-norm is defined as

$$\|\mathbf{x}\|_p = \left(\sum_{i=1}^{\infty} |x_i|^p \right)^{1/p}, \qquad (1.5\text{-}9)$$

when $1 \le p < \infty$; and

$$\|\mathbf{x}\|_p = \sup_{i=1,2,\cdots} \{|x_i|\}, \qquad (1.5\text{-}10)$$

when $p = \infty$. It turns out that none of these l^p norms except for the case of $p = 2$ can be induced by an inner product. Thus, the l^p space when $p \ne 2$ is not a Hilbert space. ∎

Example 1.5-3 *The L^2 space.* On the L^2 space, for $\mathbf{f} = f(t), \mathbf{g} = g(t) \in L^2$ their inner product is defined as

$$\langle \mathbf{f}, \ \mathbf{g} \rangle = \langle f(t), \ g(t) \rangle = \int_{-\infty}^{\infty} f(t)\overline{g(t)} \, dt. \qquad (1.5\text{-}11)$$

The induced L^2 norm is

$$\|\mathbf{f}\|_2 = \left(\int_{-\infty}^{\infty} |f(t)|^2 \, dt \right)^{1/2}, \qquad (1.5\text{-}12)$$

for all $\mathbf{f} = f(t) \in L^2$. Also, the L^2 space is complete with respect to the L^2 norm. Hence, the L^2 space is a Hilbert space.

From Section 1.3, a generalization of the L^2 space is the L^p space on which the L^p norm is defined as

$$\|\mathbf{f}\|_p = \left(\int_{-\infty}^{\infty} |f(t)|^p \, dt \right)^{1/p}, \qquad (1.5\text{-}13)$$

for all $\mathbf{f} = f(t) \in L^p$. It turns out that none of these L^p norms except for the case of $p = 2$ can be induced by an inner product. Therefore, the L^p space when $p \ne 2$ is not a Hilbert space. ∎

Example 1.5-4 *The space of pairs of L^2 functions.* It can be shown that the space of pairs of L^2 functions discussed in Example 1.4-3 in the last section is also a Hilbert space. The details of this demonstration is left to the reader.

Later in Chapter 7 (Section 7.6) this space will be found useful in solving an important problem in astronomy called *blind deconvolution*. There, the source function and the system distortion function, both unknown, are treated natually as an element in the space of pairs of L^2 functions. ∎

Orthonormal Basis

In a Hilbert space **H**, an orthonormal set is said to be an *orthonormal basis* for **H** if it is maximal in the sense that it cannot be extended to a larger orthonormal set in **H**. [†] An orthonormal basis is said to be *countable* if we can enumerate its elements using the integer sequence $n = 1, 2, 3, \cdots$. In other words, a countable orthonormal basis for a Hilbert space can be described by the set notation $\{\mathbf{u}_n, n = 1, 2, \cdots\} \in$ **H** . There can be many orthonormal bases for a Hilbert space. However, if one of them is countable then all of them are countable. From the point view of practical applications, most of the Hilbert spaces such as the R^n, l^2, and L^2 spaces admit to countable orthonormal bases. Therefore, in the following we will mainly focus on Hilbert spaces that admit to countable orthonormal bases.

The concept of orthonormal basis plays a major role in the analysis of Hilbert spaces. One of the most significant results regarding orthonormal basis in Hilbert spaces is the following:

Theorem 1.5-1 *Let* $\{\mathbf{u}_n, n = 1, 2, \cdots\}$ *be a countable orthonormal basis in a Hilbert space* **H**. *Then the following hold:*

(i) *For every* $\mathbf{x} \in$ **H**,

$$\mathbf{x} = \sum_{n=1}^{\infty} \langle \mathbf{x}, \mathbf{u}_n \rangle \mathbf{u}_n. \tag{1.5-14}$$

(ii) *For every* $\mathbf{x} \in$ **H**,

$$\|\mathbf{x}\|^2 = \sum_{n=1}^{\infty} |\langle \mathbf{x}, \mathbf{u}_n \rangle|^2. \tag{1.5-15}$$

The proof is given in several places, for example, see [4].

Theorem 1.5-1 reveals some important characteristics of Hilbert spaces. Equation (1.5-14) states that any vector in a Hilbert space can be decomposed into a summation of mutually orthogonal components, while Eq. (1.5-15) relates the norm of the vector with those of its decomposed components. In the literature, Eq. (1.5-14) is known as the *Fourier expansion formula*, and Eq. (1.5-15) as *Parseval's formula*.

These properties of an orthonormal basis for a Hilbert space are not only important for developing Hilbert space theory, but are also quite useful in practical applications. For example, in signal processing where signals are treated as vectors in a Hilbert space we often choose an orthonormal basis to represent a set of elementary signals. According to Eq. (1.5-14) we can decompose an arbitrary signal into a summation of elementary signals. Moreover Eq. (1.5-15) implies that such a decomposition has the so-called energy preservation property—the energy of the original signal is the same as the sum of those of the decomposed elementary

---

[†] By applying Zorn's lemma [4], one can show that every Hilbert space does admit an orthonormal basis.

signals. Note that Eq. (1.5-14) can be viewed from two different aspects. The first is, as pointed out above, that any signal can be decomposed [†] into elementary signals, i.e., the $\mathbf{u}_n$'s. One application of this fact is that in signal analysis we use the so-called spectrum analyzer to examine the frequency contents of a signal. The spectrum analyzer yields $\langle \mathbf{x}, \mathbf{u}_n \rangle$ or $|\langle \mathbf{x}, \mathbf{u}_n \rangle|^2$ for each $\mathbf{u}_n$ of interest. The second view of Eq. (1.5-14) is that any signal of interest can be reproduced from elementary signals. One application of this fact is that in signal synthesis, we use the so-called signal synthesizer to generate arbitrary waveforms from available elementary waveforms. [‡]

In the following we give several examples of orthonormal bases.

Example 1.5-5 For the Euclidean space R^n, the set of standard basis vectors $\{\mathbf{e}_1, \mathbf{e}_2, \cdots, \mathbf{e}_n\}$ with $\mathbf{e}_1 = (1, 0, 0, \cdots, 0, 0)$, $\mathbf{e}_2 = (0, 1, 0, \cdots, 0, 0)$, $\cdots$, and $\mathbf{e}_n = (0, 0, 0, \cdots, 0, 1)$ is an orthonormal basis.   ∎

Example 1.5-6 For the l^2 space, the set of vectors $\{\mathbf{u}_n, n = 1, 2, \cdots\}$ with $\mathbf{u}_n = (0, \cdots, 0, 1, 0, \cdots)$, where a 1 occurs only in the nth entry, is an orthonormal basis.   ∎

Example 1.5-7 Let $L^2([-\pi, \pi])$ be the set of all square-integrable functions that are defined on the interval $[-\pi, \pi]$. That is, a function $\mathbf{f} = f(t), t \in [-\pi, \pi]$, belongs to $L^2([-\pi, \pi])$ if and only if

$$\int_{-\pi}^{\pi} |f(t)|^2 \, dt \leq \infty. \tag{1.5-16}$$

It is straightforward to verify that the set $L^2([-\pi, \pi])$ forms a vector space under the pointwise addition and scalar multiplication. As a matter of fact, the space $L^2([-\pi, \pi])$ is a subspace of the function space L^2. Therefore, we can directly define the same inner product for $L^2([-\pi, \pi])$ as that of L^2. More specifically, for $\mathbf{f} = f(t), \mathbf{g} = g(t) \in L^2([-\pi, \pi])$ their inner product is defined as

$$\langle \mathbf{f}, \mathbf{g} \rangle = \langle f(t), g(t) \rangle = \int_{-\pi}^{\pi} f(t)\overline{g(t)} \, dt. \tag{1.5-17}$$

And the induced $L^2([-\pi, \pi])$ norm is

$$\|\mathbf{f}\|_2 = \left(\int_{-\pi}^{\pi} |f(t)|^2 \, dt \right)^{1/2}, \tag{1.5-18}$$

for all $\mathbf{f} = f(t) \in L^2([-\pi, \pi])$. Also, the space $L^2([-\pi, \pi])$ is a Hilbert space.

[†] Some may prefer the term "resolved" to "decomposed".
[‡] Not unlike music synthesizers that generate music from a set of basic tones.

For the space $L^2([-\pi, \pi])$, the set of functions

$$\left\{ \frac{1}{\sqrt{2\pi}}, \frac{1}{\sqrt{\pi}}\cos t, \frac{1}{\sqrt{\pi}}\sin t, \frac{1}{\sqrt{\pi}}\cos 2t, \frac{1}{\sqrt{\pi}}\sin 2t, \frac{1}{\sqrt{\pi}}\cos 3t, \frac{1}{\sqrt{\pi}}\sin 3t, \cdots \right\}$$

is an orthonormal basis. For $\mathbf{f} = f(t) \in L^2([-\pi, \pi])$, from the Fourier expansion and Parseval's formula we have

$$f(t) = a_0 + \sum_{n=1}^{\infty} \left(a_n \cos nt + b_n \sin nt \right), \tag{1.5-19}$$

and

$$\|f(t)\|^2 = \pi \left[2|a_0|^2 + \sum_{n=1}^{\infty} \left(|a_n|^2 + |b_n|^2 \right) \right], \tag{1.5-20}$$

where

$$a_0 = \frac{1}{2\pi} \int_{-\pi}^{\pi} f(t) \, dt, \quad a_n = \frac{1}{\pi} \int_{-\pi}^{\pi} f(t) \cos nt \, dt, \quad b_n = \frac{1}{\pi} \int_{-\pi}^{\pi} f(t) \sin nt \, dt, \tag{1.5-21}$$

for $n = 1, 2, \cdots$.

Clearly, this last example is the classical Fourier series expansion formula which is widely used in all branches of science and engineering. ∎

Weak Convergence

Earlier in Section 1.3 we introduced the concept of convergence in a normed vector space. By definition, a sequence $\{\mathbf{x}_n\}$ in a Hilbert space $\mathbf{H}$ converges to $\mathbf{x} \in \mathbf{H}$ if $\lim_{n \to \infty} \|\mathbf{x}_n - \mathbf{x}\| = 0$. In such a case, we write $\mathbf{x}_n \longrightarrow \mathbf{x}$. Such a convergence is also called convergence in norm.

We now introduce another concept of convergence which is important for the theory of vector space analysis. In a Hilbert space $\mathbf{H}$, a sequence $\{\mathbf{x}_n\}$ is said to *converge weakly* to $\mathbf{x} \in \mathbf{H}$ if $\langle \mathbf{x}_n, \mathbf{y} \rangle$ converges to $\langle \mathbf{x}, \mathbf{y} \rangle$ for every $\mathbf{y} \in \mathbf{H}$; $\mathbf{x}$ is then called a *weak limit* of $\{\mathbf{x}_n\}$. In such a case, we write $\mathbf{x}_n \overset{w}{\longrightarrow} \mathbf{x}$. The following theorem reveals an important fact of weak convergence: a weakly convergent sequence has a unique limit.

Theorem 1.5-2 *If* $\mathbf{x}_n \overset{w}{\longrightarrow} \mathbf{x}$, *and* $\mathbf{x}_n \overset{w}{\longrightarrow} \mathbf{y} \in \mathbf{H}$, *then* $\mathbf{x} = \mathbf{y}$.

Proof: Assume that $\mathbf{x}_n \overset{w}{\longrightarrow} \mathbf{x}$, and $\mathbf{x}_n \overset{w}{\longrightarrow} \mathbf{y}$. Then by definition, for every $\mathbf{z} \in \mathbf{H}$ we have

$$\lim_{n \to \infty} \langle \mathbf{x}_n, \mathbf{z} \rangle = \langle \mathbf{x}, \mathbf{z} \rangle, \tag{1.5-22}$$

and

$$\lim_{n \to \infty} \langle \mathbf{x}_n, \mathbf{z} \rangle = \langle \mathbf{y}, \mathbf{z} \rangle. \tag{1.5-23}$$

Hence, we have

$$\langle \mathbf{x}, \mathbf{z} \rangle = \langle \mathbf{y}, \mathbf{z} \rangle \qquad (1.5\text{-}24)$$

for every $\mathbf{z} \in \mathbf{H}$. By letting $\mathbf{z} = \mathbf{x} - \mathbf{y}$, we obtain from Eq. (1.5-24)

$$\langle \mathbf{x} - \mathbf{y}, \mathbf{x} - \mathbf{y} \rangle = 0, \qquad (1.5\text{-}25)$$

which implies $\mathbf{x} = \mathbf{y}$. ∎

In contrast to weak convergence, convergence in norm is also termed *strong convergence*. An interesting question is how these two different types of convergence are connected. The answer to this question leads to the following result, which reveals another important property of weak convergence.

Theorem 1.5-3 *In a Hilbert space* **H**,

(i) *If* $\mathbf{x}_n \longrightarrow \mathbf{x}$, *then* $\mathbf{x}_n \xrightarrow{w} \mathbf{x}$.

(ii) *If* **H** *is finite-dimensional, then* $\mathbf{x}_n \xrightarrow{w} \mathbf{x}$ *implies* $\mathbf{x}_n \longrightarrow \mathbf{x}$.

Proof: (i) Assume that $\mathbf{x}_n \longrightarrow \mathbf{x}$, i.e., $\lim_{n \to \infty} \|\mathbf{x}_n - \mathbf{x}\| = 0$. Then for every $\mathbf{y} \in \mathbf{H}$, by the Schwarz inequality we have

$$|\langle \mathbf{x}_n - \mathbf{x}, \mathbf{y} \rangle| \le \|\mathbf{x}_n - \mathbf{x}\| \, \|\mathbf{y}\|. \qquad (1.5\text{-}26)$$

It follows that

$$\lim_{n \to \infty} |\langle \mathbf{x}_n - \mathbf{x}, \mathbf{y} \rangle| = 0, \qquad (1.5\text{-}27)$$

which yields

$$\lim_{n \to \infty} \langle \mathbf{x}_n, \mathbf{y} \rangle = \langle \mathbf{x}, \mathbf{y} \rangle. \qquad (1.5\text{-}28)$$

Hence, $\mathbf{x}_n \xrightarrow{w} \mathbf{x}$.

(ii) Without loss of generality, assume that **H** is of dimension $m < \infty$, and that $\{\mathbf{e}_1, \mathbf{e}_2, \cdots, \mathbf{e}_m\}$ is an orthonormal basis for **H**. Assume also that $\mathbf{x}_n \xrightarrow{w} \mathbf{x} \in \mathbf{H}$, we want to show that $\mathbf{x}_n \longrightarrow \mathbf{x}$. According to Theorem 1.5-1, we have

$$\mathbf{x}_n = \langle \mathbf{x}_n, \mathbf{e}_1 \rangle \mathbf{e}_1 + \langle \mathbf{x}_n, \mathbf{e}_2 \rangle \mathbf{e}_2 + \cdots + \langle \mathbf{x}_n, \mathbf{e}_m \rangle \mathbf{e}_m, \qquad (1.5\text{-}29)$$

for every $\mathbf{x}_n$. Similarly for the limit $\mathbf{x}$,

$$\mathbf{x} = \langle \mathbf{x}, \mathbf{e}_1 \rangle \mathbf{e}_1 + \langle \mathbf{x}, \mathbf{e}_2 \rangle \mathbf{e}_2 + \cdots + \langle \mathbf{x}, \mathbf{e}_m \rangle \mathbf{e}_m. \qquad (1.5\text{-}30)$$

By the definition of weak convergence, we have

$$\lim_{n \to \infty} \langle \mathbf{x}_n, \mathbf{y} \rangle = \langle \mathbf{x}, \mathbf{y} \rangle \qquad (1.5\text{-}31)$$

for every $\mathbf{y} \in \mathbf{H}$. Letting $\mathbf{y} = \mathbf{e}_1$, Eqs. (1.5-29) through (1.5-31) yield

$$\lim_{n \to \infty} \langle \mathbf{x}_n, \mathbf{e}_1 \rangle = \langle \mathbf{x}, \mathbf{e}_1 \rangle. \qquad (1.5\text{-}32)$$

Thus, we have

$$\lim_{n\to\infty} \langle \mathbf{x}_n - \mathbf{x}, \ \mathbf{e}_1 \rangle = 0. \tag{1.5-33}$$

In a similar fashion, we obtain

$$\lim_{n\to\infty} \langle \mathbf{x}_n - \mathbf{x}, \ \mathbf{e}_2 \rangle = 0, \cdots, \ \lim_{n\to\infty} \langle \mathbf{x}_n - \mathbf{x}, \ \mathbf{e}_m \rangle = 0. \tag{1.5-34}$$

By applying Parseval's formula in Theorem 1.5-1, we have

$$\|\mathbf{x}_n - \mathbf{x}\|^2 = \sum_{k=1}^{m} |\langle \mathbf{x}_n - \mathbf{x}, \ \mathbf{e}_k \rangle|^2. \tag{1.5-35}$$

It follows that

$$\lim_{n\to\infty} \|\mathbf{x}_n - \mathbf{x}\| = 0. \tag{1.5-36}$$

Thus, $\mathbf{x}_n \longrightarrow \mathbf{x}$.   ∎

The above theorem states that in a finite-dimensional Hilbert space, weak convergence is equivalent to strong convergence. However, in infinite-dimensional Hilbert spaces, strong convergence implies weak convergence, but not the other way around. In the following, we give two examples to demonstrate that strong convergence is indeed a stronger condition of the two in infinite-dimensional Hilbert spaces.

Example 1.5-8 In the l^2 space, the set of vectors $\{\mathbf{u}_n, n = 1, 2, \cdots\}$ with $\mathbf{u}_n = (0, \cdots, 0, 1, 0, \cdots)$, where a 1 occurs only in the nth entry, is an orthonormal basis for l^2. By Parseval's formula, for every $\mathbf{y} \in l^2$ we have

$$\|\mathbf{y}\|^2 = \sum_{n=1}^{\infty} |\langle \mathbf{y}, \ \mathbf{u}_n \rangle|^2 < \infty, \tag{1.5-37}$$

which implies

$$\lim_{n\to\infty} \langle \mathbf{y}, \ \mathbf{u}_n \rangle = 0. \tag{1.5-38}$$

Thus, we have

$$\lim_{n\to\infty} \langle \mathbf{u}_n, \ \mathbf{y} \rangle = \langle \mathbf{0}, \ \mathbf{y} \rangle \tag{1.5-39}$$

for every $\mathbf{y} \in l^2$. Hence, $\mathbf{u}_n \xrightarrow{w} \mathbf{0}$. On the other hand, the sequence $\{\mathbf{u}_n\}$ does not converge to $\mathbf{0}$, since $\|\mathbf{u}_n - \mathbf{0}\| = \|\mathbf{u}_n\| = 1$ for every n.   ∎

Example 1.5-9 In the space $L^2([-\pi, \pi])$, the set of functions

$$\left\{ \frac{1}{\sqrt{2\pi}}, \frac{1}{\sqrt{\pi}}\cos t, \frac{1}{\sqrt{\pi}}\sin t, \frac{1}{\sqrt{\pi}}\cos 2t, \frac{1}{\sqrt{\pi}}\sin 2t, \frac{1}{\sqrt{\pi}}\cos 3t, \frac{1}{\sqrt{\pi}}\sin 3t, \cdots \right\},$$

is an orthonormal basis for $L^2([-\pi, \pi])$. By using an argument similar to that in the last example, we obtain $\cos nt \xrightarrow{w} \mathbf{0}$, and $\sin nt \xrightarrow{w} \mathbf{0}$, where $\mathbf{0}$ represents the zero function in $L^2([-\pi, \pi])$. On the other hand, neither of them converges to $\mathbf{0}$ in norm. ∎

1.6 SUMMARY

In this chapter we furnished the basic theory of vector spaces. We began by defining what a vector space is and then introduced the idea of *normed* spaces and *inner product* spaces. We pointed out that the norm was closely tied to the idea of length and that the inner product of two vectors was closely related to the angular difference between vectors. We then defined and examined the properties of Hilbert spaces, and showed that three commonly used vectors and signals can be organized into Hilbert spaces. We concluded the chapter with a discussion of the difference between strong and weak convergence and proved that the two are identical in finite-dimensional spaces but could be different in infinite-dimensional spaces.

REFERENCES

1. H. Anton, *Elementary Linear Algebra,* 6th edition, John Wiley & Sons, New York, 1991.

2. S. H. Friedberg, A. J. Insel, and L. E. Spence, *Linear Algebra,* Prentice-Hall, Englewood Cliffs, NJ, 1979.

3. P. R. Halmos, *Finite-Dimensional Vector Spaces,* Springer-Verlag, New York, 1974.

4. B. V. Limaye, *Functional Analysis*, John Wiley & Sons, New York, 1981.

5. N. I. Akhiezer and I. M. Glazman, *Theory of Linear Operators in Hilbert Space,* Vol. 1, Frederick Ungar Publishing Company, New York, 1978.

EXERCISES

1-1. Verify that vector space R^n defined in Example 1.2-1 indeed forms a valid space. Verify also that the set of vectors of the form $(0, a_2, \cdots, a_n)$ is a subspace in this space.

1-2. Show that the set of all 2×2 real-coefficient matrices of the form

$$\begin{bmatrix} a_{11} & a_{12} \\ a_{21} & a_{22} \end{bmatrix}$$

forms a vector space **V** under the operations of component-wise addition and scalar multiplication; that is, for $\mathbf{x} = \begin{bmatrix} x_{11} & x_{12} \\ x_{21} & x_{22} \end{bmatrix}$, and

$\mathbf{y} = \begin{bmatrix} y_{11} & y_{12} \\ y_{21} & y_{22} \end{bmatrix}$ both in **V**, and α a real number,

$$\mathbf{x} + \mathbf{y} = \begin{bmatrix} x_{11} + y_{11} & x_{12} + y_{12} \\ x_{21} + y_{21} & x_{22} + y_{22} \end{bmatrix},$$

and

$$\alpha\mathbf{x} = \begin{bmatrix} \alpha x_{11} & \alpha x_{12} \\ \alpha x_{21} & \alpha x_{22} \end{bmatrix}.$$

1-3. Show that in a vector space **V** the span of a non-empty set S forms a subspace in **V**.

1-4. Prove that if a vector space has a basis that contains a finite number of vectors, then all the other bases of the vector space will contain exactly the same number of vectors.

1-5. Find a basis for the vector space in Exercise 1-2. Also, what is its dimension?

1-6. Find a subspace of the vector space in Exercise 1-2.

1-7. Show that every set with more than n vectors is necessarily linearly dependent in an n-dimensional vector space.

1-8. Show that a convergent sequence has a unique limit.

1-9. Give an example of a non-convergent Cauchy sequence in a vector space.

1-10. Show that a set is closed if and only if it contains all of its limit points.

1-11. Show that for any two 2×2 matrices $\mathbf{x} = \begin{bmatrix} x_{11} & x_{12} \\ x_{21} & x_{22} \end{bmatrix}$, and $\mathbf{y} = \begin{bmatrix} y_{11} & y_{12} \\ y_{21} & y_{22} \end{bmatrix}$, the following formula

$$\langle \mathbf{x},\ \mathbf{y} \rangle \overset{\triangle}{=} x_{11}y_{11} + x_{12}y_{12} + x_{21}y_{21} + x_{22}y_{22}$$

defines an inner product on the vector space in Exercise 1-2.

1-12. Let $\mathbf{x} = (x_1, x_2, \cdots, x_n) \in R^n$, and $\mathbf{y} = (y_1, y_2, \cdots, y_n) \in R^n$. Show that

$$\langle \mathbf{x},\ \mathbf{y} \rangle \overset{\triangle}{=} \sum_{i=1}^{n} w_i x_i y_i$$

defines an inner product on R^n if $w_1, w_2, \cdots, w_n$ are positive real numbers. This inner product is called the *weighted Euclidean inner product* on R^n. Also, show that the weighted norm in Eq. (1.3-5) (Section 1.3) is induced by the weighted Euclidean inner product with a proper choice of weights $w_1, w_2, \cdots, w_n$ in Eq. (1.3-5).

1-13. Extend the Euclidean inner product in R^n to the space of n-tuple complex numbers C^n, i.e., a vector $\mathbf{x}$ in C^n has the form $(c_1, c_2, \cdots, c_n)$, where each $C_i, i = 1, 2, \cdots, n$, is a complex number.

1-14. Show that an orthonormal set in an inner product space is automatically an independent set.

1-15. Use the Gram-Schmidt process to transform the basis $\{\mathbf{u}_1, \mathbf{u}_2, \mathbf{u}_3\}$ into an orthonormal one in the Euclidean space R^3.

    (a) $\mathbf{u}_1 = (1, 1, 1), \mathbf{u}_2 = (-1, 1, 0), \mathbf{u}_3 = (1, 2, 1)$

    (b) $\mathbf{u}_1 = (1, 0, 0), \mathbf{u}_2 = (3, 7, -2), \mathbf{u}_3 = (0, 4, 1)$.

1-16. Show that Eq. (1.4-35) defines a valid inner product on the space of L^2 function pairs.

2

Projections Onto Convex Sets

2.1 INTRODUCTION

Having reviewed some of the key concepts of vector spaces in Chapter 1, we are now ready to discuss a main result of vector-space projections, namely, the theory of *projections onto convex sets* (POCS). The concept of a convex set is first introduced and the projection onto a convex set is discussed. The properties associated with projections are examined in great detail. Then the fundamental theory of POCS is presented and demonstrated rigorously using an important result from *fixed point theory* of non-expansive mappings by Opial [1]. The iterative POCS algorithm and its variants are discussed and their numerical issues are studied. The proof of Opial's theorem is furnished near the end of the chapter for completeness, although it is not needed to understand the material in the rest of the book. We complete the chapter by discussing projections in *product spaces* and the interesting results that follow from this concept. Throughout this chapter, we will restrict ourselves to the framework of Hilbert spaces unless otherwise specified.

2.2 CONVEX SETS

Consider the set $C_1 = \{ \mathbf{x} = (x_1, x_2) \ : \ x_1^2 + x_2^2 \leq 4 \}$ in the space R^2. This set consists of all the points *on and inside* the circle shown in Fig. 2.2-1. The set C_1 has the property that every point on the line segment l connecting any two points $\mathbf{x}_1$ and $\mathbf{x}_2$ in C_1 is also in C_1. How do we describe this mathematically? Consider a point $\mathbf{x}$ on the line segment l connecting $\mathbf{x}_2$ to $\mathbf{x}_1$. Clearly the vector $\mathbf{x} - \mathbf{x}_2$ is

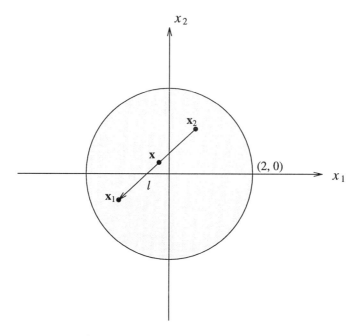

Fig. 2.2-1 The $C_1 = \left\{\, \mathbf{x} = (x_1, x_2) \;:\; x_1^2 + x_2^2 \leq 4 \,\right\}$ in R^2 consists of all the points *on and inside* the circle which is of radius 2 and centered at origin.

parallel to the vector $\mathbf{x}_1 - \mathbf{x}_2$; we write this as

$$\mathbf{x} - \mathbf{x}_2 = \mu(\mathbf{x}_1 - \mathbf{x}_2), \qquad (2.2\text{-}1)$$

where μ is a real constant of proportionality. Now if $\mathbf{x}$ lies on the line segment l then the length of $\mathbf{x} - \mathbf{x}_2$ must not exceed that of $\mathbf{x}_1 - \mathbf{x}_2$. Thus, $0 \leq \mu \leq 1$. Note that when $\mu = 1$ (i.e., $\mathbf{x} = \mathbf{x}_1$) the vector $\mathbf{x} - \mathbf{x}_2$ achieves its maximum length which is $\|\mathbf{x}_1 - \mathbf{x}_2\|$; and when $\mu = 0$ (i.e., $\mathbf{x} = \mathbf{x}_2$) it achieves its minimum length which is zero. A set with the property that all points on the line segment joining *any* two points in the set are also in the set is said to be *convex* or exhibit *convexity*. Thus for a convex set C the following holds true: Let $\mathbf{x}_1, \mathbf{x}_2 \in C$ and let $0 \leq \mu \leq 1$ then $\mathbf{x} \overset{\triangle}{=} \mu(\mathbf{x}_1 - \mathbf{x}_2) + \mathbf{x}_2 \in C$. Or, equivalently, a set C is convex if

$$\mathbf{x} = \mu\,\mathbf{x}_1 + (1 - \mu)\,\mathbf{x}_2 \in C \qquad (2.2\text{-}2)$$

for all $\mathbf{x}_1, \mathbf{x}_2 \in C$ and $0 \leq \mu \leq 1$. The set $C_1 = \left\{\, \mathbf{x} \in R^2 \;:\; x_1^2 + x_2^2 \leq 4 \,\right\}$ is easily demonstrated to satisfy this property. Let $\mathbf{x}_1, \mathbf{x}_2 \in C$, then $\|\mathbf{x}_1\| \leq 2$, and $\|\mathbf{x}_2\| \leq 2$. Thus, for $\mathbf{x} = \mu\,\mathbf{x}_1 + (1 - \mu)\,\mathbf{x}_2 \in C$ with $0 \leq \mu \leq 1$, we have

$$\|\mathbf{x}\| = \|\mu\,\mathbf{x}_1 + (1 - \mu)\,\mathbf{x}_2\| \leq \mu\,\|\mathbf{x}_1\| + (1 - \mu)\,\|\mathbf{x}_2\| \leq 2. \qquad (2.2\text{-}3)$$

Indeed, C_1 is convex.

Are all sets convex? Most assurely not. Consider the set

$$C_2 = \left\{ \mathbf{x} = (x_1, x_2) \ : \ x_1^2 + x_2^2 = 4 \right\} \qquad (2.2\text{-}4)$$

in R^2. This set is a circle. Now consider $\mathbf{x}_1 = (0, 2) \in C_2$ and $\mathbf{x}_2 = (0, -2) \in C_2$ and let $\mu = 0.5$, then $\mathbf{x} = \mu\,\mathbf{x}_1 + (1 - \mu)\,\mathbf{x}_2 = (0, 0)$. Clearly this point is not in the set. Indeed, while a great many sets of engineering and scientific importance are convex, many are not. In the theory of POCS, all sets are assumed convex. Chapter 5 considers vector-space projections in which sets are non-convex.

2.3 PROJECTION

Consider the following geometry problems (Fig. 2.3-1): In the two-dimensional (2-D) plane, the perpendicular from an arbitrary point $\mathbf{x}$ to a line l intersects the line at exactly one point, say $\mathbf{x}^*$. This point $\mathbf{x}^*$ has an important property in that among all the points on the line l it is the closest (in the Euclidean distance) to the point $\mathbf{x}$ (Fig. 2.3-1(a)). Next, assume that D is a disc and $\mathbf{y}$ an arbitrary point outside the disc, as shown in Fig. 2.3-1(b). Among all circles centered at the point $\mathbf{y}$, only one touches the disc D at exactly one point, say $\mathbf{y}^*$. This point $\mathbf{y}^*$ has the property that it is the closest to the point $\mathbf{y}$ among all the points on the disc D. How do we find $\mathbf{x}^*$ and $\mathbf{y}^*$ analytically? The answer is furnished in this section.

Note that geometric entities such as a line or a disc can be viewed as sets of points in the space R^2. More importantly, the set of points on either a line or a disc is a closed convex set in R^2. The two problems examined above reveal a very important property of closed convex sets in a Hilbert space, which is summarized in the following theorem:

Theorem 2.3-1 *Let C be a closed convex set in a Hilbert space $\mathbf{H}$, then for each $\mathbf{x}$ in $\mathbf{H}$, there exists a unique $\mathbf{x}^*$ in C that is closest to $\mathbf{x}$. In short,*

$$\|\mathbf{x} - \mathbf{x}^*\| = \min_{\mathbf{y} \in C} \|\mathbf{x} - \mathbf{y}\|, \qquad (2.3\text{-}1)$$

and $\mathbf{x}^* \in C$.

Proof: For each $\mathbf{x} \in \mathbf{H}$ we need to show both the *existence* and the *uniqueness* of such an $\mathbf{x}^*$ satisfying Eq. (2.3-1).

First, the easier part (uniqueness): assume that such an $\mathbf{x}^*$ exists; we show that it is unique by contradiction. Suppose there exists another $\mathbf{z}$ in C other than $\mathbf{x}^*$ satisfying Eq. (2.3-1). Then, we have

$$\|\mathbf{x}^* - \mathbf{z}\| > 0 \ \ \text{and} \ \ \|\mathbf{x} - \mathbf{z}\| = \|\mathbf{x} - \mathbf{x}^*\| = \min_{\mathbf{y} \in C} \|\mathbf{x} - \mathbf{y}\|. \qquad (2.3\text{-}2)$$

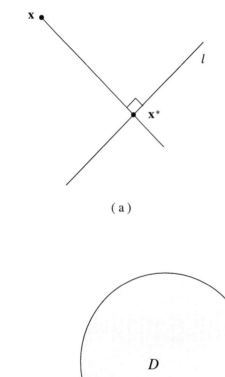

(a)

(b)

Fig. 2.3-1   (a) In the 2-D plane, the perpendicular from an arbitrary point **x** to a line l intersects the line at exactly one point, say $\mathbf{x}^*$. This point is the *projection* of **x** onto l because it is the closest point to **x** among all points on l. (b) Also in the 2-D plane, among all circles centered at the point **y**, only one touches the disc D at exactly one point, say $\mathbf{y}^*$. This point is the projection of **y** onto D.

Invoking the parallelogram law of the norm in a Hilbert space

$$\|u + v\|^2 + \|u - v\|^2 = 2\|u\|^2 + 2\|v\|^2 \qquad (2.3\text{-}3)$$

with $u = x - z$ and $v = x - x^*$, we obtain

$$\|2x - (z + x^*)\|^2 + \|x^* - z\|^2 = 2\|x - z\|^2 + 2\|x - x^*\|^2. \qquad (2.3\text{-}4)$$

Combining Eq. (2.3-2) with Eq. (2.3-4), we obtain

$$\left\|x - \tfrac{z + x^*}{2}\right\|^2 < \|x - x^*\|^2, \qquad (2.3\text{-}5)$$

which is impossible since, by the definition of x^*, it is closest to x. Note that the set C is convex and $(z + x^*)/2$ belongs to C. Thus, the uniqueness of x^* follows.

Next, we demonstrate that for each $x \in H$ there exists an x^* satisfying Eq. (2.3-1) (existence). Let δ be the largest lower bound on the distance between the point x and any point $y \in C$. That is,

$$\delta = \inf_{y \in C} \|x - y\|. \qquad (2.3\text{-}6)$$

Then for each $\delta_n = \delta + \frac{1}{n}$, $n = 1, 2, \cdots$, there exists a point $x_n \in C$ such that $\|x - x_n\| < \delta_n$. Therefore, we obtain a sequence $\{x_n\}$ such that

$$\lim_{n \to \infty} \|x - x_n\| = \delta. \qquad (2.3\text{-}7)$$

Invoking the parallelogram law in Eq. (2.3-3) with $u = x - x_m$ and $v = x - x_n$, we obtain

$$\|x_m - x_n\|^2 = 2\|x - x_m\|^2 + 2\|x - x_n\|^2 - \|2x - (x_m + x_n)\|^2. \qquad (2.3\text{-}8)$$

On the other hand, the set C is convex, so that $(x_m + x_n)/2 \in C$ and

$$\left\|x - \tfrac{x_m + x_n}{2}\right\| \geq \delta. \qquad (2.3\text{-}9)$$

Thus, Eq. (2.3-8) yields

$$\|x_m - x_n\|^2 \leq 2\|x - x_m\|^2 + 2\|x - x_n\|^2 - 4\delta^2, \qquad (2.3\text{-}10)$$

from which we obtain

$$\lim_{m,n \to \infty} \|x_m - x_n\|^2 \leq 2 \lim_{m \to \infty} \|x - x_m\|^2 + 2 \lim_{n \to \infty} \|x - x_n\|^2 - 4\delta^2 = 0.$$
$$(2.3\text{-}11)$$

Thus, the sequence $\{x_n\}$ is a Cauchy sequence and hence has a limit, say x^*, in

H. [†] Since the set C is closed, the limit $\mathbf{x}^*$ also belongs to C. Note that

$$
\begin{aligned}
\|\mathbf{x} - \mathbf{x}^*\| &= \|\mathbf{x} - \mathbf{x}_n + \mathbf{x}_n - \mathbf{x}^*\| \\
&\leq \|\mathbf{x} - \mathbf{x}_n\| + \|\mathbf{x}_n - \mathbf{x}^*\|.
\end{aligned} \tag{2.3-12}
$$

By letting $n \to \infty$, we obtain $\|\mathbf{x} - \mathbf{x}^*\| \leq \delta$. Since $\mathbf{x}^* \in C$, $\|\mathbf{x} - \mathbf{x}^*\| \geq \delta$. Therefore, the point $\mathbf{x}^*$ satisfies Eq. (2.3-1). By the first half of the theorem, such an $\mathbf{x}^*$ is also unique. ∎

Theorem 2.3-1 reveals a fundamental characteristic associated with closed convex sets in a Hilbert space. For a closed convex set C in a Hilbert space $\mathbf{H}$, Eq. (2.3-1) defines a rule which assigns to every $\mathbf{x} \in \mathbf{H}$ its unique nearest neighbor in the set C which we shall call the *projection* of $\mathbf{x}$ onto C. This rule is completely determined by the set C and is called the *projector* or *projection operator* onto C, which we shall denote by P_C. That is, for every $\mathbf{x} \in \mathbf{H}$ its projection $P_C\mathbf{x}$ onto C is defined by

$$
\|\mathbf{x} - P_C\mathbf{x}\| = \min_{\mathbf{y} \in C} \|\mathbf{x} - \mathbf{y}\|. \tag{2.3-13}
$$

Now, let's look back at the examples considered at the beginning of this section. In Fig. 2.3-1(a) the point $\mathbf{x}^*$ has the shortest distance to the point $\mathbf{x}$ among all the points on the line l (which is a closed convex set). Thus, $\mathbf{x}^*$ is the projection of $\mathbf{x}$ onto the line l. In Fig. 2.3-1(b) the point $\mathbf{y}^*$ has the shortest distance to the point $\mathbf{y}$ among all the points on the disc D (which later on is shown to a closed convex set in this section). Thus $\mathbf{y}^*$ is the projection of $\mathbf{y}$ onto the disc D.

Equation (2.3-13) gives an implicit form of the projection of an arbitrary point onto a closed convex set. In practice it is often desired to derive the explicit form of the projection for a given closed convex set. For example, in Fig. 2.3-1(b) we may want to express $\mathbf{y}^*$ explicitly in terms of $\mathbf{y}$ and the disc D. For such a purpose, we find the following result very helpful:

Corollary 2.3-1 *Let C be a closed convex set in a Hilbert space $\mathbf{H}$, then for each point $\mathbf{x} \in \mathbf{H}$: (i) if $\mathbf{x} \in C$ then $P_C\mathbf{x} = \mathbf{x}$; and (ii) if $\mathbf{x} \notin C$, then $P_C\mathbf{x}$ is on the boundary of C.*

Proof: Obviously, if $\mathbf{x} \in C$, then the closest point to $\mathbf{x}$ in C is $\mathbf{x}$ itself, i.e., $P_C\mathbf{x} = \mathbf{x}$. In what follows we show by contradiction that for $\mathbf{x} \notin C$, $P_C\mathbf{x}$ is a boundary point of C. Suppose that $P_C\mathbf{x}$ is not on the boundary of C, then it is an interior point of C so that there exists an $\epsilon > 0$ such that the set $S_\epsilon = \{\mathbf{y} : \|\mathbf{y} - P_C\mathbf{x}\| \leq \epsilon\}$ is contained in C. By assumption, $\mathbf{x} \notin C$, so $\|\mathbf{x} - P_C\mathbf{x}\| > 0$. Let

$$
\mathbf{z} = P_C\mathbf{x} + \frac{\epsilon}{2\|\mathbf{x} - P_C\mathbf{x}\|} (\mathbf{x} - P_C\mathbf{x}), \tag{2.3-14}
$$

[†]Recall that in a Hilbert space every Cauchy sequence has a limit.

then $\mathbf{z} \in S_\epsilon$. Clearly, if ϵ is chosen small enough, say $\epsilon < \|\mathbf{x} - P_C\mathbf{x}\|$, then

$$\|\mathbf{x} - \mathbf{z}\| = \left(1 - \frac{\epsilon}{2\|\mathbf{x} - P_C\mathbf{x}\|}\right)\|\mathbf{x} - P_C\mathbf{x}\| < \|\mathbf{x} - P_C\mathbf{x}\|, \qquad (2.3\text{-}15)$$

which is impossible, since $\mathbf{z} \in C$. Thus, $P_C\mathbf{x}$ has to be a boundary point of C. ∎

To rephrase: Corollary 2.3-1 states that the projection of a point onto a closed convex set is itself if the point is already in the set; on the other hand, if the point is outside the set, then its projection onto the set is always on the boundary of the set. This immediately leads to the following result:

Corollary 2.3-2 *Let C be a closed convex set in a Hilbert space $\mathbf{H}$, and P_C the projector onto C. Then $P_C\mathbf{x} = \mathbf{x}$ if and only if $\mathbf{x} \in C$. That is, P_C has the set C as its set of fixed points.*[†]

Proof: If $\mathbf{x} \in C$, then from Corollary 2.3-1 $P_C\mathbf{x} = \mathbf{x}$. On the other hand, if $P_C\mathbf{x} = \mathbf{x}$, then $\mathbf{x} \in C$ since $P_C\mathbf{x} \in C$.   ∎

From the discussion above it is clear that in a Hilbert space a closed convex set is completely characterized by its associated projector. Indeed, given a closed convex set, its projector is well defined from Eq. (2.3-13). On the other hand, given a valid projector onto a closed convex set, its associated closed convex set is simply the set of fixed points of the projector.

In the following we give some examples that illustrate that the results furnished above are useful for finding the projectors onto closed convex sets. For simplicity, let's first consider a problem raised at the beginning of this section.

Example 2.3-1 In R^2, consider the unit-disc defined by the set

$$C = \{\mathbf{x} : \|\mathbf{x}\| \le 1\}. \qquad (2.3\text{-}16)$$

We claim that the set C is closed and convex. Indeed, let $\mathbf{x}_1, \mathbf{x}_2 \in C$, and $\alpha \in [0, 1]$, then for $\mathbf{x}_3 = \alpha\,\mathbf{x}_1 + (1 - \alpha)\,\mathbf{x}_2$, we have $\|\mathbf{x}_3\| \le \|\alpha\mathbf{x}_1\| + \|(1 - \alpha)\mathbf{x}_2\| = \alpha\|\mathbf{x}_1\| + (1 - \alpha)\|\mathbf{x}_2\| \le \alpha + (1 - \alpha) = 1$. Hence C is convex. Next, let $\{\mathbf{y}_n\}$ be a converging sequence contained in C. We want to show that its limit, say $\mathbf{y}^*$, is also contained in C.[‡] Indeed,

$$\|\mathbf{y}^*\| = \|(\mathbf{y}^* - \mathbf{y}_n) + \mathbf{y}_n\| \le \|\mathbf{y}^* - \mathbf{y}_n\| + \|\mathbf{y}_n\| \le \|\mathbf{y}^* - \mathbf{y}_n\| + 1 \quad (2.3\text{-}17)$$

for every n. By letting $n \to \infty$, we obtain $\|\mathbf{y}^*\| \le 1$. Thus $\mathbf{y}^* \in C$. This proof also reveals that, in general, the unit-disc defined in Eq. (2.3-16) is always closed and convex in an arbitrary normed vector space.

[†]A point $\mathbf{x}$ is said to be a *fixed point* of an operator T if $T\mathbf{x} = \mathbf{x}$.
[‡]Recall that if $\mathbf{y}^*$ is the limit of the sequence $\{\mathbf{y}_n\}$ then $\lim_{n\to\infty} \|\mathbf{y}_n - \mathbf{y}^*\| = 0$ by definition.

The question of concern then is to derive the projector P_C for the set C. To start with, let $\mathbf{x}$ be an arbitrary point in R^2. According to Corollary 2.3-1, if $\mathbf{x} \in C$, then $P_C\mathbf{x} = \mathbf{x}$. On the other hand, if $\mathbf{x} \notin C$, then $P_C\mathbf{x}$ lies on the boundary of C, which is the unit-circle on which every point $\mathbf{x}$ satisfies $\|\mathbf{x}\| = 1$. Thus, we have

$$\|\mathbf{x} - P_C\mathbf{x}\| = \min_{\|\mathbf{y}\| \leq 1} \|\mathbf{x} - \mathbf{y}\| = \min_{\|\mathbf{y}\| = 1} \|\mathbf{x} - \mathbf{y}\|. \qquad (2.3\text{-}18)$$

To find $P_C\mathbf{x}$, we apply the method of Lagrange multipliers. We write the Lagrange functional

$$J = \|\mathbf{x} - \mathbf{y}\|^2 + \lambda \left(\|\mathbf{y}\|^2 - 1 \right) = \langle \mathbf{x} - \mathbf{y},\ \mathbf{x} - \mathbf{y} \rangle + \lambda \left(\langle \mathbf{y},\ \mathbf{y} \rangle - 1 \right), \qquad (2.3\text{-}19)$$

and then take the gradient with respect to $\mathbf{y}$ and set it to zero to obtain

$$\nabla J = 2(\mathbf{y} - \mathbf{x}) + 2\lambda \mathbf{y} = \mathbf{0}, \qquad (2.3\text{-}20)$$

which yields

$$\mathbf{y} = \frac{1}{1 + \lambda}\mathbf{x}. \qquad (2.3\text{-}21)$$

Invoking the condition that $\|\mathbf{y}\| = 1$, we obtain

$$|1 + \lambda| = \|\mathbf{x}\|. \qquad (2.3\text{-}22)$$

When $1 + \lambda = \|\mathbf{x}\|$, $\mathbf{y} = \mathbf{x}/\|\mathbf{x}\|$; and when $1 + \lambda = -\|\mathbf{x}\|$, $\mathbf{y} = -\mathbf{x}/\|\mathbf{x}\|$. Clearly, of the two $\mathbf{y} = \mathbf{x}/\|\mathbf{x}\|$ has a shorter distance to $\mathbf{x}$, so we have $P_C\mathbf{x} = \mathbf{x}/\|\mathbf{x}\|$. In fact, $\mathbf{y} = -\mathbf{x}/\|\mathbf{x}\|$ has the longest distance to $\mathbf{x}$ among all the points in C. So finally we have

$$P_C\mathbf{x} = \begin{cases} \mathbf{x} & \text{if } \mathbf{x} \in C \\ \mathbf{x}/\|\mathbf{x}\| & \text{if } \mathbf{x} \notin C. \end{cases} \qquad (2.3\text{-}23)$$

Note that the projector P_C here is not a linear operator. In general, the projection operator onto a closed convex set is nonlinear, as we shall see from examples furnished throughout this book. ∎

Before moving on, we would like to emphasize that in the above development the projection or the projector is defined for *closed, convex* sets in Hilbert spaces. If we attempt to extend these concepts to other situations such as non-closed sets or non-convex sets (the latter being discussed in detail in Chapter 5) in a Hilbert space, the properties[†] of the projection may not hold most of the time. In such a case, either the existence or the uniqueness of the "projection" may become questionable[‡]. In the following, we present two examples to clarify this point.

[†]Additional properties of projection are discussed in the next section.
[‡]This important observation suggests that in the world of problem solving, a more logical taxonomy than "linear versus non-linear" is "convex versus non-convex" constraints.

Example 2.3-2 In this example, we demonstrate that when the set is not closed, the "projection" of a point outside the set *may not even exist*. In Example 2.3-1, we derived the projection operator onto the unit-disc in the space R^2. Let's consider the open unit-disc defined by

$$C' = \{\mathbf{x} : \|\mathbf{x}\| < 1\} \tag{2.3-24}$$

in R^2. Clearly, the point $\mathbf{x}_0 = (0,1) \notin C'$. For a point $\mathbf{y} = (y_1, y_2) \in C'$, we have

$$\|\mathbf{x}_0 - \mathbf{y}\| = \sqrt{y_1^2 + (y_2 - 1)^2}. \tag{2.3-25}$$

Now, consider the sequence $\{\mathbf{y}_n\}$ with $\mathbf{y}_n = (0, 1 - 1/n)$ for each $n = 1, 2, \cdots$. Clearly, $\mathbf{y}_n \in C'$ for each n since $\|\mathbf{y}_n\| = 1 - \frac{1}{n} < 1$. But

$$\lim_{n\to\infty} \|\mathbf{x}_0 - \mathbf{y}_n\| = \lim_{n\to\infty} \frac{1}{n} = 0. \tag{2.3-26}$$

Thus we have

$$\inf_{\mathbf{y}\in C'} \|\mathbf{x}_0 - \mathbf{y}\| = 0. \tag{2.3-27}$$

On the other hand, no point in C' can achieve this distance because $\mathbf{x}_0 \notin C'$. This implies that in the set C' no point possesses the *shortest* distance to $\mathbf{x}_0$. Therefore, the concept of projection, as defined, becomes meaningless in such a case.   ∎

In practical applications, we may encounter some useful sets that are not closed. In such cases, to avoid the theoretical pitfalls in applying projection theory, we can enlarge a non-closed set by including its limit points. In other words, we modify a non-closed set by using its closure set. For example, in the above example we may approximate the open unit-disc by the closed unit-disc. This approach has been applied in several practical problems that we shall discuss in later chapters. Alternatively we can use so-called "relaxed projections" discussed in the next section.

Example 2.3-3 This example is used to demonstrate that the projection onto a closed and convex set *may not be unique* when the underlying space is not a Hilbert space. As demonstrated in the previous chapter, the linear space R^n, associated with the p-norm, is not a Hilbert space when $p \neq 2$. Let's consider the unit-disc

$$C = \{\mathbf{x} : \|\mathbf{x}\|_1 \le 1\}, \tag{2.3-28}$$

in the R^2 space under the 1-norm. From the result in Example 2.3-1 the set C is closed and convex. Let's consider the "projection" of the point $\mathbf{x}_0 = (1,1)$ onto this set. Clearly, $\|\mathbf{x}_0\|_1 = 1 + 1 = 2$ so $\mathbf{x}_0 \notin C$. For $\mathbf{y} = (y_1, y_2) \in C$, $\|\mathbf{y}\|_1 = |y_1| + |y_2| \le 1$, and we have

$$\|\mathbf{x}_0 - \mathbf{y}\|_1 = |1 - y_1| + |1 - y_2| \ge (1 - |y_1|) + (1 - |y_2|) = 2 - \|\mathbf{y}\|_1 \ge 1. \tag{2.3-29}$$

Thus the l^1 distance between the point $\mathbf{x}_0$ and any point in the set C is bounded from below by 1. On the other hand, consider the point $\mathbf{y} = (1/2, 1/2) \in C$,

$$\|\mathbf{x}_0 - \mathbf{y}\|_1 = \left|1 - \frac{1}{2}\right| + \left|1 - \frac{1}{2}\right| = 1. \qquad (2.3\text{-}30)$$

Therefore there exists at least one point in C that achieves the minimum distance 1 to the point $\mathbf{x}_0$. Moreover *all the points* on the line segment $y_1 + y_2 = 1$ in the first quadrant of the y_1-y_2 plane achieve this distance. Therefore, the "projection" of the point $\mathbf{x}_0$ onto the set C is no longer unique in such a case.   ∎

2.4 PROPERTIES OF PROJECTIONS

In this section, we study some useful properties of projections onto closed convex sets in a Hilbert spaces. These properties are central to the development of POCS theory in later sections of this chapter.

Theorem 2.4-1 *Let C be a closed convex set in a Hilbert space* **H**, *and P_C the projector onto C. Then,*

(i) For every $\mathbf{x} \in \mathbf{H}$, *and every* $\mathbf{y} \in C$,

$$\Re\langle \mathbf{x} - P_C\mathbf{x},\ \mathbf{y} - P_C\mathbf{x}\rangle \leq 0. \qquad (2.4\text{-}1)$$

(ii) For all $\mathbf{x}, \mathbf{y} \in \mathbf{H}$,

$$\|P_C\mathbf{x} - P_C\mathbf{y}\|^2 \leq \Re\langle \mathbf{x} - \mathbf{y},\ P_C\mathbf{x} - P_C\mathbf{y}\rangle. \qquad (2.4\text{-}2)$$

Proof: (i) Since $P_C\mathbf{x} \in C$, $\alpha\,\mathbf{y} + (1 - \alpha)\,P_C\mathbf{x} \in C$ for every $\alpha \in (0, 1)$. Thus

$$
\begin{aligned}
\|\mathbf{x} - P_C\mathbf{x}\|^2 &\leq\ \|\mathbf{x} - [\alpha\,\mathbf{y} + (1 - \alpha)\,P_C\mathbf{x}]\|^2 \\
&=\ \|\mathbf{x} - P_C\mathbf{x} - \alpha(\mathbf{y} - P_C\mathbf{x})\|^2.
\end{aligned}
\qquad (2.4\text{-}3)
$$

Applying the expansion formula of the norm

$$\|\mathbf{u} + \mathbf{v}\|^2 = \|\mathbf{u}\|^2 + \|\mathbf{v}\|^2 + 2\Re\langle \mathbf{u},\ \mathbf{v}\rangle \qquad (2.4\text{-}4)$$

to the last term of Eq. (2.4-3) with $\mathbf{u} = \mathbf{x} - P_C\mathbf{x}$ and $\mathbf{v} = -\alpha(\mathbf{y} - P_C\mathbf{x})$, we obtain

$$
\begin{aligned}
\|\mathbf{x} - P_C\mathbf{x}\|^2 \leq\ &\|\mathbf{x} - P_C\mathbf{x}\|^2 + \alpha^2\|\mathbf{y} - P_C\mathbf{x}\|^2 \\
&-2\alpha\Re\langle \mathbf{x} - P_C\mathbf{x},\ \mathbf{y} - P_C\mathbf{x}\rangle.
\end{aligned}
\qquad (2.4\text{-}5)
$$

Thus,

$$\Re\langle \mathbf{x} - P_C\mathbf{x},\ \mathbf{y} - P_C\mathbf{x}\rangle \leq \alpha\|\mathbf{y} - P_C\mathbf{x}\|^2, \qquad (2.4\text{-}6)$$

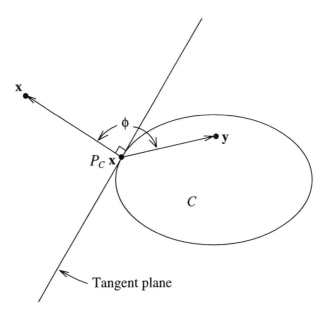

Fig. 2.4-1 The projection of a point $\mathbf{x}$ onto a convex closed set C has the geometric property that the vector $\mathbf{x} - P_C\mathbf{x}$ is "normal" to the "tangent plane" of C at the point $P_C\mathbf{x}$.

and letting $\alpha \to 0$, we obtain Eq. (2.4-1).

(ii) Since $P_C\mathbf{y} \in C$, it follows from Eq. (2.4-1) that

$$\Re\langle\mathbf{x} - P_C\mathbf{x},\ P_C\mathbf{y} - P_C\mathbf{x}\rangle \leq 0. \tag{2.4-7}$$

Similarly,

$$\Re\langle\mathbf{y} - P_C\mathbf{y},\ P_C\mathbf{x} - P_C\mathbf{y}\rangle \leq 0. \tag{2.4-8}$$

Thus

$$\Re\langle\mathbf{x} - P_C\mathbf{x},\ P_C\mathbf{y} - P_C\mathbf{x}\rangle + \Re\langle\mathbf{y} - P_C\mathbf{y},\ P_C\mathbf{x} - P_C\mathbf{y}\rangle \leq 0, \tag{2.4-9}$$

which, after re-arranging the terms, gives rise to Eq. (2.4-2). ∎

Note that in a real Hilbert space, Eq. (2.4-1) simply reduces to

$$\langle\mathbf{x} - P_C\mathbf{x},\ y - P_C\mathbf{x}\rangle \leq 0. \tag{2.4-10}$$

This inequality has an interesting geometric interpretation. As shown in Fig. 2.4-1, the vector $\mathbf{x} - P_C\mathbf{x}$ is "normal" to the "tangent plane" of C at the point $P_C\mathbf{x}$. Equation (2.4-10) implies that this tangent plane separates the point $\mathbf{x}$ from the set C so that they "sit" on opposite sides of the tangent plane. In other words, the angle ϕ between the vectors $\mathbf{x} - P_C\mathbf{x}$ and $\mathbf{y} - P_C\mathbf{x}$ is never less than 90° for every $\mathbf{y} \in C$.

It is easy to show that by its definition a subspace is also a convex set (see Exercise 2-1). A central theorem regarding subspaces is the following:

Theorem 2.4-2 (*Projection Theorem*) *Let* $\mathbf{W}$ *denote a non-empty closed subspace in a Hilbert space* $\mathbf{H}$. *If* $\mathbf{W}^{\perp}$ *denotes the set of all elements of* $\mathbf{H}$ *which are orthogonal to* $\mathbf{W}$, *then* $\mathbf{W}^{\perp}$ *is a closed subspace in* $\mathbf{H}$, *and we have* $\mathbf{W} \cap \mathbf{W}^{\perp} = \{\mathbf{0}\}$, *and* $\mathbf{H} = \mathbf{W} + \mathbf{W}^{\perp}$, *by which we mean that each* $\mathbf{x} \in \mathbf{H}$ *possesses a unique decomposition of the form* $\mathbf{x} = \mathbf{x}_1 + \mathbf{x}_2$ *with* $\mathbf{x}_1 \in \mathbf{W}$ *and* $\mathbf{x}_2 \in \mathbf{W}^{\perp}$.

Proof: Let $\mathbf{x}_1, \mathbf{x}_2 \in \mathbf{W}^{\perp}$, then for every $\mathbf{y} \in \mathbf{W}$, $\langle \mathbf{x}_1, \mathbf{y} \rangle = \langle \mathbf{x}_2, \mathbf{y} \rangle = 0$. Thus, for α, β arbitrary $\langle \alpha \mathbf{x}_1 + \beta \mathbf{x}_2, \mathbf{y} \rangle = 0$, which implies that $\alpha \mathbf{x}_1 + \beta \mathbf{x}_2 \in \mathbf{W}^{\perp}$. Hence, the set $\mathbf{W}^{\perp}$ is a subspace in $\mathbf{H}$. Since $\langle \mathbf{x}, \mathbf{x} \rangle = 0$ if and only if $\mathbf{x} = \mathbf{0}$, it follows that $\mathbf{W} \cap \mathbf{W}^{\perp} = \{\mathbf{0}\}$.

Next, we show that $\mathbf{W}^{\perp}$ is also closed. Let $\{\mathbf{x}_n\}$ be a converging sequence in $\mathbf{W}^{\perp}$ with limit $\mathbf{x}^*$. Then for every $\mathbf{y} \in \mathbf{W}$, $\langle \mathbf{x}_n, \mathbf{y} \rangle = 0$ and there results

$$
\begin{aligned}
|\langle \mathbf{x}^*, \mathbf{y} \rangle| &= |\langle \mathbf{x}_n, \mathbf{y} \rangle - \langle \mathbf{x}^*, \mathbf{y} \rangle| \\
&= |\langle \mathbf{x}_n - \mathbf{x}^*, \mathbf{y} \rangle| \le \|\mathbf{x}_n - \mathbf{x}^*\| \, \|\mathbf{y}\| \to 0, \quad (2.4\text{-}11)
\end{aligned}
$$

which yields $\langle \mathbf{x}^*, \mathbf{y} \rangle = 0$ so that $\mathbf{x}^* \in \mathbf{W}^{\perp}$. Hence, the subspace $\mathbf{W}^{\perp}$ is closed.

We show next that $\mathbf{H} = \mathbf{W} + \mathbf{W}^{\perp}$. Let P_W denote the projection operator onto $\mathbf{W}$. By Theorem 2.4-1, for each $\mathbf{x} \in \mathbf{H}$ and every $\mathbf{y} \in \mathbf{W}$, we have

$$
\Re \langle \mathbf{x} - P_W \mathbf{x}, \, \mathbf{y} - P_W \mathbf{x} \rangle \le 0. \quad (2.4\text{-}12)
$$

Now, since $\mathbf{W}$ is a subspace it also contains $\alpha \mathbf{y}$ for every real α, and Eq. (2.4-12) yields

$$
\alpha \Re \langle \mathbf{x} - P_W \mathbf{x}, \, \mathbf{y} \rangle \le \Re \langle \mathbf{x} - P_W \mathbf{x}, \, P_W \mathbf{x} \rangle, \quad (2.4\text{-}13)
$$

which can be true only if

$$
\Re \langle \mathbf{x} - P_W \mathbf{x}, \, \mathbf{y} \rangle = 0. \quad (2.4\text{-}14)
$$

If $\mathbf{H}$ is a complex vector space, we can replace $\mathbf{y}$ by $j\mathbf{y}$ in Eq. (2.4-14) and obtain

$$
\Im \langle \mathbf{x} - P_W \mathbf{x}, \, \mathbf{y} \rangle = 0, \quad (2.4\text{-}15)
$$

where $\Im$ is used to denote the imaginary part of a complex number. Thus,

$$
\langle \mathbf{x} - P_W \mathbf{x}, \, \mathbf{y} \rangle = 0, \quad (2.4\text{-}16)
$$

for every $\mathbf{y} \in \mathbf{W}$. Thus, we have $\mathbf{x} - P_W \mathbf{x} \in \mathbf{W}^{\perp}$.

Let $\mathbf{x}_1 = P_W \mathbf{x}$, and $\mathbf{x}_2 = \mathbf{x} - P_W \mathbf{x}$. Then $\mathbf{x}_1 \in \mathbf{W}$, $\mathbf{x}_2 \in \mathbf{W}^{\perp}$, and $\mathbf{x} = \mathbf{x}_1 + \mathbf{x}_2$. Suppose that there is another decomposition such that $\mathbf{x} = \mathbf{x}_3 + \mathbf{x}_4$ with $\mathbf{x}_3 \in \mathbf{W}$ and $\mathbf{x}_4 \in \mathbf{W}^{\perp}$, then we have $\mathbf{x}_1 - \mathbf{x}_3 = \mathbf{x}_2 - \mathbf{x}_4$, which implies that $\mathbf{x}_1 - \mathbf{x}_3 \in \mathbf{W}$ and $\mathbf{x}_1 - \mathbf{x}_3 \in \mathbf{W}^{\perp}$. Clearly, $\mathbf{x}_1 - \mathbf{x}_3 = \mathbf{0}$ because $\mathbf{W} \cap \mathbf{W}^{\perp} = \{\mathbf{0}\}$. Therefore, such a decomposition is unique and $\mathbf{H} = \mathbf{W} + \mathbf{W}^{\perp}$. ∎

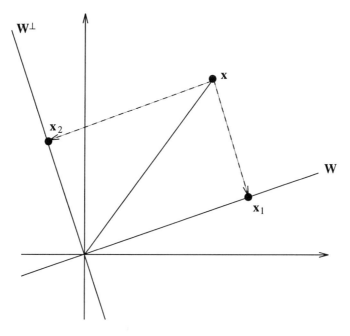

Fig. 2.4-2 For a given closed subspace **W** of a Hilbert space **H**, every $\mathbf{x} \in \mathbf{H}$ can be decomposed uniquely as $\mathbf{x} = \mathbf{x}_1 + \mathbf{x}_2$, where $\mathbf{x}_1 \in \mathbf{W}$ and $\mathbf{x}_2 \in \mathbf{W}^\perp$.

The above theorem says that for a given closed subspace **W** of a Hilbert space **H**, every $\mathbf{x} \in \mathbf{H}$ can be decomposed uniquely as $\mathbf{x} = \mathbf{x}_1 + \mathbf{x}_2$, where $\mathbf{x}_1 \in \mathbf{W}$ and $\mathbf{x}_2 \in \mathbf{W}^\perp$. The proof of the above theorem reveals that $\mathbf{x}_1$ is simply the projection of **x** onto **W**. It can also be shown that $\mathbf{x}_2$ is the projection of **x** onto $\mathbf{W}^\perp$. In the space R^2, this result is illustrated in Fig. 2.4-2. The subspace $\mathbf{W}^\perp$ is called the *orthogonal complement* of the subspace **W**. It is easy to show that $\mathbf{W}^{\perp\perp} = \mathbf{W}$; that is, the subspace **W** is the orthogonal complement of $\mathbf{W}^\perp$. Therefore, **W** and $\mathbf{W}^\perp$ are orthogonal complement of each other.

The above theorem also reveals a special property of the projection operator onto a closed subspace, which we summarize below:

Lemma 2.4-1 *Let P_W be the projection operator onto a closed subspace **W** in a Hilbert space **H**. Then for all $\mathbf{x}, \mathbf{y} \in \mathbf{H}$:*

(i) The following identity holds:

$$\langle \mathbf{x}, \ P_W \mathbf{y} \rangle = \langle P_W \mathbf{x}, \ \mathbf{y} \rangle = \langle P_W \mathbf{x}, \ P_W \mathbf{y} \rangle. \tag{2.4-17}$$

(ii) For α, β arbitrary,

$$P_W(\alpha \mathbf{x} + \beta \mathbf{y}) = \alpha P_W \mathbf{x} + \beta P_W \mathbf{y}. \tag{2.4-18}$$

*In words, P_W is a linear operator on **H**.*

Proof: (i) From the proof of Theorem 2.4-2, we have $\mathbf{x} = P_W\mathbf{x} + (\mathbf{x} - P_W\mathbf{x})$. Thus

$$\langle \mathbf{x}, \ P_W\mathbf{y} \rangle = \langle P_W\mathbf{x} + (\mathbf{x} - P_W\mathbf{x}), \ P_W\mathbf{y} \rangle = \langle P_W\mathbf{x}, \ P_W\mathbf{y} \rangle. \qquad (2.4\text{-}19)$$

Similarly, we have $\langle P_W\mathbf{x}, \ \mathbf{y} \rangle = \langle P_W\mathbf{x}, \ P_W\mathbf{y} \rangle$.

(ii) Note that

$$\alpha\mathbf{x} + \beta\mathbf{y} = (\alpha P_W\mathbf{x} + \beta P_W\mathbf{y}) + (\alpha\mathbf{x} - \alpha P_W\mathbf{x} + \beta\mathbf{y} - \beta P_W\mathbf{y}), \qquad (2.4\text{-}20)$$

where $\alpha P_W\mathbf{x} + \beta P_W\mathbf{y} \in \mathbf{W}$, and $\alpha\mathbf{x} - \alpha P_W\mathbf{x} + \beta\mathbf{y} - \beta P_W\mathbf{y} \in \mathbf{W}^\perp$. It follows that $P_W(\alpha\mathbf{x} + \beta\mathbf{y}) = \alpha P_W\mathbf{x} + \beta P_W\mathbf{y}$. ∎

From the discussion in Section 2.3 it follows that the projector P_C for a closed convex set C in a Hilbert space H is uniquely defined. We now introduce an extension of P_C called a *relaxed projector*. For each constant λ in the range of $(0, 2)$, we define an operator T_C as

$$T_C = I + \lambda(P_C - I), \qquad (2.4\text{-}21)$$

where I is the identity operator on $\mathbf{H}$. For each $\mathbf{x} \in C$, the operator T_C acts in the following fashion

$$T_C\mathbf{x} = \mathbf{x} + \lambda(P_C\mathbf{x} - \mathbf{x}) = (1 - \lambda)\mathbf{x} + \lambda P_C\mathbf{x}. \qquad (2.4\text{-}22)$$

Clearly, when the constant $\lambda = 1$, T_C is simply the projector P_C.

It is interesting to examine the behavior of the operator T_C. First, if $\mathbf{x} \in C$, then from Eq. (2.4-22) $T_C\mathbf{x} = \mathbf{x}$. Thus, when the operator T_C is applied to points inside the set C, it does not change them. In other words, each point in the set C is a fixed point of T_C. Furthermore, from Eq. (2.4-22), $T_C\mathbf{x} = \mathbf{x}$ if and only $P_C\mathbf{x} = \mathbf{x}$. Thus, the operator T_C has the same set of fixed points as the projector P_C, which is exactly the set C.

On the other hand, if $\mathbf{x} \notin C$, Eq. (2.4-22) has an instructive geometric interpretation. As illustrated in Fig. 2.4-3, when λ varies from 0 to 2, $T_C\mathbf{x}$ traces out the line segment from the point $\mathbf{x}$ to the point $\mathbf{x} + 2(P_C\mathbf{x} - \mathbf{x})$. More importantly, for each $\lambda \in (0, 2)$, the vector $T_C\mathbf{x} - \mathbf{x}$ always points to the same direction as that of the projection. As a matter of fact, the point $T_C\mathbf{x}$ is always *closer* to the set C than the point $\mathbf{x}$ itself is. More precisely, we have the following result:

Corollary 2.4-1 *Let C be a closed convex set in a Hilbert space $\mathbf{H}$, and P_C the projector onto C. Then for each $\mathbf{x} \in \mathbf{H}$,*

$$\|T_C\mathbf{x} - \mathbf{y}\|^2 \leq \|\mathbf{x} - \mathbf{y}\|^2 - \lambda(2 - \lambda)\|\mathbf{x} - P_C\mathbf{x}\|^2 \qquad (2.4\text{-}23)$$

for all $\mathbf{y} \in C$.

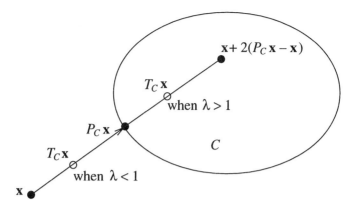

Fig. 2.4-3 Geometric interpretation of the operator T_C: when λ varies from 0 to 2, $T_C\mathbf{x}$ traces out the line segment from the point $\mathbf{x}$ to the point $\mathbf{x} + 2(P_C\mathbf{x} - \mathbf{x})$.

Proof: By invoking the expansion formula $\|\mathbf{u} + \mathbf{v}\|^2 = \|\mathbf{u}\|^2 + \|\mathbf{v}\|^2 + 2\Re\langle \mathbf{u},\ \mathbf{v}\rangle$, we obtain

$$
\begin{aligned}
\|T_C\mathbf{x} - \mathbf{y}\|^2 &= \|\mathbf{x} - \mathbf{y} + \lambda(P_C\mathbf{x} - \mathbf{x})\|^2 \\
&= \|\mathbf{x} - \mathbf{y}\|^2 + \lambda^2\|\mathbf{x} - P_C\mathbf{x}\|^2 - 2\lambda\Re\langle \mathbf{x} - \mathbf{y},\ \mathbf{x} - P_C\mathbf{x}\rangle \\
&= \|\mathbf{x} - \mathbf{y}\|^2 + \lambda^2\|\mathbf{x} - P_C\mathbf{x}\|^2 \\
&\quad - 2\lambda\Re\langle \mathbf{x} - P_C\mathbf{x} + P_C\mathbf{x} - \mathbf{y},\ \mathbf{x} - P_C\mathbf{x}\rangle \\
&= \|\mathbf{x} - \mathbf{y}\|^2 - \lambda(2 - \lambda)\|\mathbf{x} - P_C\mathbf{x}\|^2 \\
&\quad + 2\lambda\Re\langle \mathbf{x} - P_C\mathbf{x},\ \mathbf{y} - P_C\mathbf{x}\rangle, \tag{2.4-24}
\end{aligned}
$$

from which Eq. (2.4-23) follows, since $\Re\langle \mathbf{x} - P_C\mathbf{x},\ \mathbf{y} - P_C\mathbf{x}\rangle \leq 0$ by Theorem 2.4-1. ∎

It is clear from Eq. (2.4-23) that for each $\mathbf{x} \notin C$, $\|T_C\mathbf{x} - \mathbf{y}\| < \|\mathbf{x} - \mathbf{y}\|$ for every $\mathbf{y} \in C$. Indeed, T_C does bring the point $\mathbf{x}$ closer to the set C.

From Eq. (2.4-22), we see that when the parameter $\lambda \in (0, 1)$ the point $T_C\mathbf{x}$ lies in-between the point $\mathbf{x}$ and its projection $P_C\mathbf{x}$. As a result, the point $T_C\mathbf{x}$ lies outside the set C. In such a case, the operator T_C seems to *under-project*[†] the point $\mathbf{x}$ toward the set C. On the other hand, when $\lambda > 1$, the point $T_C\mathbf{x}$ lies farther away from the point $\mathbf{x}$ than the projection $P_C\mathbf{x}$ does. In such a case, the operator T_C seems to *over-project* the point $\mathbf{x}$ toward the set C. Due to this behavior, the operator T_C is called the *relaxed projector* for the set C, and its associated constant λ is called the *relaxation parameter*. Despite its name, however, the relaxed projector T_C is not a projector unless the relaxation parameter has value

---

[†]Under- and over-project are not terms that indicate a failure of the operator T_c.

1. As we will see in later sections of this chapter, these properties of the relaxed projector make it very useful for speeding up the convergence of projection-type algorithms.

In addition, the relaxed projector has another important property which we summarize below:

Theorem 2.4-3 *Let C be a closed convex set in a Hilbert space* **H***, then its projector P_C is non-expansive;[†] furthermore, for each $\lambda \in (0, 2)$ its associated relaxed projector T_C is also non-expansive.*

Proof: First, we show that the projector P_C is non-expansive. According to Theorem 2.4-1, we have for every **x** and **y** $\in$ **H**,

$$\|P_C\mathbf{x} - P_C\mathbf{y}\|^2 \leq \Re\langle \mathbf{x} - \mathbf{y},\ P_C\mathbf{x} - P_C\mathbf{y}\rangle. \tag{2.4-25}$$

Invoking the Schwarz inequality, we obtain

$$|\langle \mathbf{x} - \mathbf{y},\ P_C\mathbf{x} - P_C\mathbf{y}\rangle| \leq \|\mathbf{x} - \mathbf{y}\|\|P_C\mathbf{x} - P_C\mathbf{y}\|. \tag{2.4-26}$$

Combining the above two equations together yields

$$\|P_C\mathbf{x} - P_C\mathbf{y}\| \leq \|\mathbf{x} - \mathbf{y}\|. \tag{2.4-27}$$

Thus, the projector P_C is non-expansive.

Next, we show that for each $\lambda \in (0, 2)$ the relaxed projector T_C is also non-expansive. Consider **x** and **y** $\in$ **H**,

$$\|T_C\mathbf{x} - T_C\mathbf{y}\| = \|(1 - \lambda)(\mathbf{x} - \mathbf{y}) + \lambda(P_C\mathbf{x} - P_C\mathbf{y})\|. \tag{2.4-28}$$

For $\lambda \in (0, 1)$, $1 - \lambda > 0$, and it follows from Eq. (2.4-27) that

$$\|T_C\mathbf{x} - T_C\mathbf{y}\| \leq (1 - \lambda)\|\mathbf{x} - \mathbf{y}\| + \lambda\|P_C\mathbf{x} - P_C\mathbf{y}\| \leq \|\mathbf{x} - \mathbf{y}\|. \tag{2.4-29}$$

On the other hand, when $\lambda \in (1, 2)$, $1 - \lambda < 0$ and $2 - \lambda > 0$. By invoking the expansion formula $\|\mathbf{u} + \mathbf{v}\|^2 = \|\mathbf{u}\|^2 + \|\mathbf{v}\|^2 + 2\Re\langle \mathbf{u},\ \mathbf{v}\rangle$, and by applying Eq. (2.4-27) and Theorem 2.4-1, we obtain

$$\begin{aligned}
\|T_C\mathbf{x} - T_C\mathbf{y}\|^2 &= \|(1 - \lambda)(\mathbf{x} - \mathbf{y}) + \lambda(P_C\mathbf{x} - P_C\mathbf{y})\|^2 \\
&= (1 - \lambda)^2\|\mathbf{x} - \mathbf{y}\|^2 + \lambda^2\|P_C\mathbf{x} - P_C\mathbf{y}\|^2 \\
&\quad + 2\lambda(1 - \lambda)\Re\langle \mathbf{x} - \mathbf{y},\ P_C\mathbf{x} - P_C\mathbf{y}\rangle \\
&\leq (1 - \lambda)^2\|\mathbf{x} - \mathbf{y}\|^2 + \lambda^2\|P_C\mathbf{x} - P_C\mathbf{y}\|^2 \\
&\quad + 2\lambda(1 - \lambda)\|P_C\mathbf{x} - P_C\mathbf{y}\|^2
\end{aligned}$$

[†]An operator G on a Hilbert space **H** is said to be *non-expansive* if for every **x** and **y** $\in$ **H**, $\|G\mathbf{x} - G\mathbf{y}\| \leq \|\mathbf{x} - \mathbf{y}\|$.

$$
\begin{aligned}
&= (1-\lambda)^2 \|\mathbf{x}-\mathbf{y}\|^2 + \lambda(2-\lambda)\|P_C\mathbf{x}-P_C\mathbf{y}\|^2 \\
&\leq (1-\lambda)^2 \|\mathbf{x}-\mathbf{y}\|^2 + \lambda(2-\lambda)\|\mathbf{x}-\mathbf{y}\|^2 \\
&= \|\mathbf{x}-\mathbf{y}\|^2.
\end{aligned}
\tag{2.4-30}
$$

Thus, for each $\lambda \in (0,2)$, we have $\|T_C\mathbf{x} - T_C\mathbf{y}\| \leq \|\mathbf{x}-\mathbf{y}\|$, and Theorem 2.4-3 follows. ∎

2.5   FUNDAMENTAL THEORY OF POCS

In this section, we discuss the fundamental result of the theory of projections onto closed convex sets in a Hilbert space. While the theory is of profound mathematical interest, it has led to extraordinary success in solving practical problems by modern-day computers. Some of the discussion parallels that of Youla [2].

Let's begin with some standard notations. Assume that $C_1, C_2, \cdots, C_m$ denote m closed convex sets in a Hilbert space $\mathbf{H}$, and C_0 denotes their intersection set $C_0 = \bigcap_{i=1}^m C_i$. For each $i = 1, 2, \cdots, m$, let P_i denote the projection operator onto the set C_i, and T_i denote the corresponding relaxed projector $T_i = I + \lambda_i(P_i - I)$ for $\lambda_i \in (0,2)$. Also let T denote the *concatenation*[†] of all these relaxed projectors, that is

$$
T = T_m T_{m-1} \cdots T_1.
\tag{2.5-1}
$$

Then we have the following:

Theorem 2.5-1 (*Fundamental Theorem of POCS*) *Assume that C_0 is non-empty. Then for every $\mathbf{x} \in \mathbf{H}$ and for every $\lambda_i \in (0,2), i = 1, 2, \cdots, m$, the sequence $\{T^n\mathbf{x}\}$ converges weakly to a point of C_0.*

This theorem was first proved by Gubin, Polyak, and Raik in 1967 [3]. Later in 1982, Youla provided an alternative proof [4]. Youla's approach is based on the following important result in the fixed-point theory of nonexpansive operators [1]:

Theorem 2.5-2 (*Opial*) *Let G be a non-expansive asymptotically regular operator[‡] on a Hilbert space $\mathbf{H}$ with a non-empty set of fixed points $\Gamma \subset \mathbf{H}$. Then for every*

[†]Also known as *composition*.

[‡]An operator G on a Hilbert space $\mathbf{H}$ is said to be *asymptotically regular* if

$$
\lim_{n \to \infty} \|G^n\mathbf{x} - G^{n+1}\mathbf{x}\| = 0
$$

for every $\mathbf{x} \in \mathbf{H}$. Not every operator is asymptotically regular. For example define the operator G by $G\mathbf{x} = -\mathbf{x}$. Then

$$
\lim_{n \to \infty} \|G^n\mathbf{x} - G^{n+1}\mathbf{x}\| = 2\|\mathbf{x}\| \neq 0
$$

for $\mathbf{x} \neq \mathbf{0}$.

$\mathbf{x} \in \mathbf{H}$, *the sequence* $\{\mathbf{x}_n = G^n\mathbf{x}\}$ *converges weakly to a point of* $\Gamma.$[†]

Clearly, if we take Opial's theorem for granted, then Theorem 2.5-1 follows immediately if we show that under the setting of Theorem 2.5-1 the operator T satisfies the following three conditions: (i) It is non-expansive; (ii) it is asymptotically regular; and (iii) it has the set C_0 as its set of fixed-points. In what follows, we show that the operator T indeed satisfies these three conditions. For the purpose of clarity, we shall rephrase and prove each of these conditions in the form of separate lemmas. The proof of Opial's theorem is somewhat less straightforward so we relegate it to Section 2.8 for the interest of brevity. Readers who are mainly interested in the application of the POCS method may skip Section 2.8 without any sacrifice in continuity in the reading of the rest of the book.

Lemma 2.5-1 *The operator* T *is non-expansive in* $\mathbf{H}$.

Proof: By Theorem 2.4-3, T_i is non-expansive for each $i = 1, 2, \cdots, m$. Hence for $\mathbf{x}, \mathbf{y} \in \mathbf{H}$,

$$
\begin{aligned}
\|T\mathbf{x} - T\mathbf{y}\| &= \|T_m(T_{m-1} \cdots T_1\mathbf{x}) - T_m(T_{m-1} \cdots T_1\mathbf{y})\| \\
&\leq \|T_{m-1} \cdots T_1\mathbf{x} - T_{m-1} \cdots T_1\mathbf{y}\| \\
&\quad \vdots \\
&\leq \|T_1\mathbf{x} - T_1\mathbf{y}\| \\
&\leq \|\mathbf{x} - \mathbf{y}\|.
\end{aligned}
\tag{2.5-2}
$$

Thus, T is non-expansive. ∎

Lemma 2.5-2 *The set of fixed points of the operator* T *is* C_0.

Proof: Let Γ denote the set of fixed points of T. Clearly, if $\mathbf{x} \in C_0$, then $T_i\mathbf{x} = \mathbf{x}$ for each $i = 1, 2, \cdots, m$, so that $T\mathbf{x} = \mathbf{x}$. Thus, $C_0 \subset \Gamma$.

On the other hand, let $\mathbf{x} \in \Gamma$, then for $\mathbf{y} \in C_0$, we have

$$
\|\mathbf{x} - \mathbf{y}\| = \|T\mathbf{x} - T\mathbf{y}\| \leq \|T_1\mathbf{x} - T_1\mathbf{y}\| = \|T_1\mathbf{x} - \mathbf{y}\|.
\tag{2.5-3}
$$

But, by Corollary 2.4-1 we have

$$
\|T_1\mathbf{x} - \mathbf{y}\| \leq \|\mathbf{x} - \mathbf{y}\|.
\tag{2.5-4}
$$

Therefore, we have

$$
\|T_1\mathbf{x} - \mathbf{y}\| = \|\mathbf{x} - \mathbf{y}\|,
\tag{2.5-5}
$$

[†]In its general form, Opial's theorem holds for an operator G whose domain is simply a closed convex set in a Hilbert space. Note that by definition a Hilbert space itself is a closed convex set. We adopt the current form purely for the sake of simplicity.

which is possible only if $\mathbf{x} \in C_1$. Thus, $\mathbf{x} \in C_1$, and we have $\mathbf{x} = T_m T_{m-1} \cdots T_2 \mathbf{x}$. A repetition of the argument leads to $\mathbf{x} \in C_0$ so that $\Gamma \subset C_0$. Hence, $\Gamma = C_0$, and the lemma follows. ∎

Lemma 2.5-3 *The operator T is asymptotically regular, i.e.,*

$$\lim_{n \to \infty} \|T^n \mathbf{x} - T^{n+1}\mathbf{x}\| = 0 \qquad (2.5\text{-}6)$$

for every $\mathbf{x} \in \mathbf{H}$.

Proof: Let $\mathbf{y} \in C_0$. By Corollary 2.4-1 for every $\mathbf{x} \in \mathbf{H}$

$$\|T_1 \mathbf{x} - \mathbf{y}\|^2 \le \|\mathbf{x} - \mathbf{y}\|^2 - \lambda_1 (2 - \lambda_1) \|\mathbf{x} - P_1 \mathbf{x}\|^2, \qquad (2.5\text{-}7)$$

which yields

$$\|\mathbf{x} - P_1 \mathbf{x}\|^2 \le \frac{1}{\lambda_1 (2 - \lambda_1)} \left(\|\mathbf{x} - \mathbf{y}\|^2 - \|T_1 \mathbf{x} - \mathbf{y}\|^2 \right). \qquad (2.5\text{-}8)$$

Note that $T_1 \mathbf{x} = \mathbf{x} + \lambda_1 (\mathbf{x} - P_1 \mathbf{x})$, and we have

$$\|\mathbf{x} - T_1 \mathbf{x}\|^2 = \lambda_1^2 \|\mathbf{x} - P_1 \mathbf{x}\|^2 \le \frac{\lambda_1}{2 - \lambda_1} \left(\|\mathbf{x} - \mathbf{y}\|^2 - \|T_1 \mathbf{x} - \mathbf{y}\|^2 \right), \quad (2.5\text{-}9)$$

from which an induction argument leads to

$$\|\mathbf{x} - T\mathbf{x}\|^2 \le b_m 2^{m-1} \left(\|\mathbf{x} - \mathbf{y}\|^2 - \|T\mathbf{x} - \mathbf{y}\|^2 \right), \qquad (2.5\text{-}10)$$

where

$$b_m = \max_{1 \le i \le m} \left\{ \frac{\lambda_i}{2 - \lambda_i} \right\}. \qquad (2.5\text{-}11)$$

Indeed, for $m = 1$, Eq. (2.5-10) reduces simply to Eq. (2.5-9). Assume that Eq. (2.5-10) holds for $m = k - 1, k = 2, 3, \cdots$. Let

$$K \triangleq T_{k-1} T_{k-2} \cdots T_1. \qquad (2.5\text{-}12)$$

Then

$$\|\mathbf{x} - K\mathbf{x}\|^2 \le b_{k-1} 2^{k-2} \left(\|\mathbf{x} - \mathbf{y}\|^2 - \|K\mathbf{x} - \mathbf{y}\|^2 \right). \qquad (2.5\text{-}13)$$

For $m = k$, $T = T_k K$, and

$$\begin{aligned}
\|\mathbf{x} - T\mathbf{x}\|^2 &= \|\mathbf{x} - K\mathbf{x} + K\mathbf{x} - T\mathbf{x}\|^2 \\
&\le (\|\mathbf{x} - K\mathbf{x}\| + \|K\mathbf{x} - T\mathbf{x}\|)^2 \\
&\le 2 \left(\|\mathbf{x} - K\mathbf{x}\|^2 + \|K\mathbf{x} - T\mathbf{x}\|^2 \right) \\
&= 2 \left(\|\mathbf{x} - K\mathbf{x}\|^2 + \|K\mathbf{x} - T_k K\mathbf{x}\|^2 \right). \qquad (2.5\text{-}14)
\end{aligned}$$

From Eq. (2.5-11), $b_{k-1} \leq b_k$, and Eq. (2.5-13) gives

$$\|\mathbf{x} - K\mathbf{x}\|^2 \leq b_k 2^{k-2} \left(\|\mathbf{x} - \mathbf{y}\|^2 - \|K\mathbf{x} - \mathbf{y}\|^2 \right). \qquad (2.5\text{-}15)$$

Furthermore, by Eq. (2.5-9)

$$
\begin{aligned}
\|K\mathbf{x} - T_k K\mathbf{x}\|^2 &\leq \frac{\lambda_k}{2 - \lambda_k} \left(\|K\mathbf{x} - \mathbf{y}\|^2 - \|T_k K\mathbf{x} - \mathbf{y}\|^2 \right) \\
&\leq b_k \left(\|K\mathbf{x} - \mathbf{y}\|^2 - \|T_k K\mathbf{x} - \mathbf{y}\|^2 \right) \\
&\leq 2^{k-2} b_k \left(\|K\mathbf{x} - \mathbf{y}\|^2 - \|T_k K\mathbf{x} - \mathbf{y}\|^2 \right) \quad (2.5\text{-}16)
\end{aligned}
$$

for $k \geq 2$. Therefore, by combining Eqs. (2.5-14)–(2.5-16), we obtain

$$
\begin{aligned}
\|\mathbf{x} - T\mathbf{x}\|^2 &\leq b_k 2^{k-1} \left(\|\mathbf{x} - \mathbf{y}\|^2 - \|K\mathbf{x} - \mathbf{y}\|^2 \right) \\
&\quad + 2^{k-1} b_k \left(\|K\mathbf{x} - \mathbf{y}\|^2 - \|T_k K\mathbf{x} - \mathbf{y}\|^2 \right) \\
&= b_k 2^{k-1} \left(\|\mathbf{x} - \mathbf{y}\|^2 - \|T_k K\mathbf{x} - \mathbf{y}\|^2 \right) \\
&= b_k 2^{k-1} \left(\|\mathbf{x} - \mathbf{y}\|^2 - \|T\mathbf{x} - \mathbf{y}\|^2 \right). \qquad (2.5\text{-}17)
\end{aligned}
$$

Thus, Eq. (2.5-10) follows by induction.

By Eq. (2.5-10), we have

$$
\begin{aligned}
\|\mathbf{x} - T\mathbf{x}\|^2 &\leq b_m 2^{m-1} \left(\|\mathbf{x} - \mathbf{y}\|^2 - \|T\mathbf{x} - \mathbf{y}\|^2 \right) \\
\|T\mathbf{x} - T^2\mathbf{x}\|^2 &\leq b_m 2^{m-1} \left(\|T\mathbf{x} - \mathbf{y}\|^2 - \|T^2\mathbf{x} - \mathbf{y}\|^2 \right) \\
&\vdots \\
\|T^n\mathbf{x} - T^{n+1}\mathbf{x}\|^2 &\leq b_m 2^{m-1} \left(\|T^n\mathbf{x} - \mathbf{y}\|^2 - \|T^{n+1}\mathbf{x} - \mathbf{y}\|^2 \right).
\end{aligned}
$$

Thus

$$
\begin{aligned}
\sum_{l=0}^{n} \|T^l\mathbf{x} - T^{l+1}\mathbf{x}\|^2 &= b_m 2^{m-1} \left(\|\mathbf{x} - \mathbf{y}\|^2 - \|T^{n+1}\mathbf{x} - \mathbf{y}\|^2 \right) \\
&\leq b_m 2^{m-1} \|\mathbf{x} - \mathbf{y}\|^2. \qquad (2.5\text{-}18)
\end{aligned}
$$

By letting $n \to \infty$, we obtain

$$\sum_{l=0}^{\infty} \|T^l\mathbf{x} - T^{l+1}\mathbf{x}\|^2 \leq b_m 2^{m-1} \|\mathbf{x} - \mathbf{y}\|^2 < \infty, \qquad (2.5\text{-}19)$$

which implies that

$$\lim_{l \to \infty} \|T^l\mathbf{x} - T^{l+1}\mathbf{x}\|^2 = 0. \tag{2.5-20}$$

Thus, the operator T is asymptotically regular.[†]   ∎

Therefore, Theorem 2.5-1 follows from the argument above. Note that Theorem 2.5-1 applies to general closed convex sets in a Hilbert space and the resulting sequence of iterates $\{T^n\mathbf{x}\}$ enjoys the weak convergence property. A reasonable conjecture then is that if we impose certain additional conditions on the closed convex sets involved, we may achieve strong convergence for the sequence $\{T^n\mathbf{x}\}$. Indeed, an interesting and important result is manifest immediately when all the sets C_i are closed subspaces:

Corollary 2.5-1 *Under the constraints of Theorem 2.5-1, if each C_i is a closed subspace for $i = 1, 2, \cdots, m$, and if their intersection C_0 is non-empty, then for every $\mathbf{x} \in \mathbf{H}$ the sequence $\{T^n\mathbf{x}\}$ converges strongly to $P_0\mathbf{x}$, the projection of $\mathbf{x}$ onto C_0.*[‡]

Proof: By Theorem 2.4-2, the projector P_i for each closed subspace C_i is linear in $\mathbf{H}$; moreover, for all $\mathbf{x}, \mathbf{y} \in \mathbf{H}$, we have $\langle \mathbf{x}, P_i\mathbf{y} \rangle = \langle P_i\mathbf{x}, \mathbf{y} \rangle$. It follows that the relaxed projector $T_i = I + \lambda_i(P_i - I)$ is also linear, and so is $T = T_m T_{m-1} \cdots T_1$. Furthermore, we have

$$\begin{aligned} \langle \mathbf{x}, T_i\mathbf{y} \rangle &= \langle \mathbf{x}, (1-\lambda)\mathbf{y} + P_i\mathbf{y} \rangle = (1-\lambda)\langle \mathbf{x}, \mathbf{y} \rangle + \langle \mathbf{x}, P_i\mathbf{y} \rangle \\ &= (1-\lambda)\langle \mathbf{x}, \mathbf{y} \rangle + \langle P_i\mathbf{x}, \mathbf{y} \rangle = \langle T_i\mathbf{x}, \mathbf{y} \rangle, \end{aligned} \tag{2.5-21}$$

for all $\mathbf{x}, \mathbf{y} \in \mathbf{H}$. Thus,

$$\begin{aligned} \langle \mathbf{x}, T\mathbf{y} \rangle &= \langle \mathbf{x}, T_m T_{m-1} \cdots T_1\mathbf{y} \rangle \\ &= \langle T_m\mathbf{x}, T_{m-1} \cdots T_1\mathbf{y} \rangle \\ &\vdots \\ &= \langle T^*\mathbf{x}, \mathbf{y} \rangle, \end{aligned} \tag{2.5-22}$$

where $T^* = T_1 \cdots T_{m-1}T_m$. Clearly, the operator T^* also has C_0 as its set of fixed points. Note that $C_0 = \bigcap_{i=1}^{m} C_i$ is a closed subspace in $\mathbf{H}$.

Let $\mathcal{R}(I-T)$ denote the set of vectors of the form $(I-T)\mathbf{x}$ for all $\mathbf{x} \in \mathbf{H}$.[§] Due to the linearity of T, it is easy to show that $\mathcal{R}(I-T)$ is a subspace in $\mathbf{H}$. Therefore, the closure of the set $\mathcal{R}(I-T)$, denoted by $\overline{\mathcal{R}(I-T)}$, is a closed subspace in $\mathbf{H}$. We claim that a vector $\mathbf{x} \in \mathbf{H}$ is orthogonal to $\overline{\mathcal{R}(I-T)}$ if it is orthogonal to

---

[†]An operator that satisfies the condition in Eq. (2.5-19) is called a *reasonable wanderer*. Clearly, a reasonable wanderer is always asymptotically regular, but the converse is not always true.

[‡]As a matter of fact, this result is an extension of I. Halperin's generalization of von Neumann's celebrated alternating projection theorem [5] from $m = 2$ to arbitrary m [6].

[§]In the literature, $\mathcal{R}(I - T)$ is called the *range* of the operator $(I - T)$.

$\mathcal{R}(I - T)$. Indeed, assume that $\mathbf{x}$ is orthogonal to every vector in $\mathcal{R}(I - T)$, then for any $\mathbf{y} \in \overline{\mathcal{R}(I - T)}$, if $\mathbf{y} \in \mathcal{R}(I - T)$, then $\mathbf{x}$ is orthogonal to $\mathbf{y}$. On the other hand, if $\mathbf{y} \notin \mathcal{R}(I - T)$, then there exists a sequence $\{\mathbf{y}_n\}$ in $\mathcal{R}(I - T)$ such that $\mathbf{y}_n \to \mathbf{y}$. Note that $\langle \mathbf{x}, \mathbf{y}_n \rangle = 0$ and we have

$$
\begin{aligned}
|\langle \mathbf{x}, \mathbf{y} \rangle| &= |\langle \mathbf{x}, \mathbf{y}_n \rangle - \langle \mathbf{x}, \mathbf{y} \rangle| \\
&= |\langle \mathbf{x}, \mathbf{y}_n - \mathbf{y} \rangle| \le \|\mathbf{x}\| \, \|\mathbf{y}_n - \mathbf{y}\| \to 0, \quad\quad (2.5\text{-}23)
\end{aligned}
$$

which yields $\langle \mathbf{x}, \mathbf{y} \rangle = 0$, so that the claim follows.

Consequently, the orthogonal complement of the closed subspace $\overline{\mathcal{R}(I - T)}$, say $\overline{\mathcal{R}(I - T)}^{\perp}$, consists of all the vectors $\mathbf{x} \in \mathbf{H}$ that are orthogonal to $\mathcal{R}(I - T)$. That is, $\mathbf{x} \in \overline{\mathcal{R}(I - T)}^{\perp}$ if and only if $\langle \mathbf{x}, (I - T)\mathbf{y} \rangle = 0$ for all $\mathbf{y} \in \mathbf{H}$. Note that by Eq. (2.5-22),

$$
\begin{aligned}
\langle \mathbf{x}, (I - T)\mathbf{y} \rangle &= \langle \mathbf{x}, \mathbf{y} \rangle - \langle \mathbf{x}, T\mathbf{y} \rangle \\
&= \langle \mathbf{x}, \mathbf{y} \rangle - \langle T^*\mathbf{x}, \mathbf{y} \rangle = \langle (I - T^*)\mathbf{x}, \mathbf{y} \rangle, \quad\quad (2.5\text{-}24)
\end{aligned}
$$

so that $\mathbf{x} \in \overline{\mathcal{R}(I - T)}^{\perp}$ if and only if $\langle (I - T^*)\mathbf{x}, \mathbf{y} \rangle = 0$ for all $\mathbf{y} \in \mathbf{H}$, which is possible only if $(I - T^*)\mathbf{x} = \mathbf{0}$ or $T^*\mathbf{x} = \mathbf{x}$. In other words, $\overline{\mathcal{R}(I - T)}^{\perp}$ consists of all the fixed points of T^*, which is C_0. Thus, C_0 is the orthogonal complement of $\overline{\mathcal{R}(I - T)}$, and we have $\mathbf{H} = C_0 + \overline{\mathcal{R}(I - T)}$.

By Theorem 2.4-2, for any $\mathbf{x} \in \mathbf{H}$, there exists a unique decomposition $\mathbf{x} = \mathbf{z} + \mathbf{y}$ with $\mathbf{z} \in C_0$ and $\mathbf{y} \in \overline{\mathcal{R}(I - T)}$. Consequently, $T^n \mathbf{z} = \mathbf{z}$, and

$$
T^n \mathbf{x} = T^n \mathbf{z} + T^n \mathbf{y} = \mathbf{z} + T^n \mathbf{y}, \quad\quad (2.5\text{-}25)
$$

because T^n is linear for every n. Next, we show that $T^n \mathbf{y} \to \mathbf{0}$. For $\mathbf{y} \in \overline{\mathcal{R}(I - T)}$, if $\mathbf{y} \in \mathcal{R}(I - T)$, then there exists a $\mathbf{w} \in \mathbf{H}$ such that $\mathbf{y} = (I - T)\mathbf{w}$, and we obtain

$$
T^n \mathbf{y} = T^n (I - T)\mathbf{w} = T^n \mathbf{w} - T^{n+1} \mathbf{w} \to \mathbf{0}, \qu\quad (2.5\text{-}26)
$$

because T is asymptotically regular according to Lemma 2.5-3. On the other hand, if $\mathbf{y} \notin \mathcal{R}(I - T)$, then there exists a sequence $\{\mathbf{y}_l\}$ in $\mathcal{R}(I - T)$ such that $\mathbf{y}_l \to \mathbf{y}$, and we have

$$
\begin{aligned}
\|T^n \mathbf{y}\| &= \|T^n \mathbf{y} - T^n \mathbf{y}_l + T^n \mathbf{y}_l\| \\
&\le \|T^n \mathbf{y} - T^n \mathbf{y}_l\| + \|T^n \mathbf{y}_l\| \\
&\le \|\mathbf{y} - \mathbf{y}_l\| + \|T^n \mathbf{y}_l\|, \quad\quad (2.5\text{-}27)
\end{aligned}
$$

because T is non-expansive. By letting $n \to \infty$, Eq. (2.5-27) yields

$$
\lim_{n \to \infty} \|T^n \mathbf{y}\| \le \|\mathbf{y} - \mathbf{y}_l\|, \quad\quad (2.5\text{-}28)
$$

because by what we just proved above $\|T^n \mathbf{y}_l\| \to 0$ as $n \to \infty$. Also, $\mathbf{y}_l \to \mathbf{y}$, so

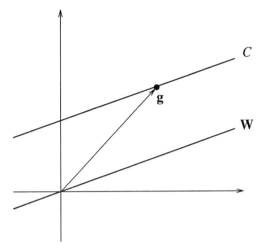

Fig. 2.5-1 In a vector space, a linear variety C is obtained by translating a closed subspace **W** by a fixed vector **g**.

we obtain

$$\lim_{n \to \infty} \|T^n \mathbf{y}\| = 0, \tag{2.5-29}$$

and Eq. (2.5-25) yields

$$T^n \mathbf{x} \to \mathbf{z} = P_0 \mathbf{x}, \tag{2.5-30}$$

where P_0 denotes the projector onto C_0. The proof is completed. ∎

To recapitulate, Corollary 2.5-1 says that when all the sets C_i in Theorem 2.5-1 are closed subspaces, the sequence of iterates $\{T^n \mathbf{x}\}$ not only converges strongly but actually converges to the projection of $\mathbf{x}$ onto C_0. In other words, the resulting convergent point is the point in C_0 that is closest to the starting point of the iteration. As will be made clear from applications described in later chapters, such a property is very desirable in certain practical problems. The condition for this desirable result, however, is that all the sets involved are closed subspaces which proves to be too strong for most practical situations. Fortunately, it turns out that this result can be extended to *linear varieties*—a close relative to closed subspace in concept.

In a vector space, a set C is said to be a *linear variety* if for some fixed $\mathbf{g} \in \mathbf{H}$ the set of vectors $\{\mathbf{x} - \mathbf{g}\}$ for all $\mathbf{x} \in C$ is a closed subspace in $\mathbf{H}$. In other words, a linear variety is obtained by translating a closed subspace by a fixed vector. This is illustrated in Fig. 2.5-1 in the space R^2.

An equivalent definition for linear variety is the following: A closed set C is a linear variety if for every $\mathbf{x}, \mathbf{y} \in C$ the vector $\alpha \mathbf{x} + (1 - \alpha) \mathbf{y} \in C$ for every constant α. Clearly a closed subspace is automatically a linear variety.

The equivalence of these two definitions can be easily demonstrated. Let C satisfy the first definition, that is, there exists a $\mathbf{g} \in \mathbf{H}$ such that all the vectors of the form $\mathbf{x} - \mathbf{g}$ for $\mathbf{x} \in C$ is a closed subspace, say $\mathbf{W}$, in $\mathbf{H}$. Then, for $\mathbf{x}, \mathbf{y} \in C$, we have $\mathbf{x} - \mathbf{g} \in \mathbf{W}$, and $\mathbf{y} - \mathbf{g} \in \mathbf{W}$, so that for α arbitrary $\alpha\,(\mathbf{x} - \mathbf{g}) + (1 - \alpha)\,(\mathbf{y} - \mathbf{g}) = \alpha\,\mathbf{x} + (1 - \alpha)\,\mathbf{y} - \mathbf{g} \in \mathbf{W}$. Thus, we have $\alpha\,\mathbf{x} + (1 - \alpha)\,\mathbf{y} \in C$. The set C is closed because $\mathbf{W}$ is closed. Hence, C also satisfies the second definition. On the other hand, let C satisfy the second definition. Let $\mathbf{g}$ be a fixed element of C, and $\mathbf{W}$ be the set of all the vectors of the form $\mathbf{x} - \mathbf{g}$ for $\mathbf{x} \in C$. Then, for each $\mathbf{u} \in \mathbf{W}$, $\mathbf{u} + \mathbf{g} \in C$. By definition, for α arbitrary, the vector $\alpha\,(\mathbf{u} + \mathbf{g}) + (1 - \alpha)\,\mathbf{g} = \mathbf{g} + \alpha\mathbf{u} \in C$. It follows that $\alpha\mathbf{u} \in \mathbf{W}$. Let also $\mathbf{v} \in \mathbf{W}$, then $\mathbf{v} + \mathbf{g} \in C$. Clearly, the vector $1/2\,(\mathbf{u} + \mathbf{g}) + (1 - 1/2)\,(\mathbf{v} + \mathbf{g}) = 1/2(\mathbf{u} + \mathbf{v}) + \mathbf{g} \in C$. Thus, $\frac{1}{2}(\mathbf{u} + \mathbf{v}) \in \mathbf{W}$. By what we just proved above, $\mathbf{u} + \mathbf{v} \in \mathbf{W}$. Thus, the set $\mathbf{W}$ is a subspace. It is closed because C is closed. Hence, C also satisfies the first definition. The equivalence of the two definitions is thus established.

The above demonstration also reveals a fact that if a linear variety is translated by any vector of its own, the resulting set of vectors is a closed subspace.

Corollary 2.5-2 *Under the constraints of Theorem 2.5-1, if each C_i is a linear variety for $i = 1, 2, \cdots, m$, and if their intersection C_0 is non-empty, then for every $\mathbf{x} \in \mathbf{H}$ the sequence $\{T^n\mathbf{x}\}$ converges strongly to $P_0\mathbf{x}$, the projection of $\mathbf{x}$ onto C_0.*

Proof: Let $\mathbf{g} \in C_0$ be fixed, then $\mathbf{W}_i$, the set of vectors of the form $\mathbf{x} - \mathbf{g}$ for all $\mathbf{x} \in C_i$ is a closed subspace for each $i = 1, 2, \cdots, m$. Let Q_i denote the projector for $\mathbf{W}_i$, then $P_i\mathbf{x}$, the projection of $\mathbf{x}$ onto C_i, is given by

$$P_i\mathbf{x} = \mathbf{g} + Q_i(\mathbf{x} - \mathbf{g}) = Q_i\mathbf{x} + (I - Q_i)\mathbf{g}. \qquad (2.5\text{-}31)$$

Then

$$T_i\mathbf{x} = (1 - \lambda_i)\mathbf{x} + \lambda_i P_i\mathbf{x} = L_i\mathbf{x} + \lambda_i(I - Q_i)\mathbf{g}, \qquad (2.5\text{-}32)$$

where $L_i = (1 - \lambda_i)I + \lambda_i Q_i$, which is clearly linear for all $i = 1, 2, \cdots, m$. Thus, by simple algebra,

$$T\mathbf{x} = T_m T_{m-1} \cdots T_1 \mathbf{x} = L\mathbf{x} + f(\mathbf{g}), \qquad (2.5\text{-}33)$$

where $L = L_m L_{m-1} \cdots L_1$ and $f(\mathbf{g})$ is a term that is not dependent on $\mathbf{x}$. For example, for $m = 3$, $f(\mathbf{g}) = [\lambda_3(I - Q_3) + \lambda_2 L_2(I - Q_2) + \lambda_1 L_3 L_2(I - Q_1)]\,\mathbf{g}$. It follows form Eq. (2.5-33) that

$$T^n\mathbf{x} = L^n\mathbf{x} + \sum_{r=0}^{n-1} L^r f(\mathbf{g}). \qquad (2.5\text{-}34)$$

In particular, for $\mathbf{x} = \mathbf{g}$, $T^n\mathbf{g} = \mathbf{g}$, so we have

$$T^n\mathbf{g} = L^n\mathbf{g} + \sum_{r=0}^{n-1} L^r f(\mathbf{g}). \qquad (2.5\text{-}35)$$

Thus,

$$T^n \mathbf{x} - T^n \mathbf{g} = T^n \mathbf{x} - \mathbf{g} = L^n (\mathbf{x} - \mathbf{g}).$$ (2.5-36)

According to Corollary 2.5-1, we have

$$L^n (\mathbf{x} - \mathbf{g}) \to Q_0 (\mathbf{x} - \mathbf{g}),$$ (2.5-37)

where Q_0 is the projector onto $\mathbf{W}_0 = \cap_{i=1}^m \mathbf{W}_i$. Therefore, Eq. (2.5-36) yields

$$T^n \mathbf{x} \to \mathbf{g} + Q_0 (\mathbf{x} - \mathbf{g}).$$ (2.5-38)

By comparing with Eq. (2.5-31), we have $\mathbf{g} + Q_0 (\mathbf{x} - \mathbf{g}) = P_0 \mathbf{x}$, which is the projection of $\mathbf{x}$ onto C_0. ∎

2.6 THE POCS ALGORITHM AND ITS NUMERICAL ASPECTS

The significance of Theorem 2.5-1—the fundamental theorem of POCS—is that it defines a systematic numerical algorithm for finding a point in the intersection of closed convex sets. To rephrase the theorem, assume that $C_1, C_2, \cdots, C_m$ are m closed convex sets in a Hilbert space whose intersection $C_0 = \cap_{i=1}^m C_i$ is non-empty. Let T_i denote the relaxed projector associated with each set C_i for $i = 1, 2, \cdots, m$. Then, the iterates $\{\mathbf{x}_k\}$ generated by

$$\mathbf{x}_{n+1} = T_m T_{m-1} \cdots T_1 \mathbf{x}_n$$ (2.6-1)

with an arbitrary starting point $\mathbf{x}_0$ will converge weakly to a point of C_0. This algorithm is generally referred as *the convex projection algorithm* or simply *the POCS algorithm*. In particular, when the projector P_i is used instead of T_i for each set in Eq. (2.6-1), the POCS algorithm reduces to

$$\mathbf{x}_{n+1} = P_m P_{m-1} \cdots P_1 \mathbf{x}_n.$$ (2.6-2)

For ease of reference, the algorithm in Eq. (2.6-2) is referred as *the pure projection algorithm*, in contrast to the algorithm in Eq. (2.6-1), which is referred as *the relaxed projection algorithm*.

Due to its widespread application in science and engineering, the POCS algorithm has been studied in the literature regarding its numerical behavior in practical implementation [7, 8]. Two important issues related to the practical use of the POCS method are: (i) the choice of initial starting point of the iteration; and (ii) the choice of relaxation parameters in the relaxed projectors. In this section, we consider these issues by way of several numerical examples. For the ease of visualization, we present these examples in the Euclidean space R^2.

Example 2.6-1 This example is to demonstrate the effect of the choice of relaxation parameters on the convergence of the POCS algorithm. Suppose we are seeking a vector $\mathbf{x} = (x_1, x_2)$ in R^2 such that $\mathbf{x}$ has simultaneous membership in

$$C_1 = \{\mathbf{x} : x_1^2 + x_2^2 \le 1\} \tag{2.6-3}$$

and

$$C_2 = \{\mathbf{x} : x_1 = 1\}. \tag{2.6-4}$$

Clearly, both C_1 and C_2 are closed and convex and their intersection contains the single point $(1, 0)$.[†] In such a case, we expect the POCS algorithm to converge to this point regardless of the choice of initial starting point and the choice of relaxation parameters.

From Example 2.3-1 in Section 2.3 we know that the projection of any point $\mathbf{x}$ onto the set C_1 is

$$P_1\mathbf{x} = \begin{cases} \mathbf{x} & \text{if } \mathbf{x} \in C_1 \\ \mathbf{x}/\|\mathbf{x}\| & \text{if } \mathbf{x} \notin C_1. \end{cases} \tag{2.6-5}$$

Then, the relaxed projection of $\mathbf{x}$ onto C_1 is

$$T_1\mathbf{x} = \begin{cases} \mathbf{x} & \text{if } \mathbf{x} \in C_1 \\ \left(1 - \lambda_1 + \dfrac{\lambda_1}{\|\mathbf{x}\|}\right)\mathbf{x} & \text{if } \mathbf{x} \notin C_1, \end{cases} \tag{2.6-6}$$

where λ_1 is the relaxation parameter of T_1.

The projection of any point $\mathbf{x}$ onto the set C_2 can be shown to be (see Exercise 2-11)

$$P_2\mathbf{x} = (1, x_2). \tag{2.6-7}$$

First, let's consider the pure projection algorithm $\mathbf{x}_{k+1} = P_2 P_1 \mathbf{x}_k$ with $\mathbf{x}_k = (x_{1k}, x_{2k})$. Assume $\mathbf{x}_0 \ne (1, 0)$. Then from Eq. (2.6-5) and Eq. (2.6-7) we have

$$\mathbf{x}_1 = \begin{cases} (1, x_{20}) & \text{if } \mathbf{x}_0 \in C_1 \\ (1, x_{20}/\|\mathbf{x}_0\|) & \text{if } \mathbf{x}_0 \notin C_1, \end{cases} \tag{2.6-8}$$

and for $k \ge 1$,

$$\mathbf{x}_{k+1} = \left(1, \frac{1}{(1 + x_{2k}^2)^{1/2}} x_{2k}\right). \tag{2.6-9}$$

Clearly, the sequence of iterates $\{\mathbf{x}_k\}$ generated by the pure projection algorithm $\mathbf{x}_{k+1} = P_2 P_1 \mathbf{x}_k$ will converge to $\mathbf{x}^* = (1, 0)$, as illustrated in Fig. 2.6-1.

Next, let's consider the relaxed projection algorithm $\mathbf{x}_{k+1} = P_2 T_1 \mathbf{x}_k$. Notice that a relaxed projector is used for the set C_1, while a true projector is used for the set C_2. This choice is made in view of the fact that C_2 is a linear subspace.[‡]

[†]A set containing a single point is sometimes called a *singleton* set.
[‡]Levi in [8] showed that it is more efficient, from the point of view of rapidity of convergence, to use pure rather than relaxed projectors when projecting onto a linear subspace.

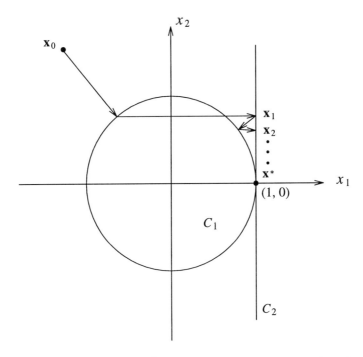

Fig. 2.6-1 In the Euclidean space R^2 the intersection of the sets $C_1 = \{\mathbf{x} : x_1^2 + x_2^2 \leq 1\}$ and $C_2 = \{\mathbf{x} : x_1 = 1\}$ contains only the point $(1, 0)$.

From Eq. (2.6-6) and Eq. (2.6-7) we have

$$\mathbf{x}_1 = \begin{cases} (1, x_{20}) & \text{if } \mathbf{x}_0 \in C_1 \\[2ex] \left(1, (1 - \lambda_1 + \frac{\lambda_1}{\|\mathbf{x}_0\|})x_{20}\right) & \text{if } \mathbf{x}_0 \notin C_1, \end{cases} \tag{2.6-10}$$

and for $k \geq 1$,

$$\mathbf{x}_{k+1} = \left(1, (1 - \lambda_1 + \frac{\lambda_1}{(1 + x_{2k}^2)^{1/2}})x_{2k}\right). \tag{2.6-11}$$

It is interesting to examine how the choice of the relaxation parameter λ_1 affects the convergence of iteration. First, let's consider $0 < \lambda_1 < 1$. In such a case we have

$$1 - \lambda_1 > \frac{1 - \lambda_1}{(1 + x_{2k}^2)^{1/2}} > 0 \tag{2.6-12}$$

for $x_{2k} \neq 0$, which yields

$$1 - \lambda_1 + \frac{\lambda_1}{(1 + x_{2k}^2)^{1/2}} > \frac{1}{(1 + x_{2k}^2)^{1/2}} . \qquad (2.6\text{-}13)$$

This implies that the iterates generated by the pure projection algorithm in Eq. (2.6-9) will converge to $(1, 0)$ faster than that by the relaxed projection algorithm in Eq. (2.6-11) for each $0 < \lambda_1 < 1$.

On the other hand, for the case of $1 < \lambda_1 < 2$,

$$1 - \lambda_1 < \frac{1 - \lambda_1}{(1 + x_{2k}^2)^{1/2}} < 0 \qquad (2.6\text{-}14)$$

for $x_{2k} \neq 0$, which yields

$$1 - \lambda_1 + \frac{\lambda_1}{(1 + x_{2k}^2)^{1/2}} < \frac{1}{(1 + x_{2k}^2)^{1/2}} . \qquad (2.6\text{-}15)$$

Thus, when the iteration is close to convergence, i.e., when x_{2k}^2 is close to 0, the resulting term on the left-hand side of Eq. (2.6-15) will be positive. This implies that for each $1 < \lambda_1 < 2$ the iterates generated by Eq. (2.6-11) will converge to $(1, 0)$ faster than that by Eq. (2.6-9).

To help illustrate the above analysis, some numerical results are furnished in Tables 2.6-1 to 2.6-3. The experiment demonstrates the convergence behaviors of the pure projection algorithm in Eq. (2.6-9) and the relaxed projection algorithm in Eq. (2.6-11) with $\lambda_1 = 0.5$ and then with $\lambda_1 = 1.5$ for the following arbitrarily chosen initial starting points: $(1, 1)$, $(100, -100)$, and $(-123, 9001)$, respectively. These results consistently show that the relaxed projection algorithm when $\lambda_1 = 1.5$ converges faster than the pure projection algorithm while it is slower than the latter when $\lambda_1 = 0.5$. ∎

Example 2.6-2 This example is to demonstrate the effect of the choice of initial starting point on the convergence behavior of the POCS algorithm. Suppose we are seeking a vector $\mathbf{x} = (x_1, x_2)$ in R^2 such that $\mathbf{x}$ has simultaneous membership in

$$C_1 = \{\mathbf{x} : x_1^2 + x_2^2 \leq 1\} \qquad (2.6\text{-}16)$$

and

$$C_2 = \{\mathbf{x} : 1/2 \leq x_1 \leq 1\}. \qquad (2.6\text{-}17)$$

Clearly, the solution set $C_0 = C_1 \cap C_2$ contains an infinite number of points, as illustrated in Fig. 2.6-2. In such a case, we expect that the choice of initial starting point will affect the convergence of the POCS algorithm.

The projection onto the set C_1 is given in Eq. (2.6-5). The set C_2 is clearly closed and convex. The projection of any point $\mathbf{x}$ onto this set can be shown to be

Table 2.6-1 Numerical results of Example 2.6-1: (i) the pure projection algorithm ($\lambda_1 = 1$); (ii) the relaxed projection algorithm with $\lambda_1 = 1.5$; and (iii) the relaxed projection algorithm with $\lambda_1 = 0.5$ when the initial starting point is $\mathbf{x}_0 = (1, 1)$.

Iterations	(i) $\lambda_1 = 1$	(ii) $\lambda_1 = 1.5$	(iii) $\lambda_1 = 0.5$
0	(1, 1)	(1, 1)	(1, 1)
1	(1, 0.707107)	(1, 0.560660)	(1, 0.853553)
20	(1, 0.218218)	(1, 0.175088)	(1, 0.314664)
40	(1, 0.156174)	(1, 0.126119)	(1, 0.224483)
60	(1, 0.128037)	(1, 0.103685)	(1, 0.183518)
80	(1, 0.111111)	(1, 0.090122)	(1, 0.158942)
100	(1, 0.099504)	(1, 0.080791)	(1, 0.142134)
140	(1, 0.084215)	(1, 0.068466)	(1, 0.120062)
160	(1, 0.078811)	(1, 0.064100)	(1, 0.112280)
180	(1, 0.074329)	(1, 0.060476)	(1, 0.105834)
200	(1, 0.070535)	(1, 0.057405)	(1, 0.100383)

(see Exercise 2-12)

$$P_2\mathbf{x} = \begin{cases} (1/2, x_2) & \text{if } x_1 < 1/2 \\ (1, x_2) & \text{if } x_1 > 1 \\ \mathbf{x} & \text{if } 1/2 \le x_1 \le 1. \end{cases} \tag{2.6-18}$$

And the relaxed projection of $\mathbf{x}$ onto C_2 is

$$T_2\mathbf{x} = \begin{cases} (x_1 + \lambda_2(1/2 - x_1), x_2) & \text{if } x_1 < 1/2 \\ (x_1 + \lambda_2(1 - x_1), x_2) & \text{if } x_1 > 1 \\ \mathbf{x} & \text{if } 1/2 \le x_1 \le 1, \end{cases} \tag{2.6-19}$$

where λ_2 is the relaxation parameter of T_2.

To investigate the effect of the choice of initial starting point on the convergence of the POCS algorithm, we consider the pure projection algorithm $\mathbf{x}_{k+1} = P_2 P_1 \mathbf{x}_k$ to isolate the effect of relaxation parameters. Trivially, if $\mathbf{x}_0 \in C_0$, i.e., the starting point is in the solution set, then $\mathbf{x}_0 = P_2 P_1 \mathbf{x}_0$ and the problem is over. Thus assume $\mathbf{x}_0 = (x_{10}, x_{20}) \notin C_0$, then if $\mathbf{x}_0$ is chosen such that $P_2 P_1 \mathbf{x}_0 \in C_1$, the algorithm will reach convergence in a single iteration. From Eq. (2.6-5) and Eq. (2.6-18), this is possible if and only if

(i) $\mathbf{x}_0 \in C_1$, and $\mathbf{x}_0$ satisfies

$$x_{10} < 1/2 \quad \text{and} \quad \|(1/2, x_{20})\| \le 1. \tag{2.6-20}$$

Table 2.6-2 Numerical results of Example 2.6-1: (i) the pure projection algorithm ($\lambda_1 = 1$); (ii) the relaxed projection algorithm with $\lambda_1 = 1.5$; and (iii) the relaxed projection algorithm with $\lambda_1 = 0.5$ when the initial starting point is $\mathbf{x}_0 = (100, -100)$.

Iterations	(i) $\lambda_1 = 1$	(ii) $\lambda_1 = 1.5$	(iii) $\lambda_1 = 0.5$
0	$(100, -100)$	$(100, -100)$	$(100, -100)$
1	$(1, -0.707107)$	$(1, 48.939340)$	$(1, -50.353553)$
20	$(1, -0.218218)$	$(1, 0.172808)$	$(1, -0.418863)$
40	$(1, -0.156174)$	$(1, 0.125263)$	$(1, -0.255455)$
60	$(1, -0.128037)$	$(1, 0.103209)$	$(1, -0.199453)$
80	$(1, -0.111111)$	$(1, 0.089809)$	$(1, -0.169001)$
100	$(1, -0.099504)$	$(1, 0.080566)$	$(1, -0.149206)$
120	$(1, -0.090909)$	$(1, 0.073695)$	$(1, -0.135032)$
140	$(1, -0.084215)$	$(1, 0.068329)$	$(1, -0.124244)$
160	$(1, -0.078811)$	$(1, 0.063987)$	$(1, -0.115680)$
180	$(1, -0.074329)$	$(1, 0.060381)$	$(1, -0.108670)$
200	$(1, -0.070535)$	$(1, 0.057324)$	$(1, -0.102793)$

(ii) $\mathbf{x}_0 \notin C_1$, and $\mathbf{x}_0$ satisfies either

$$\frac{1}{2} \le \frac{x_{10}}{\|\mathbf{x}_0\|} \le 1, \tag{2.6-21}$$

or

$$\frac{x_{10}}{\|\mathbf{x}_0\|} < \frac{1}{2} \quad \text{and} \quad \|(1/2, x_{20}/\|\mathbf{x}_0\|)\| \le 1. \tag{2.6-22}$$

In the case of Eq. (2.6-20), $\mathbf{x}_0$ lies in the following region

$$\Gamma_1 = \left\{ \mathbf{x} : \|\mathbf{x}\| \le 1, x_1 < 1/2, \text{ and } |x_2| \le \sqrt{3}/2 \right\}. \tag{2.6-23}$$

The convergent point of the algorithm for this case is $\mathbf{x}^* = (1/2, x_{20})$. In the case of Eq. (2.6-21), $\mathbf{x}_0$ lies in the following region

$$\Gamma_2 = \left\{ \mathbf{x} : \|\mathbf{x}\| > 1, \text{ and } |x_2| \le \sqrt{3}x_1 \right\}. \tag{2.6-24}$$

The convergent point for this case is $\mathbf{x}^* = \mathbf{x}_0/\|\mathbf{x}_0\|$. In the case of Eq. (2.6-22), $\mathbf{x}_0$ lies in the following region

$$\Gamma_3 = \left\{ \mathbf{x} : \|\mathbf{x}\| > 1, \text{ and } |x_2| \le -\sqrt{3}x_1 \right\}. \tag{2.6-25}$$

The convergent point for this case is $\mathbf{x}^* = (1/2, x_{20}/\|\mathbf{x}_0\|)$.

Single-iteration convergence is not possible if $\mathbf{x}_0$ lies in either of the following

Table 2.6-3 Numerical results of Example 2.6-1: (i) the pure projection algorithm ($\lambda_1 = 1$); (ii) the relaxed projection algorithm with $\lambda_1 = 1.5$; and (iii) the relaxed projection algorithm with $\lambda_1 = 0.5$ when the initial starting point is $\mathbf{x}_0 = (-123, 9001)$.

Iterations	(i) $\lambda_1 = 1$	(ii) $\lambda_1 = 1.5$	(iii) $\lambda_1 = 0.5$
0	$(-123, 9001)$	$(-123, 9001)$	$(-123, 9001)$
1	$(1, 0.999907)$	$(1, -4499)$	$(1, 4501)$
20	$(1, 0.223606)$	$(1, -0.247307)$	$(1, 0.605385)$
40	$(1, 0.158114)$	$(1, -0.146250)$	$(1, 0.286814)$
60	$(1, 0.129099)$	$(1, -0.114004)$	$(1, 0.213525)$
80	$(1, 0.111803)$	$(1, -0.096644)$	$(1, 0.177350)$
100	$(1, 0.100000)$	$(1, -0.085386)$	$(1, 0.154873)$
120	$(1, 0.091287)$	$(1, -0.077328)$	$(1, 0.139196)$
140	$(1, 0.084515)$	$(1, -0.071194)$	$(1, 0.127469)$
160	$(1, 0.079057)$	$(1, -0.066321)$	$(1, 0.118272)$
180	$(1, 0.074536)$	$(1, -0.062330)$	$(1, 0.110811)$
200	$(1, 0.070711)$	$(1, -0.058983)$	$(1, 0.104601)$

two regions in the R^2:

$$\Gamma_4 = \left\{ \mathbf{x} : x_2 > \sqrt{3}/2, \text{ and } x_2 > \sqrt{3}|x_1| \right\} \qquad (2.6\text{-}26)$$

and

$$\Gamma_5 = \left\{ \mathbf{x} : x_2 < -\sqrt{3}/2, \text{ and } x_2 < -\sqrt{3}|x_1| \right\}. \qquad (2.6\text{-}27)$$

By a similar argument to that leading to Eq. (2.6-9) in Example 2.6-1, we can show that the POCS algorithm $\mathbf{x}_{k+1} = P_2 P_1 \mathbf{x}_k$ always converges to $\mathbf{x}^* = (1/2, \sqrt{3}/2)$ when it is started with an arbitrary point $\mathbf{x}_0 \in \Gamma_4$. In such a case, it takes infinite number of iterations for convergence to occur. Similarly, for each $\mathbf{x}_0 \in \Gamma_5$ the POCS algorithm always converges to $\mathbf{x}^* = (1/2, -\sqrt{3}/2)$ after infinite number of iterations.

To help clarify the point, we summarize the above analysis result in Fig. 2.6-3 where the effect of the choice of initial starting point on the POCS algorithm is shown. When the initial starting point is chosen in the region Γ_1 or Γ_3, the algorithm converges to a point on the line segment connecting point $A : (1/2, \sqrt{3}/2)$ and point $B : (1/2, -\sqrt{3}/2)$; when it is chosen in the region Γ_2, the algorithm converges to a point on the arc segment $\|\mathbf{x}\| = 1$ with $1/2 \le x_1 \le 1$. Note the convergence for the above cases takes merely a single iteration. On the other hand, when the algorithm is started from a point in the region Γ_4 or Γ_5 it takes infinite number of iterations to reach the point A or B, respectively. Finally, when the initial starting point is inside the solution region Γ_0, it is unaffected by the algorithm. ∎

Example 2.6-3 In this example, we illustrate the effect of the choice of relaxation

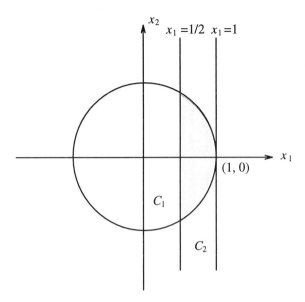

Fig. 2.6-2 In R^2 the intersection of the sets $C_1 = \{\mathbf{x} : x_1^2 + x_2^2 \leq 1\}$ and $C_2 = \{\mathbf{x} : 1/2 \leq x_1 \leq 1\}$ contains infinite number of points.

parameters on the convergence behavior of the POCS algorithm. Suppose we are seeking a vector $\mathbf{x} = (x_1, x_2)$ in R^2 such that $\mathbf{x}$ has simultaneous membership in

$$C_1 = \{\mathbf{x} : x_1^2 + x_2^2 \leq 1\} \tag{2.6-28}$$

and

$$C_2 = \{\mathbf{x} : 1/2 < x_1 \leq 1\}. \tag{2.6-29}$$

The solution set is $C_0 = C_1 \cap C_2$, which contains infinite number of points.

Here the set C_2 is not a closed set and its projection is not well defined. Consider the set $\overline{C_2}$, the closure of C_2. In such a case the POCS algorithm will converge to a point of the set $C_1 \cap \overline{C_2}$. This set is different from the true solution set $C_0 = C_1 \cap C_2$ in that it also contains the boundary points on the line segment AB, as shown in Fig. 2.6-3, which are obviously erroneous solutions. Unfortunately, from the discussion in Example 2.6-2 it is clear that a pure projection algorithm will converge to such an erroneous solution as long as the starting point is chosen to be within the regions Γ_1, Γ_2, or Γ_3. This is caused by the fact that all the erroneous solutions are on the boundary of the set $\overline{C_2}$ and the projector P_2 always reaches a boundary point of $\overline{C_2}$ from a point outside $\overline{C_2}$.

A possible remedy for the erroneous solution in such a case is to replace the POCS solution $\mathbf{x}^*$ by a point in its neighborhood which is inside the true solution set C_0. In other words, we can modify the final convergent point, say $\mathbf{x}^*$, by a small vector $\mathbf{e}$ such that $\mathbf{x}^* + \mathbf{e} \in C_0$.

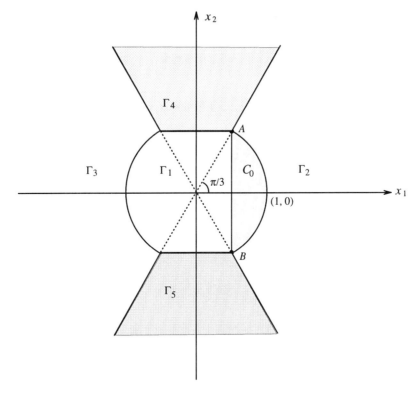

Fig. 2.6-3 The effect of the choice of the initial starting point on the convergence of the POCS algorithm $\mathbf{x}_{k+1} = P_2 P_1 \mathbf{x}_k$. Starting points in Γ_1 or Γ_3 lead to convergence to points on the line segment $\overline{AB}$. Points A and B have coordinates $(1/2, \sqrt{3}/2)$ and $(1/2, -\sqrt{3}/2)$, respectively.

Another way of avoiding an erroneous solution is to use the relaxed projector for the set $\overline{C_2}$ in the POCS algorithm. From our discussion in Section 2.4, a relaxed projector tends to reach a point inside a set rather on its boundary. Therefore, the use of relaxed projectors in the POCS algorithm will greatly reduce the likelihood of converging to a boundary point of the solution set.

Shown in Tables 2.6-4 and 2.6-5 are some numerical results of the relaxed projection algorithm $\mathbf{x}_{k+1} = T_2 T_1 \mathbf{x}_k$ with $\lambda_1 = \lambda_2 = 1.5$ when the initial starting point $\mathbf{x}_0$ is chosen arbitrarily in different regions of Fig. 2.6-3. It is interesting to note that in these results the relaxed projection algorithm not only converges to a point inside the correct solution set but also converges in a finite number of iterations even when the initial starting point is chosen in the regions Γ_4 and Γ_5.

This example reveals the interesting fact that a POCS algorithm need not take an infinite number of iterations to achieve convergence. In a situation where speed of convergence is critical, the use of the relaxed projectors can help to speed up the convergence. ∎

Table 2.6-4 Numerical results of the relaxed projection algorithm $x_{k+1} = T_2 T_1 x_k$ with $\lambda_1 = \lambda_2 = 1.5$ for the following starting points: (i) $x_0 = (8, -5) \in \Gamma_2$; and (ii) $x_0 = (-1234, -100) \in \Gamma_3$. The asterisk indicates the convergence point.

Iterations	(i) $x_0 = (8, -5) \in \Gamma_2$	(ii) $x_0 = (-1234, -100) \in \Gamma_3$
1	(2.1140, 1.7050)	(−306.2525, 49.8788)
2	(0.6947, 0.0892)*	(−74.3229, −24.6983)
3	——————	(−16.3690, 11.8761)
4		(−1.9852, −5.0572)
5		(0.5278, 1.1323)
6		(0.5651, 0.7934)*

We remark that the elementary examples furnished on the previous pages were selected only for the purpose of easy demonstration. Without a doubt the reader could have solved them by observation. In practical applications, however, this is rarely the case and the POCS algorithm must be realized on a computer to obtain a solution. In later chapters of this book, problems involving hundreds of constraints in high-dimensional spaces will be solved by this method.

Before we end this section, let us examine the practical aspect of the *weak* convergence nature of the POCS algorithm. Clearly, the POCS algorithm is an iterative algorithm. Theoretically, possibly an infinite number of iterations need to be done in order to achieve convergence. On the other hand, in practice we definitely do not want to run an iterative algorithm indefinitely. As a result, we simply terminate an iterative algorithm when we believe that the iterates are *close* enough to convergence. A common criterion to judge *closeness* to convergence is to compute the distance between two successive iterates, say $\|x_{n+1} - x_n\|$, and terminate the algorithm when it is small enough. But, the weak convergence nature of the algorithm in Eq. (2.6-1) seems to prevent the use of this criterion because it is not necessarily true that the distance between two successive elements converges to zero for a weakly convergent sequence. For example in the l^2 space the sequence $\{u_n\}$ with $u_n = (0, \cdots, 0, 1, 0, \cdots)$, where 1 only occurs in the nth entry, converges weakly to 0 (see Example 1.5-8 in Section 1.5), while the distance $\|u_m - u_n\| = \sqrt{2}$ for any distinct m and n. Fortunately, this is not an issue in practice because the algorithm in Eq. (2.6-1) is implemented on a computer where for the ease of numerical implementation a problem formulated in an infinite-dimensional space is often, if not always, approximated in a finite-dimensional space[†]. In other words, the practical, i.e., numerical version of the algorithm in Eq. (2.6-1) is carried out in a finite-dimensional space. In such a case, the weak convergence is equivalent to strong convergence and the seemingly obstacle no longer exists.

[†]After all, a digital computer is a finite-state machine and therefore is limited to representing finite-dimensional spaces.

Table 2.6-5 Numerical results of the relaxed projection algorithm $\mathbf{x}_{k+1} = T_2 T_1 \mathbf{x}_k$ with $\lambda_1 = \lambda_2 = 1.5$ for the following starting points: (i) $\mathbf{x}_0 = (-1, 8654) \in \Gamma_4$; and (ii) $\mathbf{x}_0 = (13, -99) \in \Gamma_5$.

Iterations	(i) $\mathbf{x}_0 = (-1, 8654) \in \Gamma_4$	(ii) $\mathbf{x}_0 = (13, -99) \in \Gamma_5$
1	(0.5001, −4325.5000)	(3.9024, 48.0128)
2	(0.8749, 2161.2500)	(1.6648, −22.5113)
3	(0.9684, −1079.1250)	(1.1109, 9.7597)
4	(0.9914, 538.0625)	(0.9429, −3.3895)
5	(0.9965, −267.5313)	(0.7847, 0.2496)*
6	(0.9963, 132.2656)	————
7	(0.9934, −64.6329)	
8	(0.9868, 30.8166)	
9	(0.9727, −13.9091)	
10	(0.9409, 5.4582)	
11	(0.8578, −1.2509)	
12	(0.5403, −0.6116)*	

2.7 APPLICATION OF POCS METHOD

In his pioneering paper Youla [4] introduced the POCS method to the signal processing community. Since then the POCS method has found a plethora of applications in practical problems. Even as this book is being written, new applications are reported in the technical literature. In these applications, the POCS method either provides an alternate to previously existing solutions or offers brand new solutions. Even in the former case, it may shed new light on our understanding of existing algorithms. One such example is the famous Gerchberg-Papoulis algorithm, which we shall discuss at the end of this section and again in Chapter 7.

In its most general form, the POCS method for solving a practical problem has the following framework: We want to recover, design, or determine an unknown quantity about which some information is known in the form of constraints. The unknown quantity is treated as a vector in a Hilbert space, and the known constraints are described in the form of closed convex sets in this space. Without loss of generality, assume that there are a total of m such sets $C_1, C_2, \cdots, C_m$ available. Each set is usually associated with a single constraint although sometimes it is convenient to include multiple constraints in a single set. Then, the intersection of all these sets, say $C_0 = \cap_{i=1}^m C_i$, will contain all the possible solutions to the problem because each solution satisfies all the available information about the unknown. To find such a solution, we apply the fundamental theorem of POCS: Let T_i denote the relaxed projector associated with each set C_i for $i = 1, 2, \cdots, m$, then the iterates generated by

$$\mathbf{x}_{n+1} = T_m T_{m-1} \cdots T_1 \mathbf{x}_n, \qquad (2.7\text{-}1)$$

with x_0 arbitrary, will converge weakly to a point of C_0, i.e., a solution to the problem.

The key to the successful application of the POCS method is how to define the appropriate constraint sets $C_1, C_2, \cdots, C_m$ to describe the available information. This is the creative part of the problem. The computation of the projection P_i may be technically challenging but can usually be achieved. In what follows, we illustrate this by briefly discussing the POCS approach to the following two broad categories of problems: (i) recovery problem, and (ii) design problem. Specific problems and their projection solutions are discussed in some detail in Chapters 6 through 9.

Loosely speaking, a *recovery problem* refers to the situation where given data g we want to determine the source f that produces g. Examples are many. In a digital communication system a waveform signal such as a voice signal at the transmitter is first converted to a series of digital symbols and then sent to the receiver. The receiver then tries to recover the waveform signal that produced the received digital symbols. In astronomical imaging, a picture is taken of an object of interest. Due to interference caused by so-called atmospheric turbulence, the picture often suffers from distortion such as blur. An interesting task then is how to remove the distortion from the picture. In other words, we want to recover the original source image that is free of distortion. In image processing, this is referred to as *image restoration*. Another example of recovery is in statistical estimation, such as probability density function (pdf) estimation, where we want to estimate the pdf that underlies the observed data samples. Yet another example is the *phase recovery* problem in astronomy, crystallography, and optics. The phase is difficult to measure directly yet is needed for reconstructing a picture of the object.

To apply the POCS method to solve a recovery problem, we treat the unknown source f as a vector in a proper Hilbert space. We then define one or more closed convex sets in this vector space to *relate* the unknown source f to the known data g. For easy reference, constraint sets of this nature are simply referred as *data sets*. From the discussion above, it is clear that in a practical problem the relation between the unknown source f and the data g can be deterministic or stochastic in nature. As a result, for some applications, defining data sets is by no means a trivial matter and becomes the key to the success of the POCS method. For sake of illustration, let us consider a simple example. Suppose that we want to recover an unknown source f from its K measurements $\mathbf{g} = \{g_1, g_2, \cdots, g_K\}$. Assume that the unknown source can be described by a function $\mathbf{f} = f(x)$ in L^2 space. Let $\mathcal{M}_i$ denote the operation for obtaining the ith measurement, i.e., $\mathcal{M}_i(\mathbf{f}) = g_i$. Then we can define the following constraint sets in L^2

$$C_i = \{\mathbf{f} : \mathcal{M}_i(\mathbf{f}) = g_i\} \qquad (2.7\text{-}2)$$

for $i = 1, 2, \cdots, K$. In words, the set C_i is the collection of all functions in L^2 that produces the ith measurement g_i. Assume that all the sets C_i are convex and closed (we will deal with non-convex closed sets in Chapter 5). Then, the intersection of all these sets will contain all the feasible sources that could have produced all the

m measurement. Clearly, the inclusion of the data sets in the POCS algorithm will enforce that the recovered source $\mathbf{f}$ satisfies the data information $\mathbf{g}$.

Note that the intersection of all the data sets in Eq. (2.7-2) can be written as

$$C_0 = \{\mathbf{f} : \mathcal{M}_i(\mathbf{f}) = g_i \text{ for } i = 1, 2, \cdots, K\}. \tag{2.7-3}$$

The reader might ask: why not include the set C_0 directly in the POCS algorithm instead of the individual sets C_i in Eq. (2.7-2). After all, we could then enforce the data constraint by using the set C_0 alone. However, this approach may suffer from the limitation that it may not be easy to compute the projection directly onto the set C_0, while it may be easy or at least easier to compute the projection onto each individual set C_i. This is an important point.

A fundamental issue in a recovery problem is that the data $\mathbf{g}$ may not be adequate for determining $\mathbf{f}$ uniquely. This is usually the case in practical applications. For example, in digital communications, a waveform is often *quantized* to produce digital symbols for transmission—a non-invertible process which means that the original waveform cannot be recovered exactly from the digital symbols. Another example is in astronomical imaging, where the picture is often corrupted by noise. As a result, it is virtually impossible to recover the original noise-free image exactly.

When the data $\mathbf{g}$ are not adequate for determining $\mathbf{f}$ uniquely, there will be more than one source $\mathbf{f}$ that produces the same data $\mathbf{g}$. In such cases, the data set C_0 in Eq. (2.7-3) will contain more than one element. Clearly, if no further knowledge is available, then there is little more we can do and we simply take any element in C_0 as our solution. However, it is often the case that in certain applications we have some sort of *prior knowledge* about the unknown source, which is independent of the observed data. In other words, the prior knowledge is known to us even before any data are collected about the unknown. For example, in pdf estimation, we know that the function that we are looking for is non-negative and has unity area. This is always true regardless of the distribution of the data. Conceivably, when prior knowledge is available, we would like to use it to eliminate as much ambiguity in the data set as possible to narrow down the possible solutions.

The POCS method offers a natural way for us to incorporate prior knowledge into the recovery process. Indeed this is one of the great advantage of POCS. Specifically, we define closed convex constraint sets to describe the prior knowledge in the same fashion as we did with the data information. For easy reference, constraint sets of this nature are referred as *property sets*. Then the solution that we seek will be a point in the intersection of all the available data sets and property sets because such a point will meet not only the data information but also the prior knowledge that we have. In such a case, the inclusion of property sets in the POCS algorithm eliminates "spurious" solutions in contrast to the situation where only data sets are used. Clearly, the POCS method offers a flexible framework for the incorporation of prior knowledge in a recovery problem.

In practice, it is often the case that we want to produce a signal or system such that desired specifications are met. Such problems are quite naturally called *design* or *synthesis* problems. For example, in signal processing, we may want to design a filter to achieve certain desired frequency response. In dynamic control engineering, we may want to design a controller so that a closed-loop system will achieve certain desired performance. Or, we might want to design a lens with certain properties.

To apply the POCS method to solve a design problem, we treat the quantity to be designed, say $\mathbf{f}$, as a vector in a proper Hilbert space. We then define closed convex sets in this space that relate $\mathbf{f}$ to the design specifications. As an illustration, assume there are a set of K design specifications $\{d_1, d_2, \cdots, d_K\}$ and we want $\mathbf{f}$ such that $\mathcal{D}_i(\mathbf{f}) \leq d_i$ for each $i = 1, 2, \cdots, K$, where $\mathcal{D}_i$ denotes the operation that relates the unknown $\mathbf{f}$ to the ith specification d_i. Then we can define the following constraint sets

$$C_i = \{\mathbf{f} : \mathcal{D}(\mathbf{f}) \leq d_i\} \tag{2.7-4}$$

for $i = 1, 2, \cdots, K$. In words, the set C_i is the collection of all possible designs that will satisfy the ith specification. As a result, the intersection of all these sets C_i will contain all the feasible designs that meet the design specifications. Of course, not all design problems involve such a type of constraint as $\mathcal{D}(\mathbf{f}) \leq d_i$. POCS can accommodate other types of constraints as well.

In a design problem we often have some prior knowledge about the unknown design in addition to having knowledge of the design specifications. For example, we may have some preference for certain design patterns. To incorporate prior knowledge into the design, we simply define additional closed convex sets as in the case of recovery problem. In such a case, a point in the intersection of all the constraint sets will satisfy both the design specifications and the prior knowledge regarding our preferences. To find such a point, we can invoke the POCS algorithm. Of course, the key to the success of this approach lies in how to define the constraint sets so that they are convex and closed and their projections are numerically computable.

Before we end this section, we furnish, for the sake of illustration, a classic example of a signal recovery problem that has received much attention in the literature.

Example 2.7-1 *Signal Recovery* In the space of L^2 signals, a signal $f(t)$ is said to be *band-limited* if its Fourier transform

$$F(\omega) = \int_{-\infty}^{\infty} f(t)e^{-j\omega t}dt \tag{2.7-5}$$

is zero outside a finite interval, say $[-2\pi B, 2\pi B]$. If B is the *smallest* number for which $F(\omega) = 0$ when $\omega \notin [-2\pi B, 2\pi B]$, then B is called the *bandwidth* of the signal $f(t)$ and is measured in cycles per second or *Hertz* (Hz).

A well-known fact is that a band-limited signal $f(t)$ is *analytic* on the entire t-axis (see Exercise 2-9), that is, it admits to the Taylor series expansion [10]

$$f(t) = \sum_{n=0}^{\infty} \frac{f^{(n)}(t_0)}{n!}(t - t_0)^n \tag{2.7-6}$$

about every $t = t_0$.

According to Eq. (2.7-6) it is clear that if a band-limited signal $f(t)$ is known within some ϵ interval around any point t_0, then it is completely determined for the entire t-axis. Furthermore, Eq. (2.7-6) offers an analytical approach to extrapolating a band-limited signal $f(t)$ from merely a single segment of it. That is, given a segment of $f(t)$, we can choose a point t_0 inside this known segment and compute the derivatives $f^{(n)}(t_0)$ for every n. Then the entire function $f(t)$ is completely recovered from Eq. (2.7-6). This approach is known as *analytic continuation*. Logical as it may be, it is rarely used in practice because numerical evaluation of a derivative is very sensitive to noise. An interesting question then is to consider an alternative approach. It turns out that the POCS method offers an interesting and more robust alternative to this problem.

Let $g(t)$ denote a known segment, say over the finite interval $[a, b]$, of a band-limited signal $f(t)$ with bandwidth B. Then the function we are seeking satisfies the following two properties: (i) it coincides with $g(t)$ over the interval $[a, b]$; and (ii) it is band-limited to B Hz. These properties can be described by the following constraint sets in the L^2 space:

$$C_1 = \{f(t) : f(t) = g(t) \text{ for every } t \in [a, b] \}, \tag{2.7-7}$$

and

$$C_2 = \{f(t) : F(\omega) = \mathcal{F}\{f(t)\} = 0 \text{ for all } |\omega| > 2\pi B \}, \tag{2.7-8}$$

where $\mathcal{F}$ denotes the Fourier transform operator, i.e.,

$$\mathcal{F}\{f(t)\} = \int_{-\infty}^{\infty} f(t)e^{-j\omega t}dt. \tag{2.7-9}$$

By the argument of analytic continuation it is clear that the solution set $C_0 = C_1 \cap C_2$ contains exactly one element which is the solution to the problem. As we will see from the examples in Chapter 3, both the sets C_1 and C_2 are closed convex sets. Furthermore, the set C_1 is a linear variety and the set C_2 is a linear subspace. Thus, according to Corollary 2.5-2 the pure projection algorithm

$$f_{n+1}(t) = P_2 P_1 f_n(t) \tag{2.7-10}$$

with $f_0(t)$ arbitrary will converge strongly to the solution.

The projectors P_1 and P_2 in Eq. (2.7-10) are given as follows (see Exercise 2-13 and Exercise 2-14): For $f(t) \notin C_1$,

$$P_1 f(t) = \begin{cases} g(t) & \text{if } t \in [a, b] \\ f(t) & \text{otherwise.} \end{cases} \tag{2.7-11}$$

And for $f(t) \notin C_2$, $P_2 f(t)$ in Fourier domain is given by

$$\mathcal{F}\{P_2 f(t)\} = \begin{cases} F(\omega) & \text{if } |\omega| \leq 2\pi B \\ 0 & \text{otherwise,} \end{cases} \tag{2.7-12}$$

where $F(\omega)$ is the Fourier transform of $f(t)$.

From the projection operations in Eq. (2.7-11) and Eq. (2.7-12) it is clear that during each iteration the POCS algorithm in Eq. (2.7-10) first imposes the known information about the signal in the time (or space) domain then it imposes the known information about the signal in the Fourier domain. It is interesting to note that this algorithm is virtually identical to the so-called Gerchberg-Papoulis algorithm which is frequently used for restoration of a signal from both its time domain and Fourier domain information [9, 11]. Further details about the Gerchberg-Papoulis algorithm will be given in Chapter 7 for its use in super-resolution image reconstruction.

Finally, we give a numerical demonstration of the algorithm in Eq. (2.7-10). Here we want to reconstruct the signal

$$g(t) = \frac{\sin 2\pi B t}{2\pi B t}, \tag{2.7-13}$$

whose Fourier transform is

$$G(\omega) = \begin{cases} \frac{1}{2B} & \text{if } |\omega| \leq 2\pi B \\ 0 & \text{otherwise.} \end{cases} \tag{2.7-14}$$

Clearly, $g(t)$ is band-limited to B Hz.

In this demonstration the information used about the signal $g(t)$ is that it is band-limited to $B = 10$ Hz and that it is assumed to be known over the time interval $[0, 40/512]$ for the set C_1 in Eq. (2.7-7). The known portion of $g(t)$ is plotted in Fig. 2.7-1(a). The reconstructed signals after the first 10, 20, and 500 iterations of the algorithm in Eq. (2.7-10) are shown in Fig. 2.7-1(b), (c), and (d), respectively. It is interesting to see that the POCS algorithm converges nearly to the true solution, even after just a few iterations. ∎

2.8   PROOF OF OPIAL'S THEOREM

In this section, we furnish a proof for the Opial's theorem which we used in the proof of Theorem 2.5-1—the Fundamental Theorem of POCS—in Section 2.5. Again, readers who are mainly interested in the application of the POCS method may skip this section without sacrifice in continuity for the reading of the rest of the book.

For easy reference, we repeat the Opial's theorem below:

Theorem 2.8-1 (*Opial*) *Let G be a non-expansive asymptotically regular operator on a Hilbert space $\mathbf{H}$ with a non-empty set of fixed points $\Gamma \in \mathbf{H}$. Then for every*

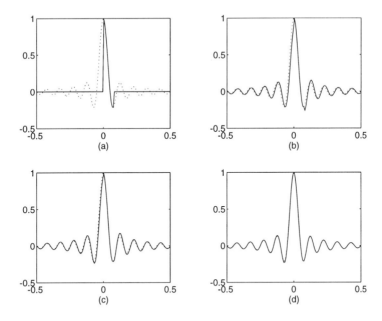

Fig. 2.7-1 The first few iterates of the algorithm in Eq. (2.7-10): (a) known portion; (b) after 10 iterations; (c) after 20 iterations; (d) after 500 iterations.

$\mathbf{x} \in \mathbf{H}$, *the sequence* $\{\mathbf{x}_n = G^n\mathbf{x}\}$ *converges weakly to a point of* Γ.

To prove the Opial's theorem, we start with a standard result on weak convergence in Hilbert spaces and three preparatory lemmas.

Theorem 2.8-2 *In a Hilbert space* **H** *every weakly convergent sequence is bounded; Also, every bounded sequence in* **H** *contains a weakly convergent subsequence.*

We omit the proof of this theorem for brevity. For reference, the interested reader is directed to a standard text on linear functional analysis, for example, see [12].

Lemma 2.8-1 *The set of fixed points of a non-expansive operator* G *in a Hilbert space* **H** *is a closed convex set.*

Proof: Let Γ denote the set of fixed points of G. Clearly, if Γ is empty the Lemma follows immediately. Assume that Γ is non-empty. We show first that Γ is closed. Let $\{\mathbf{x}_n\}$ be a sequence in Γ such that $\mathbf{x}_n \to \mathbf{x}$. Since G is non-expansive,

$$\begin{aligned}
\|G\mathbf{x} - \mathbf{x}\| &= \|G\mathbf{x} - G\mathbf{x}_n + G\mathbf{x}_n - \mathbf{x}\| \\
&= \|G\mathbf{x} - G\mathbf{x}_n + \mathbf{x}_n - \mathbf{x}\| \\
&\leq \|G\mathbf{x} - G\mathbf{x}_n\| + \|\mathbf{x}_n - \mathbf{x}\| \\
&\leq \|\mathbf{x} - \mathbf{x}_n\| + \|\mathbf{x}_n - \mathbf{x}\| \to 0.
\end{aligned} \qquad (2.8\text{-}1)$$

Thus, $G\mathbf{x} = \mathbf{x}$ and Γ is closed.

Next, we show that the set Γ is convex. We claim that Γ is identical to the set

$$\Gamma' = \left\{ \mathbf{z} : 2\Re\langle \mathbf{x} - G\mathbf{x}, \, \mathbf{z}\rangle \leq \|\mathbf{x}\|^2 - \|G\mathbf{x}\|^2 \text{ for every } \mathbf{x} \in \mathbf{H}\right\}. \qquad (2.8\text{-}2)$$

Let $\mathbf{y} \in \Gamma$, then by invoking the expansion formula $\|\mathbf{u} + \mathbf{v}\|^2 = \|\mathbf{u}\|^2 + \|\mathbf{v}\|^2 + 2\Re\langle \mathbf{u}, \, \mathbf{v}\rangle$ we obtain

$$\begin{aligned}
0 \;\leq\; \|\mathbf{x} - \mathbf{y}\|^2 - \|G\mathbf{x} - G\mathbf{y}\|^2 &= \|\mathbf{x} - \mathbf{y}\|^2 - \|G\mathbf{x} - \mathbf{y}\|^2 \\
&= \|\mathbf{x}\|^2 + \|\mathbf{y}\|^2 - 2\Re\langle \mathbf{x}, \, \mathbf{y}\rangle - \|G\mathbf{x}\|^2 - \|\mathbf{y}\|^2 + 2\Re\langle G\mathbf{x}, \, \mathbf{y}\rangle \\
&= \|\mathbf{x}\|^2 - \|G\mathbf{x}\|^2 - 2\Re\langle \mathbf{x} - G\mathbf{x}, \, \mathbf{y}\rangle, \qquad (2.8\text{-}3)
\end{aligned}$$

for every $\mathbf{x} \in \mathbf{H}$. Thus, $\mathbf{y} \in \Gamma'$ so that $\Gamma \subset \Gamma'$.

On the other hand, let $\mathbf{y} \in \Gamma'$, then

$$2\Re\langle \mathbf{x} - G\mathbf{x}, \, \mathbf{y}\rangle \leq \|\mathbf{x}\|^2 - \|G\mathbf{x}\|^2 \qquad (2.8\text{-}4)$$

for every $\mathbf{x} \in \mathbf{H}$. In particular, for $\mathbf{x} = \mathbf{y}$, we obtain

$$2\Re\langle \mathbf{y} - G\mathbf{y}, \, \mathbf{y}\rangle \leq \|\mathbf{y}\|^2 - \|G\mathbf{y}\|^2, \qquad (2.8\text{-}5)$$

which yields

$$\|\mathbf{y} - G\mathbf{y}\| \leq 0, \qquad (2.8\text{-}6)$$

so that $\mathbf{y} = G\mathbf{y}$. Thus $\mathbf{y} \in \Gamma$, and $\Gamma' \subset \Gamma$. Indeed, we have $\Gamma' = \Gamma$. Obviously the set Γ' is convex. ∎

Lemma 2.8-2 *In a Hilbert space* $\mathbf{H}$, *if* $\mathbf{x}_n \xrightarrow{w} \mathbf{x}_0$, *then*

$$\varliminf \|\mathbf{x}_n - \mathbf{x}\| > \varliminf \|\mathbf{x}_n - \mathbf{x}_0\| \qquad (2.8\text{-}7)$$

for every $\mathbf{x} \neq \mathbf{x}_0$ *in* $\mathbf{H}$.[†]

Proof: By Theorem 2.8-2, the weakly convergent sequence $\{\mathbf{x}_n\}$ is bounded, so both the limit inferiors in Eq. (2.8-7) are finite. Note that for $\mathbf{x} \neq \mathbf{x}_0$

$$\begin{aligned}
\|\mathbf{x}_n - \mathbf{x}\|^2 &= \|\mathbf{x}_n - \mathbf{x}_0 + \mathbf{x}_0 - \mathbf{x}\|^2 \\
&= \|\mathbf{x}_n - \mathbf{x}_0\|^2 + \|\mathbf{x}_0 - \mathbf{x}\|^2 + 2\Re\langle \mathbf{x}_n - \mathbf{x}_0, \, \mathbf{x}_0 - \mathbf{x}\rangle
\end{aligned}$$

[†]Let $\{x_n\}$ be a sequence of real numbers, then its *limit inferior* $\varliminf x_n$ and *limit superior* $\varlimsup x_n$ are defined respectively by

$$\varliminf x_n = \sup_n \inf_{k \geq n} x_k, \quad \text{and} \quad \varlimsup x_n = \inf_n \sup_{k \geq n} x_k.$$

For a bounded sequence both its limit inferior and limit superior exist as finite numbers. Moreover, the limit of a sequence exists if and only if its limit inferior equals to its limit superior.

$$> \quad \|\mathbf{x}_n - \mathbf{x}_0\|^2 + 2\Re\langle\mathbf{x}_n - \mathbf{x}_0, \mathbf{x}_0 - \mathbf{x}\rangle \qquad (2.8\text{-}8)$$

for every n. Since $\mathbf{x}_n \xrightarrow{w} \mathbf{x}_0$, we have $\langle\mathbf{x}_n - \mathbf{x}_0, \mathbf{x}_0 - \mathbf{x}\rangle \to 0$, and Eq. (2.8-7) follows. ∎

Lemma 2.8-3 *Let G be a non-expansive operator in a Hilbert space* **H**. *If $\mathbf{x}_n \xrightarrow{w}$ $\mathbf{x}_0$, and $(I - G)\mathbf{x}_n \to \mathbf{y}_0 \in$ **H**, then $(I - G)\mathbf{x}_0 = \mathbf{y}_0$.*

Proof: Since G is non-expansive, and $(I - G)\mathbf{x}_n \to \mathbf{y}_0$,

$$
\begin{aligned}
\underline{\lim} \, \|\mathbf{x}_n - \mathbf{x}_0\| &\geq \underline{\lim} \, \|G\mathbf{x}_n - G\mathbf{x}_0\| \\
&= \underline{\lim} \, \|\mathbf{y}_0 - (I - G)\mathbf{x}_n + \mathbf{x}_n - \mathbf{y}_0 - G\mathbf{x}_0\| \\
&\geq \underline{\lim} \, [\, \|\mathbf{y}_0 - (I - G)\mathbf{x}_n\| - \|\mathbf{x}_n - \mathbf{y}_0 - G\mathbf{x}_0\| \,] \\
&= \underline{\lim} \, \|\mathbf{x}_n - \mathbf{y}_0 - G\mathbf{x}_0\|. \qquad (2.8\text{-}9)
\end{aligned}
$$

By Lemma 2.8-2 if $\mathbf{x}_0 \neq \mathbf{y}_0 + G\mathbf{x}_0$, we would have

$$\underline{\lim} \, \|\mathbf{x}_n - \mathbf{x}_0\| < \underline{\lim} \, \|\mathbf{x}_n - \mathbf{y}_0 - G\mathbf{x}_0\|, \qquad (2.8\text{-}10)$$

which contradicts Eq. (2.8-9). Thus, $\mathbf{x}_0 = \mathbf{y}_0 + G\mathbf{x}_0$ or $(I - G)\mathbf{x}_0 = \mathbf{y}_0$ and the Lemma follows. ∎

Proof of Opial's Theorem

Let $\mathbf{y}$ be a fixed point of G. Then for every $\mathbf{x} \in$ **H**, the sequence $\{d_n\}$ with $d_n = \|G^n\mathbf{x} - \mathbf{y}\|$ is non-increasing because

$$d_{n+1} = \|G^{n+1}\mathbf{x} - \mathbf{y}\| = \|G^{n+1}\mathbf{x} - G\mathbf{y}\| \leq \|G^n\mathbf{x} - \mathbf{y}\| = d_n. \qquad (2.8\text{-}11)$$

Thus, the limit

$$d(\mathbf{y}) = \lim_{n\to\infty} d_n(\mathbf{y}) = \lim_{n\to\infty} \|G^n\mathbf{x} - \mathbf{y}\| \qquad (2.8\text{-}12)$$

exists as a finite number for every $\mathbf{y} \in \Gamma$. We claim that there exists a unique $\mathbf{y}^* \in \Gamma$ at which $d(\mathbf{y})$ assumes its minimum value among all $\mathbf{y} \in \Gamma$.

According to Lemma 2.8-1, Γ is a closed convex set in **H**. Let

$$\delta = \inf_{\mathbf{y}\in\Gamma} d(\mathbf{y}). \qquad (2.8\text{-}13)$$

Then for each $\delta_m = \delta + 1/m$, $m = 1, 2, \cdots$, there exist a point $\mathbf{y}_m \in \Gamma$ such that $d(\mathbf{y}_m) < \delta_m$. Therefore, we obtain a sequence $\{\mathbf{y}_m\}$ such that

$$\lim_{m\to\infty} d(\mathbf{y}_m) = \delta. \qquad (2.8\text{-}14)$$

Invoking the parallelogram law $\|\mathbf{u} + \mathbf{v}\|^2 + \|\mathbf{u} - \mathbf{v}\|^2 = 2\|\mathbf{u}\|^2 + 2\|\mathbf{v}\|^2$ with

$\mathbf{u} = G^n\mathbf{x} - \mathbf{y}_m$ and $\mathbf{v} = G^n\mathbf{x} - \mathbf{y}_l$, we obtain

$$\|\mathbf{y}_l - \mathbf{y}_m\|^2 = 2\|G^n\mathbf{x} - \mathbf{y}_l\|^2 + 2\|G^n\mathbf{x} - \mathbf{y}_m\|^2 - \|2G^n\mathbf{x} - (\mathbf{y}_l + \mathbf{y}_m)\|^2. \tag{2.8-15}$$

The set Γ is convex, so that $(\mathbf{y}_l + \mathbf{y}_m)/2 \in \Gamma$ and

$$\|G^n\mathbf{x} - \tfrac{\mathbf{y}_l + \mathbf{y}_m}{2}\| \geq d(\frac{\mathbf{y}_l + \mathbf{y}_m}{2}) \geq \delta. \tag{2.8-16}$$

Thus, Eq. (2.8-15) yields

$$\|\mathbf{y}_l - \mathbf{y}_m\|^2 \leq 2\|G^n\mathbf{x} - \mathbf{y}_l\|^2 + 2\|G^n\mathbf{x} - \mathbf{y}_m\|^2 - 4\delta^2. \tag{2.8-17}$$

By letting $n \to \infty$,

$$\|\mathbf{y}_l - \mathbf{y}_m\|^2 \leq 2d^2(\mathbf{y}_l) + 2d^2(\mathbf{y}_m) - 4\delta^2. \tag{2.8-18}$$

Thus,

$$\lim_{l,m\to\infty} \|\mathbf{y}_l - \mathbf{y}_m\|^2 \leq 2 \lim_{l\to\infty} d^2(\mathbf{y}_l) + 2 \lim_{m\to\infty} d^2(\mathbf{y}_m) - 4\delta^2 = 0, \tag{2.8-19}$$

which implies that the sequence $\{\mathbf{y}_m\}$ is a Cauchy sequence and has a limit, say $\mathbf{y}^*$, in $\mathbf{H}$. Since the set Γ is closed, $\mathbf{y}^*$ also belongs to Γ. Note that

$$\|G^n\mathbf{x} - \mathbf{y}^*\| = \|G^n\mathbf{x} - \mathbf{y}_m + \mathbf{y}_m - \mathbf{y}^*\| \leq \|G^n\mathbf{x} - \mathbf{y}_m\| + \|\mathbf{y}_m - \mathbf{y}^*\|. \tag{2.8-20}$$

By letting $n \to \infty$, we obtain

$$d(\mathbf{y}^*) \leq d(\mathbf{y}_m) + \|\mathbf{y}_m - \mathbf{y}^*\| \tag{2.8-21}$$

for every m. Since $\mathbf{y}_m \to \mathbf{y}^*$, we have

$$d(\mathbf{y}^*) \leq \lim_{m\to\infty} d(\mathbf{y}_m) = \delta. \tag{2.8-22}$$

Also, $\mathbf{y}^* \in \Gamma$ so that $d(\mathbf{y}^*) \geq \delta$. Thus, $d(\mathbf{y}^*) = \delta$. Indeed, $d(\mathbf{y})$ achieves its minimum value at $\mathbf{y}^*$. Furthermore, such a point $\mathbf{y}^*$ is unique. Suppose in addition there is another point $\mathbf{y}' \neq \mathbf{y}^*$ such that $d(\mathbf{y}') = d(\mathbf{y}^*)$. Then $(\mathbf{y}^* + \mathbf{y}')/2 \in \Gamma$ since Γ is convex. Thus, by the parallelogram law,

$$\begin{aligned} \|G^n\mathbf{x} - \tfrac{\mathbf{y}^* + \mathbf{y}'}{2}\|^2 &= \frac{1}{2}\|G^n\mathbf{x} - \mathbf{y}^*\|^2 + \frac{1}{2}\|G^n\mathbf{x} - \mathbf{y}'\|^2 - \|\tfrac{\mathbf{y}^* - \mathbf{y}'}{2}\|^2 \\ &< \frac{1}{2}\|G^n\mathbf{x} - \mathbf{y}^*\|^2 + \frac{1}{2}\|G^n\mathbf{x} - \mathbf{y}'\|^2. \end{aligned} \tag{2.8-23}$$

Letting $n \to \infty$ yields $d((\mathbf{y}^* + \mathbf{y}')/2) < d(\mathbf{y}^*)$, which is impossible by the definition of $d(\mathbf{y}^*)$. Indeed, $\mathbf{y}^*$ is unique.

To summarize, we have proved that for each $\mathbf{x} \in \mathbf{H}$, there exists a unique $\mathbf{y}^* \in \Gamma$

such that $d(\mathbf{y}) = \lim_{n \to \infty} \|G^n \mathbf{x} - \mathbf{y}\|$ assumes its minimum value at $\mathbf{y}^*$.

Next, we show that the sequence $\{G^n \mathbf{x}\}$ converges weakly to $\mathbf{y}^*$. Note that for $\mathbf{y}_0 \in \Gamma$, the sequence $\{d_n\}$ with $d_n = \|G^n \mathbf{x} - \mathbf{y}_0\|$ is non-increasing so that $d_n \le d_0 \le \infty$ for every n. It follows that the sequence $\{G^n \mathbf{x}\}$ is bounded because

$$\|G^n \mathbf{x}\| = \|G^n \mathbf{x} - \mathbf{y}_0 + \mathbf{y}_0\| \le \|G^n \mathbf{x} - \mathbf{y}_0\| + \|\mathbf{y}_0\| \le \|\mathbf{x} - \mathbf{y}_0\| + \|\mathbf{y}_0\| \quad (2.8\text{-}24)$$

for every n. By Theorem 2.8-2, the sequence $\{G^n \mathbf{x}\}$ contains at least one weakly convergent subsequence, say $\{\mathbf{w}_k\}$, with $\mathbf{w}_k = G^{n_k} \mathbf{x}$ for some index sequence $n_1 < n_2 < n_3 < \cdots$. Assume that the weak limit of this subsequence is $\mathbf{y}' \in \mathbf{H}$, that is

$$G^{n_k} \mathbf{x} \xrightarrow{w} \mathbf{y}' \text{ as } k \to \infty. \quad (2.8\text{-}25)$$

By assumption G is asymptotically regular, so we have

$$\lim_{k \to \infty} \|(I - G)\mathbf{w}_k\| = \lim_{k \to \infty} \|G^{n_k} \mathbf{x} - G^{n_k+1} \mathbf{x}\| = 0. \quad (2.8\text{-}26)$$

In short, we have $\mathbf{w}_k \xrightarrow{w} \mathbf{y}'$ and $(I - G)\mathbf{w}_k \to \mathbf{0}$. By Lemma 2.8-3 we have

$$(I - G)\mathbf{y}' = \mathbf{0}, \text{ or } G\mathbf{y}' = \mathbf{y}'. \quad (2.8\text{-}27)$$

Thus, $\mathbf{y}'$ is a fixed point G, i.e., $\mathbf{y}' \in \Gamma$.

Since $\{\mathbf{w}_k\}$ is a subsequence of $\{G^n \mathbf{x}\}$, it follows that

$$\lim_{k \to \infty} \|\mathbf{w}_k - \mathbf{y}'\| = d(\mathbf{y}') \text{ and } \lim_{k \to \infty} \|\mathbf{w}_k - \mathbf{y}^*\| = d(\mathbf{y}^*). \quad (2.8\text{-}28)$$

On the other hand, by Lemma 2.8-2 if $\mathbf{y}' \ne \mathbf{y}^*$ we would have

$$\lim_{k \to \infty} \|\mathbf{w}_k - \mathbf{y}'\| < \lim_{k \to \infty} \|\mathbf{w}_k - \mathbf{y}^*\|, \quad (2.8\text{-}29)$$

which is impossible by Eq. (2.8-28), since $\mathbf{y}' \in \Gamma$. Thus, $\mathbf{y}' = \mathbf{y}^*$.

From the above demonstration it is clear that all the possible weakly convergent subsequences of $\{G^n \mathbf{x}\}$ have $\mathbf{y}^*$ as their weak limit. This leads to the conclusion that the sequence $\{G^n \mathbf{x}\}$ itself converges weakly to $\mathbf{y}^*$. Indeed, suppose this is not the case, then it is true that for some $\mathbf{z} \in \mathbf{H}$ the sequence $\{\langle G^n \mathbf{x}, \mathbf{z}\rangle\}$ fails to converge to $\langle \mathbf{y}^*, \mathbf{z}\rangle$. Because the sequence $\{G^n \mathbf{x}\}$ is bounded, the scalar sequence $\{\langle G^n \mathbf{x}, \mathbf{z}\rangle\}$ is also bounded. Therefore, there exists a subsequence of $\{\langle G^n \mathbf{x}, \mathbf{z}\rangle\}$ which converges to a number, say α, which is different from $\langle \mathbf{y}^*, \mathbf{z}\rangle$. For notational simplicity, let $\left\{G^{n'_k} \mathbf{x}\right\}$ denote this subsequence. Since the sequence $\left\{G^{n'_k} \mathbf{x}\right\}$ is bounded, it has a weakly convergent subsequence, say $\left\{G^{n''_k} \mathbf{x}\right\}$. Note that $\left\{G^{n''_k} \mathbf{x}\right\}$ is also a subsequence of $\{G^n \mathbf{x}\}$. According to what we just proved above $\left\{G^{n''_k} \mathbf{x}\right\}$ converges weakly to $\mathbf{y}^*$. Thus, $\langle G^{n''_k} \mathbf{x}, \mathbf{z}\rangle \to \langle \mathbf{y}^*, \mathbf{z}\rangle$. On the other hand, because $\langle G^{n'_k} \mathbf{x}, \mathbf{z}\rangle \to \alpha$, we have $\langle G^{n''_k} \mathbf{x}, \mathbf{z}\rangle \to \alpha \ne \langle \mathbf{y}^*, \mathbf{z}\rangle$, which is clearly impossible. Therefore, the sequence $\{G^n\mathbf{x}\}$ converges weakly to $\mathbf{y}^*$. ∎

2.9 POCS ALGORITHM FORMULATED IN A PRODUCT SPACE

The fundamental theorem of POCS (Theorem 2.5-1) defines a successive projection algorithm (Eq. (2.6-1)) for solving the problem of finding a point in the intersection of several closed convex sets. In [13] Pierra introduced a formulation of this problem in a product vector space. The application of the fundamental theorem of POCS to this product vector space automatically leads to an interesting alternative algorithm for finding a point in the intersection of several closed convex sets.

Assume that $C_1, C_2, \cdots, C_m$ denote m closed convex sets in a Hilbert space $\mathbf{H}$, and C_0 denotes their intersection set $C_0 = \bigcap_{i=1}^{m} C_i$ which is non-empty. For each $i = 1, 2, \cdots, m$, let P_i denote the projection operator onto the set C_i. Then we have the following result:

Theorem 2.9-1 *For every $\mathbf{x}_0 \in \mathbf{H}$ and every choice of positive constants w_1, w_2, $\cdots, w_m$ such that $\sum_{i=1}^{m} w_i = 1$, the sequence $\{\mathbf{x}_n\}$ generated by*

$$\mathbf{x}_{n+1} = \sum_{i=1}^{m} w_i P_i \mathbf{x}_n \qquad (2.9\text{-}1)$$

converges weakly to a point of C_0.

Proof: The proof is a direct result of the formalism developed by Pierra [13]. Let us introduce the product space defined by $\mathcal{H} \triangleq \mathbf{H}^m$, i.e., the m-fold Cartesian product of the Hilbert space H; its elements are ordered m-tuples $(\mathbf{x}_1, \mathbf{x}_2, \cdots, \mathbf{x}_m)$, where $\mathbf{x}_i \in \mathbf{H}$ for each $i = 1, 2, \cdots, m$, and will be simply denoted by $\mathbf{X}$. The inner product on $\mathcal{H}$ and its induced norm are denoted, respectively, by $<< \cdot, \cdot >>$ and $\||| \cdot \|||$, with

$$<< \mathbf{X}, \mathbf{Y} >> \triangleq \sum_{i=1}^{m} w_i \langle \mathbf{x}_i, \mathbf{y}_i \rangle, \qquad (2.9\text{-}2)$$

and

$$\|||\mathbf{X}|||^2 = << \mathbf{X}, \mathbf{X} >> = \sum_{i=1}^{m} w_i \|\mathbf{x}_i\|^2, \qquad (2.9\text{-}3)$$

where $\mathbf{X} = (\mathbf{x}_1, \mathbf{x}_2, \cdots, \mathbf{x}_m) \in \mathcal{H}$, $\mathbf{Y} = (\mathbf{y}_1, \mathbf{y}_2, \cdots, \mathbf{y}_m) \in \mathcal{H}$, $\langle \cdot, \cdot \rangle$ and $\| \cdot \|$ are, respectively, the inner product and the norm defined on $\mathbf{H}$. As an exercise the reader is urged to verify that Eq. (2.9-2) indeed defines a valid norm on $\mathcal{H}$; also, to verify that the vector space $\mathcal{H}$ is a Hilbert space.

In the space $\mathcal{H}$ we define the following sets:

$$\mathbf{C} = \{ \mathbf{X} : \mathbf{X} = (\mathbf{x}_1, \mathbf{x}_2, \cdots, \mathbf{x}_m) \in \mathcal{H}, \text{ and } \mathbf{x}_i \in C_i \text{ for each } i = 1, 2, \cdots, m \},$$
$$(2.9\text{-}4)$$

and

$$\mathbf{D} = \{\, \mathbf{X} :\ \mathbf{X} = (\mathbf{x}_1, \mathbf{x}_2, \cdots, \mathbf{x}_m) \in \mathcal{H}, \ \text{and}\ \mathbf{x}_1 = \mathbf{x}_2 = \cdots = \mathbf{x}_m \,\}. \quad (2.9\text{-}5)$$

In words, the set $\mathbf{C}$ is the set of all ordered m-tuples $(\mathbf{x}_1, \mathbf{x}_2, \cdots, \mathbf{x}_m)$ such that $\mathbf{x}_i, i = 1, 2, \cdots, m$, has membership in C_i. Likewise, the set $\mathbf{D}$ is the set of all m-tuples $(\mathbf{x}_1, \mathbf{x}_2, \cdots, \mathbf{x}_m)$ such that each component is equal to all others.

It can be shown that the set $\mathbf{C}$ is convex and closed since each set C_i is closed and convex in $\mathbf{H}$. Also, the set $\mathbf{D}$ is convex and closed in $\mathcal{H}$. In fact, $\mathbf{D}$ is a closed subspace. Therefore, by the fundamental theorem of POCS, the sequence $\{\mathbf{X}_n\}$ generated by the following algorithm

$$\mathbf{X}_{n+1} = P_{\mathbf{D}} P_{\mathbf{C}} \mathbf{X}_n, \quad (2.9\text{-}6)$$

where $P_{\mathbf{D}}, P_{\mathbf{C}}$ denote, respectively, the projectors onto $\mathbf{D}$ and $\mathbf{C}$, will converge weakly to a point $\mathbf{X}^* \in \mathcal{H}$ under the inner product $<< \cdot, \cdot >>$ in $\mathcal{H}$. We denote the starting point by $\mathbf{X}_0 \overset{\triangle}{=} (\mathbf{x}_0, \mathbf{x}_0, \cdots, \mathbf{x}_0)$ where $\mathbf{x}_0 \in \mathbf{H}$.

Now, let us consider the computation of the projectors $P_{\mathbf{C}}$ and $P_{\mathbf{D}}$ in Eq. (2.9-6). The computation of $P_{\mathbf{C}}$ proceeds as follows. For an arbitrary $\mathbf{X} = (\mathbf{x}_1, \mathbf{x}_2, \cdots, \mathbf{x}_m) \in \mathcal{H}$, and $\mathbf{Y} = (\mathbf{y}_1, \mathbf{y}_2, \cdots, \mathbf{y}_m) \in \mathbf{C}$, we seek the point $\mathbf{Y}^* \in \mathbf{C}$ *nearest* to $\mathbf{X}$. Hence we begin with

$$\|\|\mathbf{X} - \mathbf{Y}\|\|^2 = \sum_{i=1}^{m} w_i \|\mathbf{x}_i - \mathbf{y}_i\|^2. \quad (2.9\text{-}7)$$

Note that for $\mathbf{y}_i \in C_i$, $\|\mathbf{x}_i - \mathbf{y}_i\|^2$ is minimized when $\mathbf{y}_i = P_i \mathbf{x}_i$. Hence, the projection of an arbitrary $\mathbf{X}$ onto $\mathbf{C}$ is given by

$$\mathbf{Y}^* = P_{\mathbf{C}} \mathbf{X} = (P_1 \mathbf{x}_1, P_2 \mathbf{x}_2, \cdots, P_m \mathbf{x}_m). \quad (2.9\text{-}8)$$

Next, we consider the computation of $P_{\mathbf{D}}$. Again for an arbitrary $\mathbf{X} \in \mathcal{H}$, we seek the point $\mathbf{Y}^* \in \mathbf{D}$ *nearest* to $\mathbf{X}$. For $\mathbf{Y} = (\mathbf{y}_1, \mathbf{y}_2, \cdots, \mathbf{y}_m) \in \mathbf{D}$, we write

$$\|\|\mathbf{X} - \mathbf{Y}\|\|^2 = \sum_{i=1}^{m} w_i \|\mathbf{x}_i - \mathbf{y}\|^2. \quad (2.9\text{-}9)$$

By taking the gradient with respect to $\mathbf{y}$ and setting it to zero we get

$$\mathbf{y} = \sum_{i=1}^{m} w_i \mathbf{x}_i. \quad (2.9\text{-}10)$$

Thus,

$$\mathbf{Y}^* = P_{\mathbf{D}} \mathbf{X} = \left(\sum_{i=1}^{m} w_i \mathbf{x}_i, \sum_{i=1}^{m} w_i \mathbf{x}_i, \cdots, \sum_{i=1}^{m} w_i \mathbf{x}_i \right). \quad (2.9\text{-}11)$$

Let us return to the iteration in Eq. (2.9-6). Let the nth iterate be denoted by $\mathbf{X}_n \overset{\triangle}{=} (\mathbf{x}_{n1}, \mathbf{x}_{n2}, \cdots, \mathbf{x}_{nm})$. Since $\mathbf{X}_n \in \mathbf{D}$, we have $\mathbf{x}_{n1} = \mathbf{x}_{n2} = \cdots = \mathbf{x}_{nm} \overset{\triangle}{=} \mathbf{x}_n$. That is, $\mathbf{X}_n = (\mathbf{x}_n, \mathbf{x}_n, \cdots, \mathbf{x}_n)$ where $\mathbf{x}_n \in \mathbf{H}$. Then, Eq. (2.9-6) yields $\mathbf{X}_{n+1} = (\mathbf{x}_{n+1}, \mathbf{x}_{n+1}, \cdots, \mathbf{x}_{n+1})$ with $\mathbf{x}_{n+1} = \sum_{i=1}^{m} w_i P_i \mathbf{x}_n$, which is the iteration in Eq. (2.9-1).

Finally, consider $\mathbf{X}^* \in \mathbf{C} \cap \mathbf{D}$. Since it is in $\mathbf{D}$, it must be of the form $\mathbf{X}^* = (\mathbf{x}^*, \mathbf{x}^*, \cdots, \mathbf{x}^*)$ for some $\mathbf{x}^* \in \mathbf{H}$. Since it is also in $\mathbf{C}$, $\mathbf{x}^* \in C_i$ for each $i = 1, 2, \cdots, m$. Thus, $\mathbf{x}^* \in C_0$. To prove Theorem 2.9-1, it remains to be shown that $\{\mathbf{x}_n\}$ converges weakly to $\mathbf{x}^*$. Considering an arbitrary $\mathbf{z} \in \mathbf{H}$, we must show that $\langle \mathbf{x}_n, \mathbf{z} \rangle$ converges to $\langle \mathbf{x}^*, \mathbf{z} \rangle$. Since $\{\mathbf{X}_n\}$ converges weakly to $\mathbf{X}^* \in \mathcal{H}$, we have, by definition, $<< \mathbf{X}_n, \mathbf{Z} >>$ converging to $<< \mathbf{X}^*, \mathbf{Z} >>$ for arbitrary $\mathbf{Z} \in \mathcal{H}$ and in particular for $\mathbf{Z} = (\mathbf{z}, \mathbf{z}, \cdots, \mathbf{z})$. Note that $<< \mathbf{X}_n, \mathbf{Z} >>= \sum_{i=1}^{m} w_i \langle \mathbf{x}_n, \mathbf{z} \rangle = \langle \mathbf{x}_n, \mathbf{z} \rangle$, and $<< \mathbf{X}^*, \mathbf{Z} >>= \sum_{i=1}^{m} w_i \langle \mathbf{x}^*, \mathbf{z} \rangle = \langle \mathbf{x}^*, \mathbf{z} \rangle$. It follows that $\langle \mathbf{x}_n, \mathbf{z} \rangle$ converges to $\langle \mathbf{x}^*, \mathbf{z} \rangle$. The proof is now completed. ∎

From the proof above is it clear that the iterative algorithm in Eq. (2.9-1) $\mathbf{x}_{n+1} = \sum_{i=1}^{m} w_i P_i \mathbf{x}_n$ is a result of reformulating the problem of finding a point in the intersection of m closed convex sets, i.e., $C_1, C_2, \cdots, C_m$, in a Hilbert space $\mathbf{H}$ as a problem of finding a point in the intersection of only two convex sets, namely $\mathbf{C}$ in Eq. (2.9-4) and $\mathbf{D}$ in Eq. (2.9-5), in a new Hilbert space, i.e., the product space $\mathcal{H} = \mathbf{H}^m$. As we will see in the next section and later in Chapter 5, such a formulation leads to important applications.

So far only the pure projectors $P_{\mathbf{D}}$ and $P_{\mathbf{C}}$ were used in the algorithm $\mathbf{X}_{n+1} = P_{\mathbf{D}} P_{\mathbf{C}} \mathbf{X}_n$ in Eq. (2.9-6). A generalization of this algorithm to include relaxed projectors is readily obtained. In particular, let $T_{\mathbf{C}}$ be the relaxed version of $P_{\mathbf{C}}$, i.e.,

$$T_{\mathbf{C}} \overset{\triangle}{=} I + \lambda(P_{\mathbf{C}} - I), \tag{2.9-12}$$

where I is the identity operator on $\mathcal{H}$ and $0 < \lambda < 2$ is the relaxation parameter. Then, we have the relaxed algorithm

$$\mathbf{X}_{n+1} = P_{\mathbf{D}} T_{\mathbf{C}} \mathbf{X}_n. \tag{2.9-13}$$

Observe that the projector $P_{\mathbf{D}}$ is a linear operator on $\mathcal{H}$—a direct result that $\mathbf{D}$ is a closed subspace. Also, from Eq. (2.9-13) we have $\mathbf{X}_n \in \mathbf{D}$ so $P_{\mathbf{D}} \mathbf{X}_n = \mathbf{X}_n$. Thus,

$$\begin{aligned} P_{\mathbf{D}} T_{\mathbf{C}} \mathbf{X}_n &= P_{\mathbf{D}} [\mathbf{X}_n + \lambda(P_{\mathbf{C}} \mathbf{X}_n - \mathbf{X}_n)] \\ &= \mathbf{X}_n + \lambda(P_{\mathbf{D}} P_{\mathbf{C}} \mathbf{X}_n - \mathbf{X}_n). \end{aligned} \tag{2.9-14}$$

From the proof of Theorem 2.9-1, for $\mathbf{X}_n = (\mathbf{x}_n, \mathbf{x}_n, \cdots, \mathbf{x}_n)$ where $\mathbf{x}_n \in \mathbf{H}$ we have

$$P_{\mathbf{D}} P_{\mathbf{C}} \mathbf{X}_n = \left(\sum_{i=1}^{m} w_i P_i \mathbf{x}_n, \sum_{i=1}^{m} w_i P_i \mathbf{x}_n, \cdots, \sum_{i=1}^{m} w_i P_i \mathbf{x}_n \right). \tag{2.9-15}$$

Therefore, with $\mathbf{X}_{n+1} = (\mathbf{x}_{n+1}, \mathbf{x}_{n+1}, \cdots, \mathbf{x}_{n+1})$, the iteration $\mathbf{X}_{n+1} = P_{\mathbf{D}} T_{\mathbf{C}} \mathbf{X}_n$ yields

$$\mathbf{x}_{n+1} = \mathbf{x}_n + \lambda \left(\sum_{i=1}^{m} w_i P_i \mathbf{x}_n - \mathbf{x}_n \right). \tag{2.9-16}$$

This result is summarized in the following:

Theorem 2.9-2 *For every* $\mathbf{x}_0 \in \mathbf{H}$, $0 < \lambda < 2$, *and every choice of positive constants* $w_1, w_2, \cdots, w_m$ *such that* $\sum_{i=1}^{m} w_i = 1$, *the sequence* $\{ \mathbf{x}_n \}$ *generated by*

$$\mathbf{x}_{n+1} = \mathbf{x}_n + \lambda \left(\sum_{i=1}^{m} w_i P_i \mathbf{x}_n - \mathbf{x}_n \right) \tag{2.9-17}$$

converges weakly to a point of C_0.

The algorithm in Eq. (2.9-17) is a more general form of the iterative algorithm $\mathbf{x}_{n+1} = \sum_{i=1}^{m} w_i P_i \mathbf{x}_n$ in Eq. (2.9-1). According to this algorithm, at every iteration the new iterate $\mathbf{x}_{n+1}$ is obtained by first projecting $\mathbf{x}_n$ *simultaneously* onto all the sets C_i to get $P_i \mathbf{x}_n, i = 1, 2, \cdots, m$, and then the resulting projections are *averaged* in a weighted fashion to form the new iterate. In particular, if the weights w_i are all equal, i.e., $w_i = 1/m$, then Eq. (2.9-17) becomes

$$\mathbf{x}_{n+1} = \mathbf{x}_n + \frac{1}{m} \left(\sum_{i=1}^{m} P_i \mathbf{x}_n - \mathbf{x}_n \right). \tag{2.9-18}$$

In contrast, the POCS algorithm in Eq. (2.6-2)

$$\mathbf{x}_{n+1} = P_m P_{m-1} \cdots P_1 \mathbf{x}_n \tag{2.9-19}$$

finds its new iterate $\mathbf{x}_{n+1}$ by sequentially projecting $\mathbf{x}_n$ onto every set C_i in a cyclic manner. As a result, the algorithm in Eq. (2.9-19) is called, variously, a *successive*, *serial*, or sequential, projections algorithm, while the one in Eq. (2.9-17) is more reasonably called a *parallel* or *simultaneous* projections algorithm.

A further generalization of the algorithm in Eq. (2.9-13) is obtained by allowing the relaxation parameter λ in $T_{\mathbf{C}}$ to vary from iteration to iteration in Eq. (2.9-12), and hence in Eq. (2.9-13). That is, we have:

Theorem 2.9-3 *For every* $\mathbf{x}_0 \in \mathbf{H}$, *and every choice of positive constants* $w_1, w_2, \cdots, w_m$ *such that* $\sum_{i=1}^{m} w_i = 1$, *the sequence* $\{ \mathbf{x}_n \}$ *generated by*

$$\mathbf{x}_{n+1} = \mathbf{x}_n + \lambda_n \left(\sum_{i=1}^{m} w_i P_i \mathbf{x}_n - \mathbf{x}_n \right) \tag{2.9-20}$$

where $\epsilon < \lambda_n < 2 - \epsilon$ *for some arbitrary* $0 < \epsilon < 1$ *converges weakly to a point of* C_0.

Pierra [13] proposed an extrapolation technique to determine the relaxation parameter λ from iteration to iteration. We cite his result without proof below.

Theorem 2.9-4 *For every $\mathbf{x}_0 \in \mathbf{H}$, and every choice of positive constants $w_1, w_2,$* $\cdots, w_m$ *such that $\sum_{i=1}^{m} w_i = 1$, the sequence $\{\mathbf{x}_n\}$ generated by*

$$\mathbf{x}_{n+1} = \mathbf{x}_n + \lambda_n \left(\sum_{i=1}^{m} w_i P_i \mathbf{x}_n - \mathbf{x}_n \right) \tag{2.9-21}$$

with $\epsilon < \lambda_n < L_n$ for some $0 < \epsilon < 1$, and

$$L_n \triangleq \frac{\sum_{i=1}^{m} w_i \|P_i \mathbf{x}_n - \mathbf{x}_n\|^2}{\|\sum_{i=1}^{m} w_i P_i \mathbf{x}_n - \mathbf{x}_n\|^2} \tag{2.9-22}$$

converges weakly to a point of C_0.

Note that since $0 < w_i < 1$ and $\sum_{i=1}^{m} w_i = 1$, we have

$$\sum_{i=1}^{m} w_i P_i \mathbf{x}_n - \mathbf{x}_n = \sum_{i=1}^{m} w_i \left(P_i \mathbf{x}_n - \mathbf{x}_n \right). \tag{2.9-23}$$

The convexity of the function $\|\mathbf{x}\|^2$ implies that

$$\left\| \sum_{i=1}^{m} w_i \left(P_i \mathbf{x}_n - \mathbf{x}_n \right) \right\|^2 \leq \sum_{i=1}^{m} w_i \|P_i \mathbf{x}_n - \mathbf{x}_n\|^2.$$

Thus, we have $L_n \geq 1$ in Eq. (2.9-22).

In the literature the algorithm in Eq. (2.9-21) is called *extrapolated parallel projection method* (EPPM). It is reported that this technique can achieve fast numerical convergence due to the fact that larger values can be used for the relaxation parameter since L_n may be larger than 2 [13–15].

2.10 PROJECTION ALGORITHMS IN THE PRESENCE OF NON-INTERSECTING SETS

So far we have been concentrating on the problem of finding a point in the *intersection* of several closed and convex sets in a Hilbert space. In summary, let $C_1, C_2, \cdots, C_m$ denote m closed and convex sets in a Hilbert space $\mathbf{H}$, and P_i denote the projection operator onto C_i, $i = 1, 2, \cdots, m$. Then the sequence $\{\mathbf{x}_n\}$ generated by either the sequential projection algorithm

$$\mathbf{x}_{n+1} = P_m \cdots P_2 P_1 \mathbf{x}_n, \tag{2.10-24}$$

or the simultaneous projection algorithm

$$\mathbf{x}_{n+1} = \sum_{i=1}^{m} w_i P_i \mathbf{x}_n, \tag{2.10-25}$$

where $w_1, w_2, \cdots, w_m$ are positive constants such that $\sum_{i=1}^{m} w_i = 1$, will converge weakly to a point in C_0, the *non-empty* intersection of the sets $C_1, C_2, \cdots, C_m$. Moreover, both the algorithms in Eq. (2.10-24) and Eq. (2.10-25) can be relaxed to achieve faster numerical convergence.

A frequently asked question is: what if the sets $C_1, C_2, \cdots, C_m$ are *non-intersecting*, i.e., when their intersection C_0 is empty? Will these algorithms still converge? and if so, to what do they converge? These matters are discussed in the rest of this section.

Sequential Projection Algorithm

The convergence properties of the sequential projection algorithm in the presence of non-intersecting sets were studied in the literature (see [3], [16], [17], for example). Without proof, we summarize the main results below:

Theorem 2.10-1 *Assume that one of the sets $C_1, C_2, \cdots, C_m$ is bounded[†]. Then the sequence $\{ \mathbf{x}_n \}$ generated by*

$$\mathbf{x}_{n+1} = P_m \cdots P_2 P_1 \mathbf{x}_n \tag{2.10-26}$$

converges weakly to a point, say, $\mathbf{x}_m^*$ in C_m. Moreover, there exist points $\mathbf{x}_i^* \in C_i, i = 1, 2, \cdots, m-1$, such that $\mathbf{x}_{i+1}^* = P_{i+1}\mathbf{x}_i^*$ and $\mathbf{x}_1^* \stackrel{\triangle}{=} P_m \mathbf{x}_m^*$.

This convergence behavior is illustrated in Fig. 2.10-1 where three sets are involved. The algorithm $\mathbf{x}_{n+1} = P_3 P_2 P_1 \mathbf{x}_n$ leads to a closed path $\mathbf{x}_1^* \rightarrow \mathbf{x}_2^* = P_2 \mathbf{x}_1^* \rightarrow \mathbf{x}_3^* = P_3 \mathbf{x}_2^* \rightarrow \mathbf{x}_1^* = P_1 \mathbf{x}_3^*$. Such a closed path is called a *greedy* path in [17], owing to the fact that the distance between every two successive points in the path is minimized.

A particularly interesting case is when $m = 2$, i.e., when only two sets C_1 and C_2 are involved. In such a case the algorithm $\mathbf{x}_{n+1} = P_2 P_1 \mathbf{x}_n$ converges weakly to a point, say, $\mathbf{x}_2^*$ in C_2 such that $P_2 P_1 \mathbf{x}_2^* = \mathbf{x}_2^*$. In other words, $\mathbf{x}_2^*$ is a point in C_2 whose distance from C_1 is a minimum among all points of C_2. That is, if we let $\mathbf{x}_1^* = P_1 \mathbf{x}_2^*$, then the distance between $\mathbf{x}_1^*$ and $\mathbf{x}_2^*$ is the distance between the sets C_1 and C_2. This observation will be of use below.

[†]A set C is bounded if there exists some $\mu < \infty$ such that $\|\mathbf{x}\| < \mu$ for every $\mathbf{x} \in C$.

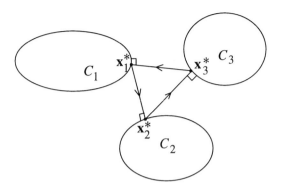

Fig. 2.10-1 When the sets C_1, C_2 and C_3 are non-intersecting, the sequential algorithm $\mathbf{x}_{n+1} = P_3 P_2 P_1 \mathbf{x}_n$ leads to a greedy path $\mathbf{x}_1^* \rightarrow \mathbf{x}_2^* = P_2 \mathbf{x}_1^* \rightarrow \mathbf{x}_3^* = P_3 \mathbf{x}_2^* \rightarrow \mathbf{x}_1^* = P_1 \mathbf{x}_3^*$.

Simultaneous Projection Algorithm

From the previous section we observe that the simultaneous projection algorithm is, in essence, a two-set sequential projection algorithm in a product space. Specifically, let $\mathcal{H}$ denote the m-fold Cartesian product space $\mathbf{H}^m$, of which the elements $\mathbf{X}$ are ordered m-tuples $(\mathbf{x}_1, \mathbf{x}_2, \cdots, \mathbf{x}_m)$, where $\mathbf{x}_i \in \mathbf{H}$ for each $i = 1, 2, \cdots, m$. Moreover, as stated earlier and repeated here for convenience, the inner product $<< \cdot, \cdot >>$ and induced norm $||| \cdot |||$ on $\mathcal{H}$ are, respectively, given by

$$<< \mathbf{X}, \mathbf{Y} >> \overset{\triangle}{=} \sum_{i=1}^{m} w_i \langle \mathbf{x}_i, \mathbf{y}_i \rangle, \qquad (2.10\text{-}27)$$

and

$$|||\mathbf{X}|||^2 = << \mathbf{X}, \mathbf{X} >> = \sum_{i=1}^{m} w_i ||\mathbf{x}_i||^2, \qquad (2.10\text{-}28)$$

where $\mathbf{X} = (\mathbf{x}_1, \mathbf{x}_2, \cdots, \mathbf{x}_m) \in \mathcal{H}$, $\mathbf{Y} = (\mathbf{y}_1, \mathbf{y}_2, \cdots, \mathbf{y}_m) \in \mathcal{H}$, and $\langle \cdot, \cdot \rangle$ and $|| \cdot ||$ are, respectively, the inner product and the norm defined on $\mathbf{H}$. In the product space $\mathcal{H}$ consider again two sets $\mathbf{C}$ and $\mathbf{D}$ defined earlier as

$$\mathbf{C} = \{ \, \mathbf{X} : \mathbf{X} = (\mathbf{x}_1, \mathbf{x}_2, \cdots, \mathbf{x}_m) \in \mathcal{H}, \text{ and } \mathbf{x}_i \in C_i \text{ for each } i = 1, 2, \cdots, m \, \}, \qquad (2.10\text{-}29)$$

and

$$\mathbf{D} = \{ \, \mathbf{X} : \mathbf{X} = (\mathbf{x}_1, \mathbf{x}_2, \cdots, \mathbf{x}_m) \in \mathcal{H}, \text{ and } \mathbf{x}_1 = \mathbf{x}_2 = \cdots = \mathbf{x}_m \, \}. \qquad (2.10\text{-}30)$$

It is easy to see that if the sets $C_1, C_2, \cdots, C_m$ are non-intersecting, then $\mathbf{C}$ and $\mathbf{D}$ are non-intersecting.

Now, consider the sequential projection algorithm

$$\mathbf{X}_{n+1} = P_{\mathbf{D}} P_{\mathbf{C}} \mathbf{X}_n. \tag{2.10-31}$$

Based on the convergence properties of the sequential projection algorithm discussed above, it is reasonable to conjecture[†] that this algorithm converges weakly to a point in **D** whose distance to **C** is minimized. Indeed, the following result is rigorously demonstrated in [18]:

Theorem 2.10-2 *Assume that one of the sets $C_1, C_2, \cdots, C_m$ is bounded. Then the sequence $\{ \mathbf{x}_n \}$ generated by the relaxed algorithm*

$$\mathbf{X}_{n+1} = \mathbf{X}_n + \lambda_n \left(P_{\mathbf{D}} P_{\mathbf{C}} \mathbf{X}_n - \mathbf{X}_n \right), \tag{2.10-32}$$

*where $\epsilon \le \lambda_n \le 2 - \epsilon$ for some arbitrary $\epsilon > 0$ converges weakly to a point, say, $\mathbf{X}^*$ in **D**, which has the minimum distance to the set **C**.*

Consider now the convergence point $\mathbf{X}^*$ of the algorithm in Eq. (2.10-32). Since $\mathbf{X}^* \in \mathbf{D}$, then $\mathbf{X}^* = (\mathbf{x}^*, \mathbf{x}^*, \cdots, \mathbf{x}^*)$ for some $\mathbf{x}^* \in \mathbf{H}$. Note that the distance of $\mathbf{X}^*$ to the set **C** is simply given by $|||\mathbf{X}^* - P_{\mathbf{C}} \mathbf{X}^*|||$. Also,

$$P_{\mathbf{C}} \mathbf{X}^* = (P_1 \mathbf{x}^*, P_2 \mathbf{x}^*, \cdots, P_m \mathbf{x}^*). \tag{2.10-33}$$

It follows that

$$|||\mathbf{X}^* - P_{\mathbf{C}} \mathbf{X}^*|||^2 = \sum_{i=1}^{m} w_i \|\mathbf{x}^* - P_i \mathbf{x}^*\|^2 = \sum_{i=1}^{m} w_i d^2(\mathbf{x}^*, C_i), \tag{2.10-34}$$

where $d(\mathbf{x}^*, C_i)$ denotes the distance of $\mathbf{x}^*$ from the set C_i, $i = 1, 2, \cdots, m$. Therefore, the minimum distance property of $\mathbf{X}^* \in \mathbf{D}$, i.e., the minimum of $|||\mathbf{X}^* - P_{\mathbf{C}} \mathbf{X}^*|||$, implies that the *weighted set-distance squares*

$$\Phi(\mathbf{x}^*) \triangleq \sum_{i=1}^{m} w_i d^2(\mathbf{x}^*, C_i) \tag{2.10-35}$$

is minimized. This result is summarized in the following:

Corollary 2.10-1 *For every $\mathbf{x}_0 \in \mathbf{H}$ and every choice of positive constants $w_1, w_2, \cdots, w_m$ such that $\sum_{i=1}^{m} w_i = 1$, the sequence $\{ \mathbf{x}_n \}$ generated by*

$$\mathbf{x}_{n+1} = \mathbf{x}_n + \lambda_n \left(\sum_{i=1}^{m} w_i \, P_i \mathbf{x}_n - \mathbf{x}_n \right), \tag{2.10-36}$$

[†]Note, however, that the set **C** is not necessarily bounded even if one of the sets $C_1, C_2, \cdots, C_m$ is bounded.

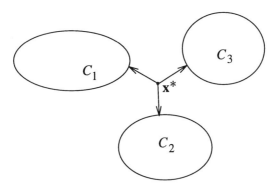

Fig. 2.10-2 When the sets C_1, C_2 and C_3 are non-intersecting, the parallel algorithm in Eq. (2.10-36) converges weakly to a point $\mathbf{x}^*$ such that the weighted set-distance squares $\Phi(\mathbf{x}^*) \triangleq \sum_{i=1}^{3} w_i d^2(\mathbf{x}^*, C_i)$ is minimized.

where $\epsilon \leq \lambda_n \leq 2 - \epsilon$ for some arbitrary $\epsilon > 0$ converges weakly to a point $\mathbf{x}^*$ such that $\Phi(\mathbf{x}^*) = \sum_{i=1}^{m} w_i d^2(\mathbf{x}^*, C_i)$ is minimized.

Note that this result applies when the sets intersect and when they do not. For the former, $\mathbf{x}^* \in C_i$ for $i = 1, 2, \cdots, m$ so that $\Phi(\mathbf{x}^*) = 0$; for the latter, $\mathbf{x}^*$ only *approximately* satisfies all the set constraints in the sense of *weighted least-squares*. This property is further illustrated in Fig. 2.10-2 when three sets are involved in contrast to the cyclic convergence of the sequential algorithm illustrated earlier in Fig. 2.10-1.

The algorithm in Eq. (2.10-36) is described in [18] for signal reconstruction from inconsistent constraints caused by imprecise signal information. In practice this imprecision can be due to causes ranging from inadequate data modeling to noise perturbations in data measurements. It is suggested that the weights $w_i, i = 1, 2, \cdots, m$, be chosen so as to reflect the influences of the constraint sets C_i on the final solution. More specifically, the larger a particular weight w_i, the closer the solution to the corresponding set C_i. Hence, if some constraints C_i are judged to be more critical than others, then their corresponding weights should be assigned larger values. If, on the other hand, all constraints are judged to be equally critical, then the weights should be taken to be all equal, i.e., $w_i = 1/m$.

2.11 SUMMARY

In this chapter we discussed the mathematical theory underlying the methods of *projections onto convex sets* (POCS), a major result of vector space projections. We defined and illustrated what a projection actually is and the properties it must have. We then stated the Fundamental Theorem of POCS (Theorem 2.5-1) and showed how it could be derived from a theorem by Opial (Theorem 2.5-2). We then proceeded to examine the POCS algorithm from a numerical point-of-view paying special attention to the role of the *relaxed projector*, whose use can greatly accelerate convergence, and the influence of the *starting point* on the time-to-convergence. Throughout the chapter we furnished numerous examples to illustrate the somewhat abstract mathematical ideas. To give readers a taste of the utility of POCS, we applied it to the restoration of a band-limited signal. For completeness we furnished a derivation of Opial's basic theorem. While this theorem is fundamental to the derivation of POCS, it is not required reading if the reader is primarily interested in learning how POCS is applied. We concluded the chapter with a discussion of the POCS algorithm formulated in a product space. This alternative algorithm may yield faster convergence, and will play an important role in Chapter 5, where we discuss non-convex sets.

REFERENCES

1. Z. Opial, Weak convergence of the sequence of successive approximation for nonexpansive mappings, *Bull. Am. Math. Soc.*, **73**:591–597, 1967.

2. D. C. Youla, Mathematical theory of image restoration by the method of convex projections, Chapter 2 in *Image Recovery: Theory and Applications*, (H. Stark, ed.), Academic Press, Orlando, FL, 1987.

3. L. G. Gubin, B. T. Polyak, and E. V. Raik, The method of projections for finding the common point in convex sets, *USSR Comput. Math. Phys.*, **7**(6):1–24, 1967.

4. D.C. Youla and H. Webb, Image restoration by the method of convex projections: Part 1–theory," *IEEE Trans. Med. Imaging*, **MI-1**:81–94, Oct. 1982.

5. J. von Neumann, *Functional Operators,* vol. II of *Ann. Math. Stud.*, **22**(55), Princeton University Press, Princeton, NJ, 1950.

6. I. Halperin, The product of projection operators, *Acta Sci. Math.*, **23**:96–99, 1960.

7. A. Levi and H. Stark, Signal restoration from phase by projections onto convex sets, *J. Opt. Soc. Am.*, **73**:810–822, 1983.

8. A. Levi, Image restoration by the method of projections with applications to the phase and magnitude retrieval problem, Ph.D. thesis, Dept. Electrical, Comp. and System Eng., Rensselaer Polytechnic Institute, Troy, NY 12181, Dec. 1983.

9. A. Papoulis, A new algorithm in spectral analysis and bandlimited extrapolation, *IEEE Trans. Circuits Syst.*, **CAS-22**:735–742, 1975.

10. H. Stark, F. B. Tuteur, and J. B. Anderson, *Analog, Digital, and Optical Communications*, 2nd ed., Prentice-Hall, Englewood Cliffs, NJ, 1988.

11. R.W. Gerchberg, "Super-resolution through error energy reduction," *Optica Acta*, **21**:709–720, 1974.

12. B. V. Limaye, *Functional Analysis*, John Wiley & Sons, New York, 1981.

13. G. Pierra, Decomposition through formalization in a product space, *Math. Programming*, **28**:96–115, Jan. 1984.

14. P. L. Combettes and H. Puh, Iterations of parallel convex projections in Hilbert spaces, *Numer. Funct. Anal. Optim.*, **15**:225–243, 1994.

15. P. L. Combettes, Convex set theoretic image recovery by extrapolated iterations of parallel subgradient projections, *IEEE Trans. Image Processing*, **6**:493–506, April 1997.

16. M. Goldburg and R. J. Marks II, Signal synthesis in the presence of inconsistent set of constraints, *IEEE Trans. Circuits Syst.*, **332**(7):647–663, 1985.

17. D. C. Youla and V. Velasco, Extensions of a result on the synthesis of signals in the presence of inconsistent constraints, *IEEE Trans. Circuits Syst.*, **33**(4):465–468, 1986.

18. P. L. Combettes, Inconsistent signal feasibility problem: Least-squares solutions in a product space, *IEEE Trans. Signal Processing*, **42**:2955–2866, Nov. 1994.

EXERCISES

2-1. Show that a subspace is also a convex set.

2-2. Show that the projection onto the unit-disc is not unique in the R^2 space under the p-norm where $p = \infty$.

2-3. Given C, a set in a vector space, define $\tilde{C}$ as the set of vectors of the form $\alpha \mathbf{x}_1 + (1 - \alpha)\mathbf{x}_2$ for all $\mathbf{x}_1, \mathbf{x}_2$ in C and all $0 \le \alpha \le 1$. Show that the set $\tilde{C}$ is convex. In fact, $\tilde{C}$ is the minimal convex set that contains C. The set $\tilde{C}$ is sometimes called the *convex hull* of C. Also, if C is closed, is $\tilde{C}$ closed?

2-4. In the L^2 space of square-integrable functions define set C as

$$C = \left\{ f(\cdot) : \int_0^b f(x)dx = 1 \right\},$$

where b is a constant. Show that:

(a) when b is finite the set C is convex and closed;

(b) when $b = \infty$ the set C is not closed.

2-5. Let $(C)_\mathbf{a}$ denote the *translation* of a set C by vector $\mathbf{a}$, i.e.,

$$(C)_\mathbf{a} \overset{\triangle}{=} \{\mathbf{x} : \mathbf{x} = \mathbf{a} + \mathbf{y} \text{ for some } \mathbf{y} \in C\}.$$

Show that $(C)_\mathbf{a}$ is also closed and convex if C is.

2-6. Let $\alpha(C)$ denote the *scaling* of a set C by a constant α (which is not necessarily positive), i.e.,

$$\alpha(C) \overset{\triangle}{=} \{\mathbf{x} : \mathbf{x} = \alpha \mathbf{y} \text{ for some } \mathbf{y} \in C\}.$$

Show that $\alpha(C)$ is also closed and convex if C is.

2-7. Let C_3 denote the *summation* of two sets C_1 and C_2, i.e.,

$$C_3 \overset{\triangle}{=} \{\mathbf{x} : \mathbf{x} = \mathbf{x}_1 + \mathbf{x}_2 \text{ for some } \mathbf{x}_1 \in C_1 \text{ and some } \mathbf{x}_2 \in C_2\}.$$

Show that C_3 is also closed convex if both C_1 and C_2 are.

2-8. Let C be an open set and $\mathbf{x} \notin C$. Show that none of the points $\mathbf{y}$ in C achieve a minimal distance to $\mathbf{x}$, hence the projection onto an open set is meaningless.

2-9. Show that a band-limited signal $f(t)$ is analytic on the entire t-axis, i.e., it admits to a Taylor series expansion.

2-10. Let $P\mathbf{x}$ denote the projection of $\mathbf{x}$ onto a closed convex set C. Show that the projection of $\mathbf{x}$ onto $(C)_{\mathbf{a}}$, the translation of C by vector $\mathbf{a}$, is generally not given by $P\mathbf{x} - \mathbf{a}$. What if the set C is a linear variety?

2-11. Show that in R^2 the projection of any point $\mathbf{x} = (x_1, x_2)$ onto the set $C = \{\mathbf{x} : x_1 = 1\}$ is given by $P\mathbf{x} = (1, x_2)$.

2-12. Consider the set $C = \{\mathbf{x} : 1/2 \leq x_1 \leq 1\}$ in R^2. Show that the projection of any point $\mathbf{x} \notin C$ onto this set is given by

$$P\mathbf{x} = \begin{cases} (1/2, x_2) & \text{if } x_1 < 1/2 \\ (1, x_2) & \text{if } x_1 > 1 \\ \mathbf{x} & \text{if } 1/2 \leq x_1 \leq 1. \end{cases}$$

2-13. Derive the projector in Eq. (2.7-11).

2-14. Derive the projector in Eq. (2.7-12).

2-15. Show that the set $\mathbf{C}$ in Eq. (2.9-4) is closed and convex if each set C_i is closed and convex in $\mathbf{H}$.

2-16. Show that the set $\mathbf{D}$ is a closed subspace in $\mathcal{H}$.

3
Elementary Projectors

3.1 INTRODUCTION

In this chapter, we define and discuss some elementary constraint sets that are frequently used in the method of convex projections. The purpose here is two-fold. First, we want to demonstrate, through examples, how to determine if a set is convex and closed, and compute its projector. Second, we want these examples to be useful for practical applications. With this in mind, we first study several constraint sets in general Hilbert spaces, and discuss their practical aspects. In the last two sections, we give several constraint sets that are proven to be useful in the context of signal processing.

To minimize notational confusion, in the following we shall consistently adopt the following notation rule unless otherwise specified: we use C to denote a constraint set under study in a Hilbert space, which is denoted by $\mathbf{H}$. We use $\mathbf{x}$ to denote an arbitrary vector in the space $\mathbf{H}$, which may or may not be in the set C. Also, we use $\mathbf{y}$ to denote an arbitrary vector in C, and $\mathbf{g}$ to denote the projection of $\mathbf{x}$ onto C.

3.2 LINEAR TYPE CONSTRAINT

Linear mathematical models that relate input to output, cause and effect, source and measurement, object and image, etc., are widely used in problems in science and engineering. A major reason for this usage is that many practical problems are either inherently linear, almost linear, or modeled as linear. Nonlinear problems are often approximated by linear ones, largely due to the fact that a linear relationship is mathematically more tractable than a nonlinear one. For example, it is generally agreed that a linear equation is much easier to solve than a nonlinear one.

Linear relationships are typically described by *linear equations*. In the context of constraint sets, a linear type relationship is easily described by a closed convex set, which we shall call a *linear type constraint*. In this section, we investigate such constraints.

In a Hilbert space a linear equation is often described by the underlying inner product. More specifically, let b be a scalar and $\mathbf{a}$ a non-zero vector in a Hilbert space $\mathbf{H}$. Then the equation

$$\langle \mathbf{y}, \mathbf{a} \rangle = b \tag{3.2-1}$$

defines a linear equation where $\mathbf{y}$ is, typically, an unknown vector [†].

Typically we are interested in the solution to Eq. (3.2-1). Namely, any vector $\mathbf{y}$ whose inner product with the vector $\mathbf{a}$ is b is taken as a solution to Eq. (3.2-1). Thus, the collection of all such solutions defines the following set

$$C = \{ \, \mathbf{y} \, : \, \langle \mathbf{y}, \mathbf{a} \rangle = b \, \}, \tag{3.2-2}$$

in the space $\mathbf{H}$. In words, the set C is the collection of vectors (or points) that satisfy the linear equation in Eq. (3.2-1). The set C offers an equivalent way of describing the linear relationship in Eq. (3.2-1).

Before we examine the explicit form of the set C in any specific Hilbert space, we first demonstrate that the set C is convex and closed, and derive the projection onto C.

CONVEXITY: Let $\mathbf{y}_1, \mathbf{y}_2 \in C$ and $\mathbf{y}_3 = \alpha \, \mathbf{y}_1 + (1 - \alpha) \, \mathbf{y}_2$ for $\alpha \in [0, 1]$. Then

$$
\begin{aligned}
\langle \mathbf{y}_3, \mathbf{a} \rangle &= \langle \alpha \, \mathbf{y}_1 + (1 - \alpha) \, \mathbf{y}_2, \mathbf{a} \rangle \\
&= \alpha \, \langle \mathbf{y}_1, \mathbf{a} \rangle + (1 - \alpha) \, \langle \mathbf{y}_2, \mathbf{a} \rangle \\
&= \alpha \, b + (1 - \alpha) \, b \\
&= b.
\end{aligned} \tag{3.2-3}
$$

Hence, $\mathbf{y}_3 \in C$ and C is convex. ∎

CLOSEDNESS: Let $\{\mathbf{y}_n\}$ be a sequence in C such that $\mathbf{y}_n \to \mathbf{y}^*$. We want to show that its limit $\mathbf{y}^*$ is also in C. By the Schwarz inequality, we obtain

$$|\langle \mathbf{y}_n, \mathbf{a} \rangle - \langle \mathbf{y}^*, \mathbf{a} \rangle| = |\langle \mathbf{y}_n - \mathbf{y}^*, \mathbf{a} \rangle \leq \|\mathbf{y}_n - \mathbf{y}^*\| \cdot \|\mathbf{a}\| \to 0, \tag{3.2-4}$$

[†]For example, $y_1 + 2y_2 = 4$, where y_1, y_2 are real, can be written as $\langle \mathbf{y}, \mathbf{a} \rangle = 4$ where $\mathbf{a} = (1, 2)$ and $\mathbf{y} = (y_1, y_2)$.

since $\mathbf{y}_n \to \mathbf{y}^*$. Thus,

$$\langle \mathbf{y}^*, \mathbf{a} \rangle = \lim_{n \to \infty} \langle \mathbf{y}_n, \mathbf{a} \rangle = b. \tag{3.2-5}$$

Indeed, $\mathbf{y}^* \in C$ and the set C is closed. ∎

Note that the argument leading to Eq. (3.2-3) still holds when α takes on an arbitrary value. Therefore, the set C is not only a closed convex set, but also a linear variety. We are going apply this result in Chapter 4.

PROJECTION: The projection of an arbitrary $\mathbf{x}$ onto the set C is given by

$$\mathbf{g} \triangleq P\mathbf{x} = \begin{cases} \mathbf{x} & \text{if } \mathbf{x} \in C \\ \mathbf{x} + \frac{b - \langle \mathbf{x}, \mathbf{a} \rangle}{\|\mathbf{a}\|^2} \mathbf{a} & \text{otherwise,} \end{cases} \tag{3.2-6}$$

where P is used to denote the projector.

DERIVATION: As always we wish to find a $\mathbf{y} \in C$, for an arbitrary $\mathbf{x} \in \mathbf{H}$, that minimizes $\|\mathbf{y} - \mathbf{x}\|$. Let $\mathbf{a}_0 = \mathbf{a}/\|\mathbf{a}\|$. Then, each vector $\mathbf{x} \in \mathbf{H}$ has the following orthogonal decomposition

$$\mathbf{x} = \langle \mathbf{x}, \mathbf{a}_0 \rangle \mathbf{a}_0 + \mathbf{d}, \tag{3.2-7}$$

where $\mathbf{d} = \mathbf{x} - \langle \mathbf{x}, \mathbf{a}_0 \rangle \mathbf{a}_0$. Clearly $\langle \mathbf{d}, \mathbf{a}_0 \rangle = 0$ so $\mathbf{d}$ is orthogonal to $\mathbf{a}_0$.

Since each $\mathbf{y} \in C$ satisfies $\langle \mathbf{y}, \mathbf{a} \rangle = b$, i.e., $\langle \mathbf{y}, \mathbf{a}_0 \rangle = b/\|\mathbf{a}\|$, we can write $\mathbf{y}$ according to Eq. (3.2-7) as

$$\mathbf{y} = \frac{b}{\|\mathbf{a}\|} \mathbf{a}_0 + \mathbf{e}, \tag{3.2-8}$$

where $\mathbf{e}$ is a vector that is orthogonal to $\mathbf{a}_0$.

Now, consider $\mathbf{x} \notin C$. From Eqs. (3.2-7) and (3.2-8) we have

$$\begin{aligned} \|\mathbf{x} - \mathbf{y}\|^2 &= \left\| [\langle \mathbf{x}, \mathbf{a}_0 \rangle \mathbf{a}_0 + \mathbf{d}] - [\frac{b}{\|\mathbf{a}\|} \mathbf{a}_0 + \mathbf{e}] \right\|^2 \\ &= \left\| [\langle \mathbf{x}, \mathbf{a}_0 \rangle - \frac{b}{\|\mathbf{a}\|}] \mathbf{a}_0 + (\mathbf{d} - \mathbf{e}) \right\|^2 \\ &= \left\| [\langle \mathbf{x}, \mathbf{a}_0 \rangle - \frac{b}{\|\mathbf{a}\|}] \mathbf{a}_0 \right\|^2 + \|\mathbf{d} - \mathbf{e}\|^2, \end{aligned} \tag{3.2-9}$$

because $\langle \mathbf{a}_0, \mathbf{d} - \mathbf{e} \rangle = 0$. Clearly, $\|\mathbf{x} - \mathbf{y}\|$ is minimized when $\mathbf{e} = \mathbf{d}$, which yields

$$\begin{aligned} \mathbf{y} &= \frac{b}{\|\mathbf{a}\|} \mathbf{a}_0 + \mathbf{d} \\ &= \frac{b}{\|\mathbf{a}\|} \mathbf{a}_0 + (\mathbf{x} - \langle \mathbf{x}, \mathbf{a}_0 \rangle \mathbf{a}_0) \\ &= \mathbf{x} + [\frac{b}{\|\mathbf{a}\|} - \langle \mathbf{x}, \mathbf{a}_0 \rangle] \mathbf{a}_0 \end{aligned}$$

$$= \mathbf{x} + \frac{b - \langle \mathbf{x}, \mathbf{a} \rangle}{\|\mathbf{a}\|^2} \mathbf{a}, \qquad (3.2\text{-}10)$$

which is the projection of $\mathbf{x}$ onto the set C. Thus, Eq. (3.2-6) follows.　∎

DISCUSSION: As pointed out earlier in this section, the linear type constraint set

$$C = \{\, \mathbf{y} \, : \, \langle \mathbf{y}, \mathbf{a} \rangle = b \,\}, \qquad (3.2\text{-}11)$$

represents different expressions in different vector spaces. From the discussions furnished above it is clear that any set that can be put into this form is guaranteed to be convex and closed, and the projection onto this set is readily given in Eq. (3.2-6). In what follows, we present a few concrete examples to illustrate the usefulness and versatility of this type of constraint sets.

Example 3.2-1 In the space R^n, a vector[†] $\mathbf{x}$ has the form $\mathbf{x} = (x_1, x_2, \cdots, x_n)$ where each x_i is a real number, and for $\mathbf{a} = (a_1, a_2, \cdots, a_n)$ in R^n the inner product of $\mathbf{x}$ and $\mathbf{a}$ is given by

$$\langle \mathbf{x}, \mathbf{a} \rangle = \sum_{i=1}^{n} a_i x_i. \qquad (3.2\text{-}12)$$

Thus, the linear constraint set C in Eq. (3.2-2) is written equivalently as

$$C = \{\, \mathbf{y} \, : \, \textstyle\sum_{i=1}^{n} a_i y_i = b \,\}. \qquad (3.2\text{-}13)$$

From linear algebra we know that in the space R^n the linear equation

$$a_1 x_1 + a_2 x_2 + \cdots + a_n x_n = b \qquad (3.2\text{-}14)$$

represents a hyperplane whose normal vector is $\mathbf{a} = (a_1, a_2, \cdots, a_n)$. Thus, the linear constraint set C in Eq. (3.2-13) actually represents the collection of the points on a hyperplane in the R^n space.

According to Eq. (3.2-6) the projection $\mathbf{g}$ of $\mathbf{x} \notin C$ onto C is given by

$$\mathbf{g} = \mathbf{x} + \beta \mathbf{a}, \qquad (3.2\text{-}15)$$

where

$$\beta = \frac{b - \sum_{i=1}^{n} a_i x_i}{\sum_{i=1}^{n} a_i^2}. \qquad (3.2\text{-}16)$$

It is clear from Eq. (3.2-15) that the projection $\mathbf{g}$ is obtained by adding to $\mathbf{x}$ a vector in the direction of the normal vector $\mathbf{a}$, whose length is $\beta\|\mathbf{a}\|$. In other words, the projection onto a hyperplane from an arbitrary point outside the hyperplane is

---

[†]We implicitly assume row vectors.

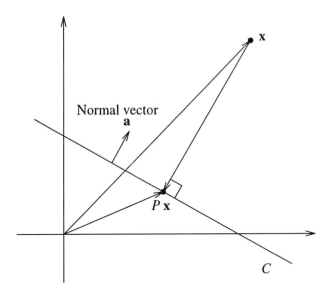

Fig. 3.2-1 The projection onto a hyperplane from an arbitrary point $\mathbf{x}$ outside the hyperplane is obtained by dropping the perpendicular from $\mathbf{x}$ to the hyperplane and their intersection point gives the projection. Note $\mathbf{x} - P\mathbf{x}$ is parallel to the normal vector $\mathbf{a}$.

obtained by dropping the perpendicular from this point to the hyperplane and their intersection point gives the projection—a rather instructive geometric interpretation that we observed when we first introduced the concept of projection onto a convex set at the beginning of Section 2.3. This is further illustrated in Fig. 3.2-1 in the space R^2.

For a specific example, let's consider the set which we defined in Eq. (2.6-4) earlier in Section 2.6 which we rewrite as

$$C_2 = \{\, \mathbf{y} \,:\, y_1 = 1 \,\}. \tag{3.2-17}$$

in the space R^2. Clearly, with $n = 2$, $\mathbf{a} = (1,0)$, and $b = 1$, the set in Eq. (3.2-13) reduces to this set.

Consider $\mathbf{x} = (x_1, x_2) \notin C_2$. Let $\mathbf{g}$ be its projection onto C_2. Then Eq. (3.2-16) yields $\beta = 1 - x_1$. From Eq. (3.2-15) we obtain

$$\mathbf{g} = \mathbf{x} + (1 - x_1)\mathbf{a} = (1, x_2) \tag{3.2-18}$$

which clearly agrees with the result given earlier in Eq. (2.6-7). ∎

Example 3.2-2 In our earlier discussion in Example 1.5-7 in Section 1.5, we introduced the space $L^2([-\pi, \pi])$, i.e., the space of all square-integrable functions defined on the interval $[-\pi, \pi]$. That is, a function $\mathbf{f} = f(t), t \in [-\pi, \pi]$, belongs to $L^2([-\pi, \pi])$ if and only if

$$\int_{-\pi}^{\pi} |f(t)|^2 \, dt \leq \infty. \tag{3.2-19}$$

For every $\mathbf{f} = f(t) \in L^2([-\pi, \pi])$, we have the Fourier series expansion

$$f(t) = a_0 + \sum_{n=1}^{\infty} (a_n \cos nt + b_n \sin nt), \tag{3.2-20}$$

where

$$a_0 = \frac{1}{2\pi} \int_{-\pi}^{\pi} f(t) \, dt, \quad a_n = \frac{1}{\pi} \int_{-\pi}^{\pi} f(t) \cos nt \, dt, \quad b_n = \frac{1}{\pi} \int_{-\pi}^{\pi} f(t) \sin nt \, dt, \tag{3.2-21}$$

for $n = 1, 2, \cdots$.

As an example of linear constraint set in this space, define the set C as the set of functions whose nth Fourier coefficient b_n has some given value b, i.e.,

$$C = \left\{ y(t) \; : \; \frac{1}{\pi} \int_{-\pi}^{\pi} y(t) \sin nt \, dt = b \right\}. \tag{3.2-22}$$

To demonstrate that this set is of the general form in Eq. (3.2-2), let $h(t) = 1/\pi \sin nt$. Then

$$\|h(t)\|^2 = \int_{-\pi}^{\pi} h(t)^2 dt = \int_{-\pi}^{\pi} \left(\frac{1}{\pi} \sin nt \right)^2 dt = \frac{1}{\pi} < \infty, \tag{3.2-23}$$

so $h(t) \in L^2[-\pi, \pi]$ and for $y(t) \in L^2[-\pi, \pi]$

$$\langle y(t), \, h(t) \rangle = \int_{-\pi}^{\pi} y(t) \overline{h(t)} dx = \frac{1}{\pi} \int_{-\pi}^{\pi} y(t) \sin nt \, dt. \tag{3.2-24}$$

Indeed, the set in Eq. (3.2-22) is of the form in Eq. (3.2-2) with $\mathbf{a} = h(t), \mathbf{y} = y(t)$ so it is both closed and convex in $L^2[-\pi, \pi]$.

Consider $f(t) \notin C$. Let $g(t)$ be its projection onto C. According to Eq. (3.2-6), we obtain

$$g(t) = f(t) + \frac{b - \frac{1}{\pi} \int_{-\pi}^{\pi} f(t) \sin nt \, dt}{\frac{1}{\pi}} h(t)$$

$$= f(t) + \left[\pi b - \int_{-\pi}^{\pi} f(t) \sin nt \, dt \right] \sin nt, \tag{3.2-25}$$

for every $t \in [-\pi, \pi]$.   ∎

Example 3.2-3 In the space L^2—the space of all square-integrable functions defined on R, let's consider the set

$$C = \left\{ \, y(x) \; : \; \int_a^b y(x)dx = p \, \right\},$$
(3.2-26)

where p is a constant scalar and a, b are finite real numbers. Without loss of generality, assume $a < b$. At the first glance, it may not be immediately clear that this set is of the general form given in Eq. (3.2-2). However, if we define

$$h(x) = \begin{cases} 1 & \text{if } a \leq x \leq b \\ 0 & \text{otherwise,} \end{cases}$$
(3.2-27)

then,

$$\|h(x)\|^2 = \int_{-\infty}^{\infty} h^2(x)dx = \int_a^b 1^2 dx = b - a < \infty.$$
(3.2-28)

Hence, $h(x) \in L^2$ and for $y(x) \in L^2$

$$\langle y(x), \; h(x) \rangle = \int_{-\infty}^{\infty} y(x)\overline{h(x)}dx = \int_a^b y(x)dx.$$
(3.2-29)

Indeed, the set in Eq. (3.2-26) is of the form in Eq. (3.2-2), so it is both closed and convex in L^2.

Consider $f(x) \notin C$. Let $g(x)$ be its projection onto C. According to Eq. (3.2-6), we obtain

$$\begin{aligned} g(x) &= f(x) + \frac{p - \int_a^b f(x)dx}{b - a} h(x) \\ &= \begin{cases} f(x) + \frac{p - \int_a^b f(x)dx}{b-a} & \text{if } a \leq x \leq b \\ f(x) & \text{otherwise.} \end{cases} \end{aligned}$$
(3.2-30)

■

3.3 SOFT LINEAR TYPE CONSTRAINT

The linear type constraint set

$$C = \{ \, \mathbf{y} \; : \; \langle \mathbf{y}, \, \mathbf{a} \rangle = b \, \},$$
(3.3-1)

furnishes a general way of describing a linear-type constraint for the method of convex projections. In practical applications the value b is either known *a priori* or is estimated or inferred from data. For some applications, the prior knowledge or measurement of b may not be 100 percent accurate due to noise or other factors. For example, in statistics it is often the case that we accept at some prescribed confidence level that the quantity of interest is within some interval. In such a case, it is more appropriate to expect the quantity $\langle \mathbf{y}, \mathbf{a} \rangle$ to be *somewhat close to* b, rather than equal b exactly. A way to incorporate this uncertainty is to replace the exact constraint

$$\langle \mathbf{y}, \mathbf{a} \rangle = b, \tag{3.3-2}$$

by

$$|\langle \mathbf{y}, \mathbf{a} \rangle - b| \leq \epsilon, \tag{3.3-3}$$

where ϵ is chosen *a priori*. In a real Hilbert space, Eq. (3.3-3) is equivalent to

$$b - \epsilon \leq \langle \mathbf{y}, \mathbf{a} \rangle \leq b + \epsilon, \tag{3.3-4}$$

or simply

$$b_1 \leq \langle \mathbf{y}, \mathbf{a} \rangle \leq b_2, \tag{3.3-5}$$

where $b_1 \stackrel{\triangle}{=} b - \epsilon$ and $b_2 \stackrel{\triangle}{=} b + \epsilon$. That is, we are now dealing with a *linear inequality* instead of a *linear equality*. The collection of solutions to this inequality automatically leads to the following set

$$C = \{ \, \mathbf{y} \; : \; b_1 \leq \langle \mathbf{y}, \mathbf{a} \rangle \leq b_2 \, \}. \tag{3.3-6}$$

Comparing Eq. (3.3-6) with Eq. (3.3-1), we find that the set C in Eq. (3.3-6) is a less restrictive version of the one in Eq. (3.3-1). Consequently, the constraint in Eq. (3.3-6) is referred as a *soft linear constraint*, while the one in Eq. (3.3-1) is referred as a *hard linear constraint*. We point out that the set C in Eq. (3.3-6) can be easily extended to a complex Hilbert space. For example, one may put constraint on the real part of the linear quantity $\langle \mathbf{y}, \mathbf{a} \rangle$. In such a case the set in Eq. (3.3-6) will have the form ($b_2 > b_1$)

$$C' = \{ \, \mathbf{y} \; : \; b_1 \leq \Re \langle \mathbf{y}, \mathbf{a} \rangle \leq b_2 \, \}. \tag{3.3-7}$$

In a similar fashion, one can put constraint on the imaginary part of the linear quantity $\langle \mathbf{y}, \mathbf{a} \rangle$. To avoid notational complexity, we are going to restrict ourselves in what follows to the setting of real Hilbert spaces. Nevertheless, the reader should find it relatively straightforward to extend these results to complex vector spaces.

Let's first demonstrate that the soft linear constraint set C in Eq. (3.3-6) is both convex and closed, and then derive the projection onto this set.

CONVEXITY: Let $\mathbf{y}_1, \mathbf{y}_2 \in C$ and $\mathbf{y}_3 = \alpha \, \mathbf{y}_1 + (1 - \alpha) \, \mathbf{y}_2$ for $\alpha \in [0, 1]$. Then

$$\begin{aligned}
\langle \mathbf{y}_3, \mathbf{a} \rangle &= \langle \alpha \, \mathbf{y}_1 + (1 - \alpha) \, \mathbf{y}_2, \mathbf{a} \rangle \\
&= \alpha \, \langle \mathbf{y}_1, \mathbf{a} \rangle + (1 - \alpha) \, \langle \mathbf{y}_2, \mathbf{a} \rangle \\
&\leq \alpha \, b_2 + (1 - \alpha) \, b_2 \\
&= b_2.
\end{aligned} \tag{3.3-8}$$

Similarly, we have $\langle \mathbf{y}_3, \mathbf{a} \rangle \geq b_1$. Hence, $\mathbf{y}_3 \in C$ and the set C is convex.   ∎

CLOSEDNESS: Let $\{\mathbf{y}_n\}$ be a sequence in C such that $\mathbf{y}_n \to \mathbf{y}^*$. From the demonstration leading to Eq. (3.2-5), we have

$$\langle \mathbf{y}^*, \, \mathbf{a} \rangle = \lim_{n \to \infty} \langle \mathbf{y}_n, \, \mathbf{a} \rangle. \tag{3.3-9}$$

It follows that $b_1 \le \langle \mathbf{y}^*, \, \mathbf{a} \rangle \le b_2$. Thus, $\mathbf{y}^* \in C$ and the set C is closed. ∎

PROJECTION: The projection of an arbitrary $\mathbf{x}$ onto the set C is given by

$$\mathbf{g} \overset{\triangle}{=} P\mathbf{x} = \begin{cases} \mathbf{x} & \text{if } \mathbf{x} \in C \\ \mathbf{x} + \dfrac{b_1 - \langle \mathbf{x}, \, \mathbf{a} \rangle}{\|\mathbf{a}\|^2}\mathbf{a} & \text{if } \langle \mathbf{x}, \, \mathbf{a} \rangle < b_1 \\ \mathbf{x} + \dfrac{b_2 - \langle \mathbf{x}, \, \mathbf{a} \rangle}{\|\mathbf{a}\|^2}\mathbf{a} & \text{if } \langle \mathbf{x}, \, \mathbf{a} \rangle > b_2, \end{cases} \tag{3.3-10}$$

where P is used to denote the projector.

DERIVATION: Consider $\mathbf{x} \notin C$. From Corollary 2.3-1, the projection $\mathbf{g} = P\mathbf{x}$ of $\mathbf{x}$ onto the set C is on the boundary of C. In other words, $\mathbf{g}$ satisfies either $\langle \mathbf{g}, \, \mathbf{a} \rangle = b_1$ or $\langle \mathbf{g}, \, \mathbf{a} \rangle = b_2$. Based on the result in Eq. (3.2-6), we have either

$$\mathbf{g} = \mathbf{x} + \frac{b_1 - \langle \mathbf{x}, \, \mathbf{a} \rangle}{\|\mathbf{a}\|^2}\mathbf{a}, \tag{3.3-11}$$

or

$$\mathbf{g} = \mathbf{x} + \frac{b_2 - \langle \mathbf{x}, \, \mathbf{a} \rangle}{\|\mathbf{a}\|^2}\mathbf{a}. \tag{3.3-12}$$

Let

$$d_1 = \left\| \mathbf{x} - \left(\mathbf{x} + \frac{b_1 - \langle \mathbf{x}, \, \mathbf{a} \rangle}{\|\mathbf{a}\|^2}\mathbf{a} \right) \right\| = |b_1 - \langle \mathbf{x}, \, \mathbf{a} \rangle| / \|\mathbf{a}\|, \tag{3.3-13}$$

and

$$d_2 = \left\| \mathbf{x} - \left(\mathbf{x} + \frac{b_2 - \langle \mathbf{x}, \, \mathbf{a} \rangle}{\|\mathbf{a}\|^2}\mathbf{a} \right) \right\| = |b_2 - \langle \mathbf{x}, \, \mathbf{a} \rangle| / \|\mathbf{a}\|. \tag{3.3-14}$$

Now, if $\langle \mathbf{x}, \, \mathbf{a} \rangle < b_1$, then

$$\begin{aligned} |b_2 - \langle \mathbf{x}, \, \mathbf{a} \rangle| &= |(b_2 - b_1) + (b_1 - \langle \mathbf{x}, \, \mathbf{a} \rangle)| \\ &= |b_2 - b_1| + |b_1 - \langle \mathbf{x}, \, \mathbf{a} \rangle| \\ &> |b_1 - \langle \mathbf{x}, \, \mathbf{a} \rangle|, \end{aligned} \tag{3.3-15}$$

so that $d_2 > d_1$. Thus, the projection $\mathbf{g}$ in such a case is given by Eq. (3.3-11).

Similarly, when $\langle \mathbf{x}, \, \mathbf{a} \rangle > b_2$, the projection $\mathbf{g}$ is given by Eq. (3.3-12). Combining the above results leads to Eq. (3.3-10). ∎

DISCUSSION: Equation (3.3-6) gives a general form of a soft linear-type constraint set in Hilbert spaces. The demonstration above shows that any set that can be put into this form is guaranteed to be a convex and closed set. The projection onto this set can be computed according to Eq. (3.3-10).

A variant of the constraint set in Eq. (3.2-2) is the unilateral constraint of the form

$$C = \{ \mathbf{y} : \langle \mathbf{y}, \mathbf{a} \rangle \leq b \}. \tag{3.3-16}$$

From the demonstration above, it is easy to see that this set C is also closed and convex. The projector onto this set is given by

$$P\mathbf{x} = \begin{cases} \mathbf{x} & \text{if } \mathbf{x} \in C \\ \mathbf{x} + \dfrac{b - \langle \mathbf{x}, \mathbf{a} \rangle}{\|\mathbf{a}\|^2} \mathbf{a} & \text{if } \langle \mathbf{x}, \mathbf{a} \rangle > b. \end{cases} \tag{3.3-17}$$

Note that a unilateral constraint set can also have the form

$$C = \{ \mathbf{y} : \langle \mathbf{y}, \mathbf{a} \rangle \geq b \}. \tag{3.3-18}$$

We can easily convert this set into the form of Eq. (3.3-16) by replacing the vector $\mathbf{a}$ by $-\mathbf{a}'$ and b by $-b'$. The details are omitted for the interest of brevity.

In the following we give several explicit examples of linear-type constraint sets to illustrate the application of these results.

Example 3.3-1 In the space R^n, the constraint set C in Eq. (3.3-6) is equivalent to

$$C = \{ \mathbf{y} : b_1 \leq \sum_{i=1}^{n} a_i y_i \leq b_2 \}. \tag{3.3-19}$$

According to Eq. (3.3-10) the projection $\mathbf{g}$ of $\mathbf{x} \notin C$ onto C is given by

$$\mathbf{g} = \mathbf{x} + \beta \mathbf{a}, \tag{3.3-20}$$

where

$$\beta = \begin{cases} \left(b_1 - \sum_{i=1}^{n} a_i x_i \right) \big/ \sum_{i=1}^{n} a_i^2 & \text{if } \sum_{i=1}^{n} a_i x_i < b_1 \\ \left(b_2 - \sum_{i=1}^{n} a_i x_i \right) \big/ \sum_{i=1}^{n} a_i^2 & \text{if } \sum_{i=1}^{n} a_i x_i > b_2. \end{cases} \tag{3.3-21}$$

It is clear form Eq. (3.3-20) that the projection $\mathbf{g}$ is obtained by adding to $\mathbf{x}$ a vector in the direction of the normal vector $\mathbf{a}$, whose length is $\beta \|\mathbf{a}\|$. The resulting vector $\mathbf{g}$ lies on the closest hyperplane in the set, a condition that is guaranteed by Eq. (3.3-21). This is further illustrated in Fig. 3.3-1 in the space R^2.   ∎

Example 3.3-2 In the L^2 space, consider the set

$$C = \left\{ y(x) : p_1 \leq \int_a^b y(x) dx \leq p_2 \right\}, \tag{3.3-22}$$

where p_1, p_2 and a, b are finite real numbers. This is the soft-constraint extension of the hard constraint C in Eq. (3.2-26). Without loss of generality, assume $a < b$.

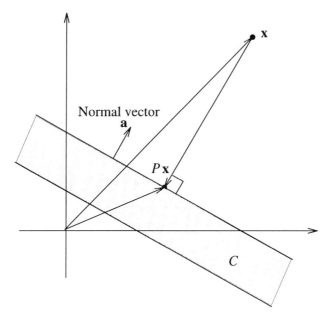

Fig. 3.3-1 The projection onto the soft linear constraint set: The projection of a point **x**, not inside the set, is obtained by adding to **x** a vector in the direction of the normal vector so that the the resulting vector lies on the closest hyperplane in the set.

Let

$$h(x) = \begin{cases} 1 & \text{if } a \leq x \leq b \\ 0 & \text{otherwise;} \end{cases} \qquad (3.3\text{-}23)$$

then we get

$$\langle y(x), \, h(x) \rangle = \int_a^b y(x)dx \; , \qquad (3.3\text{-}24)$$

so the set in Eq. (3.3-22) indeed is of the form in Eq. (3.3-6). Consider $f(x) \notin C$. Let $g(x)$ be its projection onto C. According to Eq. (3.3-10), we obtain

$$g(x) = f(x) + \beta h(x) = \begin{cases} f(x) + \beta & \text{if } a \leq x \leq b \\ f(x) & \text{otherwise,} \end{cases} \qquad (3.3\text{-}25)$$

where

$$\beta = \begin{cases} \dfrac{p_1 - \int_a^b y(x)dx}{b-a} & \text{if } \int_a^b y(x)dx < p_1 \\[2ex] \dfrac{p_2 - \int_a^b y(x)dx}{b-a} & \text{if } \int_a^b y(x)dx > p_2. \end{cases} \qquad (3.3\text{-}26)$$

■

3.4 ERROR TYPE CONSTRAINT

As already pointed out earlier in this book, the concept of norm is very important in vector spaces—not only theoretically but also for practical purposes. The norm is related directly to many real-world quantities such as distance, length, and energy. Consequently, constraint sets that involve bounds on norms are very useful in problem-solving by projection methods.

In the first part of this section, we examine constraint sets in Hilbert spaces resulting from the relation

$$\|\mathbf{x} - \mathbf{r}\| \leq \epsilon. \tag{3.4-1}$$

In Eq. (3.4-1), both the vector $\mathbf{r}$ and the constant ϵ are given, and the vector $\mathbf{x}$ is unknown. Mathematically, Eq. (3.4-1) describes the fact that the vector $\mathbf{x}$ is within an ϵ distance of the known vector $\mathbf{r}$. As a result, this type of constraint is referred as *distance-type* constraint. Also, because in some applications, the vector $\mathbf{r}$ represents an *ideal* while $\mathbf{x}$ represents the *achievable*, constraints of the form $\|\mathbf{x} - \mathbf{r}\| \leq \epsilon$ are also called *error-type constraints*.

In practice, the constraint in Eq. (3.4-1) may come about in several ways. For example, $\mathbf{r}$ may denote a reference image and $\mathbf{x}$ may be our reconstructed image and we have prior knowledge that $\mathbf{x}$ cannot differ from $\mathbf{r}$ by more than ϵ. Or, $\mathbf{r}$ may denote a spoken word and $\mathbf{x}$ the same word synthesized by machine and we want $\|\mathbf{x} - \mathbf{r}\| < \epsilon$, where ϵ is some level of fidelity.

In the second part of this section, we examine a variant of the distance type constraint in the Euclidean space R^n—weighted-error type constraint – which is found useful for signal processing. The numerical issues involved in computing the projection onto its corresponding constraint set is investigated in detail.

Distance Type Constraint

Let $\mathbf{H}$ be a Hilbert space. For $\mathbf{r} \in \mathbf{H}$ and ϵ a scalar, consider the set

$$C = \{\, \mathbf{y} \; : \; \|\mathbf{y} - \mathbf{r}\| \leq \epsilon \,\}. \tag{3.4-2}$$

Clearly, any point in the set C is a solution to Eq. (3.4-1) and *vice versa*, that is, C is the set equivalent of the constraint in Eq. (3.4-1). In the following we demonstrate that C is a convex and closed set in $\mathbf{H}$, and derive the projection onto C.

CONVEXITY: Let $\mathbf{y}_1, \mathbf{y}_2 \in C$ and $\mathbf{y}_3 = \alpha\,\mathbf{y}_1 + (1 - \alpha)\,\mathbf{y}_2$ for $\alpha \in [0, 1]$. Then by the triangle inequality and the linearity of the norm, we get

$$
\begin{aligned}
\|\mathbf{y}_3 - \mathbf{r}\| &= \|\alpha\,\mathbf{y}_1 + (1 - \alpha)\,\mathbf{y}_2 - \mathbf{r}\| \\
&= \|\alpha\,(\mathbf{y}_1 - \mathbf{r}) + (1 - \alpha)\,(\mathbf{y}_2 - \mathbf{r})\| \\
&\leq \alpha\,\|\mathbf{y}_1 - \mathbf{r}\| + (1 - \alpha)\,\|\mathbf{y}_2 - \mathbf{r}\| \\
&\leq \alpha\,\epsilon + (1 - \alpha)\,\epsilon = \epsilon.
\end{aligned}
\tag{3.4-3}
$$

Indeed, $\mathbf{y}_3 \in C$ and the set C is convex. ∎

CLOSEDNESS: Let $\{\mathbf{y}_n\}$ be a sequence in C such that $\mathbf{y}_n \to \mathbf{y}^*$. We want to show that its limit $\mathbf{y}^*$ is also in C. Note that

$$
\begin{aligned}
\|\mathbf{y}^* - \mathbf{r}\| &= \|(\mathbf{y}^* - \mathbf{y}_n) + (\mathbf{y}_n - \mathbf{r})\| \\
&\leq \|\mathbf{y}^* - \mathbf{y}_n\| + \|\mathbf{y}_n - \mathbf{r}\| \\
&\leq \|\mathbf{y}^* - \mathbf{y}_n\| + \epsilon.
\end{aligned}
\tag{3.4-4}
$$

Letting $n \to \infty$, we obtain $\|\mathbf{y}^* - \mathbf{r}\| \leq \epsilon$. Thus, $\mathbf{y}^* \in C$ and the set C is closed. ∎

PROJECTION: The projection of an arbitrary $\mathbf{x}$ onto the set C is given by

$$
P\mathbf{x} = \begin{cases} \mathbf{x} & \text{if } \mathbf{x} \in C \\ \mathbf{r} + \dfrac{\epsilon}{\|\mathbf{x} - \mathbf{r}\|}(\mathbf{x} - \mathbf{r}) & \text{otherwise,} \end{cases}
\tag{3.4-5}
$$

where P is used to denote the projector.

DERIVATION: For $\mathbf{x} \notin C$ and $\mathbf{y} \in C$, we have

$$
\|\mathbf{x} - \mathbf{r}\| > \epsilon \text{ and } \|\mathbf{y} - \mathbf{r}\| \leq \epsilon.
\tag{3.4-6}
$$

By invoking the Schwarz inequality we obtain

$$
\begin{aligned}
\|\mathbf{x} - \mathbf{y}\|^2 &= \|(\mathbf{x} - \mathbf{r}) - (\mathbf{y} - \mathbf{r})\|^2 \\
&= \|\mathbf{x} - \mathbf{r}\|^2 + \|\mathbf{y} - \mathbf{r}\|^2 - 2\Re\langle \mathbf{x} - \mathbf{r}, \mathbf{y} - \mathbf{r}\rangle \\
&\geq \|\mathbf{x} - \mathbf{r}\|^2 + \|\mathbf{y} - \mathbf{r}\|^2 - 2\,|\langle \mathbf{x} - \mathbf{r}, \mathbf{y} - \mathbf{r}\rangle| \\
&\geq \|\mathbf{x} - \mathbf{r}\|^2 + \|\mathbf{y} - \mathbf{r}\|^2 - 2\,\|\mathbf{x} - \mathbf{r}\| \cdot \|\mathbf{y} - \mathbf{r}\| \\
&= (\|\mathbf{x} - \mathbf{r}\| - \|\mathbf{y} - \mathbf{r}\|)^2 \\
&\geq (\|\mathbf{x} - \mathbf{r}\| - \epsilon)^2.
\end{aligned}
\tag{3.4-7}
$$

Thus, $\|\mathbf{x} - \mathbf{y}\|$ is bounded from below by $\|\mathbf{x} - \mathbf{r}\| - \epsilon > 0$. On the other hand, let

$$
\mathbf{y} = \mathbf{r} + \frac{\epsilon}{\|\mathbf{x} - \mathbf{r}\|}(\mathbf{x} - \mathbf{r}).
\tag{3.4-8}
$$

Then, $\|\mathbf{x} - \mathbf{y}\|$ assumes this lower bound. Note that $\|\mathbf{y} - \mathbf{r}\| = \epsilon$, so $\mathbf{y} \in C$. Thus, the projection in Eq. (3.4-5) follows. ∎

DISCUSSION: The projection in Eq. (3.4-5) has an interesting geometrical structure. Let $\gamma = \epsilon/\|\mathbf{x} - \mathbf{r}\|$. Then for $\mathbf{x} \notin C$, $0 < \gamma < 1$. Equation (3.4-5) can be rewritten as

$$
P\mathbf{x} = \gamma\,\mathbf{x} + (1 - \gamma)\,\mathbf{r}.
\tag{3.4-9}
$$

That is, the projection $P\mathbf{x}$ is a convex combination of $\mathbf{x}$ and the reference vector

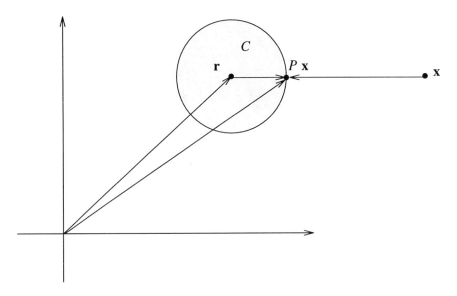

Fig. 3.4-1   In R^n, the constraint set C in Eq. (3.4-2) represents a spheroid that is centered at the reference vector **r**.

r. Thus, $P\mathbf{x}$ always lies in-between **x** and **r** so it is always closer to the reference signal **r** than **x** is itself. This result is better illustrated in the Euclidean space R^n. In R^n, the constraint set C in Eq. (3.4-2) represents a spheroid of radius ϵ that is centered at the reference vector **r**. The projection from a point **x** outside the spheroid is the point on the surface of the spheroid that lies on the line segment between **x** and **r**. This is illustrated in Fig. 3.4-1.

The constraint set C in Eq. (3.4-2) has a noteworthy special form when the reference vector **r** is the zero vector. In such a case, it simply reduces to

$$C = \{\, \mathbf{y} \,:\, \|\mathbf{y}\| \le \epsilon \,\}. \tag{3.4-10}$$

In words, the set C now becomes a constraint on the size or *length* of vectors.

According to Eq. (3.4-5), the projector for the set in Eq. (3.4-10) becomes

$$P\mathbf{x} = \begin{cases} \mathbf{x} & \text{if } \mathbf{x} \in C \\ \frac{\epsilon}{\|\mathbf{x}\|}\mathbf{x} & \text{otherwise.} \end{cases} \tag{3.4-11}$$

That is, the projection $P\mathbf{x}$ for $\mathbf{x} \notin C$ is obtained by a proper scaling of **x**.

In signal processing, it is often the case that we have some knowledge about the energy (or power) of signals. The energy of a signal (or noise) **f** is often taken to be $\|\mathbf{f}\|^2$. Therefore, we can directly translate the knowledge on $\|\mathbf{f}\|^2$ into that on $\|\mathbf{f}\|$, and then apply the constraints in Eq. (3.4-2) or Eq. (3.4-10). For instance, if it is known that the energy of a signal is bounded from above by ϵ^2, then we have $\|\mathbf{f}\|^2 \le \epsilon^2$, which is equivalently to $\|\mathbf{f}\| \le \epsilon$, and the constraint set C in

Eq. (3.4-10) can be applied. Consequently, the constraint sets in Eq. (3.4-2) and Eq. (3.4-10) are also frequently referred as *energy type constraints*.

As a simple example of this type of constraints, consider in R^2 the unit disk

$$C = \{ \mathbf{x} : \|\mathbf{x}\| \leq 1 \}. \tag{3.4-12}$$

According to Eq. (3.4-11), the projection of a point $\mathbf{x}$ onto this set is given by

$$P\mathbf{x} = \begin{cases} \mathbf{x} & \text{if } \mathbf{x} \in C \\ \frac{\mathbf{x}}{\|\mathbf{x}\|} & \text{otherwise.} \end{cases} \tag{3.4-13}$$

Clearly, this result agrees with the one that we derived earlier in Eq. (2.3-23) using the method of Lagrange multipliers.

Weighted-Error Type Constraint

In the above we considered the constraint set

$$C = \{ \mathbf{y} : \|\mathbf{y} - \mathbf{r}\| \leq \epsilon \}, \tag{3.4-14}$$

where the quantity $\|\mathbf{y} - \mathbf{r}\|$ is used to characterize the distance or error between the unknown $\mathbf{y}$ and the reference vector $\mathbf{r}$. For some applications, another type error measure called *weighted error*, which we shall define shortly, is also of practical importance. For simplicity, in the following we are going to restrict our discussions to the Euclidean space R^n. The extension to the space L^2 is left as an exercise for the reader (Exercise 3-9).

In the space R^n, let $\mathbf{w} = (w_1, w_2, \cdots, w_n)$ be a known vector. Then for $\mathbf{x} = (x_1, x_2, \cdots, x_n)$ and $\mathbf{r} = (r_1, r_2, \cdots, r_n)$, we define

$$e_w(\mathbf{x}, \mathbf{r}) = \left[\sum_{i=1}^{n} w_i^2 (x_i - r_i)^2 \right]^{1/2} \tag{3.4-15}$$

as the *weighted error* between the vector $\mathbf{x}$ and the reference vector $\mathbf{r}$. The vector $\mathbf{w}$ is called the *weighting vector*. Clearly, when $w_1 = w_2 = \cdots = w_n = 1$, the weighted error $e_w(\mathbf{x}, \mathbf{r})$ simply becomes to the regular distance between $\mathbf{x}$ and $\mathbf{r}$.

An appealing feature of the weighted error in Eq. (3.4-15) is that it allows differential treatment of the individual components of $\mathbf{x} - \mathbf{r}$. A practical example of this is in statistical pattern recognition where a process called "whitening" is often used to weigh the error components differently when they exhibit different noise characteristics [1].

Based on the weighted-error measure, the distance type constraint set in Eq. (3.4-2) is accordingly modified as

$$C = \{ \mathbf{y} : \sum_{i=1}^{n} w_i^2 (y_i - r_i)^2 \leq \epsilon^2 \}. \tag{3.4-16}$$

It is straightforward to show that the set C defined in Eq. (3.4-16) is both closed and convex. For brevity the details are omitted (see Exercise 3-13). In the following we derive the projection onto this set.

Consider $\mathbf{x} \notin C$. From Corollary 2.3-1, the projection $\mathbf{g} = P\mathbf{x}$ of $\mathbf{x}$ onto C is on the boundary of C. Note that a boundary point $\mathbf{y} = (y_1, y_2, \cdots, y_n)$ of C must satisfy

$$\sum_{i=1}^{n} w_i^2 (y_i - r_i)^2 = \epsilon^2. \tag{3.4-17}$$

Thus, the projection $\mathbf{g}$ minimizes the distance $\|\mathbf{x} - \mathbf{y}\|$ among all $\mathbf{y}$ under the condition that $\mathbf{y}$ satisfies Eq. (3.4-17). To find $\mathbf{g}$, we use the Lagrange multiplier's method. We construct the Lagrange functional

$$
\begin{aligned}
J(\mathbf{y}) &= \|\mathbf{x} - \mathbf{y}\|^2 + \lambda \left[\sum_{i=1}^{n} w_i^2 (y_i - r_i)^2 - \epsilon^2 \right] \\
&= \sum_{i=1}^{n} (x_i - y_i)^2 + \lambda \left[\sum_{i=1}^{n} w_i^2 (y_i - r_i)^2 - \epsilon^2 \right]. \tag{3.4-18}
\end{aligned}
$$

Taking the partial derivative of $J(\mathbf{y})$ with respect to a particular y_i and setting it to zero, we obtain

$$\frac{\partial J(\mathbf{y})}{\partial y_i} = 2(y_i - x_i) + 2\lambda w_i^2 (y_i - r_i) = 0, \tag{3.4-19}$$

which yields

$$y_i = \frac{x_i + \lambda w_i^2 r_i}{1 + \lambda w_i^2}. \tag{3.4-20}$$

Invoking the condition in Eq. (3.4-17) yields

$$\sum_{i=1}^{n} w_i^2 (y_i - r_i)^2 = \sum_{i=1}^{n} \frac{w_i^2 (x_i - r_i)^2}{(1 + \lambda w_i^2)^2} = \epsilon^2. \tag{3.4-21}$$

Thus, the projection $\mathbf{g} = (g_1, g_2, \cdots, g_n)$ is given by

$$g_i = \frac{x_i + \lambda w_i^2 r_i}{1 + \lambda w_i^2}, \quad i = 1, 2, \cdots, n \tag{3.4-22}$$

where λ satisfies

$$\sum_{i=1}^{n} \frac{w_i^2 (x_i - r_i)^2}{(1 + \lambda w_i^2)^2} = \epsilon^2. \tag{3.4-23}$$

Clearly, Eq. (3.4-23) is a non-linear equation in λ. To compute the projection $\mathbf{g}$, one needs to solve Eq. (3.4-23) for λ. With some algebra it can be shown that Eq. (3.4-23) can be rewritten as a $2n$th order polynomial in λ so it has up to $2n$

roots. Due to the uniqueness of the projection **g** in Eq. (3.4-22), only one of these $2n$ roots is associated with the true projection. A natural question then is which one is the correct one. It turns out that for $\mathbf{x} \notin C$ Eq. (3.4-22) has a unique positive root, and this positive root results in a **g** that has a smaller distance to **x** than that furnished by use of any other root. Thus, the unique positive root of Eq. (3.4-22) gives the correct projection **g** in Eq. (3.4-22) when $\mathbf{x} \notin C$. This result is further clarified in the following theorem. To avoid deviation from the main theme, its proof is delayed until Appendix: Part A.

Theorem 3.4-1 *Let* $\mathbf{x} = (x_1, x_2, \cdots, x_n)$ *be a point outside the weighted-error constraint set*

$$C = \{\, \mathbf{y} \; : \; \textstyle\sum_{i=1}^{n} w_i^2 (y_i - r_i)^2 \leq \epsilon^2 \,\}. \qquad (3.4\text{-}24)$$

Then, its projection onto C, *denoted by* $\mathbf{g} = (g_1, g_2, \cdots, g_n)$, *is given by*

$$g_i = \frac{x_i + \lambda w_i^2 r_i}{1 + \lambda w_i^2}, \qquad (3.4\text{-}25)$$

where λ *is the only positive root of the equation*

$$\sum_{i=1}^{n} \frac{w_i^2 (x_i - r_i)^2}{(1 + \lambda w_i^2)^2} = \epsilon^2. \qquad (3.4\text{-}26)$$

According to this theorem, in order to compute the projection **g**, one needs to solve only for the positive root of Eq. (3.4-26). Due to the inherent non-linearity of Eq. (3.4-26), the positive root is usually found by applying an iterative numerical algorithm. An immediate issue then is how do we guarantee that the numerical algorithm always converge to the correct root—since the equation has more than one root. Another issue is the speed of convergence of the algorithm. It turns out that for Eq. (3.4-26), Newton's method (see Appendix: Part B) resolves these issues very nicely. The numerical properties of this algorithm is summarized in the following theorem. Its proof is given in Appendix: Part C.

Theorem 3.4-2 *Let*

$$\psi(\lambda) \triangleq \sum_{i=1}^{n} \frac{w_i^2 (x_i - r_i)^2}{(1 + \lambda w_i^2)^2} - \epsilon^2, \qquad (3.4\text{-}27)$$

be such that $\psi(0) > 0$. *Then, with* $\lambda_0 = 0$ *the iterates generated by Newton's method*

$$\lambda_{k+1} = \lambda_k - \frac{\psi(\lambda_k)}{\psi'(\lambda_k)}, \qquad k = 0, 1, 2, \cdots \qquad (3.4\text{-}28)$$

will converge increasingly to λ_+, *the unique positive root of* $\psi(\lambda) = 0$. *In other words, we have* $\lambda_k < \lambda_{k+1} < \lambda_+$ *for every* k.

DISCUSSION: Geometrically, the weighted error set C represents an ellipsoid in R^n. To help visualize this fact, let's consider the case of R^2. The set C in R^2 has the form

$$C = \{ (y_1, y_2) : w_1^2(y_1 - r_1)^2 + w_2^2(y_2 - r_2)^2 \leq \epsilon^2 \}. \tag{3.4-29}$$

Clearly, the set C represents the collection of the points inside the ellipse

$$w_1^2(y_1 - r_1)^2 + w_2^2(y_2 - r_2)^2 = \epsilon^2, \tag{3.4-30}$$

on the 2-D plane.

Consider a point $\mathbf{x} = (x_1, x_2) \notin C$. Its projection $\mathbf{g} = (g_1, g_2)$ according to Theorem 3.4-1, is given by

$$g_i = \frac{x_i + \lambda w_i^2 r_i}{1 + \lambda w_i^2} \tag{3.4-31}$$

for $i = 1, 2$. The parameter λ is the positive root of

$$\frac{w_1^2(x_1 - r_1)^2}{(1 + \lambda w_1^2)^2} + \frac{w_2^2(x_2 - r_2)^2}{(1 + \lambda w_2^2)^2} = \epsilon^2. \tag{3.4-32}$$

Clearly, even in the space R^2, the parameter λ does not have an easy explicit solution so a numerical approach has to be adopted.

To illustrate the convergence behavior of the numerical algorithm given in Theorem 3.4-1, let's consider the set C in Eq. (3.4-29) with $(r_1, r_2) = (0, 0), w_1 = 1, w_2 = 2$ and $\epsilon = 1$, which is rewritten as

$$C = \{ \mathbf{y} : y_1^2 + 4y_2^2 \leq 1 \}. \tag{3.4-33}$$

Clearly, an arbitrarily chosen point $\mathbf{x} = (10, -10)$ is outside this set. To find its projection onto C, let's first solve for the positive root of Eq. (3.4-26), which is rewritten as

$$\frac{100}{(1 + \lambda)^2} + \frac{400}{(1 + 4\lambda)^2} = 1. \tag{3.4-34}$$

Let

$$\psi(\lambda) = \frac{100}{(1 + \lambda)^2} + \frac{400}{(1 + 4\lambda)^2} - 1. \tag{3.4-35}$$

Then

$$\psi'(\lambda) = -2 \left[\frac{100}{(1 + \lambda)^3} + \frac{1600}{(1 + 4\lambda)^3} \right]. \tag{3.4-36}$$

The iteration in Eq. (3.4-28) can then be carried out, even using a hand calculator. The first few iterates of the algorithm are listed in Table 3.4-1. Also listed in Table 3.4-1 are the corresponding values of the function $\psi(\lambda)$ for each iterate of λ. Clearly, the numerical result confirms what is described in Theorem 3.4-2 about the convergence behavior of the algorithm in Eq. (3.4-28). That is, the iterates of λ converge increasingly to its final value after only 13 iterations. Once λ is computed the projection of the point $\mathbf{x} = (10, -10)$ onto the set C in Eq. (3.4-33) can be computed according to Eq. (3.4-25), which is found to be $(0.881595, -0.236003)$.

Table 3.4-1 Numerical results of the first few iterates of the algorithm in Eq. (3.4-28).

Iterations	λ	$\psi(\lambda)$
1	0.146765	499.000000
2	0.397371	233.850194
3	0.823882	109.865688
4	1.515743	50.739606
5	2.570050	22.818732
6	4.077934	9.989647
7	6.042667	4.212846
8	8.173388	1.647513
9	9.750907	0.540683
10	10.296761	0.115141
11	10.342788	0.008347
12	10.343076	0.000052
13	10.343076	0.000000

One can quickly confirm that this point is indeed on the boundary of the set C (to the extent of numerical accuracy, of course).

3.5 SIMILARITY CONSTRAINT

In the last section, we discussed constraint sets that restrict vectors to lie within some bound of a reference vector. The criteria used there for the judgment of closeness between two vectors are based on their geometric distance. In certain applications, however, the geometric distance may not be the best measure for closeness between two vectors. One such example is the digital waveform detection problem that was discussed in Section 1.4. As argued there, the relative correlation defined by

$$\Upsilon(\mathbf{f}, \mathbf{g}) = \frac{\langle \mathbf{f}, \mathbf{g} \rangle}{\|\mathbf{f}\| \cdot \|\mathbf{g}\|} \tag{3.5-1}$$

serves as a better measure on the closeness or *similarity* between the two vectors **f** and **g**. The relative correlation stresses similarity in shape and size rather than absolute differences in amplitudes. In this section we define and study a constraint set that characterizes the closeness between a vector and a known reference vector using this similarity measure.

Let **r** be a known non-zero vector. Then the closeness between a vector **y** and the known vector **r** is reflected by the similarity measure $\Upsilon(\mathbf{y}, \mathbf{r})$, whose *magnitude* is between 0 and 1, where a value of 0 indicates that they are most dissimilar while a value of 1 indicates that they are wholly similar. A constraint set that captures the idea of similarity is:

$$C = \{\, \mathbf{y} \,:\, \langle \mathbf{y},\, \mathbf{r} \rangle \geq \rho \, \|\mathbf{r}\| \cdot \|\mathbf{y}\| \,\}, \tag{3.5-2}$$

where $\mathbf{r}$ is a known non-zero reference vector, and ρ is a known constant such that $0 \leq \rho \leq 1$.

The constant ρ in Eq. (3.5-2) determines the level of similarity between a vector in C and the reference vector $\mathbf{r}$. Indeed, for any non-zero vector $\mathbf{y} \in C$, we have

$$\Upsilon(\mathbf{y}, \mathbf{r}) \geq \rho. \tag{3.5-3}$$

Thus, to say that a vector $\mathbf{x}$ is very close to the reference vector $\mathbf{r}$ is equivalent to saying that the vector $\mathbf{x}$ is a member of the set C for some high similarity level ρ.

For notational simplicity, let $\mathbf{r}_0 = \mathbf{r}/\|\mathbf{r}\|$. Then the set C in Eq. (3.5-2) can be written equivalently as

$$C = \{\, \mathbf{y} \,:\, \langle \mathbf{y},\, \mathbf{r}_0 \rangle \geq \rho \, \|\mathbf{y}\| \,\}. \tag{3.5-4}$$

For the rest of the development we are going to work with the set C in this form. In the following we demonstrate that C is a convex and closed set, and derive the projection onto this set.

CONVEXITY: Let $\mathbf{y}_1, \mathbf{y}_2 \in C$ and $\mathbf{y}_3 = \alpha\,\mathbf{y}_1 + (1-\alpha)\,\mathbf{y}_2$ for $\alpha \in [0,1]$. Then

$$
\begin{aligned}
\langle \mathbf{y}_3,\, \mathbf{r}_0 \rangle &= \langle \alpha\,\mathbf{y}_1 + (1-\alpha)\,\mathbf{y}_2,\, \mathbf{r}_0 \rangle \\
&= \alpha\,\langle \mathbf{y}_1,\, \mathbf{r}_0 \rangle + (1-\alpha)\,\langle \mathbf{y}_2,\, \mathbf{r}_0 \rangle \\
&\geq \alpha\rho\,\|\mathbf{y}_1\| + (1-\alpha)\rho\,\|\mathbf{y}_2\| \\
&= \rho\,[\alpha\|\mathbf{y}_1\| + (1-\alpha)\|\mathbf{y}_2\|] \\
&\geq \rho\,\|\mathbf{y}_3\|.
\end{aligned}
\tag{3.5-5}
$$

Hence, $\mathbf{y}_3 \in C$ and the set C is convex. ∎

CLOSEDNESS: Let $\{\mathbf{y}_n\}$ be a sequence in C such that $\mathbf{y}_n \rightarrow \mathbf{y}^*$. From the argument leading to Eq. (3.2-5), we have

$$\lim_{n \to \infty} \langle \mathbf{y}_n,\, \mathbf{r}_0 \rangle = \langle \mathbf{y}^*,\, \mathbf{r}_0 \rangle. \tag{3.5-6}$$

On the other hand, $\mathbf{y}_n \in C$ so that

$$\langle \mathbf{y}_n,\, \mathbf{r}_0 \rangle \geq \rho \, \|\mathbf{y}_n\|. \tag{3.5-7}$$

By letting $n \to \infty$, we obtain

$$\langle \mathbf{y}^*,\, \mathbf{r}_0 \rangle = \lim_{n \to \infty} \langle \mathbf{y}_n,\, \mathbf{r}_0 \rangle \geq \lim_{n \to \infty} \rho \, \|\mathbf{y}_n\| = \rho \|\mathbf{y}^*\|. \tag{3.5-8}$$

Thus, $\mathbf{y}^* \in C$ and the set C is closed. ∎

PROJECTION: The projection of an arbitrary $\mathbf{x}$ onto the set C is given by

$$Px = \begin{cases} \mathbf{x} & \text{if } \mathbf{x} \in C \\ \mathbf{0} & \text{if } \gamma \leq 0 \\ \gamma(\rho\,\mathbf{r}_0 + \sqrt{1-\rho^2}\,\mathbf{d}_0) & \text{if } \gamma > 0, \end{cases} \tag{3.5-9}$$

where

$$\mathbf{d}_0 \triangleq \frac{\mathbf{x} - \langle \mathbf{x},\, \mathbf{r}_0 \rangle \mathbf{r}_0}{\|\mathbf{x} - \langle \mathbf{x},\, \mathbf{r}_0 \rangle \mathbf{r}_0\|}, \tag{3.5-10}$$

γ is a constant given by

$$\gamma \triangleq \rho\,\langle \mathbf{x},\, \mathbf{r}_0 \rangle + \sqrt{1-\rho^2}\,\langle \mathbf{x},\, \mathbf{d}_0 \rangle, \tag{3.5-11}$$

and, as usual, P is used to denote the projector.

DERIVATION: For any vector $\mathbf{x} \in \mathbf{H}$, we can write

$$\mathbf{x} = \langle \mathbf{x},\, \mathbf{r}_0 \rangle \mathbf{r}_0 + \mathbf{d}, \tag{3.5-12}$$

where $\mathbf{d} = \mathbf{x} - \langle \mathbf{x},\, \mathbf{r}_0 \rangle \mathbf{r}_0$. Clearly, $\langle \mathbf{d},\, \mathbf{r}_0 \rangle = 0$, so $\mathbf{d}$ is orthogonal to $\mathbf{r}_0$.

From Corollary 2.3-1 the projection of $\mathbf{x}$ onto the set C for $\mathbf{x} \notin C$ is always on the boundary of C. Note that a boundary point $\mathbf{y}$ of C satisfies

$$\langle \mathbf{y},\, \mathbf{r}_0 \rangle = \rho\,\|\mathbf{y}\|. \tag{3.5-13}$$

Write $\mathbf{y}$ as

$$\mathbf{y} = \langle \mathbf{y},\, \mathbf{r}_0 \rangle \mathbf{r}_0 + \mathbf{e} = \beta \mathbf{r}_0 + \mathbf{e}, \tag{3.5-14}$$

where $\beta = \langle \mathbf{y},\, \mathbf{r}_0 \rangle$ which is clearly non-negative, and $\mathbf{e}$ is a vector that is orthogonal to $\mathbf{r}_0$. Then,

$$\|\mathbf{y}\|^2 = \beta^2 + \|\mathbf{e}\|^2. \tag{3.5-15}$$

Note that Eq. (3.5-13) yields

$$\|\mathbf{y}\|^2 = \frac{\beta^2}{\rho^2}. \tag{3.5-16}$$

Thus,

$$\|\mathbf{e}\|^2 = \|\mathbf{y}\|^2 - \beta^2 = \frac{1-\rho^2}{\rho^2}\beta^2. \tag{3.5-17}$$

Note that for $\mathbf{H}$ a real Hilbert space,

$$\begin{aligned} \|\mathbf{x} - \mathbf{y}\|^2 &= \|\mathbf{x}\|^2 + \|\mathbf{y}\|^2 - 2\langle \mathbf{x},\, \mathbf{y} \rangle \\ &= \|\mathbf{x}\|^2 + \frac{\beta^2}{\rho^2} - 2\langle \langle \mathbf{x},\, \mathbf{r}_0 \rangle \mathbf{r}_0 + \mathbf{d},\, \beta \mathbf{r}_0 + \mathbf{e} \rangle \\ &= \|\mathbf{x}\|^2 + \frac{\beta^2}{\rho^2} - 2(\beta\langle \mathbf{x},\, \mathbf{r}_0 \rangle + \langle \mathbf{d},\, \mathbf{e} \rangle) \end{aligned}$$

$$\geq \quad \|\mathbf{x}\|^2 + \frac{\beta^2}{\rho^2} - 2\beta \langle \mathbf{x}, \mathbf{r}_0 \rangle - 2\|\mathbf{d}\| \cdot \|\mathbf{e}\|$$

$$= \quad \|\mathbf{x}\|^2 + \frac{\beta^2}{\rho^2} - 2\beta \langle \mathbf{x}, \mathbf{r}_0 \rangle - 2\frac{\sqrt{1-\rho^2}}{\rho}\beta\|\mathbf{d}\|, \quad (3.5\text{-}18)$$

where the last line is obtained with the help of Eq. (3.5-17).

With $\gamma = \rho \langle \mathbf{x}, \mathbf{r}_0 \rangle + \sqrt{1-\rho^2}\|\mathbf{d}\|$, and

$$\psi(\beta) \triangleq \frac{\beta^2}{\rho^2} - 2\beta \langle \mathbf{x}, \mathbf{r}_0 \rangle - 2\frac{\sqrt{1-\rho^2}}{\rho}\beta\|\mathbf{d}\| = \frac{\beta^2}{\rho^2} - 2\frac{\gamma}{\rho}\beta. \quad (3.5\text{-}19)$$

It is straightforward to show that for $\beta \geq 0$ the function $\psi(\beta)$ is minimized when

$$\beta = \begin{cases} 0 & \text{if } \gamma \leq 0 \\ \rho\gamma & \text{if } \gamma > 0. \end{cases} \quad (3.5\text{-}20)$$

Thus, Eq. (3.5-18) yields

$$\|\mathbf{x} - \mathbf{y}\|^2 = \|\mathbf{x}\|^2 + \psi(\beta) \geq \begin{cases} \|\mathbf{x}\|^2 & \text{if } \gamma \leq 0 \\ \|\mathbf{x}\|^2 - \gamma^2 & \text{if } \gamma > 0, \end{cases} \quad (3.5\text{-}21)$$

for every $\mathbf{y}$ on the boundary of the set C.

Equation (3.5-21) gives the lower bound on the distance $\|\mathbf{x} - \mathbf{y}\|$ from $\mathbf{x}$ to any point $\mathbf{y}$ on the boundary of C. Any point $\mathbf{y}$ on the boundary of C that assumes this lower bound will no doubt be the projection of $\mathbf{x}$ onto C. In the following we show that such a point indeed exists. First, when $\gamma \leq 0$, let $\mathbf{y} = \mathbf{0}$. Then $\mathbf{y} \in C$ and $\|\mathbf{x} - \mathbf{y}\|^2 = \|\mathbf{x}\|^2$. Next, when $\gamma > 0$, let

$$\mathbf{y} = \gamma \left[\rho \mathbf{r}_0 + \sqrt{1-\rho^2}\frac{\mathbf{d}}{\|\mathbf{d}\|} \right]. \quad (3.5\text{-}22)$$

Then $\|\mathbf{y}\| = \gamma$ and $\langle \mathbf{y}, \mathbf{r}_0 \rangle = \rho\|\mathbf{y}\|$ so that $\mathbf{y} \in C$. Furthermore,

$$\begin{aligned}
\|\mathbf{x} - \mathbf{y}\|^2 &= \|\mathbf{x}\|^2 + \|\mathbf{y}\|^2 - 2\langle \mathbf{x}, \mathbf{y} \rangle \\
&= \|\mathbf{x}\|^2 + \gamma^2 - 2\gamma \left[\rho \langle \mathbf{x}, \mathbf{r}_0 \rangle + \sqrt{1-\rho^2}\langle \mathbf{x}, \tfrac{\mathbf{d}}{\|\mathbf{d}\|} \rangle \right] \\
&= \|\mathbf{x}\|^2 + \gamma^2 - 2\gamma \left[\rho \langle \mathbf{x}, \mathbf{r}_0 \rangle + \sqrt{1-\rho^2}\|\mathbf{d}\| \right] \\
&= \|\mathbf{x}\|^2 + \gamma^2 - 2\gamma^2 \\
&= \|\mathbf{x}\|^2 - \gamma^2.
\end{aligned} \quad (3.5\text{-}23)$$

Thus, the lower bound given in Eq. (3.5-21) is achieved in either case. The projection in Eq. (3.5-9) follows.

Finally, we point out that under the condition that $\gamma > 0$, the quantity $\|\mathbf{d}\|$ in Eq. (3.5-22) is never zero so that the vector $\mathbf{y}$ is well defined. This is so because

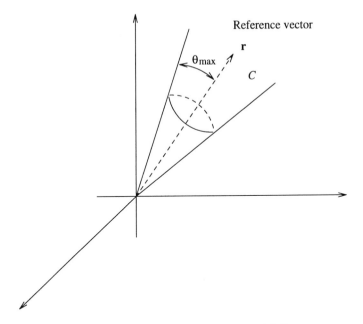

Fig. 3.5-1 In R^n, the similarity constraint set C in Eq. (3.5-2) represents a cone whose vertex is at the origin (i.e., the zero vector) and whose central axis is the reference vector $\mathbf{r}$.

otherwise we would have

$$\mathbf{x} = \langle \mathbf{x}, \, \mathbf{r}_0 \rangle \mathbf{r}_0. \tag{3.5-24}$$

Under the condition that $\gamma > 0$, Eq. (3.5-24) yields

$$\|\mathbf{x}\| = \langle \mathbf{x}, \, \mathbf{r}_0 \rangle, \tag{3.5-25}$$

which indicates that $\mathbf{x} \in C$.   ∎

DISCUSSION: In the Euclidean space R^n, the similarity constraint set

$$C = \{ \, \mathbf{y} \, : \, \langle \mathbf{y}, \, \mathbf{r} \rangle \geq \rho \, \|\mathbf{r}\| \cdot \|\mathbf{y}\| \, \} \tag{3.5-26}$$

has an interesting geometric interpretation: It is the collection of vectors whose angle θ with the reference vector $\mathbf{r}$ is no larger than $\theta_{\max} = \cos^{-1} \rho$. In other words, the set C represents a *cone* whose vertex is at the origin (i.e., the zero vector) and whose central axis is the reference vector $\mathbf{r}$. The angle formed by any vector on the surface of the cone with its central axis vector $\mathbf{r}$ is exactly $\theta_{\max}$. This is illustrated further in Fig. 3.5-1.

3.6   TIME-DOMAIN CONSTRAINTS

As mentioned earlier in this book, the L^2 function space is extraordinarily useful in solving problems in science and engineering. In signal processing for example, an electrical signal $x(t)$ such as a voltage or current is often treated as an element of the L^2 space. As will be discussed in Section 3.7, an L^2 signal $x(t)$ can also be represented in the frequency domain by its Fourier transform. Data and prior information regarding a signal can be described either in the time-domain or in the frequency domain. In this section, we present a few examples to illustrate how to construct constraint sets to characterize the time-domain properties of a signal. The resulting sets are accordingly called *time-domain constraints* in the L^2 space.

To avoid notational confusion, in the rest of the development we are going to use $f(\cdot)$ to represent a whole function vector $\mathbf{f} = f(t)$ for $-\infty < t < \infty$ in the space L^2, while reserving $f(t)$ exclusively for its value at t.

Non-negativity Constraint

In some applications it is known that certain quantities of interest never assume negative values. For example, the instantaneous magnitude of a voltage signal, the spectral intensity of a light, and the density function of a probability distribution, etc. are known to be non-negative. It turns out that this kind of information can be handily described by a closed convex set in L^2. More specifically, this set is defined as

$$C = \{ \, y(\cdot) \; : \; y(t) \geq 0 \text{ for } t \in R \, \}. \tag{3.6-1}$$

In words, C is the set of non-negative functions in L^2. Clearly, any function that is non-negative is a member of the set C. In the following, we demonstrate that this set is convex and closed, and derive the projection onto it.

CONVEXITY: Let $y_1(\cdot), y_2(\cdot) \in C$ and $y_3(\cdot) = \alpha \, y_1(\cdot) + (1 - \alpha) \, y_2(\cdot)$ for $\alpha \in [0, 1]$. Then, for $t \in R$

$$y_3(t) = \alpha \, y_1(t) + (1 - \alpha) \, y_2(t) \geq 0. \tag{3.6-2}$$

Hence, $y_3(\cdot) \in C$ and C is convex.   ∎

CLOSEDNESS: Let $\{y_n(\cdot)\}$ be a sequence in C such that $y_n(\cdot) \to y^*(\cdot)$. We want to show that $y^*(\cdot) \in C$ also. Let E be the set such that

$$E = \{ \, t \; : \; y^*(t) < 0 \, \}, \tag{3.6-3}$$

and E^c be its complement. Then

$$\|y_n(\cdot) - y^*(\cdot)\|^2 \;\; = \;\; \int_E |y_n(t) - y^*(t)|^2 \; dt + \int_{E^c} |y_n(t) - y^*(t)|^2 \; dt$$

$$\geq \int_E |y_n(t) - y^*(t)|^2 \, dt$$

$$\geq \int_E |y^*(t)|^2 \, dt. \tag{3.6-4}$$

In going from line 2 to line 3 above we use the fact that over E, $|y_n(t) - y^*(t)| > |y^*(t)|$ since $y_n(t) \geq 0$ by assumption. By letting $n \to \infty$, $\|y_n(\cdot) - y^*(\cdot)\| \to 0$, and we obtain

$$\int_E |y^*(t)|^2 \, dt = 0, \tag{3.6-5}$$

which implies that the set E is empty.[†] It follows that $y^*(\cdot) \in C$, and C is closed. ∎

PROJECTION: For $x(\cdot) \notin C$, its projection $g(\cdot) = Px(\cdot)$ onto C is given by

$$g(t) = \begin{cases} x(t) & \text{if } x(t) \geq 0 \\ 0 & \text{if } x(t) < 0. \end{cases} \tag{3.6-6}$$

In words, the projection of $x(\cdot)$ onto C is simply the *rectification* of $x(\cdot)$. Thus, the constraint C plays the role of a rectifier in electronics.

DERIVATION: For $x(\cdot) \notin C$, let E be the set such that

$$E = \{\, t \,:\, x(t) < 0 \,\}, \tag{3.6-7}$$

and E^c be its complement. Then, for $y(\cdot) \in C$,

$$\|x(\cdot) - y(\cdot)\|^2 = \int_E |x(t) - y(t)|^2 \, dt + \int_{E^c} |x(t) - y(t)|^2 \, dt. \tag{3.6-8}$$

It is readily seen that the distance $\|x(\cdot) - y(\cdot)\|$ is minimized when the function $y(\cdot)$ satisfies

$$y(t) = \begin{cases} x(t) & \text{if } t \in E \\ 0 & \text{if } t \in E^c. \end{cases} \tag{3.6-9}$$

The projection in Eq. (3.6-6) follows immediately. ∎

Known Support Constraint

In some applications it is often the case that the unknown function of interest assumes non-zero values only when its argument takes on certain values. For example, the output of an initially uncharged *causal* linear time-invariant system has non-zero values only after an input is applied. This information can easily be

---

[†] Strictly speaking, the set E has *measure* 0. In other words, $y^*(t)$ is non-negative *almost everywhere*.

described by the following set in the space L^2:

$$C = \{ y(\cdot) \ : \ y(t) = 0 \text{ for } t \notin S \}, \tag{3.6-10}$$

where S is the set of points on which the function may take on non-zero values. In the literature this set S is called the *support* of the function. For example, if a function is known to assume non-zero values only when t is in the range of 1 and 2, then S is the interval $[1, 2]$.

In what follows we demonstrate that C is convex and closed, and derive the projection onto this set.

CONVEXITY: Let $y_1(\cdot), y_2(\cdot) \in C$ and $y_3(\cdot) = \alpha \, y_1(\cdot) + (1 - \alpha) \, y_2(\cdot)$ for $\alpha \in [0, 1]$. Then, for $t \notin S$

$$y_3(t) = \alpha \, y_1(t) + (1 - \alpha) \, y_2(t) = 0. \tag{3.6-11}$$

Hence, $y_3(\cdot) \in C$ and C is convex.   ∎

CLOSEDNESS: Let $\{y_n(\cdot)\}$ be a sequence in C such that $y_n(\cdot) \to y^*(\cdot)$. We want to show that $y^*(\cdot) \in C$ also. Let S^c be the complement of the support set S. Noting that $y_n(t) = 0$ for $t \in S^c$, we get

$$\begin{aligned}
\|y_n(\cdot) - y^*(\cdot)\|^2 &= \int_S |y_n(t) - y^*(t)|^2 \ dt + \int_{S^c} |y_n(t) - y^*(t)|^2 \ dt \\
&= \int_S |y_n(t) - y^*(t)|^2 \ dt + \int_{S^c} |y^*(t)|^2 \ dt \\
&\geq \int_{S^c} |y^*(t)|^2 \ dt.
\end{aligned} \tag{3.6-12}$$

By letting $n \to \infty$, we obtain

$$\int_{S^c} |y^*(t)|^2 \ dt = 0. \tag{3.6-13}$$

It follows that $y^*(t) = 0$ for $t \in S^c$.[†] Hence, $y^*(\cdot) \in C$, and C is closed.   ∎

PROJECTION: For $x(\cdot) \notin C$, its projection $Px(\cdot)$ onto C is given by

$$Px(t) = \begin{cases} x(t) & \text{if } t \in S \\ 0 & \text{if } t \notin S. \end{cases} \tag{3.6-14}$$

In words, the projection of a function $x(t)$ onto C is obtained by truncating $x(t)$ within the support S.

[†] Strictly speaking, $y^*(t) = 0$ for almost every $t \in S^c$.

DERIVATION: Let S^c be the complement of S. Then for $x(\cdot) \notin C$ and $y(\cdot) \in C$,

$$\|x(\cdot) - y(\cdot)\|^2 = \int_S |x(t) - y(t)|^2 \, dt + \int_{S^c} |x(t) - y(t)|^2 \, dt$$

$$= \int_S |x(t) - y(t)|^2 \, dt + \int_{S^c} |x(t)|^2 \, dt. \qquad (3.6\text{-}15)$$

Observe that the distance $\|x(\cdot) - y(\cdot)\|$ is minimized when $y(t) = x(t)$ for $t \in S$. Thus, the projection given in Eq. (3.6-14) follows. ∎

Note that in demonstrating the convexity of the set C, Eq. (3.6-11) also holds for an arbitrary α. Hence the set C is also a linear variety. In fact, it can be also easily demonstrated that C is also a closed subspace.

Known Segment Constraint

In a number of applications it may be the case that we know part of an unknown signal that is described by a function in L^2. For example, our observation data may be incomplete and the signal is observed only during a specified time interval. Our prior knowledge, then, about the signal can be described by a constraint set of the following form:

$$C = \{ y(\cdot) \ : \ y(t) = h(t) \text{ for } t \in S \}, \qquad (3.6\text{-}16)$$

where S is a set of points on the real line R,[†] and $h(\cdot)$ represents the known (observed) portion of the signal over the interval S.

The set C so-defined above is the collection of functions that coincide with the known segment function $h(\cdot)$ over the set S. Clearly, any function that coincides with $h(\cdot)$ over the set S has to be member of C. Thus, the set C provides an equivalent way for describing that a function assumes certain known values over a certain interval. In what follows we demonstrate that C is convex and closed, and derive the projection onto this set.

CONVEXITY: Let $y_1(\cdot), y_2(\cdot) \in C$ and $y_3(\cdot) = \alpha \, y_1(\cdot) + (1 - \alpha) \, y_2(\cdot)$ for $\alpha \in [0, 1]$. Then, for $t \in S$,

$$y_3(t) = \alpha \, y_1(t) + (1 - \alpha) \, y_2(t) = \alpha \, h(t) + (1 - \alpha) \, h(t) = h(t). \qquad (3.6\text{-}17)$$

Hence, $y_3(\cdot) \in C$ and C is convex. ∎

CLOSEDNESS: Let $\{y_n(\cdot)\}$ be a sequence in C such that $y_n(\cdot) \to y^*(\cdot)$. We want to show that $y^*(\cdot) \in C$ also. Let S^c be the complement of the support set S. Then,

$$\|y_n(\cdot) - y^*(\cdot)\|^2 = \int_S |y_n(t) - y^*(t)|^2 \, dt + \int_{S^c} |y_n(t) - y^*(t)|^2 \, dt$$

[†] Strictly speaking, S should be a *measurable set*.

$$\geq \int_S |y_n(t) - y^*(t)|^2 \, dt. \tag{3.6-18}$$

By letting $n \to \infty$, we obtain

$$\int_S |y_n(t) - y^*(t)|^2 \, dt = 0. \tag{3.6-19}$$

It follows that $y_n(t) = y^*(t)$ for $t \in S$.[†] Hence, $y^*(\cdot) \in C$, and C is closed.   ■

PROJECTION: For $x(\cdot) \notin C$, its projection $g(\cdot) = Px(\cdot)$ onto C is given by

$$g(t) = \begin{cases} h(t) & \text{if } t \in S \\ x(t) & \text{if } t \notin S. \end{cases} \tag{3.6-20}$$

In words, the projection of a function $x(t)$ onto C is obtained by replacing $x(t)$ with $g(t)$ on the known segment $t \in S$.

DERIVATION: Let S^c be the complement of S. Then for $x(\cdot) \notin C$ and $y(\cdot) \in C$,

$$\begin{aligned} \|x(\cdot) - y(\cdot)\|^2 &= \int_S |x(t) - y(t)|^2 \, dt + \int_{S^c} |x(t) - y(t)|^2 \, dt \\ &= \int_S |x(t) - h(t)|^2 \, dt \\ &\quad + \int_{S^c} |x(t) - y(t)|^2 \, dt. \end{aligned} \tag{3.6-21}$$

Observe that the distance $\|x(\cdot) - y(\cdot)\|$ is minimized when $y(t) = x(t)$ for every $t \in S^c$. The projection given in Eq. (3.6-20) follows.   ■

The reader may have noticed that Eq. (3.6-17) still holds for an arbitrary α. Hence the set C in Eq. (3.6-16) is a linear variety. Finally, note that the known support constraint set in Eq. (3.6-10) is a special case of the known segment constraint set in Eq. (3.6-16). Indeed, the fact that a function $f(\cdot)$ has a known support S, which is described by the set in Eq. (3.6-10), is equivalent to the fact the function is known to be zero outside the known support S, which can be described equivalently by a constraint set in the form of Eq. (3.6-16). Based on this fact, one can immediately derive the projection in Eq. (3.6-14) from that in Eq. (3.6-20).

Bounded Value Constraint

In some applications we may have prior knowledge about the constrained range of amplitudes of an unknown signal that is described by a function in L^2. For example, signals in certain logic circuits may be restricted to the range of 0 to 5

[†]Strictly speaking, $y_n(t) = y^*(t)$ for *almost every* $t \in S$.

volts during certain intervals for proper operation. This kind of information can be described by the following constraint set in the L^2 space:

$$C = \{ \, y(\cdot) \; : \; l(t) \le y(t) \le u(t) \text{ for } t \in S \, \}, \qquad (3.6\text{-}22)$$

where $l(t), u(t)$ are, respectively, the *known* lower and upper bounds over the set of points S.

Note that in Eq. (3.6-22) the known bounds $l(t), u(t)$ may or may not be constant, depending on the applications. Also, the set C can be easily extended to define constraint set on functions that are bounded unilaterally either from below or from above.

It can be shown that the set C is both convex and closed. The derivation of its projection is left as an exercise (Exercise 3-10).

3.7 FREQUENCY-DOMAIN CONSTRAINTS

The *Fourier transform* of a waveform, sometimes called its *frequency-domain representation*, plays a key role in solving science and engineering problems. In this section we present a few examples to illustrate how to construct constraint sets to characterize the frequency-domain properties of a signal in L^2 space. The resulting sets are accordingly called *frequency-domain constraints*.

First, let's briefly review a few important concepts related to Fourier transforms. In the space of L^2 the Fourier transform of a function $f(\cdot)$ is defined by

$$F(\omega) = \int_{-\infty}^{\infty} f(t) e^{-j\omega t} dt. \qquad (3.7\text{-}1)$$

On the other hand, a function $f(\cdot) \in L^2$ is uniquely determined from its Fourier transform $F(\cdot)$ by

$$f(t) = \frac{1}{2\pi} \int_{-\infty}^{\infty} F(\omega) e^{j\omega t} d\omega. \qquad (3.7\text{-}2)$$

The operations defined in Eq. (3.7-1) and Eq. (3.7-2) are called, respectively, the *forward Fourier transform* and the *inverse Fourier transform*, and they are denoted respectively by operators $\mathcal{F}$ and $\mathcal{F}^{-1}$. That is, we have

$$F(\cdot) = \mathcal{F} f(\cdot) \text{ and } f(\cdot) = \mathcal{F}^{-1} F(\cdot). \qquad (3.7\text{-}3)$$

Also, $f(\cdot)$ and $F(\cdot)$ so-defined are called a *Fourier transform pair*, which is frequently denoted by

$$f(\cdot) \leftrightarrow F(\cdot). \qquad (3.7\text{-}4)$$

A useful identity between a Fourier transform pair $f(\cdot)$ and $F(\cdot)$ is

$$\int_{-\infty}^{\infty} |f(t)|^2 \, dt = \frac{1}{2\pi} \int_{-\infty}^{\infty} |F(\omega)|^2 \, d\omega. \tag{3.7-5}$$

This identity is one form of the Parseval's relation. Its significance in signal processing is that it relates the energy of the signal in time-domain and frequency domain.

The Fourier transform is a linear transform, that is, it satisfies the superposition principle. To be specific, let $f_1(\cdot), f_2(\cdot) \in L^2$, then

$$\mathcal{F}[\alpha f_1(\cdot) + \beta f_2(\cdot)] = \alpha \, \mathcal{F} f_1(\cdot) + \beta \, \mathcal{F} f_2(\cdot), \tag{3.7-6}$$

for α, β arbitrary.

For a function $f(\cdot) \in L^2$, its Fourier transform $F(\cdot)$ can be written in its polar coordinate form as

$$F(\omega) = |F(\omega)| \exp\left[j \angle F(\omega)\right], \tag{3.7-7}$$

where $|F(\omega)|$ is called the *magnitude* and $\angle F(\omega)$ the *angle* or *phase* of $F(\omega)$.

As described in Example 2.7-1 in Chapter 2, a function $f(\cdot) \in L^2$ is said to be *band-limited* if its Fourier transform $F(\cdot)$ is zero outside a finite interval, say $[-2\pi B, 2\pi B]$, where B is called the *bandwidth* of $f(\cdot)$ and is usually measured in Hertz (Hz). Here B is the smallest number such that $F(\omega) = 0$ for $|\omega| > 2\pi B$.

In signal processing, it is often the case that we have some knowledge about the signal in the frequency domain. In the following we study several constraint sets which are used to describe the signals in the frequency domain.

Bandwidth Constraint

An important class of signals in practice is the class of band-limited signals. For example the signals generated from a communication system are all, ideally, band-limited because a practical communication system always has a finite bandwidth. The fact that a signal is band-limited can be described by the following set:

$$C = \{ y(\cdot) \; : \; y(\cdot) \leftrightarrow Y(\cdot), Y(\omega) = 0, \text{ for } |\omega| > 2\pi B \}, \tag{3.7-8}$$

where B is the bandwidth which is known most of the time from the specific applications.

In words, the set C is the collection of signals whose Fourier transform is zero at frequencies higher than B Hz. In the following, we demonstrate that this set is convex and closed, and derive the projection onto it.

CONVEXITY: Let $y_1(\cdot), y_2(\cdot) \in C$ and $y_3(\cdot) = \alpha \, y_1(\cdot) + (1 - \alpha) \, y_2(\cdot)$ for $\alpha \in [0, 1]$. Then, by the linearity property of the Fourier transform we have

$$Y_3(\cdot) = \alpha \, Y_1(\cdot) + (1 - \alpha) \, Y_2(\cdot), \tag{3.7-9}$$

where $Y_i(\cdot)$ denote the Fourier transform of the functions $y_i(\cdot)$ for $i = 1, 2, 3$. It follows that $Y_3(\omega) = 0$ for $|\omega| > 2\pi B$. Thus, $y_3(\cdot) \in C$, and C is convex. ∎

It should be pointed out that the argument surrounding Eq. (3.7-9) is still valid even when α takes on an arbitrary value in R. Since C is closed (see below), it follows that C is a linear variety. Moreover, it can be further shown that C is a closed subspace in L^2 (see Exercise 3-14).

CLOSEDNESS: Let $\{y_n(\cdot)\}$ be a sequence in C such that $y_n(\cdot) \rightarrow y^*(\cdot)$. We want to show that $y^*(t) \in C$ also. Let $Y^*(\cdot)$ and $Y_n(\cdot)$ denote respectively the Fourier transform of $y^*(t)$ and $y_n(\cdot)$. For convenience, let $E = [-2\pi B, 2\pi B]$, and E^c be its complement. Then, by the Parseval's identity in Eq. (3.7-5) we obtain

$$
\begin{aligned}
\|y_n(\cdot) - y^*(\cdot)\|^2 &= \int_{-\infty}^{\infty} |y_n(t) - y^*(t)|^2 \, dt \\
&= \int_E |Y_n(\omega) - Y^*(\omega)|^2 \, d\omega + \int_{E^c} |Y_n(\omega) - Y^*(\omega)|^2 \, d\omega \\
&\geq \int_{E^c} |Y_n(\omega) - Y^*(\omega)|^2 \, d\omega \\
&= \int_{E^c} |Y^*(\omega)|^2 \, d\omega, \quad\quad\quad (3.7\text{-}10)
\end{aligned}
$$

since $Y_n(\omega) = 0$ for $\omega \in E^c$. By letting $n \rightarrow \infty$, we obtain

$$
\int_{E^c} |Y^*(\omega)|^2 \, d\omega = 0. \quad\quad\quad (3.7\text{-}11)
$$

It follows that $Y^*(\omega) = 0$ for $\omega \in E^c$. Hence, $y^*(\cdot) \in C$, and C is closed. ∎

PROJECTION: For $x(\cdot) \notin C$, its projection $g(\cdot) = Px(\cdot)$ onto C in the frequency domain, $G(\cdot) = \mathcal{F}g(\cdot)$, is given by

$$
G(\omega) = \begin{cases} X(\omega) & \text{if } \omega \in [-2\pi B, 2\pi B] \\ 0 & \text{otherwise.} \end{cases} \quad\quad\quad (3.7\text{-}12)
$$

In words, in the frequency domain the projection of a function $x(t) \notin C$ onto C is obtained by truncating the Fourier transform of $x(t)$ within the frequency band $[-2\pi B, 2\pi B]$. To obtain the projection $g(\cdot)$ in the time domain, one can simply compute the inverse Fourier transform of $G(\cdot)$ in Eq. (3.7-12).

The reader may notice that the action of the projection $Px(\cdot)$ is equivalent to passing the signal $x(\cdot)$ through an ideal lowpass filter with bandwidth B Hz. Hence in the time-domain, $Px(\cdot)$ can be expressed as

$$
Px(t) = x(t) * 2B \frac{\sin(2\pi Bt)}{2\pi Bt}, \quad\quad\quad (3.7\text{-}13)
$$

where $*$ denotes the convolution operation.

DERIVATION: Let $E = [-2\pi B, 2\pi B]$, and E^c be its complement. Then for $x(\cdot) \notin C$ and an arbitrary $y(\cdot) \in C$,

$$
\begin{aligned}
\|x(\cdot) - y(\cdot)\|^2 &= \int_{-\infty}^{\infty} |x(t) - y(t)|^2 \, dt \\
&= \int_E |X(\omega) - Y(\omega)|^2 \, d\omega + \int_{E^c} |X(\omega) - Y(\omega)|^2 \, d\omega \\
&= \int_E |X(\omega) - Y(\omega)|^2 \, d\omega + \int_{E^c} |X(\omega)|^2 \, d\omega, \qquad (3.7\text{-}14)
\end{aligned}
$$

since $Y(\omega) = 0$ for $\omega \in E^c$. Clearly, the distance $\|x(\cdot) - y(\cdot)\|$ is minimized when $Y(\omega) = X(\omega)$ for $\omega \in E$. The projection in Eq. (3.7-12) follows immediately. ∎

Known Phase Constraint

In the Fourier transform, the relative displacement of the infinitesimal sinusoids (i.e., the Fourier components) with respect to each other is described by the *phase*. The concept of phase is critical to understanding the composition of waveforms. Indeed the phase, along with amplitude and frequency, is one of the three variables that can be controlled in a sinusoidal component of a signal. In some applications it may be the case that we have knowledge about the phase of the Fourier transform of a signal. To describe this knowledge, we use the following constraint set:

$$
C = \{\, y(\cdot) \; : \; y(\cdot) \leftrightarrow Y(\cdot), \angle Y(\omega) = \phi(\omega) \,\}, \qquad (3.7\text{-}15)
$$

where $\phi(\omega)$ is a known real function of ω.

In words, the set C is the collection of functions (i.e., signals) whose phase is some known function $\phi(\omega)$. Clearly, any function that has $\phi(\omega)$ as its phase will be a member of C. In the following, we demonstrate that this set is convex and closed, and derive the projection onto it.

CONVEXITY: Let $y_1(\cdot), y_2(\cdot) \in C$. Then their respective Fourier transforms $Y_1(\cdot)$ and $Y_2(\cdot)$ have the same phase function $\phi(\omega)$. That is,

$$
Y_1(\omega) = |Y_1(\omega)| \exp[j\phi(\omega)], \text{ and } Y_2(\omega) = |Y_2(\omega)| \exp[j\phi(\omega)]. \qquad (3.7\text{-}16)
$$

Let $y_3(\cdot) = \alpha\, y_1(\cdot) + (1 - \alpha)\, y_2(\cdot)$ for $\alpha \in [0, 1]$. Then, its Fourier transform $Y_3(\cdot)$ satisfies

$$
\begin{aligned}
Y_3(\omega) &= \alpha\, Y_1(\omega) + (1 - \alpha)\, Y_2(\omega) \\
&= \alpha\, |Y_1(\omega)| \exp[j\phi(\omega)] + (1 - \alpha)\, |Y_2(\omega)| \exp[j\phi(\omega)] \\
&= [\alpha\, |Y_1(\omega)| + (1 - \alpha)\, |Y_2(\omega)|] \exp[j\phi(\omega)] \\
&= |Y_3(\omega)| \exp[j\phi(\omega)]. \qquad (3.7\text{-}17)
\end{aligned}
$$

Thus, $Y_3(\cdot)$ has the same phase function $\phi(\cdot)$, and $y_3(\cdot) \in C$. Thus, C is convex.
∎

Before we demonstrate that the set C is also closed, let's first consider the following lemma:

Lemma 3.7-1 *Let $x \geq 0, y \geq 0$. Then for arbitrary real numbers α and β the following holds*

$$|x \exp(j\alpha) - y \exp(j\beta)|^2 \geq x^2 \sin^2(\alpha - \beta). \tag{3.7-18}$$

Furthermore, with x, α, and β held fixed, the function

$$\gamma(y) = |x \exp(j\alpha) - y \exp(j\beta)|^2$$

is minimized when

$$y = \begin{cases} x \cos(\alpha - \beta) & \text{if } \cos(\alpha - \beta) > 0 \\ 0 & \text{otherwise.} \end{cases} \tag{3.7-19}$$

PROOF: Note that

$$
\begin{aligned}
|x \exp(j\alpha) - y \exp(j\beta)|^2 &= x^2 + y^2 - 2xy \cos(\alpha - \beta) \\
&= x^2 \sin^2(\alpha - \beta) + [x \cos(\alpha - \beta) - y]^2 \\
&\geq x^2 \sin^2(\alpha - \beta). \tag{3.7-20}
\end{aligned}
$$

Thus, Eq. (3.7-18) follows immediately.

Next, consider the function $\gamma(y) = |x \exp(j\alpha) - y \exp(j\beta)|^2$ for x, α, and β fixed. From the second row of the above equation we see that $\gamma(y)$ is minimized by $y = x \cos(\alpha - \beta)$ when $\cos(\alpha - \beta) > 0$ and by $y = 0$ when $\cos(\alpha - \beta) \leq 0$. Hence, Lemma 3.7-1 follows. ∎

Now, we demonstrate that the set C is also closed.

CLOSEDNESS: Let $\{y_n(\cdot)\}$ be a sequence in C such that $y_n(\cdot) \to y^*(\cdot)$. We want to show that $y^*(\cdot) \in C$ also. Clearly, $y^*(\cdot) \in C$ if $y^*(\cdot)$ is the zero function. Without loss of generality, assume that $y^*(t)$ is not the zero function. Let $Y^*(\cdot)$ and $Y_n(\cdot)$ denote, respectively, the Fourier transform of $y^*(t)$ and $y_n(\cdot)$. Let $\psi(\omega) = \angle Y^*(\omega)$. Then, by the Parseval's identity and Lemma 3.7-1, used in the third line below, we obtain

$$
\begin{aligned}
\|y_n(\cdot) - y^*(\cdot)\|^2 &= \frac{1}{2\pi} \int_R |Y_n(\omega) - Y^*(\omega)|^2 \, d\omega \\
&= \frac{1}{2\pi} \int_R ||Y_n(\omega)| \exp[j\phi(\omega)] - |Y^*(\omega)| \exp[j\psi(\omega)]|^2 \, d\omega
\end{aligned}
$$

$$\geq \frac{1}{2\pi} \int_R |Y^*(\omega)|^2 \, \sin^2 [\phi(\omega) - \psi(\omega)] \, d\omega. \qquad (3.7\text{-}21)$$

By letting $n \to \infty$, we obtain

$$\frac{1}{2\pi} \int_R |Y^*(\omega)|^2 \, \sin^2 [\phi(\omega) - \psi(\omega)] \, d\omega = 0, \qquad (3.7\text{-}22)$$

which gives rise to $\sin^2 [\phi(\omega) - \psi(\omega)] = 0.$[†] Hence, $\phi(\omega) = \psi(\omega) + k\pi$ for some integer k. In fact, k can only be even since $y_n(\cdot) \to y^*(\cdot)$. Hence, $y^*(\cdot) \in C$, and C is closed.   ∎

PROJECTION: For $x(\cdot) \notin C$, its projection $g(\cdot) = Px(\cdot)$ onto C is best described in the frequency domain. With $G(\cdot) = \mathcal{F}g(\cdot)$, we obtain

$$G(\omega) = \begin{cases} |X(\omega)| \cos[\angle X(\omega) - \phi(\omega)] \exp [j\phi(\omega)] & \text{if } \cos[\angle X(\omega) - \phi(\omega)] \geq 0 \\ 0 & \text{otherwise.} \end{cases}$$
$$(3.7\text{-}23)$$

DERIVATION: For $x(\cdot) \notin C$ and an arbitrary $y(\cdot) \in C$, we have

$$\begin{aligned} \|x(\cdot) - y(\cdot)\|^2 &= \frac{1}{2\pi} \int_R |X(\omega) - Y(\omega)|^2 \, d\omega \\ &= \frac{1}{2\pi} \int_R ||X(\omega)| \exp [j\angle X(\omega)] - |Y(\omega)| \exp [j\phi(\omega)]|^2 \, d\omega. \end{aligned}$$

From Lemma 3.7-1, the distance $\|x(\cdot) - y(\cdot)\|$ is minimized when $Y(\omega)$ takes on the value of $G(\omega)$ in Eq. (3.7-23). Thus, the projection given above follows.   ∎

DISCUSSION: Phase constraints can be applied in the time or frequency domains. In electrical engineering and optics, time-domain signals called *phasors* and space-domain signals called *complex amplitudes*, respectively, are used to model voltages, currents, and optical fields. For a complex valued signal $y(t)$ we can express it as

$$y(t) = |y(t)| \exp [j\angle y(t)], \qquad (3.7\text{-}24)$$

where, as in the case of the Fourier representation, $|y(t)|$ and $\angle y(t)$ are called the magnitude and phase of the signal $y(t)$, respectively.

If we have knowledge of the phase of a time-domain signal we can enforce the constraint that follows from this knowledge via the set

$$C = \{ \, y(\cdot) \, : \, \angle y(t) = \phi(t) \text{ for } t \in R \, \}, \qquad (3.7\text{-}25)$$

where $\phi(\cdot)$ is a known phase function. That C is closed and convex can be demonstrated using the methods of the proceeding discussion; likewise with the compu-

---

[†] Strictly speaking, $\sin^2 [\phi(\omega) - \psi(\omega)] = 0$ for almost every $\omega \in R$.

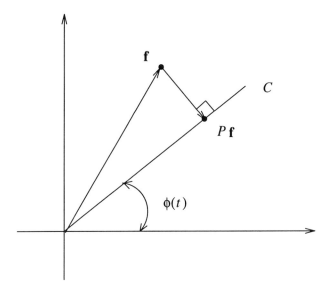

Fig. 3.7-1 Illustration that in the case of the phase constraint of Eq. (3.7-25) the projection can be deduced from a simple diagram.

tation of its projection. Nevertheless to show that sometimes a projection can be deduced from a simple diagram, refer to Fig. 3.7-1. The arbitrary complex signal $f(\cdot)$ is shown in relation to the set C. Since $\angle y(\cdot) = \phi(\cdot)$ for a typical element $y(t)$ in C, the only adjustable property of $y(\cdot)$ is its length. Form the diagram it is clear that the projection $g(\cdot)$ of $f(\cdot)$ on the direction indicated by $\phi(\cdot)$ is given by

$$g(t) = \begin{cases} |f(t)| \cos[\angle f(t) - \phi(t)] \exp[j\phi(t)] & \text{if } \cos[\angle f(t) - \phi(t)] \geq 0 \\ 0 & \text{otherwise.} \end{cases}$$

$$(3.7\text{-}26)$$

As expected, there exists a striking parallelism between this result and that given in Eq. (3.7-23).

3.8 SUMMARY

In this chapter we discussed a number of elementary but practically useful constraint sets in a Hilbert space setting. For each set we demonstrated convexity and closedness, and computed its associated projector.

The first part of the chapter emphasized linear-type constraints, error-type constraints, and similarity-type constraints. Numerous examples of constraint sets of these types were furnished in several vector spaces.

The latter part of the chapter dealt primarily with functional constraints in L^2 space. It was convenient to group time and frequency domain constraints separately.

While the associated projectors were quite elementary and usually easy to compute, their importance in problem-solving is evident from gleaming the periodical literature.

APPENDIX

Part A: Proof of Theorem 3.4-1

Proof: What needs to be shown is that for $\mathbf{x} \notin C$ Eq. (3.4-26) has a unique positive root, and the resulting $\mathbf{g}$ in Eq. (3.4-25) is closer to $\mathbf{x}$ than that resulting from any other root.

Since $\mathbf{g} \in R^n$, the parameter λ must be a real number. Therefore, we need to consider only the real roots of Eq. (3.4-26).

Define, as in Eq. (3.4-27),

$$\psi(\lambda) \triangleq \sum_{i=1}^{n} \frac{w_i^2 (x_i - r_i)^2}{(1 + \lambda w_i^2)^2} - \epsilon^2. \tag{3.A-1}$$

Then the equation $\psi(\lambda) = 0$ is equivalent to Eq. (3.4-26). By assumption $\mathbf{x} \notin C$ and we have

$$\psi(0) = \sum_{i=1}^{n} w_i^2 (x_i - r_i)^2 - \epsilon^2 > 0. \tag{3.A-2}$$

On the other hand,

$$\lim_{\lambda \to +\infty} \psi(\lambda) = -\epsilon^2 < 0. \tag{3.A-3}$$

Thus, by the continuity of $\psi(\lambda)$ for $\lambda \geq 0$, there must exist a $\lambda \in (0, \infty)$ such that $\psi(\lambda) = 0$. In other words, the equation $\psi(\lambda) = 0$ must have a positive root in λ. Since the function $\psi(\lambda)$ is strictly decreasing for all $\lambda > 0$, such a positive root is also unique. Thus, Eq. (3.4-26) has a unique positive root in λ. For convenience, we denote this root by λ_+.

Next, we show that the resulting $\mathbf{g}$ from this positive root λ_+ is closer to the point $\mathbf{x}$ than that resulting from a negative root of Eq. (3.4-26) , if it has any[†].

To be specific, let

$$D(\lambda) = \|\mathbf{x} - \mathbf{g}\|^2 = \sum_{i=1}^{n} (x_i - g_i)^2, \tag{3.A-4}$$

[†]One can show easily that Eq. (3.4-23) has at least one negative root.

and substituting for g_i from Eq. (3.4-25) yields

$$D(\lambda) = \sum_{i=1}^{n} \left[\frac{\lambda w_i^2 (x_i - r_i)}{1 + \lambda w_i^2} \right]^2. \tag{3.A-5}$$

We want to show that if Eq. (3.4-26) has a negative root, say λ_-, then $D(\lambda_+) < D(\lambda_-)$.

Note that for $\lambda \geq 0$, $|1 - \lambda w_i^2| < 1 + \lambda w_i^2$ for each i. Thus, for all $\lambda > 0$, we have $\psi(-\lambda) > \psi(\lambda)$. Since $\psi(\lambda)$ is strictly decreasing for all $\lambda > 0$, it follows that $\psi(-\lambda) > \psi_{(\lambda)} > 0$ for all $\lambda \in (0, \lambda_+)$. Hence, $\psi(\lambda) > 0$ for all $\lambda \in [-\lambda_+, 0)$. Therefore, any negative root of $\psi(\lambda) = 0$ must lie in the interval $(-\infty, -\lambda_+)$. Thus, $|\lambda_-| > \lambda_+$.

For $\lambda_- < 0$ and $|\lambda_-| > \lambda_+$, we have

$$|1 + \lambda_- w_i^2| < 1 + |\lambda_- w_i^2| = 1 - \lambda_- w_i^2 = |1 - \lambda_- w_i^2|, \tag{3.A-6}$$

and

$$1 - \lambda_- w_i^2 > 1 + \lambda_+ w_i^2 > 1. \tag{3.A-7}$$

Thus,

$$\left| \frac{\lambda_- w_i^2}{1 + \lambda_- w_i^2} \right| > \left| \frac{\lambda_- w_i^2}{1 - \lambda_- w_i^2} \right| = 1 - \frac{1}{1 - \lambda_- w_i^2}$$
$$> 1 - \frac{1}{1 + \lambda_+ w_i^2} = \frac{\lambda_+ w_i^2}{1 + \lambda_+ w_i^2}. \tag{3.A-8}$$

Therefore, Eq. (3.A-5) yields

$$D(\lambda_-) > D(\lambda_+). \tag{3.A-9}$$

Thus, Theorem 3.4-1 follows since λ_- is an arbitrary negative root. ∎

Part B: Introduction to Newton's Method

Newton's method (or sometimes called the Newton-Raphson method) is one of the most useful and best-known algorithms for finding the root of an equation $f(x) = 0$, which is likely nonlinear in x. In short, this method is summarized in the following theorem:

Theorem 3.0-1 (*Newton-Raphson Method*) *Assume that $f(x), f'(x)$, and $f''(x)$ are continuous near the root p of $f(x) = 0$. If p_0 is chosen close enough to p, then the sequence $\{p_k\}$ generated by the iteration*

$$p_{k+1} = p_k - f(p_k)/f'(p_k) \quad for \ k = 0, 1, 2, \cdots, \tag{3.B-1}$$

will converge to p.

For a rigorous proof of this theorem the interested reader is referred to any standard texts on numerical analysis, for example, [2]. In the following we furnish a brief discussion to give some insight on this method.

Consider the Taylor polynomial representation of $f(x)$ around the initial starting point p_0 with remainder term

$$f(x) = f(p_0) + f'(p_0)(x - p_0) + f''(c)\frac{(x - p_0)^2}{2} , \qquad (3.B\text{-}2)$$

where c lies somewhere between p_0 and x. Substituting $x = p$ in Eq. (3.B-2) and using that fact that $f(p) = 0$ results in

$$0 = f(p_0) + f'(p_0)(p - p_0) + f''(c)\frac{(p - p_0)^2}{2} . \qquad (3.B\text{-}3)$$

If p_0 is close enough to p, then the last term on the right side of Eq. (3.B-3) will be small and can be neglected and we obtain

$$0 \approx f(p_0) + f'(p_0)(p - p_0). \qquad (3.B\text{-}4)$$

Solving for p in Eq. (3.B-4), we get $p = p_0 - f(p_0)/f'(p_0)$. This value is used to define p_1, the next approximation to the root p. When p_k is used in place of p_0 above, the general rule in Eq. (3.B-1) follows.

The Newton's method has an informative graphical interpretation. Consider the graph of the function $f(x)$ which intersects the x-axis at the point $(p, 0)$. Let $(p_0, f(p_0))$ be the initial starting point lying on the curve near the point $(p, 0)$, as illustrated in Fig. 3.A-1. If we approximate the function $f(x)$ by the line tangent to the curve at point$(p_0, 0)$, then this tangent line will meet the x-axis at point $(p_1, f(p_1))$, where p_1 is found to be $p_1 = p_0 - f(p_0)/f'(p_0)$. The point $x = p_1$ is then used as the next approximation to the root p and a repetition of the above process will leads to the iteration in Eq. (3.B-1).

Part C: Proof of Theorem 3.4-2

Proof: The function $\psi(\lambda)$ has derivative

$$\psi'(\lambda) = -2 \sum_{i=1}^{n} \frac{w_i^4 (x_i - r_i)^2}{(1 + \lambda w_i^2)^3} .$$

Clearly, $\psi'(\lambda) < 0$ for $\lambda \geq 0$. Also, it is a continuous and strictly increasing function of λ for $\lambda \geq 0$.

Assume that $\lambda_k \geq 0$ is such that $\psi(\lambda_k) > 0$, then

$$\lambda_{k+1} = \lambda_k - \frac{\psi(\lambda_k)}{\psi'(\lambda_k)} > \lambda_k, \qquad (3.C\text{-}1)$$

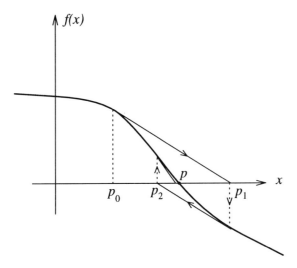

Fig. 3.A-1   Graphical illustration of the Newton-Raphson's method.

because $\psi'(\lambda_k) < 0$.

On the other hand,

$$\psi(\lambda_{k+1}) = \psi(\lambda_k) + \int_{\lambda_k}^{\lambda_{k+1}} \psi'(\lambda)\, d\lambda.$$

For $\lambda \in (\lambda_k, \lambda_{k+1}]$, $\psi'(\lambda) > \psi'(\lambda_k)$ so that

$$
\begin{aligned}
\psi(\lambda_{k+1}) \;\;>\;\; & \psi(\lambda_k) + \int_{\lambda_k}^{\lambda_{k+1}} \psi'(\lambda_k)\, d\lambda \\
=\;\; & \psi(\lambda_k) + \psi'(\lambda_k)(\lambda_{k+1} - \lambda_k).
\end{aligned}
\qquad (3.\text{C-}2)
$$

By Eq. (3.C-1) we obtain $\psi(\lambda_{k+1}) > 0$.

By induction, with $\lambda_0 = 0$ the Newton's iterates generated by

$$\lambda_{k+1} = \lambda_k - \frac{\psi(\lambda_k)}{\psi'(\lambda_k)}\,,
\qquad (3.\text{C-}3)$$

will satisfy

$$0 = \lambda_0 < \lambda_1 < \lambda_2 < \cdots.
\qquad (3.\text{C-}4)$$

Because the function $\psi(\lambda)$ is continuous and strictly decreasing for $\lambda > 0$ and $\psi(\lambda_k) > 0 = \psi(\lambda_+)$, it follows that $\lambda_k < \lambda_+$ for all $k = 0, 1, 2, \cdots$. Hence, the sequence $\{\lambda_k\}$ must converge to some number, say λ^*. By letting $k \to \infty$, Eq. (3.C-3) yields $\psi(\lambda^*) = 0$, which implies $\lambda^* = \lambda_+$.   ∎

REFERENCES

1. R. O. Duda and P. E. Hart, *Pattern Classification and Scene Analysis*, John Wiley & Sons, New York, 1973.

2. J. H. Mathews, *Numerical Methods for Computer Science, Engineering, and Mathematics*, Prentice-Hall, Englewood Cliffs, NJ, 1987.

EXERCISES

3-1. Define a set C to describe the points in the shaded fan region in the 2-D plane shown in Fig. P3.1.(a).

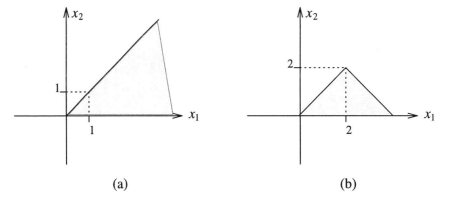

(a)   (b)

Fig. P3.1   Point sets in the 2-D plane.

    (a) Show that the set C is closed and convex.

    (b) Derive the projector onto C.

3-2. Define a set C to describe the points in the shaded triangular region in the 2-D plane shown in Fig. P3.1.(b).

    (a) Show that the set C is closed and convex.

    (b) Derive the projector onto C.

3-3. A signal $f(t)$ is known to be *symmetric* about the origin $t = 0$, i.e. $f(t) = f(-t)$ for $-\infty < t < \infty$.

    (a) Define a constraint set C to describe such a property of $f(t)$.

    (b) Is the set C in (a) closed and convex?

    (c) If the answer is yes, derive the projector onto C.

3-4. Repeat Exercise 3-3 for a signal $f(t)$ that is known to be *anti-symmetric* about the origin $t = 0$, i.e., $f(t) = -f(-t)$ for $-\infty < t < \infty$.

3-5. Show that the following set is convex and closed and derive its projector:

$$C = \{\, \mathbf{y} \; : \; |\langle \mathbf{y}, \, \mathbf{a} \rangle| \leq b \,\}.$$

3-6. Show that the following set is non-convex:

$$C = \{\, \mathbf{y} \,:\, |\langle \mathbf{y},\, \mathbf{a}\rangle| \geq b \,\}.$$

3-7. The amplitude of a voltage signal $f(t)$ is known to be bounded by 5 volts.
   (a) Define a constraint set C to describe such a property of $f(t)$.
   (b) Is the set C in (a) closed and convex?
   (c) If the answer is yes, derive the projector onto C.

3-8. A voltage signal $f(t)$ is applied across a resistor of $100\ \Omega$ from $t = 5$ seconds to $t = 10$ seconds. It is measured that the heat generated by the resistor is no larger than 5 joules.
   (a) Define a constraint set C to describe such a property of $f(t)$.
   (b) Is the set C in (a) closed and convex?
   (c) If the answer is yes, derive the projector onto C.

3-9. In the space L^2, let $\mathbf{w}$ be a known function vector and define the weighted error between a vector $\mathbf{x}$ and a reference vector $\mathbf{r}$ as

$$e_w(\mathbf{x}, \mathbf{r}) = \left[\, \int_{-\infty}^{\infty} |w(x)(y(x) - r(x)|^2\, dx \,\right]^{1/2}.$$

   (a) Define a error type constraint set C using this weighted error measure.
   (b) Show that the set C is both convex and closed.
   (c) Derive the projector onto C.

3-10. Consider the constraint set C given in Eq. (3.6-22),
   (a) Show that C is convex and closed.
   (b) Derive the projector onto C.

3-11. The magnitude of the Fourier transform of a voice waveform $f(t)$ is known to be bounded by 0.1 Volt-Sec in the frequency range of 100 Hz to 200 Hz.
   (a) Define a constraint set C to describe such a property of $f(t)$.
   (b) Is the set C in (a) closed and convex?
   (c) If the answer is yes, derive the projector onto C.

3-12. The power of an audio waveform $f(t)$ within the frequency range of 100 Hz to 200 Hz is measured to be bounded by 0.1 Watts.
   (a) Define a constraint set C to describe such a property of $f(t)$.
   (b) Is the set C in (a) closed and convex?
   (c) If the answer is yes, derive the projector onto C.

3-13. Show that the set $C = \{\, \mathbf{y} \,:\, \sum_{i=1}^{n} w_i^2 (y_i - r_i)^2 \leq \epsilon^2 \,\}$ in the Euclidean space R^n is both closed and convex.

3-14. Show that the set

$$C = \{\, y(\cdot) \,:\, y(\cdot) \leftrightarrow Y(\cdot), Y(\omega) = 0,\ \text{for } \omega \notin [-2\pi B, 2\pi B] \,\}$$

is a closed subspace in L^2.

3-15. Let X be a random variable with probability density function (pdf) $f(x)$. It is known that X is twice more likely to assume a positive value than a negative one.

   (a) Define a constraint set C to describe such a property of $f(x)$.

   (b) Is the set C in (a) closed and convex?

   (c) If the answer is yes, derive the projector onto C.

3-16. A random variable X with pdf $f(x)$ is known to have mean μ.

   (a) Define a constraint set C to describe such a property of $f(x)$.

   (b) Is the set C in (a) closed and convex?

   (c) If the answer is yes, derive the projector onto C.

4

Solutions of Linear Equations

4.1 INTRODUCTION

One of the most frequently encountered problems in all branches of analysis is the need to solve a system of simultaneous linear equations. Such a system can be written in the form

$$
\begin{aligned}
a_{11}x_1 + a_{12}x_2 + \cdots + a_{1n}x_n &= b_1 \\
a_{21}x_1 + a_{22}x_2 + \cdots + a_{2n}x_n &= b_2 \\
&\;\;\vdots \\
a_{m1}x_1 + a_{m2}x_2 + \cdots + a_{mn}x_n &= b_m,
\end{aligned}
\tag{4.1-1}
$$

where $x_1, x_2, \cdots, x_n$ are the unknown variables and the subscripted a's and b's denote constants.

A sequence of n numbers $s_1, s_2, \cdots, s_n$ is called a *solution* of the system in Eq. (4.1-1) if every equation in this system is satisfied when we substitute $x_1 = s_1, x_2 = s_2, \cdots, x_n = s_n$. The problem of concern, of course, is how to find such a solution provided that this system has at least one.

Several situations need to be classified. First, when there are as many equations as unknowns, i.e, when $m = n$ in Eq. (4.1-1), which is often the case, the system is said to be *critically determined*. Second, when there are fewer equations than unknowns, i.e, when $m < n$, the system is said to be *under-determined*. Lastly, when there are more equations than unknowns, i.e, when $m > n$, the system is said to be *over-determined*.

Not all systems of simultaneous linear equations have solutions. A system of linear equations that has no solution is said to be *inconsistent*. This occurs when some of its equations are contradictory to each other. For example, it is evident that no two real numbers would simultaneously satisfy the following equations

$$\begin{aligned} 2x + 2y &= 4 \\ x + y &= 0 \end{aligned}$$

(4.1-2)

since they contradict each other. Inconsistent systems of linear equations are most likely to occur when the system is over-determined. Yet in practical applications, an *approximate* solution is still desired for an inconsistent system of linear equations. Such an approximate solution is typically obtained by minimizing some error criterion so that it *approximately* satisfies the system of linear equations. One famous example is the so-called *least-squares solution* [1].

On the other hand, a system of linear equations that has at least one solution is said to be *consistent*. A consistent system of linear equations can have either exactly one or infinitely many solutions. For example, both the following two systems of linear equations

$$\begin{aligned} x - y &= 0 \\ x + y &= 2 \end{aligned}$$

(4.1-3)

and

$$\begin{aligned} x + y &= 1 \\ 2x + 2y &= 2 \end{aligned}$$

(4.1-4)

are consistent. More specifically, the first system has $(x = 1, y = 1)$ as its only solution, while the second system has infinitely many solutions of which examples are $(x = 1, y = 0)$, $(x = 0, y = 1)$, and $(x = 1/2, y = 1/2)$, etc. In general, assuming that no equation is a linear combination of the others[†], a critically determined consistent system has exactly one solution, while a under-determined consistent system always has infinitely many solutions.

There exist many methods for solving consistent systems of linear equations. Perhaps the most popular approach which one learns from a first course in linear algebra is *Cramer's rule*, in which each component $x_i, i = 1, 2, \cdots, n$, of the solution is computed as the quotient of a different *numerator determinant* over a common *denominator determinant*. Another popular yet computationally more efficient approach is the Gaussian elimination algorithm, in which the set of simultaneous equations in Eq. (4.1-1) is reduced to *triangular form*, which is easier to solve through systematic operations. The Gaussian elimination approach is perhaps the most popular in use today for computer-aided solution of linear equations. There exist, of course, many other more efficient algorithms for solving systems of linear equations. Those algorithms gain their efficiency by taking advantage of certain inherent structures of a system of linear equations. A detailed discussion of these algorithms will force us to deviate from the main goal of this book and thus

---

[†]In such cases we say that the equations are *independent*.

we omit such a discussion. Interested readers are directed to [1]–[6] for details.

In this chapter, we show that a system of simultaneous linear equations can be solved by vector-space projections methods. The convergence behavior of the projection-based algorithm is analyzed in detail. Moreover, a suggestion for accelerating this algorithm, based on the Gram-Schmidt process, is provided. Finally, we give two practical application examples of this approach for solving important problems in imaging science.

First, let's adopt some notation from linear algebra. Let $\mathbf{x}$ denote the column vector of unknowns, i.e., $\mathbf{x} = (x_1, x_2, \cdots, x_n)^T$, and similarly let $\mathbf{b} = (b_1, b_2, \cdots, b_m)^T$. Also let $\mathbf{A}$ be the matrix of the coefficient constants a's, i.e.,

$$\mathbf{A} = \begin{pmatrix} a_{11} & a_{12} & \cdots & a_{1n} \\ a_{21} & a_{22} & \cdots & a_{2n} \\ \vdots & \vdots & \vdots & \vdots \\ a_{m1} & a_{m2} & \cdots & a_{mn} \end{pmatrix}.$$

Then the system of simultaneous linear equations in Eq. (4.1-1) can be conveniently written in the following handy form

$$\mathbf{A}\mathbf{x} = \mathbf{b}. \tag{4.1-5}$$

In such a formulation, the unknowns are treated as a vector in the space R^n. Equation (4.1-1) is associated with a geometric meaning: an unknown vector $\mathbf{x}$ is transformed by matrix $\mathbf{A}$ into a vector $\mathbf{b}$, which is known. Then the question of concern is how to find such a vector $\mathbf{x}$ that makes this possible.

For the space R^n, the Euclidean inner product is that which was introduced in Chapter 1. Thus consider two vectors $\mathbf{x} = (x_1, x_2, \cdots, x_n)^T$ and $\mathbf{y} = (y_1, y_2, \cdots, y_n)^T$ both in R^n; their Euclidean inner product, denoted by $\langle \mathbf{x}, \mathbf{y} \rangle$, is defined as [†]

$$\langle \mathbf{x}, \mathbf{y} \rangle = \sum_{i=1}^{n} x_i y_i. \tag{4.1-6}$$

Alternatively, the inner product $\langle \mathbf{x}, \mathbf{y} \rangle$ is denoted by $\mathbf{x}^T \mathbf{y}$ or $\mathbf{x} \cdot \mathbf{y}$. The nomenclature *dot product* is widely used in physics, where the underlying vectors are forces, fields, velocities etc.

Associated with the Euclidean inner product, the Euclidean norm for a vector $\mathbf{x}$ is defined as

$$\|\mathbf{x}\| = \langle \mathbf{x}, \mathbf{x} \rangle^{1/2}. \tag{4.1-7}$$

Geometrically, the Euclidean norm serves as a measure of the length of a vector, which agrees with physical intuition. Furthermore, the Euclidean distance between

---

[†] Sometimes we repeat ourselves to avoid having the reader flip back and forth between the chapters.

two vectors (points) $\mathbf{x}$ and $\mathbf{y}$, which is denoted by $d(\mathbf{x}, \mathbf{y})$, is naturally induced as

$$d(\mathbf{x}, \mathbf{y}) = \|\mathbf{x} - \mathbf{y}\|. \tag{4.1-8}$$

With the inner product notation, Eq. (4.1-1) can be written into another useful form. Let $\mathbf{a}_i$ denote the ith row vector of the matrix $\mathbf{A}$, i.e.,

$$\mathbf{a}_i = \left(a_{i1}, a_{i2}, \cdots, a_{in}\right)^T \tag{4.1-9}$$

for $i = 1, 2, \cdots, m$. Then Eq. (4.1-1) can be rewritten as

$$
\begin{aligned}
\langle \mathbf{a}_1, \mathbf{x} \rangle &= b_1 \\
\langle \mathbf{a}_2, \mathbf{x} \rangle &= b_2 \\
&\vdots \\
\langle \mathbf{a}_m, \mathbf{x} \rangle &= b_m.
\end{aligned}
\tag{4.1-10}
$$

Clearly, Eqs. (4.1-1), (4.1-5) and (4.1-10) are simply alternate ways for describing a system of linear equations. Later on we will use them interchangeably, whichever happens to be more convenient.

4.2   CONVEX-SET FORMULATION

Consider the system of linear equations in Eq. (4.1-3), which consists of two equations of two variables x and y. From elementary algebra, we know that each equation in Eq. (4.1-3) is represented by a straight line in R^2. The graphs of these two equations, denoted by l_1 and l_2, respectively, are plotted in Fig. 4.2-1(a). Similarly, the graphs for the equations in Eq. (4.1-4) and Eq. (4.1-2) are given in Fig. 4.2-1(b) and (c), respectively.

In these plots, each line represents an equation and each point on the line represents a solution to that equation. In other words, the line for an equation is the collection of solutions to that equation. In Fig. 4.2-1(a), the lines l_1 and l_2 *intersect* at only one point, i.e., $(1, 1)$, which is the unique solution to the problem. In Fig. 4.2-1(b), the lines l_1 and l_2 coincide with each other, in which case they *intersect* at infinitely many points: $(1, 0)$, $(0, 1)$, and $(1/2, 1/2)$, to name a few, all of which are solutions to the problem. On the other hand, in Fig. 4.2-1(c), the lines l_1 and l_2 are parallel to each other, in which case they *do not intersect* and consequently the system has no solution.

The observation above also extends to higher dimensional vector spaces. Consider an equation in a system of linear equations in Eq. (4.1-10), say its ith equation,

$$\langle \mathbf{a}_i, \mathbf{y} \rangle = b_i. \tag{4.2-1}$$

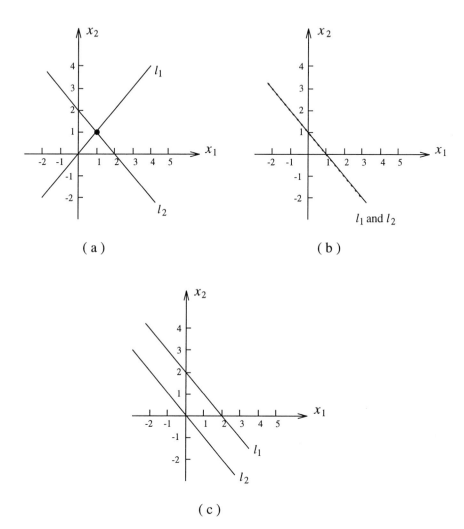

Fig. 4.2-1 A system of linear equations may have no solution, one solution, or infinitely many solutions: (a) lines l_1 and l_2, which represent two linear equations, intersect at one point, which is the only solution to the system; (b) lines l_1 and l_2 coincide with each other, thus there are infinite number of solutions to the system; and (c) lines l_1 and l_2 are parallel and do not intersect, therefore the system has no solution.

A point (vector) $\mathbf{y}^{\dagger}$ is a solution if it satisfies this equation. If we call the collection of all the solutions as the *solution set*, then the solution set to Eq. (4.2-1), denoted by C_i, is handily expressed in R^n as

$$C_i = \{\mathbf{y} : \langle \mathbf{a}_i, \mathbf{y} \rangle = b_i\}. \tag{4.2-2}$$

For m equations of the form of Eq. (4.1-1), there are m such solution sets: namely, $C_1, C_2, \cdots,$ and C_m.

Consider a solution, say $\mathbf{y}^*$, to the system of linear equations $\langle \mathbf{a}_i, \mathbf{y} \rangle = b_i$, $i = 1, 2, \cdots, m$. Since it satisfies each equation in the system, $\mathbf{y}^*$ has membership in each of the sets C_i, i.e., $\mathbf{y}^* \in C_i$ for $i = 1, 2, \cdots, m$. Consequently, we have

$$\mathbf{y}^* \in \bigcap_{i=1}^{m} C_i. \tag{4.2-3}$$

If we use C_0 to denote the set of all the solutions for the system of linear equations, then we have

$$C_0 = \bigcap_{i=1}^{m} C_i. \tag{4.2-4}$$

In other words, the set of solutions to a system of linear equations is simply the set resulting from the intersection of the solution sets for each equation in the system.

Clearly, when the system of linear equations is inconsistent, the intersection set C_0 in Eq. (4.2-4) will be empty and the system has no solution. On the other hand, when the system is consistent, this intersection set will contain either only one element, which is the unique solution to the system, or infinitely many elements, all of which are valid solutions to the system (see Exercise 4-1).

The problem of concern, of course, is how to find an element in this intersection set when it is non-empty. It would seem that the projections onto convex sets (POCS) is a natural vehicle for solving this problem. Before we apply this algorithm, however, two questions remain to be addressed: (i) Is the set C_i in Eq. (4.2-2) convex and closed? and (ii) If it is, how do we compute its projection P_i?

Note that the constraint set C_i belongs to the class of linear-type constraints in a Hilbert space that we discussed earlier in Section 3.2. Thus the above questions are answered immediately by the results developed there. Nevertheless, in the following we demonstrate in detail how to answer these questions by exploiting the properties of the Euclidean space R^n. We first demonstrate that each set C_i is both convex and closed in the space R^n. Then we derive its projector P_i.

[†] In keeping with notation used in earlier chapters, we switch the notation $\mathbf{x}$ to $\mathbf{y}$ to denote the unknown solution vector.

Convexity and Closedness of C_i

First, let's consider the convexity of C_i. Assume that $\mathbf{y}_1$ and $\mathbf{y}_2$ are two arbitrary elements of C_i, we show that their convex combination $\mathbf{y}_3 = \alpha \mathbf{y}_1 + (1-\alpha)\mathbf{y}_2$ also belongs to this set for all values of $0 \leq \alpha \leq 1$. This comes as a direct result of the linear property of the inner product. Indeed,

$$
\begin{aligned}
\langle \mathbf{a}_i, \mathbf{y}_3 \rangle &= \langle \mathbf{a}_i, \alpha \mathbf{y}_1 + (1-\alpha)\mathbf{y}_2 \rangle \\
&= \alpha \langle \mathbf{a}_i, \mathbf{y}_1 \rangle + (1-\alpha)\langle \mathbf{a}_i, \mathbf{y}_2 \rangle \\
&= \alpha b_i, +(1-\alpha)b_i = b_i.
\end{aligned} \tag{4.2-5}
$$

Hence $\mathbf{y}_3 \in C_i$, and it follows that the set C_i is convex.

Next, we show that C_i is closed. Assume that $\{\mathbf{y}_k\}$ is a converging sequence contained in C_i, we want to show that its limit, say $\mathbf{y}^*$, also belongs to C_i. Again, this comes as a result of the Schwarz inequality. Indeed,

$$
|\langle \mathbf{a}_i, \mathbf{y}_k - \mathbf{y}^* \rangle| \leq \|\mathbf{a}_i\| \, \|\mathbf{y}_k - \mathbf{y}^*\| \longrightarrow 0, \tag{4.2-6}
$$

since $\|\mathbf{y}_k - \mathbf{y}^*\| \longrightarrow 0$ as $k \longrightarrow \infty$. Note also that

$$
\langle \mathbf{a}_i, \mathbf{y}_k - \mathbf{y}^* \rangle = \langle \mathbf{a}_i, \mathbf{y}_k \rangle - \langle \mathbf{a}_i, \mathbf{y}^* \rangle \longrightarrow 0. \tag{4.2-7}
$$

Hence it follows that $\langle \mathbf{a}_i, \mathbf{y}^* \rangle = \langle \mathbf{a}_i, \mathbf{y}_k \rangle = b_i$ and therefore the set C_i is closed.

Before we move on, let's examine the geometric structure of the set C_i. From the proof of its convexity, it is clear that Eq. (4.2-5) still holds when α takes on an arbitrary value outside the interval $[0, 1]$. Therefore, the combination $\mathbf{y}_3 = \alpha \mathbf{y}_1 + (1-\alpha)\mathbf{y}_2$ always belongs to C_i as long as $\mathbf{y}_1$ and $\mathbf{y}_2$ do. This implies that the set C_i allows for a more refined categorization: it is also a linear variety. This result does not come as a surprise. From elementary linear algebra we know that each linear equation in Eq. (4.1-1) represents a hyperplane in R^n. For example, when $n = 2$, i.e., when there are two unknowns, C_i represents a straight line in R^2; when $n = 3$, C_i is a plane in R^3; for larger values of n, C_i is a *hyperplane*.

Projection P_i onto C_i

Here we present two approaches to compute the projection onto C_i. First, let us consider an intuitive approach. By its definition, the projection $P_i\mathbf{x}$ of an arbitrary vector $\mathbf{x}$ onto C_i is the point in C_i that is closest to $\mathbf{x}$. As pointed out earlier, the set C_i geometrically represents a hyperplane with $\mathbf{a}_i$ as its normal vector in the Euclidean space R^n. From geometry we know that the line connecting $\mathbf{x}$ and its projection $P_i\mathbf{x}$ is orthogonal to this hyperplane, as illustrated in Fig. 4.2-2. Therefore, we have

$$
\mathbf{x} - P_i\mathbf{x} = \alpha \mathbf{a}_i \tag{4.2-8}
$$

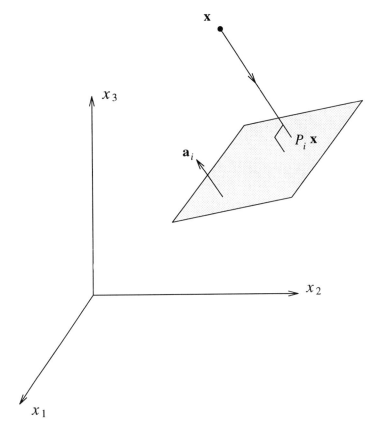

Fig. 4.2-2 In an Euclidean space the projection from a point to a hyperplane is always orthogonal to the hyperplane itself.

for some constant α. The constant α is determined from the condition that $P_i\mathbf{x}$ has to be in C_i, i.e., on the hyperplane. In other words,

$$\langle \mathbf{a}_i, P_i\mathbf{x} \rangle = b_i. \tag{4.2-9}$$

From Eq. (4.2-8), we obtain that $P_i\mathbf{x} = \mathbf{x} - \alpha\mathbf{a}_i$ so that

$$\langle \mathbf{a}_i, \mathbf{x} - \alpha\mathbf{a}_i \rangle = b_i, \tag{4.2-10}$$

which yields

$$\alpha = \frac{\langle \mathbf{a}_i, \mathbf{x} \rangle - b_i}{\|\mathbf{a}_i\|^2}. \tag{4.2-11}$$

Therefore, we have[†]

$$P_i \mathbf{x} = \mathbf{x} - \frac{\langle \mathbf{a}_i, \mathbf{x} \rangle - b_i}{\|\mathbf{a}_i\|^2} \mathbf{a}_i . \qquad (4.2\text{-}12)$$

In the derivation above, we have taken for granted that $P_i \mathbf{x} - \mathbf{x}$ is orthogonal to the hyperplane. In the following we present a rigorous mathematical approach to computing the projection onto C_i that does not require such geometrical considerations.

As stated earlier, by definition the projection $P_i \mathbf{x}$ of an arbitrary $\mathbf{x}$ onto the set C_i is the vector $\mathbf{y}$ in C_i that minimizes the distance $\|\mathbf{y} - \mathbf{x}\|$. This vector $\mathbf{y}$ can be solved by using the Lagrange multiplier's method. Indeed, write the Lagrange functional

$$\begin{aligned} J &= \|\mathbf{y} - \mathbf{x}\|^2 + \lambda \left(\langle \mathbf{a}_i, \mathbf{y} \rangle - b_i \right) \\ &= \mathbf{y}^T \mathbf{y} - \mathbf{y}^T \mathbf{x} - \mathbf{x}^T \mathbf{y} + \mathbf{x}^T \mathbf{x} + \lambda \left(\mathbf{a}_i^T \mathbf{y} - b_i \right) . \end{aligned} \qquad (4.2\text{-}13)$$

By taking the gradient with respect to $\mathbf{y}$, we get

$$\nabla J = 2\mathbf{y} - 2\mathbf{x} + \lambda \mathbf{a}_i = \mathbf{0}, \qquad (4.2\text{-}14)$$

which yields

$$\mathbf{y} = \mathbf{x} - \frac{\lambda}{2} \mathbf{a}_i. \qquad (4.2\text{-}15)$$

Since $\langle \mathbf{a}_i, \mathbf{y} \rangle = b_i$ it follows from Eq. (4.2-15) that

$$\left\langle \mathbf{a}_i, \left(\mathbf{x} - \frac{\lambda}{2} \mathbf{a}_i \right) \right\rangle = b_i. \qquad (4.2\text{-}16)$$

Solving for $\lambda/2$ and using this value in Eq. (4.2-15) yields Eq. (4.2-12).

Having answered the two questions raised above, we are ready to apply the projection algorithm as described in Section 2.6 to solve the problem[‡]. More specifically, starting with an arbitrary initial value $\mathbf{x}_0$, the iteration [§]

$$\mathbf{x}_{k+1} = P_m \cdots P_2 P_1 \, \mathbf{x}_k, \qquad k = 0, 1, 2, \cdots \qquad (4.2\text{-}17)$$

will converge to a point in the solution set C_0 defined in Eq. (4.2-4), provided that it is not empty, i.e., when the system of linear equations is consistent. The final convergence point, say $\mathbf{x}^*$, will be a solution to the system of linear equations. We remark that the iteration in Eq. (4.2-17) does converge strongly, since the Euclidean space R^n is finite dimensional.

When the iteration in Eq. (4.2-17) reaches its convergence point $\mathbf{x}^*$, the following

---

[†]The result given in Eq. (4.2-12) was first derived by Kaczmarz and published in 1937. See [3, 4] for references and further details.
[‡]See Eqs. (2.6-1) and (2.6-2).
[§]The order of the operators is immaterial; we could have just as well written $\mathbf{x}_{k+1} = P_1 P_2 \cdots P_m \, \mathbf{x}_k$.

identity holds

$$\mathbf{x}^* = P_1\mathbf{x}^* = P_2\mathbf{x}^* = \cdots = P_m\mathbf{x}^*. \qquad (4.2\text{-}18)$$

This is true since the projection theory guarantees that $\mathbf{x}^*$ belongs to C_0, which implies that $\mathbf{x}^*$ belongs to each one of the sets $C_i, i = 1, 2, \cdots, m$, which is the case when the set C_0 is not empty, i.e., when the system of linear equations is consistent. On the other hand, if the system of linear equations is inconsistent, no point in R^n will satisfy the relation in Eq. (4.2-18). Therefore, the algorithm in Eq. (4.2-17) will fail to converge to a point in C_0.

4.3   CONVERGENCE ANALYSIS

In this section we study the numerical behavior of the algorithm in Eq. (4.2-17) for solving a system of linear equations. To best visualize the result, let's start with the following simultaneous equations in two variables

$$\begin{aligned} a_{11}x_1 + a_{12}x_2 &= b_1 \\ a_{21}x_1 + a_{22}x_2 &= b_2. \end{aligned} \qquad (4.3\text{-}1)$$

As pointed earlier, these two equations represent two straight lines in R^2. Furthermore, these two lines can be described in the form of solution sets

$$C_1 = \{\mathbf{x} : a_{11}x_1 + a_{12}x_2 = b_1\}, \qquad (4.3\text{-}2)$$

and

$$C_2 = \{\mathbf{x} : a_{21}x_1 + a_{22}x_2 = b_2\}, \qquad (4.3\text{-}3)$$

where $\mathbf{x} = (x_1, x_2)^T$.

First, if the system in Eq. (4.3-1) is inconsistent, then the two lines of the equations are parallel to each other in R^2. In this case, the two solution sets C_1 and C_2 will have no point in common, that is, $C_1 \cap C_2 = \emptyset$. The algorithm in Eq. (4.2-17) would fail to converge in the sense that Eq. (4.2-18) is not satisfied, as illustrated in Fig. 4.3-1(a). On the other hand, if the system in Eq. (4.3-1) is consistent with infinitely many solutions, then the two lines coincide with each other. In this case, the algorithm in Eq. (4.2-17) will find a solution in a single step, as illustrated in Fig. 4.3-1(b).

The most interesting case, however, is when the system has exactly one solution. Then, the algorithm in Eq. (4.2-17) will ultimately converge to this solution no matter what the initial starting point is, as illustrated in Fig. 4.3-1(c). In the following, we analyze the convergence behavior for this case.

Let α denote the acute angle between these two straight lines, and $\mathbf{x}^*$ denote the solution point, as illustrated in Fig. 4.3-2. From Fig. 4.3-2,

$$\|\mathbf{x}_{k+1} - \mathbf{x}^*\| = \|\mathbf{x}'_{k+1} - \mathbf{x}^*\| \cos \alpha = \|\mathbf{x}_k - \mathbf{x}^*\| \cos^2 \alpha. \qquad (4.3\text{-}4)$$

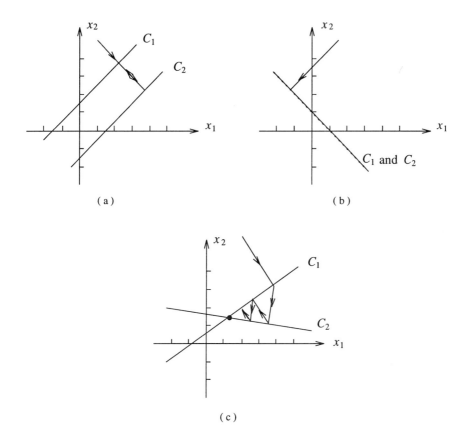

Fig. 4.3-1 Behaviors of the projection algorithm for solving a system of linear equations: (a) since C_1 doesn't intersect C_2 there is no solution; (b) when C_1 coincides with C_2, the solution depends on the starting point and is obtained in a single iteration; (c) when the system has only one solution, convergence is independent of the starting point.

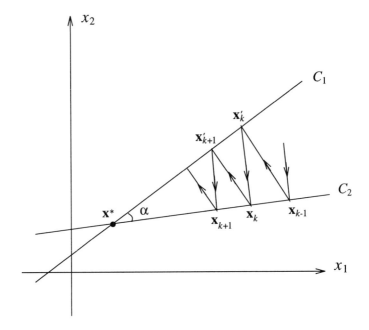

Fig. 4.3-2   The relation between angle α and the iterates. A small value of α leads to slow convergence.

Or, equivalently,

$$\frac{\|\mathbf{x}_{k+1} - \mathbf{x}^*\|}{\|\mathbf{x}_k - \mathbf{x}^*\|} = \cos^2 \alpha. \tag{4.3-5}$$

Furthermore,

$$
\begin{aligned}
\|\mathbf{x}_{k+1} - \mathbf{x}_k\| &= \|\mathbf{x}_k - \mathbf{x}^*\| - \|\mathbf{x}_{k+1} - \mathbf{x}^*\| \\
&= (1 - \cos^2 \alpha)\|\mathbf{x}_k - \mathbf{x}^*\|,
\end{aligned}
\tag{4.3-6}
$$

and

$$
\begin{aligned}
\|\mathbf{x}_k - \mathbf{x}_{k-1}\| &= \|\mathbf{x}_{k-1} - \mathbf{x}^*\| - \|\mathbf{x}_k - \mathbf{x}^*\| \\
&= \left(\frac{1}{\cos^2 \alpha} - 1\right)\|\mathbf{x}_k - \mathbf{x}^*\|.
\end{aligned}
\tag{4.3-7}
$$

Combining Eqs. (4.3-6) and (4.3-7) yields

$$\frac{\|\mathbf{x}_{k+1} - \mathbf{x}_k\|}{\|\mathbf{x}_k - \mathbf{x}_{k-1}\|} = \cos^2 \alpha. \tag{4.3-8}$$

Equations (4.3-5) and (4.3-8) show that the iteration converges at a constant linear rate which is determined only by α. Indeed, Eq. (4.3-5) indicates that the distance between the iterates and the converging point is always reduced by $\cos^2 \alpha$ times after each iteration. Furthermore, the convergence speed increases as the angle α increases from $0°$ to $90°$, as illustrated by examples in Fig. 4.3-3.

Notice that the acute angle α is determined by the coefficient vectors $\mathbf{a}_1 = (a_{11}, a_{12})^T$ and $\mathbf{a}_2 = (a_{21}, a_{22})^T$ through the following relation:

$$\cos \alpha = \frac{|\langle \mathbf{a}_1, \mathbf{a}_2 \rangle|}{\|\mathbf{a}_1\| \, \|\mathbf{a}_2\|} . \tag{4.3-9}$$

It is clear that when the two coefficient vectors are nearly parallel, i.e., when α is close to $0°$, the algorithm in Eq. (4.2-17) will converge very slowly, since $\cos^2 \alpha$ is close to 1. We can see this from Eq. (4.3-5), where the remaining error after $k + 1$ iterations is essentially the same as that after k iterations, i.e., there is little progress toward the solution $\mathbf{x}^*$ when α is close to $0°$. In this case, the iteration will go through a long "tunnel" before the final solution is reached. This behavior, of course, is undesirable in practical applications. On the other hand, when the two coefficient vectors are nearly orthogonal, i.e., when α is close to $90°$, the algorithm will converge very rapidly, since its convergence rate $\cos^2 \alpha$ is close to 0.

An interesting case is that when α is exactly equal to $90°$, Eq. (4.3-5) indicates that the solution will be reached in just a single iteration! This is indeed the case, as illustrated in Fig. 4.3-4. Moreover, this result holds in a higher-dimensional vector space. Indeed, we can state the following theorem:

Theorem 4.3-1 *If all the row vectors in a system of linear equations are mutually orthogonal, the projection algorithm in Eq. (4.2-17) will reach its solution in a single iteration.*

PROOF: Let $\mathbf{x}_1', \mathbf{x}_2', \cdots, \mathbf{x}_m'$ denote the iterates generated by successively projecting onto the sets $C_1, C_2, \cdots,$ and C_m, starting from an arbitrary starting point $\mathbf{x}_0$. From Eq. (4.2-12) it is clear that the projection $P_i \mathbf{y}$ of an arbitrary vector $\mathbf{y}$ onto C_i can be written as $P_i \mathbf{y} = \mathbf{y} - \beta_i \mathbf{a}_i$ for some scalar β_i which depends on $\mathbf{y}$. Thus, we have the following

$$
\begin{aligned}
\mathbf{x}_1' &= P_1 \mathbf{x}_0 = \mathbf{x}_0 - \beta_1 \mathbf{a}_1 \\
\mathbf{x}_2' &= P_2 \mathbf{x}_1' = \mathbf{x}_1' - \beta_2 \mathbf{a}_2 \\
&= \mathbf{x}_0 - \beta_1 \mathbf{a}_1 - \beta_2 \mathbf{a}_2 \\
\mathbf{x}_3' &= P_3 \mathbf{x}_2' = \mathbf{x}_2' - \beta_3 \mathbf{a}_3 \\
&= \mathbf{x}_0 - \beta_1 \mathbf{a}_1 - \beta_2 \mathbf{a}_2 - \beta_3 \mathbf{a}_3 \\
&\quad\vdots \\
\mathbf{x}_m' &= P_m \mathbf{x}_{m-1}' = \mathbf{x}_{m-1}' - \beta_m \mathbf{a}_m \\
&= \cdots = \mathbf{x}_0 - \sum_{i=1}^{m} \beta_i \mathbf{a}_i .
\end{aligned}
$$

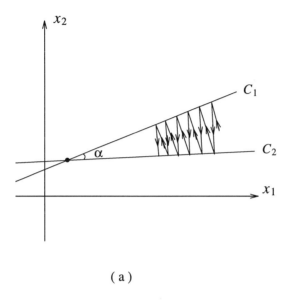

(a)

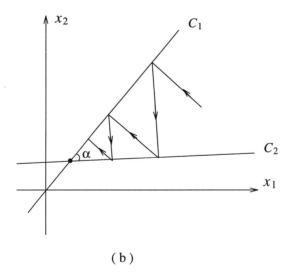

(b)

Fig. 4.3-3 The rate of convergence of the projection algorithm depends on the acute angle α between the set C_1 and C_2: (a) a small α requires many iterations to get close to the solution; (b) a large α requires few iterations.

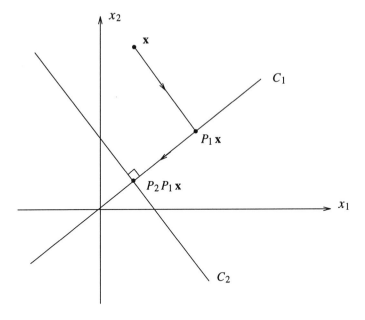

Fig. 4.3-4 When the two sets C_1 and C_2 are orthogonal to each other, the projection algorithm can reach the solution after a single iteration.

Let $\mathbf{x}_1 = \mathbf{x}'_m$. We claim that $\mathbf{x}_1$ is a solution to the system of linear equations. Indeed, consider the ith equation

$$\langle \mathbf{x}, \mathbf{a}_i \rangle = b_i \qquad (4.3\text{-}10)$$

in the system. Since all the row vectors are mutually orthogonal to each other, i.e., $\langle \mathbf{a}_i, \mathbf{a}_j \rangle = 0$ for $i \neq j$, we have

$$
\begin{aligned}
\langle \mathbf{x}_1, \mathbf{a}_i \rangle &= \left\langle \mathbf{x}_0 - \sum_{k=1}^{m} \beta_k \mathbf{a}_k, \mathbf{a}_i \right\rangle \\
&= \left\langle \mathbf{x}_0 - \sum_{k=1}^{i} \beta_k \mathbf{a}_k, \mathbf{a}_i \right\rangle - \left\langle \sum_{k=i+1}^{m} \beta_k \mathbf{a}_k, \mathbf{a}_i \right\rangle \\
&= \left\langle \mathbf{x}_0 - \sum_{k=1}^{i} \beta_k \mathbf{a}_k, \mathbf{a}_i \right\rangle \\
&= \langle \mathbf{x}'_i, \mathbf{a}_i \rangle = b_i,
\end{aligned}
$$

since $\mathbf{x}'_i$ itself is an element of C_i. In the second line the term on the extreme right is zero because of orthogonality of the $\mathbf{a}_i$'s. Repeating the argument for $i = 1, 2, \cdots, m$ leads to the theorem. Since $\mathbf{x}_1 = P_m \cdots P_2 P_1 \, \mathbf{x}_0$, convergence occurs after only one full iteration. ∎

4.4 CONVERGENCE ACCELERATION

Theorem 4.3-1 gives a promising result for applying the projection algorithm to solve systems of linear equations. It suggests that in practical applications, where the speed of convergence is usually of concern, every effort be made such that row vectors in a system of linear equations are mutually orthogonal. Unfortunately, such orthogonality rarely happens by chance. The question then is: can we still benefit from this result, especially when some of the row vectors are nearly parallel to each other? Before we answer this question, let us review the *Gram-Schmidt process* first introduced in Chapter 1. We restate the algorithm here for the reader's convenience.

The Gram-Schmidt process is a step-by-step procedure for constructing an orthogonal basis from an existing non-orthogonal basis for a vector space. Let $\{\mathbf{u}_1, \mathbf{u}_2, \cdots, \mathbf{u}_n\}$ be a basis for R^n. According to the Gram-Schmidt process the following sequence of steps will produce an orthogonal basis $\{\mathbf{v}_1, \mathbf{v}_2, \cdots, \mathbf{v}_n\}$ for R^n.

1. Let $\mathbf{v}_1 = \mathbf{u}_1$.

2. For $i = 2, 3, \cdots, n$, let

$$\mathbf{v}_i = \mathbf{u}_i - \sum_{j=1}^{i-1} \frac{\langle \mathbf{u}_i, \mathbf{v}_j \rangle}{\|\mathbf{v}_j\|^2} \mathbf{v}_j. \tag{4.4-1}$$

The fact that the sequence $\{\mathbf{v}_1, \mathbf{v}_2, \cdots, \mathbf{v}_n\}$ generated through the Gram-Schmidt process are mutually orthogonal can be easily established. First, from Eq. (4.4-1), we have

$$\langle \mathbf{v}_i, \mathbf{v}_1 \rangle = \langle \mathbf{u}_i, \mathbf{v}_1 \rangle - \sum_{j=1}^{i-1} \frac{\langle \mathbf{u}_i, \mathbf{v}_j \rangle}{\|\mathbf{v}_j\|^2} \langle \mathbf{v}_j, \mathbf{v}_1 \rangle \tag{4.4-2}$$

for $i = 2, 3, \cdots, n$. Letting $i = 2$ in Eq. (4.4-2) clearly establishes that $\langle \mathbf{v}_2, \mathbf{v}_1 \rangle = 0$, from which it can be induced that $\langle \mathbf{v}_3, \mathbf{v}_1 \rangle = 0, \cdots, \langle \mathbf{v}_n, \mathbf{v}_1 \rangle = 0$. Thus the vector $\mathbf{v}_1$ is orthogonal to the rest of $\mathbf{v}_i$'s. Second, consider that

$$
\begin{aligned}
\langle \mathbf{v}_i, \mathbf{v}_2 \rangle &= \langle \mathbf{u}_i, \mathbf{v}_2 \rangle - \sum_{j=1}^{i-1} \frac{\langle \mathbf{u}_i, \mathbf{v}_j \rangle}{\|\mathbf{v}_j\|^2} \langle \mathbf{v}_j, \mathbf{v}_2 \rangle \\
&= \langle \mathbf{u}_i, \mathbf{v}_2 \rangle - \sum_{j=2}^{i-1} \frac{\langle \mathbf{u}_i, \mathbf{v}_j \rangle}{\|\mathbf{v}_j\|^2} \langle \mathbf{v}_j, \mathbf{v}_2 \rangle
\end{aligned}
$$

for $i = 3, 4, \cdots, n$. It follows that $\langle \mathbf{v}_3, \mathbf{v}_2 \rangle = 0$, from which it can be induced that $\langle \mathbf{v}_4, \mathbf{v}_2 \rangle = 0, \cdots, \langle \mathbf{v}_n, \mathbf{v}_2 \rangle = 0$. Thus the vector $\mathbf{v}_2$ is orthogonal to the rest of $\mathbf{v}_i$'s. Similarly, it can be established that $\mathbf{v}_3, \mathbf{v}_4, \cdots, \mathbf{v}_{n-1}$ are all orthogonal to the rest of the $\mathbf{v}_i$'s. Therefore, the sequence $\{\mathbf{v}_1, \mathbf{v}_2, \cdots, \mathbf{v}_n\}$ produced as in Eq. (4.4-2) indeed forms an orthogonal basis for R^n.

We now return to our earlier problem. Consider a system of linear equations with row vectors $\{\mathbf{a}_1, \mathbf{a}_2, \cdots, \mathbf{a}_m\}$, as in Eq. (4.1-10). It is true that these vectors, in general, will not be a basis for the space R^n. From the proof given above it is clear the Gram-Schmidt process nonetheless will still produce a set of orthogonal vectors, say $\{\mathbf{a}_1', \mathbf{a}_2', \cdots, \mathbf{a}_m'\}$. Therefore we can generate a new system of linear equations

$$
\begin{aligned}
\langle \mathbf{a}_1', \mathbf{x} \rangle &= b_1' \\
\langle \mathbf{a}_2', \mathbf{x} \rangle &= b_2' \\
&\ \ \vdots \\
\langle \mathbf{a}_m', \mathbf{x} \rangle &= b_m',
\end{aligned}
\tag{4.4-3}
$$

through the following process:

1. Let $\mathbf{a}_1' = \mathbf{a}_1$, and $b_1' = b_1$.

2. For $i = 2, 3, \cdots, m$, let

$$
\mathbf{a}_i' = \mathbf{a}_i - \sum_{j=1}^{i-1} \frac{\langle \mathbf{a}_i, \mathbf{a}_j' \rangle}{\|\mathbf{a}_j'\|^2} \mathbf{a}_j',
\tag{4.4-4}
$$

and

$$
b_i' = b_i - \sum_{j=1}^{i-1} \frac{\langle \mathbf{a}_i, \mathbf{a}_j' \rangle}{\|\mathbf{a}_j'\|^2} b_j'.
\tag{4.4-5}
$$

The operations carried out in Eqs. (4.4-4) and (4.4-5) are elementary row operations (i.e., scaling, addition or subtraction) on the system of linear equations in Eq. (4.1-1). Therefore, the resulting system of linear equations in Eq. (4.4-3) is equivalent to that in Eq. (4.1-1), in the sense that they have the same solution set in R^n.

Note that if the set of vectors $\{\mathbf{a}_1, \mathbf{a}_2, \cdots, \mathbf{a}_m\}$ is not linearly independent, then the operation in Eq. (4.4-4) will produce a zero vector (see Exercise 4-2). If for some i the resulting vector $\mathbf{a}_i'$ is zero while b_i' is non-zero, the system of linear equations must be inconsistent. On the other hand, if both $\mathbf{a}_i'$ and b_i' are zero, the system of linear equations must have a redundant equation. When this happens we can simply drop it out without changing the system.

Based on the *modified system of linear equations* in Eq. (4.4-3), we obtain a new formulation of the problem in the form of convex constraint sets. More specifically, define

$$
C_i' = \{\mathbf{y} : \langle \mathbf{a}_i', \mathbf{y} \rangle = b_i'\},
\tag{4.4-6}
$$

for $i = 1, 2, \cdots, m$. Then the new solution set

$$
C_0' = \bigcap_{i=1}^{m} C_i'
\tag{4.4-7}
$$

will be identical to that defined in Eq. (4.2-4), as reasoned above. Therefore, the projection algorithm in Eq. (4.2-17) can be applied to find the solution. The difference this time, however, is that it will always find the solution after merely a single iteration, no matter what its starting point x_0 is!

Comparing the sets in Eq. (4.2-2) with those in Eq. (4.4-6), we see that what the Gram-Schmidt process does in Eq. (4.4-4) and Eq. (4.4-5) is simply to change the geometrical configuration of the constraint sets such that the normal vectors of the resulting new sets are mutually orthogonal to each other. More specifically, each set C_i is modified to generate a new set C_i' such that its normal vector is orthogonal to those of all the other new sets. The Gram-Schmidt process guarantees that the new sets C_i' have the same intersection set as the old ones, so that the new formulation solves the same problem as the old one does.

One important class of problems encountered in practice is the so-called large-scale system, where n, the number of unknowns, is very large, say in the order of thousands or even tens of thousands. Such large numbers of unknowns occur quite frequently in image processing, especially medical image processing such as *positron-emission tomography, computer-aided tomography,* or *magnetic resonance imaging.* It is typical in such problems that the elements of the row vectors in the matrix $\mathbf{A}$, are known by some formula and so can be generated as needed. Moreover, often only a small portion of the n coefficients in a row vector $\mathbf{a}_i$ have non-zero values. In other words, the coefficient matrix $\mathbf{A}$ is a *large-scale sparse matrix,* for which it is inefficient, if not impossible, to store its elements in memory in a computer-aided algorithm. In such a case, the projection algorithm can be applied to its advantage. The computation of the projection onto each set can be done by taking advantage of the fact that only a small portion of the components in each vector $\mathbf{a}_i$ are non-zero. Indeed, in obtaining $P_i\mathbf{x}$ in Eq. (4.2-12), only those components of $\mathbf{x}$ corresponding to non-zero components of the vector $\mathbf{a}_i$ need to be updated. Furthermore, in computing the inner product $\langle \mathbf{a}_i, \mathbf{y} \rangle$ multiplications are only needed for non-zero components of $\mathbf{a}_i$.

From the analysis in Section 4.3, it is seen that if some of the row vectors in a system of equations are nearly parallel, the convergence of the projection algorithm in Eq. (4.2-17) will be very slow. In such a case, a partial Gram-Schmidt process can be applied only to those vectors that are nearly parallel (if they can be identified). The convergence rate for the new resulting system of linear equations should be significantly improved.

We conclude this section by presenting a numerical example to illustrate the effect of relative angles between row vectors, i.e., the angles between the normal vectors of the constraint sets, on the convergence rate of the projection algorithm. For this example, consider the system of linear equations given in the following:

$$
\begin{aligned}
x_1 - 2x_2 + x_3 + 4x_4 + x_5 &= -536 \\
3x_1 + x_2 + 2x_3 - 6x_4 + x_5 &= -1089 \\
100x_1 - 7x_2 - 60x_3 + 2x_4 - x_5 &= 9343 \\
5x_1 - 2x_2 - 16x_3 + 8x_4 + 3x_5 &= -2294 \\
99x_1 - 8x_2 - 59x_3 - 3x_4 + x_5 &= 6924 \ ,
\end{aligned}
\tag{4.4-8}
$$

which has exactly one solution $\mathbf{x}^* = (100, -1, 30, 78, -980)^T$. The coefficient matrix of Eq. (4.4-8) is

$$
\mathbf{A} = \begin{pmatrix}
1 & -2 & 1 & 4 & 1 \\
3 & 1 & 2 & -6 & 1 \\
100 & -7 & -60 & 2 & -1 \\
5 & -2 & -16 & 8 & 3 \\
99 & -8 & -59 & -3 & 1
\end{pmatrix}.
\tag{4.4-9}
$$

It is clear that the 3rd and the 5th row vectors are very close to each other. As a matter of fact the angle between them is merely 2.717 degrees. We can expect the convergence of a projection algorithm for this system to be very slow. On the other hand, we can apply the Gram-Schmidt process to obtain a new system of equations, such that the projection can reach the solution in a single step. Here, however, we illustrate that even a partial application of the Gram-Schmidt process can improve the convergence significantly.

In this problem, we could apply the Gram-Schmidt process to the 3rd and 5th row vectors in Eq. (4.4-9) to obtain two orthogonal vectors. However, since they are nearly the same in length and direction, their difference vector would be nearly orthogonal to either of them. Therefore, we can simply use their difference to obtain a new equation with coefficient vector nearly orthogonal to either. By subtracting the 3rd equation from the 5th, we obtain a new system of equations:

$$
\begin{aligned}
x_1 - 2x_2 + x_3 + 4x_4 + x_5 &= -536 \\
3x_1 + x_2 + 2x_3 - 6x_4 + x_6 &= -1089 \\
100x_1 - 7x_2 - 60x_3 + 2x_4 - x_5 &= 9343 \\
5x_1 - 2x_2 - 16x_3 + 8x_4 + 3x_5 &= -2294 \\
-x_1 - x_2 + x_3 - 5x_4 + 2x_5 &= -2419 \,,
\end{aligned}
\tag{4.4-10}
$$

which, of course, has the same solution as in Eq. (4.4-8).

To compare the convergence rate, the projection algorithm is applied to both the system in Eq. (4.4-8) and the system in Eq. (4.4-10). To measure the accuracy of the iterate $\mathbf{x}_k$ during the kth iteration, the following relative error is used:

$$
e(\mathbf{x}_k, \mathbf{x}^*) = \frac{\|\mathbf{x}_k - \mathbf{x}^*\|}{\|\mathbf{x}^*\|}.
\tag{4.4-11}
$$

To minimize the effect of initial starting point on the convergence, the algorithm is started using the following three arbitrarily chosen initial points for both systems: (i) $\mathbf{x}_0 = (0, 0, 0, 0, 0)^T$; (ii) $\mathbf{x}_0 = (1000, 1000, 1000, 1000, 1000)^T$; and (iii) $\mathbf{x}_0 = (-20, 50, 391, 3211, 2)^T$.

The projection algorithm in Eq. (4.2-17) is first tested on the system in Eq. (4.4-8). In order to demonstrate the convergence of the algorithm, the relative error $e(\mathbf{x}_k, \mathbf{x}^*)$ is computed as the iteration progresses. The results are summarized in Table 4.4-1 for iterates obtained at every other 1000 iterations for each initial condition. Since the relative errors do not directly reflect where the actual iterates are, we also furnish, in Table 4.4-2, the values of the iterate x_k at certain iterations, To save space, only the result for the first initial condition $x_0 = (0,0,0,0,0)^T$ is given.

Table 4.4-1 List of relative errors of the numerical iterates for the system in Eq. (4.4-8) for three different initial conditions: (i) $x_0 = (0,0,0,0,0)^T$; (ii) $x_0 = (1000,1000,1000,1000,1000)^T$; and (iii) $x_0 = (-20,50,391,3211,2)^T$

Iterations	(i)	(ii)	(iii)
0	1.000000	2.774430	3.343677
1000	0.148052	1.474023	0.986837
2000	0.111165	1.106769	0.740966
3000	0.083468	0.831017	0.556354
4000	0.062672	0.623968	0.417738
5000	0.047057	0.468506	0.313658
6000	0.035333	0.351777	0.235510
7000	0.026530	0.264132	0.176832
8000	0.019920	0.198323	0.132774

Then the algorithm is tested on the modified system in Eq. (4.4-10). Similar results are summarized in Table 4.4-3 and Table 4.4-4. Clearly, a simple modification in the second case has dramatically improved the convergence rate of the algorithm, as expected. Furthermore, these results substantially agree with the analysis given in this section.

By examining the results furnished in these examples, it seems that it often takes many iterations for the projection algorithm to reach convergence. It is useful to recall that the coefficients in A and b are rarely known exactly in many practical problems. Even if we have an exact system of linear equations, it is almost inevitable that roundoff errors will be introduced during the numerical solution in the computer. Therefore, it may be pointless to run the algorithm till its hypothetical convergence point is found. For example, one can simply stop the iteration if the distance between two successively generated iterates is found to be less than some prescribed level.

Before ending this section, we raise a question which a careful reader may have considered. In Section 4.2 the iterative algorithm in $x_{k+1} = P_n \cdots P_2 P_1 x_k$ was introduced to solve a system of linear equations directly. In this section, it has been suggested that the Gram-Schmidt process be applied, wholly or partially, to modify a system of linear equations so that faster convergence is possible. Clearly, as long as a system of linear equations admits to one solution, both approaches should yield the same solution, since only one solution is admissible. On the other hand, when a system of linear equations has many solutions, which is often the case in practical applications, will these two approaches still give the same solution? Surprisingly, the answer is yes, a result that we demonstrate in the next section.

Table 4.4-2 List of the numerical values of the iterates for the system in Eq. (4.4-8) with the first initial condition $x_0 = (0,0,0,0,0)^T$. The correct solution to the system is $x^* = (100, -1, 30, 78, -980)^T$

Iterations	x_1	x_2	x_3	x_4	x_5
0	0.000000	0.000000	0.000000	0.000000	0.000000
2000	117.016275	97.026146	44.011322	112.117893	−951.380373
4000	109.593340	54.264630	37.899225	97.234794	−963.864987
6000	105.408479	30.156782	34.453380	88.844084	−970.903493
8000	103.049162	16.565396	32.510702	84.113617	−974.871623
10000	101.719039	8.902920	31.415469	81.446700	−977.108752
12000	100.969150	4.583013	30.798005	79.943161	−978.369989
14000	100.546382	2.147559	30.449895	79.095505	−979.081041
16000	100.308036	0.774513	30.253639	78.617617	−979.481915
18000	100.173663	0.000425	30.142995	78.348197	−979.707917

Table 4.4-3 List of relative errors of the numerical iterates for the system in Eq. (4.4-10) for three different initial conditions: (i) $x_0 = (0,0,0,0,0)^T$; (ii) $x_0 = (1000, 1000, 1000, 1000, 1000)^T$; and (iii) $x_0 = (-20, 50, 391, 3211, 2)^T$

Iterations	(i)	(ii)	(iii)
0	1.000000	2.774430	3.343677
50	0.168992	0.570323	0.341822
100	0.042551	0.143602	0.086068
150	0.010714	0.036158	0.021671
200	0.002698	0.009104	0.005457
250	0.000679	0.002292	0.001374
300	0.000171	0.000577	0.000346
350	0.000043	0.000145	0.000087
400	0.000011	0.000037	0.000022

Table 4.4-4   List of the numerical values of the iterates for the system in Eq. (4.4-10) with the first initial condition $\mathbf{x}_0 = (0, 0, 0, 0, 0)^T$

Iterations	x_1	x_2	x_3	x_4	x_5
0	0.000000	0.000000	0.000000	0.000000	0.000000
100	102.019318	28.455413	37.308334	84.496716	−951.675011
200	100.128022	0.867435	30.463339	78.411883	−978.204232
300	100.008116	−0.881607	30.029375	78.026113	−979.886151
400	100.000515	−0.992494	30.001862	78.001656	−979.992782
500	100.000033	−0.999524	30.000118	78.000105	−979.999542
600	100.000002	−0.999970	30.000007	78.000007	−979.999971
700	100.000000	−0.999998	30.000000	78.000000	−979.999998
800	100.000000	−1.000000	30.000000	78.000000	−980.000000

4.5   MINIMUM DISTANCE PROPERTY

As discussed in Section 4.1, the equation $\mathbf{Ax} = \mathbf{b}$ offers a handy way of describing a system of linear equations. From the point of view of linear transformation, it states that the matrix $\mathbf{A}$ transforms an unknown vector $\mathbf{x}$ in R^n into the vector $\mathbf{b}$. Therefore, if we define a set C in R^n as

$$C = \{\mathbf{y} : \mathbf{Ay} = \mathbf{b}\}, \tag{4.5-1}$$

then the set C contains all the solutions to the problem. On the other hand, from the convex set formulation in Section 4.2 it is clear that the set C is identical to the set $C_0 = \bigcap_{i=1}^m C_i$, where the C_i's are the solution sets for each individual equations. Since each set C_i is both convex and closed, it follows that C is also closed and convex. In fact, C is a closed linear variety (see Exercise 4-3). An interesting question arises: Can we directly project onto this set so that the solution can be reached in a single step? The answer is yes, provided that the projection onto C can be computed.

Notice that the set C is empty if a system of linear equations is inconsistent. In such a case the projection onto C is simply meaningless. Therefore, in the following, only consistent systems of linear equations are considered.

To derive the projection $P\mathbf{x}$ of an arbitrary vector $\mathbf{x}$ onto C, we can apply the Lagrange multipliers technique (see Exercise 4-4). Here, however, we consider a more intuitive approach. Recall that each equation $\langle \mathbf{a}_i, \mathbf{y} \rangle = b_i$ represents a hyperplane in R^n with normal vector $\mathbf{a}_i$. For any two vectors $\mathbf{y}_1$ and $\mathbf{y}_2$ on the hyperplane, i.e., $\langle \mathbf{a}_i, \mathbf{y}_1 \rangle = \langle \mathbf{a}_i, \mathbf{y}_2 \rangle = b_i$, we have $\langle \mathbf{a}_i, \mathbf{y}_1 - \mathbf{y}_2 \rangle = 0$. That is, the normal vector $\mathbf{a}_i$ is orthogonal to the difference vector $\mathbf{y}_1 - \mathbf{y}_2$ for any two vectors $\mathbf{y}_1$ and $\mathbf{y}_2$ in the set C_i. Therefore, if we take any two vectors $\mathbf{y}_1$ and $\mathbf{y}_2$ in the set C_0, we have $\langle \mathbf{a}_i, \mathbf{y}_1 - \mathbf{y}_2 \rangle = 0$ for every $i = 1, 2, \cdots, m$. That is, the normal vector $\mathbf{a}_i$ of each hyperplane is now also normal to the set C_0, which is a linear variety. On the other hand, it can be shown that any vector that is normal to C_0 should be a linear combination of the vectors $\mathbf{a}_i$ (see Exercise 4-5). Geometrically, then, if $P\mathbf{x}$ is the projection of $\mathbf{x}$ onto C_0, the vector $P\mathbf{x} - \mathbf{x}$ should be normal to C_0. Thus, we have

$$P\mathbf{x} - \mathbf{x} = \sum_{i=1}^{m} \alpha_i \mathbf{a}_i, \qquad (4.5\text{-}2)$$

for some constants $\alpha_i, i = 1, 2, \cdots, m$. These constants are to be determined from the condition that $P\mathbf{x} \in C_0$.

Let $\boldsymbol{\alpha}$ be the vector[†] of these unknown constants α_i's, i.e.,

$$\boldsymbol{\alpha} = (\alpha_1, \alpha_2, \cdots, \alpha_m)^T. \qquad (4.5\text{-}3)$$

Then the following identity holds:

$$\sum_{i=1}^{m} \alpha_i \mathbf{a}_i = \mathbf{A}^T \boldsymbol{\alpha}. \qquad (4.5\text{-}4)$$

Substituting this result into Eq. (4.5-2), we obtain

$$P\mathbf{x} = \mathbf{x} + \mathbf{A}^T \boldsymbol{\alpha}, \qquad (4.5\text{-}5)$$

which yields

$$\mathbf{A}\left(\mathbf{x} + \mathbf{A}^T \boldsymbol{\alpha}\right) = \mathbf{b}, \qquad (4.5\text{-}6)$$

since $P\mathbf{x} \in C$. Therefore,

$$\mathbf{A}\mathbf{A}^T \boldsymbol{\alpha} = \mathbf{b} - \mathbf{A}\mathbf{x}. \qquad (4.5\text{-}7)$$

Here the matrix product $\mathbf{A}\mathbf{A}^T$ is an $m \times m$ matrix. Assuming that it is invertible, we can solve for $\boldsymbol{\alpha}$ from Eq. (4.5-7) as follows

$$\boldsymbol{\alpha} = -\left[\mathbf{A}\mathbf{A}^T\right]^{-1}(\mathbf{A}\mathbf{x} - \mathbf{b}), \qquad (4.5\text{-}8)$$

which gives the projection

$$P\mathbf{x} = \mathbf{x} - \mathbf{A}^T\left[\mathbf{A}\mathbf{A}^T\right]^{-1}(\mathbf{A}\mathbf{x} - \mathbf{b}). \qquad (4.5\text{-}9)$$

Note that in deriving this result we assumed that the inverse of the matrix $\mathbf{A}\mathbf{A}^T$ exists. A natural question is what happens if it does not. To answer this question, let's first examine when this could happen. It can be shown that the matrix $\mathbf{A}\mathbf{A}^T$ is invertible if and only if all the m row vectors of $\mathbf{A}$ are linearly independent (see Exercise 4-6). In other words, $\mathbf{A}\mathbf{A}^T$ is non-invertible if and only if the row vectors of $\mathbf{A}$ are linearly dependent. If the row vectors of $\mathbf{A}$ are not linearly independent,

---

[†]Note that $\boldsymbol{\alpha}$ here is meant to have no relation to the angle between two hyperplanes introduced in Section 4.4.

Eq. (4.5-2) will hold for many solutions of α_i's, all of which will give the same vector $P\mathbf{x} - \mathbf{x}$. This is true because the projection $P\mathbf{x}$ is unique. To find one of these many solutions, one could simply use its pseudo-inverse such as the Penrose inverse [4, 6] in Eq. (4.5-8). Thus whether $\mathbf{A}\mathbf{A}^T$ is invertible or not, the projection onto C is readily computable. The non-invertible case is discussed in [3].

Having obtained the projection onto the set C, we can apply the projection algorithm to find the solution. Since only one set is involved in the algorithm, the solution will be reached in just a single step. More specifically, we have the following algorithm:

1. Choose an arbitrary initial starting point $\mathbf{x}_0$.

2. Find the solution $\mathbf{x}^*$ by computing

$$\mathbf{x}^* = \mathbf{x}_0 - \mathbf{A}^T \left[\mathbf{A}\mathbf{A}^T\right]^{-1} (\mathbf{A}\mathbf{x}_0 - \mathbf{b}). \tag{4.5-10}$$

The algorithm in Eq. (4.5-10) seems rather attractive, since it always yields the solution in a single step, unlike the algorithm introduced in Section 4.2, in which one-step convergence is only possible when all the row vectors are orthogonal, which happens rather rarely in practice. Unfortunately, it does have a limitation in practical applications; namely, in a large-scale problem the dimension of the matrix $\mathbf{A}\mathbf{A}^T$ is so large that it is hardly practical, if not impossible, to numerically compute the inverse. This does not, however, imply that the algorithm suggested in Eq. (4.5-10) is useless. Quite the contrary, this algorithm helps us to gain invaluable insight into some interesting properties of projection algorithms for solving systems of linear equations. These properties might be difficult to discern otherwise. We discuss this point in greater detail in what follows.

First, let's answer the question that we raised at the end of the previous section, namely, does modifying the original system of equations cause the POCS solution to vary if the system is underdetermined? Consider the iterative POCS algorithm $\mathbf{x}_{k+1} = P_m \cdots P_2 P_1 \mathbf{x}_k$. As pointed out earlier, each set C_i involved in this algorithm is a closed linear variety. From Corollary 2.5-1, it is clear that this algorithm will converge to the same point as the one-step projection onto their intersection set C_0. More specifically, for an arbitrary initial point $\mathbf{x}_0$, the algorithm will converge to the point $\mathbf{x}^*$ in Eq. (4.5-10). On the other hand, if a system of linear equations is modified by either the Gram-Schmidt process or other elementary operations, the resulting new system of linear equations will always have the same solution set as its original system. Therefore, the projection algorithms will always yield the same solution, no matter whether the system is modified or not.

This observation reveals a rather remarkable property of the projection algorithms for solving a system of linear equations, that is, the algorithm converges to the point $\mathbf{x}^*$ in the solution set, which is always closest to its initial starting point $\mathbf{x}_0$, a property typically enjoyed only by a one-step projection algorithm. In the following, we will refer to this as the *minimum distance property*. We say that this property is remarkable because it proves useful in practical applications, as exemplified in the following.

First, if the starting point $\mathbf{x}_0$ for the projection algorithm is chosen to be $\mathbf{x}_0 = \mathbf{0}$, then the minimum distance property implies that the algorithm will converge to the vector with the minimum norm, i.e., minimum distance from the origin. More specifically, we have [†]

$$\mathbf{x}^* = \mathbf{A}^T \left[\mathbf{A}\mathbf{A}^T \right]^{-1} \mathbf{b}, \tag{4.5-11}$$

i.e., let $\mathbf{x}_0 = \mathbf{0}$ in Eq. (4.5-10). In practical applications such as signal processing, a vector $\mathbf{x}$ typically represents a signal and its norm is directly related to the energy of the signal. In such a case, the projection algorithm finds the signal given in Eq. (4.5-11), which is characterized by having the minimum energy of all the signals that satisfy the desired property that is imposed upon by a system of linear equations. Conceivably, this property is important in practical applications.

Another reason that the minimum distance property is remarkable is that in some applications minimum distance translates to minimum adjustment to the system. In these applications, the variable vector $\mathbf{x}$ typically represents the status of the system. Possibly due to a change of specification, we might want to change the status of the system such that new specification is met. Suppose that the new specification is described by a system of linear equations in $\mathbf{x}$. Let $\mathbf{x}_{old}$ and $\mathbf{x}^*$ represent the old status and the new status of the system respectively. Then the change in system status is described by the difference between these two vectors, i.e.,

$$\Delta \mathbf{x} = \mathbf{x}^* - \mathbf{x}_{old}. \tag{4.5-12}$$

In such a case, if we start with $\mathbf{x}_0 = \mathbf{x}_{old}$, the projection algorithm will find the new status $\mathbf{x}^*$ such that the change in system status $\Delta \mathbf{x}$ will have the minimum norm, which translates into the projection algorithm finding the minimum adjustment in the system status such that the new specification is satisfied.

Finally, it is interesting to point out that when the matrix $\mathbf{A}$ in the constraint set C in Eq. (4.5-1) is invertible, then the projection solution in Eq. (4.5-10) simply reduces to $\mathbf{x}^* = \mathbf{A}^{-1}\mathbf{b}$, as it should, because, in such a case, the system $\mathbf{A}\mathbf{x} = \mathbf{b}$ has a unique solution.

4.6 A HYBRID ALGORITHM

Having established that the one-step algorithm in Eq. (4.5-10) and the iterative algorithm in Eq. (4.2-17) yield identical solutions, we realize that the former enforces all the constraints imposed by the totality of linear equations in one shot. The advantage of the one-step approach is that the solution can be found in a single operation, while the iterative algorithm enforces the constraints imposed by each equation, one at a time, which has the advantage that the projection can be com-

---

[†]Readers will recognize Eq. (4.5-11) as the *least-squares solution* to the system $\mathbf{A}\mathbf{x} = \mathbf{b}$.

puted easily without matrix inversion. In a sense, the latter approach has employed a "divide and conquer" strategy to break all the constraints into pieces so that each piece is much easier to deal with. Naturally, one wonders if there exists an approach that takes advantage of the best of the both. In this section, we present a hybrid of these two approaches that does so.

Observe that the iterative algorithm in Eq. (4.2-17) arises from the description of a system of linear equations in its "line-by-line" form

$$
\begin{aligned}
\langle \mathbf{a}_1, \mathbf{x} \rangle &= b_1 \\
\langle \mathbf{a}_2, \mathbf{x} \rangle &= b_2 \\
&\vdots \\
\langle \mathbf{a}_m, \mathbf{x} \rangle &= b_m,
\end{aligned}
\tag{4.6-1}
$$

while the one-step algorithm in Eq. (4.5-10) arises from its matrix description form $\mathbf{A}\mathbf{x} = \mathbf{b}$. Mathematically, the "line-by-line" form can be viewed as a result of a row-vector decomposition of the matrix $\mathbf{A}$, i.e.,

$$
\mathbf{A} = \begin{bmatrix} \mathbf{a}_1^T \\ \mathbf{a}_2^T \\ \vdots \\ \mathbf{a}_m^T \end{bmatrix}.
\tag{4.6-2}
$$

In a more general form, the matrix $\mathbf{A}$ can be decomposed into smaller row matrices, each of which consists of the row vectors of the matrix $\mathbf{A}$. That is,

$$
\mathbf{A} = \begin{bmatrix} \mathbf{A}_1 \\ \mathbf{A}_2 \\ \vdots \\ \mathbf{A}_M \end{bmatrix},
\tag{4.6-3}
$$

where each matrix $\mathbf{A}_i$ represents a matrix formed by a group of row vectors of matrix $\mathbf{A}$. Strictly speaking, possible row permutation is also allowed on the matrix $\mathbf{A}$ in obtaining each sub-matrix $\mathbf{A}_i$. To avoid confusion in notation, the same symbol $\mathbf{A}$ is used for the coefficient matrix. In a similar fashion, the vector $\mathbf{b}$ is decomposed into

$$
\mathbf{b} = \begin{bmatrix} \mathbf{b}_1 \\ \mathbf{b}_2 \\ \vdots \\ \mathbf{b}_M \end{bmatrix}.
\tag{4.6-4}
$$

Then the system of linear equations $\mathbf{A}\mathbf{x} = \mathbf{b}$ can be written in equivalent form as

$$
\begin{aligned}
\mathbf{A}_1 \mathbf{x} &= \mathbf{b}_1 \\
\mathbf{A}_2 \mathbf{x} &= \mathbf{b}_2
\end{aligned}
$$

$$\vdots \tag{4.6-5}$$

$$\mathbf{A}_M \mathbf{x} = \mathbf{b}_M.$$

In words, the original system is decomposed into a collection of sub-systems of linear equations. Indeed, if each matrix $\mathbf{A}_i$ contains a single row vector, then Eq. (4.6-5) is identical to its line-by-line form in Eq. (4.6-1). In general, each equation in Eq. (4.6-5) now represents a new sub-system of linear equations. Therefore, we can define a solution set, say C_i, for each one of them, that is,

$$C_i = \{\mathbf{y} : \mathbf{A}_i \mathbf{y} = \mathbf{b}_i\}, \tag{4.6-6}$$

for $i = 1, 2, \cdots, M$. Then their intersection set will be the solution set to the original system. From the argument in the last section, we know that each set C_i is a linear variety. Let P_i denote the projection operator onto C_i. Then the projection algorithm

$$\mathbf{x}_{k+1} = P_M \cdots P_2 P_1 \mathbf{x}_k, \qquad k = 0, 1, 2, \cdots \tag{4.6-7}$$

with $\mathbf{x}_0$ arbitrary will converge to a point in the solution set of the problem. Clearly, the algorithm still preserves that minimum distance property.

We use Eq. (4.5-9) to compute the projection onto the set C_i. Considering $\mathbf{y} \notin C_i$, its projection $P_i \mathbf{y}$ onto C_i is given by

$$
\begin{aligned}
P_i \mathbf{y} &= \mathbf{y} - \mathbf{A}_i^T \left[\mathbf{A}_i \mathbf{A}_i^T \right]^{-1} (\mathbf{A}_i \mathbf{y} - \mathbf{b}_i) \\
&= \mathbf{y} + \mathbf{A}_i^T \left[\mathbf{A}_i \mathbf{A}_i^T \right]^{-1} \mathbf{b}_i - \mathbf{A}_i^T \left[\mathbf{A}_i \mathbf{A}_i^T \right]^{-1} \mathbf{A}_i \mathbf{y}. \tag{4.6-8}
\end{aligned}
$$

As pointed out in last section, a major drawback of the one-step algorithm is that in large-scale problems one is faced with the numerical difficulty of computing the inverse of the matrix product $\left[\mathbf{A}\mathbf{A}^T \right]^{-1}$. The hybrid algorithm in Eq. (4.6-7) will alleviate this difficulty, since we can decompose a large-scale matrix $\mathbf{A}$ into a number of moderate-scale matrices so that the quantity $\left[\mathbf{A}_i \mathbf{A}_i^T \right]^{-1}$ is much easier to handle in computing the projection in Eq. (4.6-8). It would seem that the algorithm $\mathbf{x}_{k+1} = P_M \cdots P_2 P_1 \mathbf{x}_k$ is rather demanding in computation, since it needs the projection in Eq. (4.6-8) in each iteration. Fortunately, for each set C_i the quantities $\mathbf{A}_i^T \left[\mathbf{A}_i \mathbf{A}_i^T \right]^{-1} \mathbf{b}_i$ and $\mathbf{A}_i^T \left[\mathbf{A}_i \mathbf{A}_i^T \right]^{-1} \mathbf{A}_i$ in Eq. (4.6-8) are needed to be computed only once and then reused for the rest of iterations.

An advantage of this approach is that one never needs to deal with matrices that are too large for comfortable inversion. The matrix $\mathbf{A}$ can be decomposed in any desired fashion and the algorithm will always yield the same solution, provided that the same starting point $\mathbf{x}_0$ is used. This is guaranteed by the minimum distance property.

An interesting observation is that we can speed up the convergence of the algorithm by decomposing $\mathbf{A}$ properly. As demonstrated in Section 4.4, an iterative algorithm may converge very slowly if the row vectors of some of the sets involved in the algorithm are nearly parallel to each other. In such a case, if we can group these nearly parallel vectors into one set and apply the one-step projector to this set, then, conceivably, the convergence of algorithm in Eq. (4.6-7) can be improved, since the quasi-parallelism between the sets is removed.

In concluding this section, we give a numerical example to demonstrate the hybrid *iterative-onestep* algorithm in Eq. (4.6-7). For easy comparison with the results presented earlier for the iterative algorithm $\mathbf{x}_{k+1} = P_m \cdots P_2 P_1 \mathbf{x}_k$, the same system of linear equations as in Eq. (4.4-8) is used. First, the system of linear equation is inefficiently decomposed into two sub-systems of linear equations, of which the first one $\mathbf{A}_1^{(1)}\mathbf{x} = \mathbf{b}_1^{(1)}$ is

$$
\begin{aligned}
x_1 - 2x_2 + x_3 + 4x_4 + x_5 &= -536 \\
3x_1 + x_2 + 2x_3 - 6x_4 + x_6 &= -1089 \\
100x_1 - 7x_2 - 60x_3 + 2x_4 - x_5 &= 9343
\end{aligned}
\tag{4.6-9}
$$

and the second one $\mathbf{A}_2^{(1)}\mathbf{x} = \mathbf{b}_2^{(1)}$ is

$$
\begin{aligned}
5x_1 - 2x_2 - 16x_3 + 8x_4 + 3x_5 &= -2294 \\
99x_1 - 8x_2 - 59x_3 - 3x_4 + x_5 &= 6924.
\end{aligned}
\tag{4.6-10}
$$

We note that the 3rd and 5th equations in Eq. (4.4-8), which are nearly parallel, are aggregated into separate sets. The quasi-parallelism between these two sets is not removed, so slow numerical convergence of the algorithm is expected. For easy reference, this approach is simply referred as *scheme-one* (inefficient) decomposition. The algorithm in Eq. (4.6-7) is tested on this scheme-one decomposition and the numerical results are summarized in Table 4.6-1. Notice that the same three initial starting points are used as before. In the interest of brevity, only the relative errors of the iterates are furnished here.

Table 4.6-1 List of relative errors of the numerical iterates using inefficient decomposition for three different initial conditions: (i) $\mathbf{x}_0 = (0,0,0,0,0)^T$; (ii) $\mathbf{x}_0 = (1000, 1000, 1000, 1000, 1000)^T$; and (iii) $\mathbf{x}_0 = (-20, 50, 391, 3211, 2)^T$

Iterations	(i)	(ii)	(iii)
0	1.000000	2.774430	3.343677
1000	0.146088	1.463374	0.980071
2000	0.109013	1.091987	0.731341
3000	0.081347	0.814853	0.545735
4000	0.060702	0.608053	0.407234
5000	0.045296	0.453737	0.303883
6000	0.033801	0.338584	0.226761
7000	0.025223	0.252655	0.169212
8000	0.018821	0.188534	0.126268

Next, the system of linear equation is efficiently decomposed into two sub-

Table 4.6-2 List of relative errors of the numerical iterates using efficient decomposition for three different initial conditions: (i) $\mathbf{x}_0 = (0,0,0,0,0)^T$; (ii) $\mathbf{x}_0 = (1000, 1000, 1000, 1000, 1000)^T$; and (iii) $\mathbf{x}_0 = (-20, 50, 391, 3211, 2)^T$

Iterations	(i)	(ii)	(iii)
0	1.000000	2.774430	3.343677
50	0.069783	0.280391	0.175549
100	0.008724	0.035053	0.021946
150	0.001091	0.004382	0.002744
200	0.000136	0.000548	0.000343
250	0.000017	0.000068	0.000043
300	0.000002	0.000009	0.000005
350	0.000000	0.000001	0.000001
370	0.000000	0.000000	0.000000

systems of linear equations, of which the first one $\mathbf{A}_1^{(2)}\mathbf{x} = \mathbf{b}_1^{(2)}$ is

$$
\begin{aligned}
x_1 - 2x_2 + x_3 + 4x_4 + x_5 &= -536 \\
3x_1 + x_2 + 2x_3 - 6x_4 + x_6 &= -1089
\end{aligned}
\tag{4.6-11}
$$

and the second one $\mathbf{A}_2^{(2)}\mathbf{x} = \mathbf{b}_2^{(2)}$ is

$$
\begin{aligned}
100x_1 - 7x_2 - 60x_3 + 2x_4 - x_5 &= 9343 \\
5x_1 - 2x_2 - 16x_3 + 8x_4 + 3x_5 &= -2294 \\
99x_1 - 8x_2 - 59x_3 - 3x_4 + x_5 &= 6924.
\end{aligned}
\tag{4.6-12}
$$

Here, the 3rd and 5th equations in Eq. (4.4-8) are aggregated into the same solution set. For easy reference we call this *scheme-two* (efficient) decomposition. The quasi-parallelism between the two solution sets is removed so faster numerical convergence of the algorithm is expected. The algorithm in Eq. (4.6-7) is tested on this scheme in a similar fashion and the numerical results are summarized in Table 4.6-2. Comparing with the results in Table 4.6-1, we conclude that the second approach indeed has dramatically improved the convergence of the numerical algorithm as expected. Also, it is interesting to note that both the results in Table 4.6-1 and Table 4.6-2 have improved convergence speed compared with the results of the iterative POCS algorithm with or without orthogonalization with the Gram-Schmidt process (see Table 4.4-1 and Table 4.4-3). This improvement is consistent with our initial expectation that the numerical behavior of the hybrid algorithm lies in-between that of the one-step and the iterative algorithms given in Eq. (4.5-10) and Eq. (4.2-17), respectively.

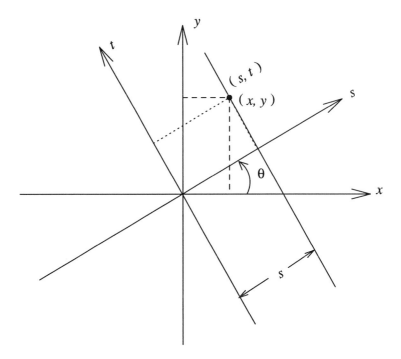

Fig. 4.7-1 The relationship between two rectangular x-y and s-t coordinate systems.

4.7 APPLICATION EXAMPLE: COMPUTED TOMOGRAPHY

Computed tomography (CT) is one of several imaging techniques that are made possible by the computing power of modern computers. Today, CT is routinely used in clinical situations such as brain lesion detection and identification, etc. Needless to say, its impact on modern medical diagnostics cannot be overemphasized.

The principle behind CT can be briefly explained by the following: Let $f_c(x,y)$ be a function of two continuous variables x and y. For example, $f_c(x,y)$ might denote the pointwise X-ray opacity or attenuation of the object being imaged. Here we use the subscript c to specify that the function is defined for *continuous* variables, to avoid any ambiguity that may arise from the notation $f(i,j)$, which will be used to represent a spatially discrete function of two integer variables i and j. Consider the geometry shown in Fig. 4.7-1, in which a rectangular x-y coordinate system is rotated about the origin through an angle θ to generate a new s-t coordinate system. Then each point in the plane has two sets of coordinates: coordinates (x,y) relative to the x-y system, and coordinates (s,t) relative to the s-t system. Given a point (s,t) in the s-t system, we can compute its coordinates in the x-y system according to

$$\begin{pmatrix} x \\ y \end{pmatrix} = \begin{pmatrix} \cos\theta & -\sin\theta \\ \sin\theta & \cos\theta \end{pmatrix} \begin{pmatrix} s \\ t \end{pmatrix}. \tag{4.7-1}$$

Note that for each given value of s in the s-t coordinate system there is a line uniquely defined that is parallel to the t-axis with a displacement value of s from the t-axis. Let $p_\theta(s)$ denote the integral of the function $f_c(x,y)$ along this line. Then we have

$$p_\theta(s) = \int_{-\infty}^{\infty} f_c(x,y)\Big|_{x=s\cos\theta-t\sin\theta, y=s\sin\theta+t\cos\theta} dt. \qquad (4.7\text{-}2)$$

This equation arises naturally in the theory of tomographic imaging. In X-ray CT, a two-dimensional (2-D) cross-section of a three-dimensional (3-D) object such as a human body is penetrated by an X-ray beam, as illustrated in Fig. 4.7-2. According to the physics of X-ray tomography, an X-ray beam is attenuated in its intensity by the object through which it passes. Furthermore, the degree of attenuation of the beam depends on the material composition of the object. For example, the attenuation caused by human soft tissue differs from that of human bone. If we use $f_c(x,y)$ to denote the pointwise opacity of the object at (x,y), then the line integral in Eq. (4.7-2) can be easily obtained from the intensity $I_\theta(s)$ recorded at a detector located at s at angle θ with displacement s (Fig. 4.7-2). Indeed, with I_0 denoting the beam intensity incident on the object, we obtain

$$p_\theta(s) = \ln \frac{I_0}{I_\theta(s)} . \qquad (4.7\text{-}3)$$

Since the value of $f_c(x,y)$ depends on the material content of the object at spatial position (x,y), it can be used to determine the material composition of the object. Thus, it is of interest to reconstruct the function $f_c(x,y)$ from its measurement $p_\theta(s)$, which is exactly the goal of CT.

In the literature the angle θ in Eq. (4.7-2), as illustrated in Fig. 4.7-2, is called the *view angle*, and the line integral $p_\theta(s)$ is called the *projection*, the *Radon transform*, or the *raysum* of $f_c(x,y)$ at view angle θ with displacement s. In the following context, however, we will refer to it specifically as the raysum of $f_c(x,y)$, while reserving the word "projection" exclusively for the projection of a vector onto a set, trying to avoid any possible ambiguity that might arise otherwise.

Due to physical and temporal limitations, the raysums $\{p_\theta(s)\}$ are measured only at a finite number of view angles, say $\{\theta_k, k = 1, 2, \cdots, K_\theta\}$, over a finite number of displacement values, say $\{s_l, l = 1, 2, \cdots, K_s\}$. Such limitation in practice comes in many ways, of which some examples are the size of of crystal detectors, the measurement precision of physical instruments, the allowed amount of exposure of the human body to radiation sources, etc. In measuring these raysums, CT systems typically utilize two major source-detector configurations, namely, the *parallel-beam* geometry and the *fan-beam* geometry, as illustrated in Fig. 4.7-3(a) and Fig. 4.7-3(b), respectively. The latter geometry is widely used in modern X-ray CT systems because of its fast scanning ability compared with the former.

The fundamental problem of CT, then, is to reconstruct a 2-D attenuation function $f_c(x,y)$ from a total of $K_\theta K_s$ raysums $p_{\theta_k}(s_l)$, where $p_{\theta_k}(s_l)$ denotes the raysum

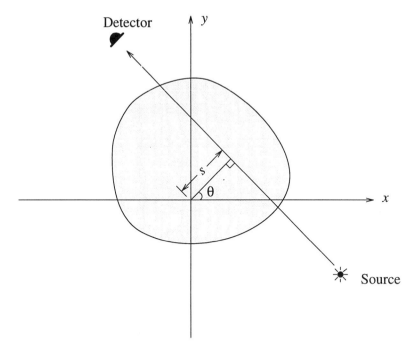

Fig. 4.7-2 In X-ray tomography, a 2-D cross-section of a 3-D object is penetrated by a series of parallel or fan-out beams. Only one such beam is shown here.

obtained at view angle θ_k with displacement value s_l. Since the reconstruction algorithm is realized on a computer, the function $f_c(x, y)$ is reconstructed only over a finite number of uniformly separated spatial positions (x_i, y_j) in the 2-D region of interest. The collection of these spatial positions (x_i, y_j) is called the *sampling grid* or *sampling lattice* and the function values $f_c(x_i, y_j)$ are called the *sampled representation* of $f_c(x, y)$. Conceivably, if the sampling grid is fine enough, the sampled representation will give us reasonably accurate information about the attenuation function $f_c(x, y)$.

Three major reconstruction algorithms have been widely studied for the CT problem in the literature. They are: (i) the algebraic reconstruction technique (ART) [7, 8], (ii) the filtered convolution back-projection algorithm [9–11], and (iii) the direct Fourier method [12, 13]. More recently, in a study reported in [14, 15], a vector-space projection approach has been proposed to solve the CT problem. A detailed discussion of each of these algorithms here is irrelevant to the purpose of this book. Instead, in what follows, only a discussion of the vector-space algorithm is furnished for the purpose of demonstrating the applicability of vector-space projections to practical problems.

First, consider the uniform rectangular sampling structure in Fig. 4.7-4, in which the region of interest is divided into square cells, of which the centers are the

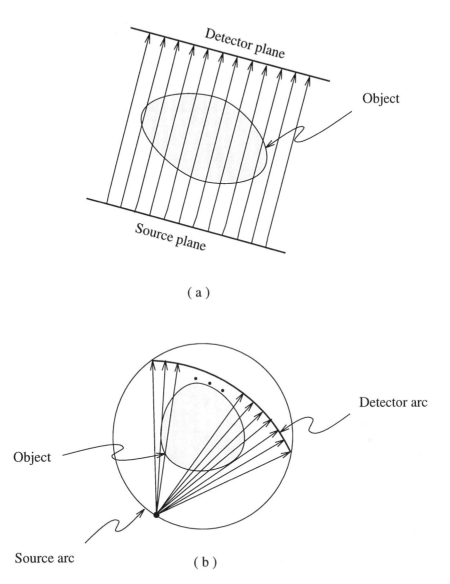

Fig. 4.7-3 In obtaining raysums, CT systems typically utilize two major source-detector configurations, namely the parallel-beam geometry in (a) and the fan-beam geometry in (b).

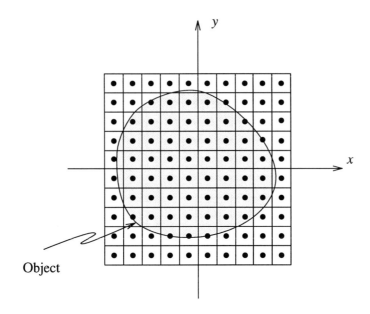

Fig. 4.7-4   A 2-D region of interest is divided into rectangular cells called pixels, of which the centers are the sampling grids which are denoted by black dots.

sampling points. The entire cell assumes the value of its sampling point. In image processing, each of these cells in the sampling region is called a *pixel*. Without loss of generality, we assume that there are a total of M pixels in each row and a total of N pixels in each column in the 2-D region of interest. Then the attenuation function $f_c(x, y)$ is represented by its $M \times N$ values $f_c(x_i, y_j)$, evaluated at the sampling grids (x_i, y_j), $i = 1, 2, \cdots, M$, $j = 1, 2, \cdots, N$. For simplicity, we define $f(i, j) \triangleq f_c(x_i, y_j)$. In other words, we are going to deal with an array of numbers $f(i, j)$ instead of the function $f_c(x, y)$. In digital image processing, this array of numbers is called an image, of which the intensity value $f(i, j)$ at pixel (i, j) represents the material attenuation factor at the spatial location (x_i, y_j).

Let $l_{\theta, s}$ denote the line along which the raysum is obtained. Let also $\mathcal{I}_{\theta, s}$ denote the set of image pixels that the line $l_{\theta, s}$ intercepts, that is,

$$\mathcal{I}_{\theta, s} = \{(i, j) : \text{pixel } (i, j) \text{ is intercepted by the line } l_{\theta, s}\}. \tag{4.7-4}$$

Furthermore, for each pixel $(i, j) \in \mathcal{I}_{\theta, s}$ we define $r_{\theta, s}(i, j)$ to be the length of the portion of the line $l_{\theta, s}$ inside the pixel (i, j), as illustrated in Fig. 4.7-5. On the assumption that the function $f_c(x, y)$ does not vary much within a pixel, the line

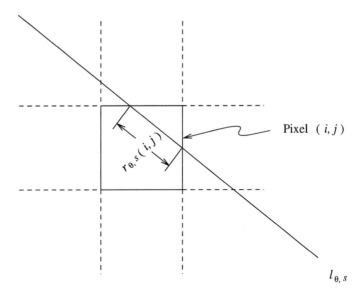

Fig. 4.7-5 The notation $r_{\theta,s}(i,j)$ is used to denote the length of the portion of the line $l_{\theta,s}$ inside the pixel (i,j).

integral can be approximated by the following:

$$
\begin{aligned}
p_\theta(s) &= \int_{-\infty}^{\infty} f_c(x,y)\bigg|_{x=s\cos\theta-t\sin\theta,\,y=s\sin\theta+t\cos\theta} \, dt \\
&\simeq \sum_{(i,j)\in\mathcal{I}_{\theta,s}} f(i,j)r_{\theta,s}(i,j).
\end{aligned}
\tag{4.7-5}
$$

Clearly, Eq. (4.7-5) relates the image pixel values $f(i,j)$ to a measured raysum through a linear equation, which is made possible by approximating an integral by a summation.

We recall that a total of $K = K_\theta K_s$ raysums are obtained, each of which corresponds to a parameter setting (θ_k, s_l), i.e., the combination of the view angle θ_k with the displacement value s_l, regardless of the source-detector geometry used. For notational simplicity, we use m to index these[†] K raysums instead of the double indices k and l for the parameters (θ_k, s_l). Also, we use R_m to denote the raysum, that is, $R_m = p_{\theta_k}(s_l)$ for some k and l. Then Eq. (4.7-5) yields

$$
R_m = \sum_{(i,j)\in\mathcal{I}_m} f(i,j)r_m(i,j)
\tag{4.7-6}
$$

[†]For example use $m = (k-1)K_s + l$ for $1 \le k \le K_\theta, 1 \le l \le K_s$. This is called *lexicographic ordering*.

for $m = 1, 2, \cdots, K$. In Eq. (4.7-6), $\mathcal{I}_m$ denotes the set of pixels intercepted by the X-ray beam line for obtaining the mth raysum, and $r_m(i, j)$ denotes the length of the portion of this line inside the pixel (i, j). In a CT system, both the set $\mathcal{I}_m$ and the associated coefficients $r_m(i, j)$ can be determined from the source-detector geometry. Note that in Eq. (4.7-6) we have replaced the sign '$\simeq$' by the equal sign '$=$', assuming that the pixels are small enough to avoid visual artifacts.

Equation (4.7-6) establishes that the relation between the unknown image and its measured raysums is governed by a system of simultaneous linear equations. Thus, the reconstruction of a CT image $f(i, j)$ from its measured raysums is equivalent to the solution of the system of linear equations in Eq. (4.7-6).

In a practical CT system, the image $f(i, j)$ can have several hundreds of pixels in each of its dimensions M and N. In other words, the system of linear equations in Eq. (4.7-6) is what we earlier called *large-scale*. It is interesting, however, to observe that the index set $\mathcal{I}_m$ of non-zero coefficients in Eq. (4.7-6) contains only a small portion of the image pixels. This is true because a single X-ray line passes through only a small portion of the total image pixels. As a result, the coefficient matrix for the system of linear equations in Eq. (4.7-6) is a sparse matrix. From the discussion in earlier sections of this chapter, the projection algorithm can be used to its advantage for solving such a kind of problem.

Finally we conclude this section by furnishing a numerical example reported in [14, 15]. Shown in Fig. 4.7-6(a) is the famous Shepp-Logan phantom, which has been widely used in the evaluation of medical imaging algorithms. In the Shepp-Logan phantom, geometric shapes such as circles and ellipses of different sizes are used to simulate the structure in a cross-section of a human head section. The size of the image in Fig. 4.7-6(a) is of 64×64 pixels. In the experiment, 4902 raysums are obtained from this image using the parallel-beam geometry. Therefore, a total of 4902 linear equations in the form of Eq. (4.7-6) are obtained. To solve this system of linear equations, the iterative projection algorithm $\mathbf{x}_{k+1} = P_K \cdots P_2 P_1 \mathbf{x}_k$ is used. In such a formulation, all the 64×64 image pixels $f(i, j)$ are treated as the unknown vector $\mathbf{x}$ and constraint sets are defined for each equation. The reconstructed image obtained after merely 20 iterations of the POCS projection algorithm is shown in Fig. 4.7-6(b). The algorithm was started with the zero vector. Clearly, the reconstruction accurately reveals the fine structure in the Shepp-Logan phantom, even thought it is still noisy.

In addition to the advantages pointed out earlier, another advantage of the projection algorithm is that it offers the opportunity to incorporate *prior* knowledge about the unknown object into the algorithm in the form of constraint sets. For example, according to the physics of X-ray tomography the attenuation factor $f(i, j)$ can never assume negative values. Such a piece of prior knowledge can be easily described by a convex constraint set and be incorporated into the iterative projection algorithm. In [14, 15] the reconstruction of the Shepp-Logan image included prior knowledge about the *size* of the image, the permissible *amplitude variation* in the image, and the maximum deviation of the image from a reference image. Space limitations preclude us from going into details, but the interested reader is directed to

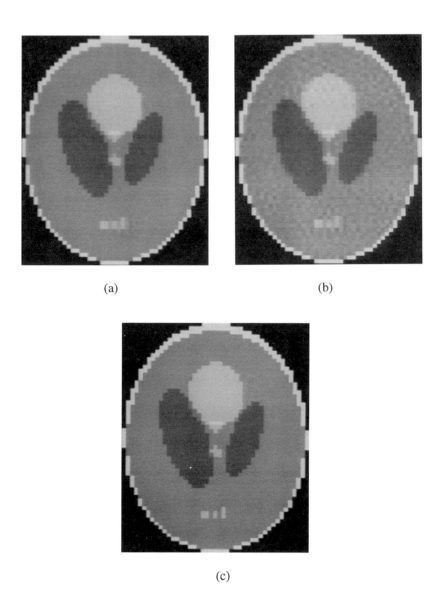

(a)   (b)

(c)

Fig. 4.7-6 Projection-based CT reconstruction example: (a) the Shepp-Logan phantom; (b) the reconstructed image after 20 iterations when only the raysum data are used; and (c) the reconstructed image after 20 iterations, when additional prior knowledge is used.

[14, 15] for the specifics. Shown in Fig. 4.7-6(c) is the reconstructed image after 20 iterations after the above prior knowledge has been used. The reconstructed image is of better quality than the image reconstructed from the measurements alone.

4.8 APPLICATION EXAMPLE: RESOLUTION ENHANCEMENT

As a second example of a practical application involving the solution of linear equations, we furnish the following. In the field of remote sensing such as high-altitude surveillance, it is often desired to obtain information about a region on the ground. An effective approach employed for such purpose today is to use a high-flying aircraft or satellite equipped with an aerial *detector* or *sensor* array to collect data directly from above the scene. Detector arrays typically consist of many individual detectors. The data obtained by a detector array using state-of-the-art technology can give us information ranging from material content to temperature distribution of the scene. It is of considerable interest for us to reconstruct the most detailed scene from the detector data using image processing technology.

For the purpose of data collection, the detectors in an image plane detector array are aligned in such a fashion that individual detectors cover only a small portion of a scene, while the field-of-view of the detector array as a whole covers the entire scene. The area covered by a detector is called the *footprint* of the detector. Because the footprint is much larger than the fine details in the image, a naive reconstruction directly from these detector readings without further processing would produce an image whose spatial resolution is limited by the size of the detector footprint and not by the high-quality optics that generate the image. Due to physical limitations, the size of the footprint for a detector cannot be made arbitrarily small. For example, it is not uncommon that in weather satellites and aircraft scanners the reading from a single detector actually corresponds to a physical area in the scale of tens or hundreds of meters on the ground. Conceivably, an image of higher resolution in such a case is more desirable since it can reveal more detailed information about the scene. In practice, the data obtained directly by a detector array are called *low-resolution* data. Thus, a natural question is: is it possible to reconstruct an image of higher resolution than that of the low-resolution data?

Another view of this problem, which may have greater appeal to more mathematically inclined readers, is that the detector array *undersamples* the image and when an image is reconstructed from the samples, it is distorted by *aliasing* artifacts.

In image processing, this problem is categorized as *resolution enhancement*. There has been a great deal written on this subject [16]. For example, it is argued in [17] that if one permits an overlap of successive scans and rescans of the same scene from different directions, there might be enough information in the acquired data to permit a higher-resolution reconstruction, i.e., one more commensurate with the imaging optics. As another example of applying the POCS algorithm for solving systems of linear equations associated with real problems, we review an approach to this problem first suggested in [18].

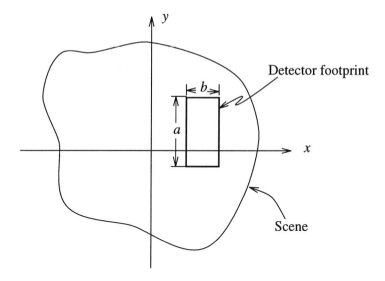

Fig. 4.8-1 In remote sensing, a detector array is superimposed upon an image of the scene of interest to collect data.

Assume that the content of the scene of interest on the ground is described by a function of two continuous variables $f_c(x, y)$. The two variables x and y here are the coordinates of a point in the scene relative to a convenient rectangular coordinate system in the image plane.[†] For example, if the object of interest is a ground image, then the value of $f_c(x, y)$ gives the monochrome reflectivity at spatial location (x, y). Now consider a detector with a *response function* $\sigma(x, y)$ superimposed upon the image. Then the detector output is given by

$$d = \int_{-\infty}^{\infty} \int_{-\infty}^{\infty} f_c(x, y)\sigma(x, y)\,dx\,dy. \tag{4.8-1}$$

As mentioned at the beginning of this section, a single detector covers only a small area in the image. As a result, the detector response function $\sigma(x, y)$ is non-zero only within a small region in the x-y plane. In fact, the region where $\sigma(x, y)$ has non-zero values is exactly the detector footprint. Therefore, the integration in Eq. (4.8-1) is carried out only over the region of the detector footprint. Equation (4.8-1) states that the output of the detector is the average of the function $f_c(x, y)$ weighted by the detector response $\sigma(x, y)$ over the detector footprint. To clarify this point, Fig. 4.8-1 illustrates that a detector with a rectangular footprint is superimposed upon an image of the scene of interest. In such a case, the detector

---

[†]Typically the ground scene is *imaged* onto the detector array. The image is of high quality so that, except for coordinate inversion and magnification, the ground scene can be assumed identical with the image scene. We use the same coordinate system for either ground or image plane for simplicity.

reading is determined by all the values of the function $f_c(x, y)$ within the whole rectangular footprint, instead of by its value at a particular point. Therefore the detector has absorbed all the fine details, and the reconstruction of $f_c(x, y)$ at a level of detail beyond that of the detectors requires additional processing.

Assume that a total of K detector readings d_k, $k = 1, 2, \cdots, K$, are collected by the detector array. Then from Eq. (4.8-1) each reading d_k is given by

$$d_k = \int_{-\infty}^{\infty} \int_{-\infty}^{\infty} f_c(x, y)\sigma_k(x, y)dxdy, \qquad (4.8\text{-}2)$$

where $\sigma_k(x, y)$ is the detector response function corresponding to the kth reading. In practice, $\sigma_k(x, y)$ can be determined from the detector configuration in a detector array. The problem then is as follows: given detector readings $d_k, k = 1, 2, \cdots, K$, how do we reconstruct the function $f_c(x, y)$ at a higher resolution than that of the detector readings?

As the reader might have noticed, there exists a strong similarity between this problem and the CT problem addressed in last section. Indeed, recall that in the CT problem a 2-D function is reconstructed from its raysum data, where each raysum is obtained by integrating the function along a parameterized line, while here a 2-D function is reconstructed from its detector data, where each detector reading is obtained by integrating the function over the detector footprint. Realizing this similarity, we find it not surprising that the resolution enhancement problem can be solved by using the POCS algorithm in a similar fashion to that for the CT problem.

In real-world situations the processing is often done by computer. In such a case, a more realistic approach is to reconstruct the function over a sampling grid in the region of interest. Assume that the uniform rectangular sampling structure, shown in Fig. 4.7-4, is used. Since the object here is to reconstruct the function at a resolution higher than that of the detector data, the rectangular sampling cell, i.e., the image pixel, has to be chosen in such a way that its size is less than that of the detector footprint. Without loss of generality, we assume further that there are a total of M pixels in each row and a total of N pixels in each column in the 2-D region of interest. Then the function $f_c(x, y)$ is represented by its $M \times N$ sample values $f_c(x_i, y_j)$ evaluated at the sampling points (x_i, y_j) for $i = 1, 2, \cdots, M$, $j = 1, 2, \cdots, N$. For simplicity, we define $f(i, j) \triangleq f_c(x_i, y_j)$. As stated earlier, in image processing this array of numbers $f(i, j)$ is called an $M \times N$ image.

Let $\mathcal{I}_k$ denote the set of image pixels that the detector footprint covers for the kth detector reading. Then, based on the assumption that the image pixel is chosen fine enough, the detector reading can be approximated by

$$d_k = \int_{-\infty}^{\infty} \int_{-\infty}^{\infty} f(x, y)\sigma_k(x, y)dxdy \qquad (4.8\text{-}3)$$

$$\simeq \sum_{(i,j)\in\mathcal{I}_k} f(i, j)\hat{\sigma}_k(i, j), \qquad (4.8\text{-}4)$$

where
$$\hat{\sigma}_k(i,j) = \sigma_k(x_i, y_j) s_k(i,j), \qquad (4.8\text{-}5)$$

in which $s_k(i,j)$ denotes the area of the portion of pixel (i,j) that lies within the detector footprint for the kth reading. More specifically, if the pixel (i,j) lies totally within the detector footprint, then $s_k(i,j)$ is the area of the pixel; otherwise, it is the fractional area of the pixel within the footprint. Clearly, Eq. (4.8-4) establishes that the relation between the unknown image and its measured detector readings is governed by a system of simultaneous linear equations. Thus, the reconstruction of the image $f(i,j)$ from its measured detector data is equivalent to solving the system of linear equations in Eq. (4.8-4).

In a practical application, it is not uncommon that the number of detector readings for a single image frame can be in the order of 10^4 or higher[†]. This implies that the system of linear equations can be very large. It is interesting to observe, however, that the index set $\mathcal{I}_k$ of non-zero coefficients in Eq. (4.8-4) contains only a very small fraction of the total pixels in the image. This is so because the detector footprint covers only a small portion of the entire scene. As a result, the coefficient matrix for the system of linear equations in Eq. (4.8-4) is a sparse matrix. Thus the situation is similar to that in the previous problem and the projection algorithm can be used to its advantage for solving such a kind of problem.

We illustrate the material in this section by furnishing a numerical example reported in [18]. Purely for demonstrational purpose, the 64×64 Shepp-Logan phantom, shown in Fig. 4.7-6, is used here instead of a real-world aerial scene. In this experiment, a rectangular detector array consisting of 8 detectors along the horizontal direction and 16 detectors along the vertical direction is used. The footprint of each detector has a geometric dimension of 8 image pixels along the horizontal direction and 4 image pixels along the vertical direction. Furthermore, the detector response function $\sigma(x,y)$ is assumed to have constant value of unity within the footprint. In such a case, the output of each detector is simply the unweighted or uniform average of the image content inside its footprint. Note that the detector array here is configured in such a way that it covers the image completely when it is perfectly aligned with the image plane, as illustrated in Fig. 4.8-2(a). Clearly, the detector array can collect only a total of $16 \times 8 = 128$ readings from the entire image. The resolution of an image produced by these 128 readings alone would be $1/32$ of that of the original image. This is all due to the fact that the detector footprint is too large so that the detector output "averages out" the details in the image.

To combat this detail-loss caused by the gross detector reading, a successive rescan scheme is used to gather more data from the same image. To be specific, the same detector array is sequentially rotated upon the image field to 32 equi-spaced angular positions in the range of $0°$ to $180°$, as shown in Fig. 4.8-2(b). The detector readings are recorded at each angular position. As a result, a total of

---

[†]For example, the number of detectors in the Amber AE-4128 infrared detector array is 16,384.

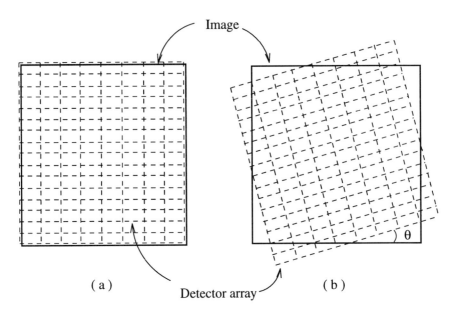

Fig. 4.8-2 In obtaining the detector data, a rectangular detector array consisting of 8 detectors along the horizontal direction and 16 detectors along the vertical direction is used. (a) Detector array in vertical position; (b) detector array is rotated through angle θ.

$128 \times 32 = 4096$ readings are taken, compared with 64×64 unknowns.

To reconstruct the image, the iterative algorithm $\mathbf{x}_{k+1} = P_K \cdots P_2 P_1 \mathbf{x}_k$ is used to solve the system of linear equations formed from the detector data. In such a formulation, all the 64×64 image pixels $f(i,j)$ are treated as components of a single unknown vector $\mathbf{x}$ and constraint sets are defined for each equation. Figure 4.8-3 shows the reconstructed images after 10, 50, and 100 iterations. In obtaining these results, the algorithm is started with the zero vector. The fine structure in the Shepp-Logan phantom becomes evident after 50 iterations and, except some noise, is reproduced exactly after 100 iterations. In [18], the study is further extended to deal with the problem of reconstruction from insufficient detector data, which happens in real-world applications when some of the detectors in a detector array are inoperative during data collection. It is demonstrated that in such a case the effect caused by insufficient data can be ameliorated when additional constraint sets based on available prior knowledge are incorporated into the reconstruction. The interested reader is directed to [18] for more details.

Finally we conclude this section by furnishing a numerical example involving *scanning*, first described in [17] and extended in [18]. Many images are scanned and often the scanning array elements are larger than the point-spread function (psf) of the imaging optics. If the effective pixel size is that of the support of the psf, the image recovered from scanning could be severely blurred. To demonstrate the recovery of such blurred images by the methods described in this chapter, we

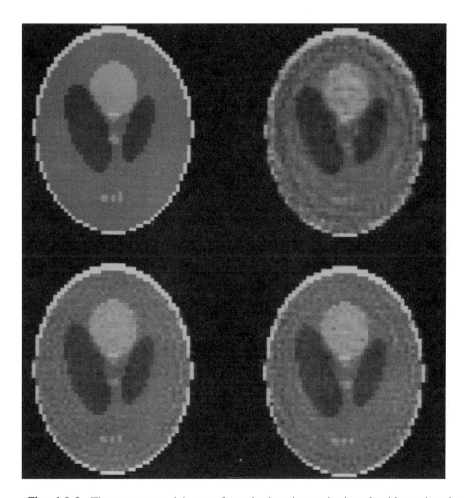

Fig. 4.8-3 The reconstructed images from the iterative projection algorithm using the detector data after 10, 50, and 100 iterations are shown in top-right, bottom-right, and bottom-left, respectively. The original image is shown in the top-left.

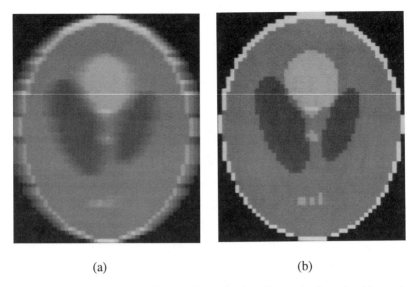

(a)   (b)

Fig. 4.8-4 The reconstructed images from the iterative projection algorithm using the detector data obtained from scanning. (a) The low-resolution image produced by scanning with a 4-pixel-by-1-pixel detector. The horizontal blur is quite obvious. The POCS image after 100 iterations is shown in (b). Prior knowledge about the finite support of the object, gray level excursions, etc., were used.

once again consider the 64×64 Shepp-Logan image shown in Fig. 4.7-6. The low-resolution image produced by scanning with a 4-pixel-by-1-pixel detector is shown in Fig. 4.8-4(a). Here the horizontal blur is quite obvious. The response of the detectors is uniform, i.e., $\sigma(x, y) = 1$ over the footprint. For each position of the scanning array, a single linear equation is obtained relating the true image values to the detector reading. As the detector is moved all over the image, enough sparse linear equations are generated to enable a high-resolution reconstruction. The POCS image after 100 iterations is shown in Fig. 4.8-4(b). Prior knowledge about the finite support of the object, gray level excursions, etc., were used. The improvement over the basic image is quite remarkable.

Additional examples of the application of POCS are given in later chapters. In particular, high-resolution image reconstruction in the presence of noise and motion between low-resolution frames, is discussed in Section 9.5 in Chapter 9.

4.9 SUMMARY

The solution to a large number of quantitative real-world problems ultimately involves solving systems of linear equations. In this chapter we showed how the vector-space projections onto convex sets (POCS) can be used to solve such systems without matrix inversion. While other methods furnish iterative non-matrix-inverting solutions to linear equations, POCS allows for the incorporation of prior knowledge to aid in finding solutions consistent with the physical laws that apply.

It may not always be a good idea to solve a given system of equations without first doing some "pre-processing". For example, when two or more equations in the system are nearly parallel, matrix inversion is very risky and iterative methods may take a long time to converge. We discussed several preprocessing techniques that significantly reduced the time-to-convergence of the POCS algorithm.

Finally we illustrated how systems of linear equations arise in the real world and their solutions by POCS. We considered two such examples: one taken from computed tomography and the other from high-resolution enhancement.

REFERENCES

1. A. W. Al-Khafaji and J. R. Tooley, *Numerical Methods in Engineering Practice,* Holt, Rinehart and Winston, Inc., New York, 1986.

2. G. E. Forsythe, M. A. Malcom, and C. B. Moler, *Computer Methods for Mathematical Computations,* Prentice-Hall, Englewood Cliffs, NJ, 1977.

3. C. I. Podilchuck and R. J. Mammone, Image recovery by convex projections using a least-squares constraints, *J. Opt. Soc. Am.,* **7**:517–521, March 1990.

4. E. K. P. Chong and S. H. Zak, *An Introduction to Optimization,* John Wiley & Sons, New York, 1996.

5. G. H. Golub and C. F. Van Loan, *Matrix Computations,* 2nd ed., Johns Hopkins University Press, Baltimore, 1989.

6. J. M. Ortega, *Matrix Theory,* Plenum Press, New York, NY, 1987.

7. G. T. Herman, A. Lent, and S. W. Rowland, ART: Mathematics and applications, *J. Theor. Biol.,* **42**:1–32, 1973.

8. G. T. Herman, On modifications to the algebraic reconstruction technique, *Comput. Biol. and Med.,* **9**:271–276, 1979.

9. R. N. Bracewell and A. C. Riddle, Inversion of fan beam scans in radio astronomy, *Astrophys. J.,* **150**:427–434, 1967.

10. G. N. Ramachandran and A. V. Lakshminarayanan, Three-dimensional reconstruction from radiographs and electron micrographs: Application of convolutions instead of Fourier transforms, *Proc. Nat. Acad. Sci. USA,* **68**:2236–2240, 1970.

11. L. A. Shepp and B. F. Logan, The Fourier reconstruction of a head section, *IEEE Trans. Nucl. Sci.,* **21**:21–43, 1974.

12. H. Stark, J. W. Woods, I. Paul, and R. Hingorani, Direct Fourier reconstruction in computer tomography, *IEEE Trans. Acoust. Speech and Sign. Pro.,* **29**:237–245, April 1981.

13. H. Stark, J. W. Woods, I. Paul, and R. Hingorani, An investigation of computerized tomography by direct Fourier inversion and optimum interpolation, *IEEE Trans. Biomed. Eng.,* **28**:496–505, July 1981.

14. P. Oskoui-Fard and H. Stark, Tomographic image reconstruction using the theory of convex projections, *IEEE Trans. Medical Imaging,* **7**:45–58, March 1988.

15. P. Oskoui-Fard and H. Stark, A comparative study of three reconstruction methods for a limited-view computer tomography problem, *IEEE Trans. Medical Imaging,* **8**:43–49, March 1988.

16. J. A. Richard, *Remote Sensing Digital Image Analysis,* Springer-Verlag, New York, 1986.

17. B. R. Frieden and H. H. G. Aumann, Image reconstruction from multiple 1-D scans using filtered localized projections, *App. Optics,* **26**:3615–3621, 1987.

18. P. Oskoui-Fard and H. Stark, High resolution image recovery from image plane arrays using convex projections, *J. Opt. Soc. Amer. A,* **6**(11):1715–1726, November 1989.

EXERCISES

4-1. Show that the solution set of a system of linear equations that is consistent can have either a single point or infinite number of points.

4-2. Show that if the set of vectors $\{a_1, a_2, \cdots, a_m\}$ is not linearly independent, then the Gram-Schmidt process in Eq. (4.4-4) will produce a zero vector.

4-3. Prove that the set $C = \{x : Ax = b\}$ is a closed linear variety in R^n.

4-4. Derive the projection onto the set $C = \{x : Ax = b\}$ using the Lagrange multipliers approach.

4-5. Show that a vector that is normal to the solution set of a system of linear equations can be written as a linear combination of its row vectors.

4-6. Show that the matrix AA^T is invertible if and only if all the m row vectors of A are linearly independent.

4-7. Write a program to test the POCS algorithm for solving the system of linear equations given in Eq. (4.4-8). Run your program and compare your results with those listed in Tables 4.4-1 and 4.4-2.

4-8. Run your program in Exercise 4-7 to solve the modified system of linear equations of Eq. (4.4-8), which is given in Eq. (4.4-10). Compare your results with those listed in Tables 4.4-3 and 4.4-4.

4-9. Write a program to test the hybrid "iterative-onestep" POCS algorithm for solving the system of linear equations given in Eq. (4.4-8). Run your program using the *scheme-one* decomposition given in Eqs. (4.6-9) and (4.6-10). Compare your results with those listed in Table 4.6-1.

4-10. Run your program in Exercise 4-9 using the *scheme-two* decomposition given in Eqs. (4.6-11) and (4.6-12). Compare your results with those listed in Table 4.6-2.

5

Generalized Projections

5.1 INTRODUCTION

Powerful as it is, the method of *projections onto convex sets* (POCS) cannot guarantee convergence to a feasible solution if any of the constraint sets are non-convex. Figure 5.1-1(a) illustrates how the non-convexity of set C_1 results in convergence to a point x_T, which is not a feasible solution, i.e., $\mathbf{x}_T \notin C_0 \triangleq C_1 \cap C_2$. Whether one converges to a feasible solution or not depends on the set configurations and the starting point. In Fig. 5.1-1(b) the set configurations are the same as in Fig. 5.1-1(a) but the starting point is nearer to the solution region. Clearly a feasible solution is achievable in this case. The point $\mathbf{x}_T$ in Fig. 5.1-1(a) is sometimes called a *trap* and is an undesirable convergence point, since it doesn't satisfy all the constraints. How easy is it to converge to a trap? Again this depends the set configuration. For example, in Fig. 5.1-1(a) it is rather easy to converge to a trap since almost any starting point in the region to the right of $\mathbf{x}_T$ will converge to one. We might call such a trap an *attractive* trap. On the other hand, consider the situation shown in Fig. 5.1-2, where the set C_1 represents the points along the vertical line and C_2 represents points along the circumference of the circle. Clearly in this case it is unlikely to converge to a trap, since only starting points along the line AB will lead to convergence to $\mathbf{x}_T$ or $\mathbf{x}'_T$. Such a trap might be called a *repulsive* trap. The point is that the use of non-convex constraints does not immediately rule out the use of projection methods. Quite the contrary, as we shall see in this and later chapters in this book, a number of important practical problems involve non-convex constraints and can be solved using a restricted form of the method of vector-space projections called *generalized projections*.

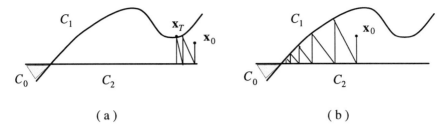

Fig. 5.1-1 (a) Non-convexity of C_1 results in convergence to a trap point $\mathbf{x}_T$, which is not a feasible solution. (b) A different starting point results in convergence to a feasible solution. A feasible solution is any point in C_0.

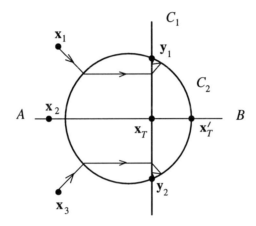

Fig. 5.1-2 All starting points, except those on line segment AB lead to feasible solutions $\mathbf{y}_1$ or $\mathbf{y}_2$. Starting points on line AB lead to traps $\mathbf{x}_T$ or $\mathbf{x}'_T$. If the starting point is chosen randomly it is unlikely to fall on AB and, hence, the algorithm is unlikely to converge to a trap.

5.2 CONVEX VERSUS NON-CONVEX SETS

We recall from Chapter 2 that in a convex set, the line connecting any two points in the set resides wholly within the set. This fundamental property of convexity is expressed mathematically as follows: with $\mathbf{x}$ denoting any point on the line segment between $\mathbf{x}_1$ and $\mathbf{x}_2$, clearly $\mathbf{x} - \mathbf{x}_2 = \mu(\mathbf{x}_1 - \mathbf{x}_2)$ (Fig. 5.2-1(a)). Then it follows that

$$\mathbf{x} = \mu \mathbf{x}_1 + (1 - \mu)\mathbf{x}_2, \qquad 0 \le \mu \le 1. \qquad (5.2\text{-}1)$$

Non-convex sets do not obey this property for every pair of points in the set. For example, in Fig. 5.2-1(b), the points $\mathbf{x}_1$ and $\mathbf{x}_2$ satisfy Eq. (5.2-1) but points $\mathbf{x}_3$ and $\mathbf{x}_4$ do not. Hence the set is not convex.

A set C consisting of a single point $\mathbf{x}$ is convex since $\mu \mathbf{x} + (1 - \mu)\mathbf{x} \in C$. Consider the set $C \overset{\triangle}{=} \{\mathbf{x} = (x_1, x_2) : 0 \le x_1 \le 2, 0 \le x_2 \le 2\}$. This convex set

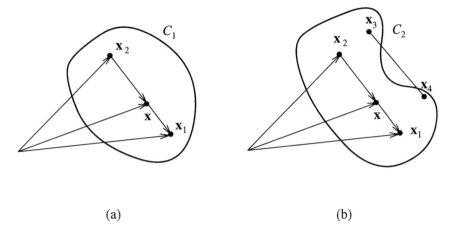

(a)   (b)

Fig. 5.2-1   (a) A convex set. (b) A non-convex set.

describes the simply connected region shown in Fig. 5.2-2(a). However, the set in Fig. 5.2-2(b) $C \triangleq \{\mathbf{x} = (x_1, x_2) : 0 \leq x_1 \leq 2 \text{ and } 0 \leq x_2 \leq 2, \text{ or, } 3 \leq x_1 \leq 4 \text{ and } 3 \leq x_2 \leq 4\}$ does not describe a simply-connected region and clearly is not convex. Sets formed from the union of several closed disjoint sets are generally not convex. Certain important sets consist of points which are n-tuples whose components take binary values such as ± 1. Such sets are also non-convex. They occur in certain types of neural nets and we shall encounter them in Chapter 8.

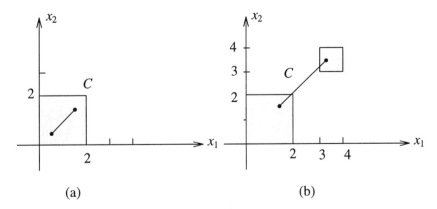

(a)   (b)

Fig. 5.2-2   (a) A convex set. (b) The union of disjoint closed convex sets is generally non-convex.

5.3 EXAMPLES OF NON-CONVEX SETS

We shall assume that all of the elements in the sets discussed below belong to the L^2 space of square-integrable functions. The following sets occur in various engineering problems and are non-convex:

$$C_1 = \{y(t) : |y(t)| = g(t), \ 0 \le t \le T\}, \tag{5.3-1}$$

$$C_2 = \{y(t) : y(t) \leftrightarrow Y(\omega), |Y(\omega)| = M(\omega), \omega \in \Omega\}, \tag{5.3-2}$$

$$C_3 = \left\{(h(t), x(t)) : \int_{-\infty}^{\infty} h(\tau)x(t-\tau)d\tau = y(t), \ 0 \le t \le T\right\}, \tag{5.3-3}$$

$$C_4 = \{y(t) : y(t) \leftrightarrow Y(\omega), \alpha \le |Y(\omega)| \le \beta, \omega \in \Omega\}. \tag{5.3-4}$$

The following set consisting of elements in R^n is also non-convex:

$$C_5 = \{(y_1, y_2, \cdots, y_n) : y_i = +1 \text{ or } -1, \ i = 1, ..., n\}. \tag{5.3-5}$$

There are many more sets of a non-convex nature of engineering interest, but we limit our discussion, for the sake of brevity, to the five cited above. Set C_1 is related to problems where we can measure the instantaneous power $|y(t)|^2$ of a signal $y(t)$. If the signal is complex we lose the phase; if the signal is bi-polar real, we lose the polarity. Such problems occur in optics, astronomy, incoherent detection of electronic signals, and elsewhere. Clearly, by measuring the instantaneous power for all $t \in [0, T]$, we can extract $g(t) = |y(t)|$ as data and require that our restored signal $y(t)$ satisfy $|y(t)| = g(t)$ the measured data. Set C_2 is similar to set C_1 but the constraint is on the magnitude of the Fourier transform. In modern astronomy one often measures not an image but the correlation function of an image. Thus if the measured data are $g(\mathbf{x}) = \int y(\mathbf{x}')y(\mathbf{x}' - \mathbf{x})d\mathbf{x}'$, then from elementary Fourier theory $G(\omega) = |Y(\omega)|^2 = M^2(\omega)$, where $M(\omega) \overset{\triangle}{=} |Y(\omega)|$. Thus the image $y(\mathbf{x})$ must satisfy the *Fourier magnitude* constraint. Set C_2 is widely used in the so-called *phase recovery problem*. This refers to the fact that we cannot recover the image from $g(\mathbf{x})$ alone or, equivalently, from $M(\omega)$ alone. However, if we could recover the *phase* of $Y(\omega)$, then we would have enough information to recover the image.

Set C_3 occurs in the *blind deconvolution* problem. Here both the original source $x(t)$ and point-spread function $h(t)$ are not known. All that is known is that the process by which the source $x(t)$ is mapped into the blurred image $y(t)$ is a linear, space-invariant operation. Set C_3 is a construct on a *pair* of functions. In general, projecting onto such sets is more difficult than projecting onto sets involving single-function constraints.

Set C_4 requires that the Fourier transform magnitude of $y(t)$ be bounded from below by α and from above by β. This constraint occurs in analog and digital *filter design*, where it is often desirable to keep the magnitude of the filter response $|Y(\omega)|$ as close to constant as feasible over the passband. Using this constraint together with the *linear-phase* constraint enables the design of filters using projection methods (see Chapter 6, Section 6.4).

Set C_5 requires that the components $y_i, i = 1, ..., n$, of a vector $\mathbf{y}$, take on binary values ± 1. This set, obviously non-convex, occurs in digital communications and pattern recognition. For example, in the discrete *Hopfield net*, all state vectors and all stored library vectors must be of binary form. Set C_5 occurs in modeling the dynamics of the Hopfield neural net by alternating pairs of projections (see Chapter 8, Section 8.3).

The above discussion hints at the fact that non-convex sets might still be useful in finding solutions by projections. Under what circumstances is this true? We consider this question in the next section.

5.4 THEORY OF GENERALIZED PROJECTION

We recall that when all sets are convex, the POCS algorithm converges *weakly* to a point $\mathbf{x}^* \in C_0$, where C_0 is the intersection of the convex constraint sets. *Weak convergence* of a sequence $\mathbf{x}_1, \mathbf{x}_2, \cdots$ to a point $\mathbf{x}$ means that

$$\lim_{n \to \infty} \langle \mathbf{x}_n, \mathbf{y} \rangle = \langle \mathbf{x}, \mathbf{y} \rangle \tag{5.4-1}$$

for all $\mathbf{y}$ in the Hilbert space. Weak convergence, also called *inner-product convergence*, is equivalent to strong convergence when the underlying space is finite-dimensional. This was demonstrated in Theorem 1.5-3 in Section 1.5. When dealing with non-convex set, it is generally not possible to achieve strong or weak convergence. Under certain conditions, however, a generalized projection algorithm can produce iterates, whose summed distance from the membership sets decrease with the iteration number. Under what circumstances this error reduction property occurs is discussed below.

Summed Distance Error

Definition 1. Let C be a set in a Hilbert space $\mathbf{H}$ and $\mathbf{x}$ a point in $\mathbf{H}$. Then

$$d(\mathbf{x}, C) = \inf_{\mathbf{y} \in C} \|\mathbf{x} - \mathbf{y}\| \tag{5.4-2}$$

is called *the distance of a point $\mathbf{x}$ to the set C*.

Definition 2. Let $C_1, C_2, .., C_m$ be m sets in a Hilbert space $\mathbf{H}$. The *summed distance error* (SDE) of a point $\mathbf{x}$ from the sets $C_1, C_2, ..., C_m$ is defined by

$$J(\mathbf{x}) \stackrel{\triangle}{=} \sum_{i=1}^{m} d(\mathbf{x}, C_i). \tag{5.4-3}$$

Remarks

1. It can be shown that if set C is closed then the distance $d(\mathbf{x}, C) = 0$ implies that $\mathbf{x} \in C$, or equivalently, $d(\mathbf{x}, C)$ is never zero for any $\mathbf{x} \notin C$ (see Exercise 5-1). As a result, the summed distance error $J(\mathbf{x})$ of a point $\mathbf{x}$ from the sets $C_1, C_2, ..., C_m$ is zero if and only if the point $\mathbf{x}$ lies in the intersection of all the sets $C_1, C_2, ..., C_m$.

2. From the discussion in Chapter 2 it is clear that when the set C is closed and convex there is a unique point in C that achieves the minimal distance $d(\mathbf{x}, C)$ for any $\mathbf{x}$ in $\mathbf{H}$. This point, of course, is the projection of $\mathbf{x}$ onto C, i.e., $P\mathbf{x}$. In short, we have $d(\mathbf{x}, C) = \|\mathbf{x} - P\mathbf{x}\|$.

3. When the set C is not convex, however, in general, there is no assurance of the existence of such a point in C that achieves the minimal distance $d(\mathbf{x}, C)$, even when C is closed. That is, such a point may not even exist. For example, after some reflection, we come up with the following example. In the l^2 space let C be the set of vectors of the following form

$$\mathbf{e}_n = \left(0, 0, \cdots, 0, 1 + \frac{1}{n}, 0, \cdots \right),$$

$n = 1, 2, 3, \cdots$, that is, only the nth component of $\mathbf{e}_n$ is nonzero and has value $1 + 1/n$. The set C is closed but nonconvex (see Exercise 5-2). It is easy to check that the distance of the zero vector $\mathbf{0}$ to the set C is 1, i.e., $d(\mathbf{0}, C) = 1$. Apparently, none of the vectors $\mathbf{e}_n$ in C achieve this distance.

At this point it seems rather discouraging to extend the concept of projection through the use of minimal distance $d(\mathbf{x}, C)$ to a nonconvex set, because the "projection" may not even exist in the first place. Fortunately, most of the sets that arise in a practical situation do admit to such a natural extension. Even more encouraging we have the following result:

Theorem 5.4-1 *In a finite-dimensional Hilbert space $\mathbf{H}$ let C be a closed set. Then for any point $\mathbf{x}$ in $\mathbf{H}$ there exists at least one point in C that achieves the distance of $\mathbf{x}$ to C, i.e., $d(\mathbf{x}, C)$.*

Proof: Let $\delta = d(\mathbf{x}, C)$, i.e.,

$$\delta = \inf_{\mathbf{y} \in C} \|\mathbf{x} - \mathbf{y}\|. \tag{5.4-4}$$

Then, for each $\delta_n = \delta + 1/n$, $n = 1, 2, \cdots$, there exists a point $\mathbf{y}_n \in C$ such that $\|\mathbf{x} - \mathbf{y}_n\| < \delta_n$. Therefore, we obtain a sequence $\{\mathbf{y}_n\}$ such that

$$\lim_{n \to \infty} \|\mathbf{x} - \mathbf{y}_n\| = \delta. \tag{5.4-5}$$

Observe that

$$\|\mathbf{y}_n\| \leq \|\mathbf{x}\| + \|\mathbf{x} - \mathbf{y}_n\| < \|\mathbf{x}\| + \delta + 1. \tag{5.4-6}$$

Therefore, the sequence $\{\mathbf{y}_n\}$ is bounded. By Theorem 2.8-2, the sequence $\{\mathbf{y}_n\}$ contains at least one weakly convergent subsequence, say $\{\mathbf{y}_{n_k}\}$, for some index sequence $n_1 < n_2 < n_3 < \cdots$. Assume that the weak limit of this subsequence is $\mathbf{y}^* \in \mathbf{H}$. Then, by Theorem 1.5-3, $\mathbf{y}^*$ is also the strong limit of the sequence $\{\mathbf{y}_{n_k}\}$ since, by assumption, $\mathbf{H}$ is finite-dimensional. That is,

$$\lim_{k \to \infty} \|\mathbf{y}_{n_k} - \mathbf{y}^*\| = 0. \tag{5.4-7}$$

Thus $\mathbf{y}^* \in C$ since C is closed. By Eq. (5.4-5), we also have

$$\lim_{k \to \infty} \|\mathbf{y}_{n_k} - \mathbf{x}\| = \delta. \tag{5.4-8}$$

Hence,

$$\|\mathbf{x} - \mathbf{y}^*\| = \|(\mathbf{x} - \mathbf{y}_{n_k}) + (\mathbf{y}_{n_k} - \mathbf{y}^*)\| \le \|\mathbf{y}_{n_k} - \mathbf{x}\| + \|\mathbf{y}_{n_k} - \mathbf{y}^*\|. \tag{5.4-9}$$

Letting $k \to \infty$, we obtain

$$\|\mathbf{x} - \mathbf{y}^*\| \le \delta. \tag{5.4-10}$$

By the definition of δ, the theorem follows. ∎

Generalized Projections

Let C be a closed set in a Hilbert space $\mathbf{H}$ and $\mathbf{x}$ a point in $\mathbf{H}$. If there exists a point $\mathbf{x}^*$ in C such that the distance from $\mathbf{x}$ to C is achieved, that is, $\|\mathbf{x} - \mathbf{x}^*\| = d(\mathbf{x}, C)$ and $\mathbf{x}^* \in C$, then $\mathbf{x}^*$ is called the *generalized projection* of $\mathbf{x}$ onto C. In short, we write $\mathbf{x}^* = P\mathbf{x}$ and P is called the *generalized projector* onto C.

Remarks

1. When the set C is convex, the generalized projection is simply the projection of $\mathbf{x}$ onto C.

2. When C is nonconvex there may be more than one point that satisfies the definition of a generalized projection. However, in practice, we may also find a procedure for uniquely choosing one of these points. For instance, in the *restoration from magnitude* problem, in projecting onto the set of functions with prescribed Fourier magnitude, the phase of the estimate at every iteration uniquely defines the projection. This is discussed in greater detail in Section 7.3.

Generalized Projection Algorithm

In anticipation of the use of f, g, h, etc. for representing two-dimensional images in later chapters, we switch at this point from vector notation to function notation. Thus, instead of seeing $\mathbf{x}, \mathbf{y}, \mathbf{z}$, etc., the reader is more likely to see f, g, h, or even $f(x), g(x), h(x)$, etc.

We want to investigate the properties of the recursive algorithm

$$f_{n+1} = T_{1,n} T_{2,n} \cdots T_{m,n} f_n, \qquad n = 0, 1, 2, \ldots, \tag{5.4-11}$$

where

$$T_{i,n} \overset{\triangle}{=} I + \lambda_{i,n}(P_i - I), \qquad i = 1, 2, \ldots, m, \tag{5.4-12}$$

I is the identity operator, and $\lambda_{i,n}$ is a relaxation parameter that can be adjusted from iteration to iteration (which explains why we add the subscript n in $\lambda_{i,n}$).

In particular, let us consider Eq. (5.4-11) when $m = 2$. We are *not* assuming that the sets C_i, $i = 1, \ldots, m$, upon which we project are convex. The problem of $m = 2$ is not trivial: a famous $m = 2$ case of Eq. (5.4-11), involving non-convex sets, is the so-called *Gerchberg-Saxton algorithm* [1] for restoring the phase of a complex optical field from image and diffraction plane pictures. This important problem has become a classic in optical information processing.

For $m = 2$, Eq. (5.4-11) becomes

$$f_{n+1} = T_{1,n} T_{2,n} f_n, \qquad f_0 \text{ arbitrary.} \tag{5.4-13}$$

To unburden the reader from trying to decipher excessive notation, let's agree to submerge the subscript n in $T_{i,n}$, so that instead of Eq. (5.4-13), we write

$$f_{n+1} = T_1 T_2 f_n, \qquad f_0 \text{ arbitrary.} \tag{5.4-14}$$

This also means that the n is submerged in $\lambda_{i,n}$ so that instead of writing $T_{i,n} \overset{\triangle}{=} I + \lambda_{i,n}(P_i - I)$, we write $T_i \overset{\triangle}{=} I + \lambda_i(P_i - I)$. However, it should be understood that λ_1, λ_2 and therefore T_1 and T_2 still depend on the iteration number n. Since the sets involved in Eq. (5.4-14) may not be convex, we cannot use the notion of weak or strong convergence to describe their performance. Instead, we use the SDE defined in Eq. (5.4-3). Let us now specialize the SDE in Eq. (5.4-3) for the generalized projection algorithm.

Let C_1, C_2 be any two sets with projectors P_1, P_2, respectively, and let T_1, T_2 be defined by Eq. (5.4-12).[†] Let f_n be the estimate of f at the nth iteration of Eq. (5.4-14); then as a *performance measure* at f_n, we take $J(f_n)$, i.e., the sum of distances between the point f_n and the sets C_1 and C_2. Thus the performance is measured by the SDE for the case $M = 2$. Specifically,

$$J(f_n) \overset{\triangle}{=} \|P_1 f_n - f_n\| + \|P_2 f_n - f_n\| \tag{5.4-15}$$

and similarly,

$$J(T_2 f_n) = \|P_1 T_2 f_n - T_2 f_n\| + \|P_2 T_2 f_n - T_2 f_n\|. \tag{5.4-16}$$

[†]Recall the dependence on iteration number n is submerged.

Note that $J(f_n) \geq 0$ and $J(f_n) = 0$ if and only if $f_n \in C_1 \cap C_2$. We are now in a position to state a theorem that describes the set-distance reduction property of the recursion given by Eq. (5.4-14). Credit for this theorem belongs to A. Levi [2].

Theorem 5.4-2 (*Fundamental Theorem of Generalized Projections*) *The recursion given by Eq. (5.4-14) has the property*

$$J(f_{n+1}) \leq J(T_2 f_n) \leq J(f_n) \tag{5.4-17}$$

for every λ_1 and λ_2 that satisfy

$$0 \leq \lambda_i \leq \frac{A_i^2 + A_i}{A_i^2 + A_i - \frac{1}{2}(A_i + B_i)}, \qquad i = 1, 2, \tag{5.4-18}$$

where

$$A_1 \triangleq \frac{\|P_1 T_2 f_n - T_2 f_n\|}{\|P_2 T_2 f_n - T_2 f_n\|}, \tag{5.4-19}$$

$$A_2 \triangleq \frac{\|P_2 f_n - f_n\|}{\|P_1 f_n - f_n\|}, \tag{5.4-20}$$

and

$$B_1 \triangleq \frac{\Re\langle P_2 T_2 f_n - T_2 f_n, P_1 T_2 f_n - T_2 f_n \rangle}{\|P_2 T_2 f_n - T_2 f_n\|^2}, \tag{5.4-21}$$

$$B_2 \triangleq \frac{\Re\langle P_1 f_n - f_n, P_2 f_n - f_n \rangle}{\|P_1 f_n - f_n\|^2}. \tag{5.4-22}$$

The result given in Eq. (5.4-17) is called the *SDE reduction property*. The proof of this theorem is given in Section 5.6. However, Section 5.6 can be omitted on a first reading without any loss of continuity.

Remarks

1. The properties of the performance measure $J(f_n)$ make $J(f_n)$ a good quantity for monitoring the performance and controlling the parameters of the algorithm given by Eq. (5.4-14).

2. It can easily be verified, by using *Schwarz inequality*, that $A_i \geq B_i, i = 1, 2$ and hence,

$$A_i \geq \frac{1}{2}(A_i + B_i) \geq 0, \qquad i = 1, 2. \tag{5.4-23}$$

Therefore, by using Eq. (5.4-23) in Eq. (5.4-18), we conclude that the value of unity will always be included in the range of $\lambda_i, i = 1, 2$, as given in Eq. (5.4-18). This means that the algorithm given by

$$f_{n+1} = P_1 P_2 f_n, \qquad f_0 \text{ arbitrary}, \tag{5.4-24}$$

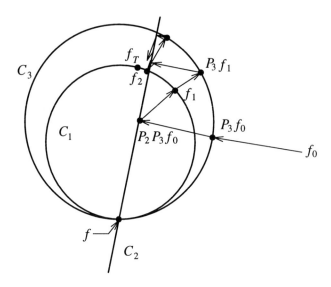

Fig. 5.4-1 An example that shows that summed-distance error reduction is not guaranteed when more than two sets are involved, of which at least one is non-convex. Note that convergence is to the non-feasible solution f_T instead of the feasible f.

will always have the property of summed distance error reduction.[†]

3. Theorem 5.4-1 is also true if $A_i \to \infty$, in which case the range of λ_i, as given in Eq. (5.4-12), becomes $0 \le \lambda_i \le 1, i = 1, 2.$[‡]

4. The SDE reduction property holds for *at least* those values of λ_i given by Eq. (5.4-18). In fact, when we examine some approximations made in the proof of the theorem, we can see that this property holds for a more extended range of λ_i than that given by Eq. (5.4-18).

The set-distance reduction property cannot be generalized to more than two sets. For two sets the assertion $J(f_{n+1}) \le J(f_n)$ follows *a fortiori* from Eq. (5.4-17). For more than two sets we cannot assert this (for any m, $J(f_n)$ is defined as the sum of distances from f_n to the sets $C_1, C_2..., C_m$). To demonstrate our inability to assert $J(f_{n+1}) \le J(f_n)$ for $m > 2$, we give the following simple counter example:

Consider the three sets described in Fig. 5.4-1. The sets C_1 and C_3 are two circles and C_2 is a line. Note that C_1 and C_3 are nonconvex. Suppose we want to find the point f which is the intersection point of C_1, C_2, and C_3 by the recursion

$$f_{n+1} = P_1 P_2 P_3 f_n, \qquad f_0 \text{ arbitrary.} \qquad (5.4\text{-}25)$$

[†]However, that does not make the algorithm in Eq. (5.4-24) optimal as far as convergence speed is concerned.

[‡]For example, $\|P_1 f_n - f_n\| \to 0$ implies $A_2 \to \infty$.

In Fig. 5.4-1 we indicate the initial point f_0 and the points generated after applying the projections $P_3, P_2, P_1, P_3, \ldots$ successively. From f_0 the iterates move in the direction of the arrows to points $f_1, f_2, \cdots$. The main conclusions drawn from this simple example are:

1. The algorithm diverges, i.e., if $\|e_n\| \overset{\triangle}{=} \|f_n - f\|$ where f is the solution, then

$$\|e_1\| \le \|e_2\| \le \ldots \; ; \tag{5.4-26}$$

2. For $n \ge 2, J(f_{n+1}) \ge J(f_n)$;

3. There exists in this case a *trap point* f_T, i.e., a fixed point of the operator $P_1 P_2 P_3$ with $f_n \to f_T$. The point f_T satisfies $P_1 P_2 P_3 f_T = f_T$. Thus for $m > 2$, the set-distance reduction property does not hold.

Despite the fact that the theorem is not valid for $m > 2$, the algorithm given by Eq. (5.4-14) is not so restrictive in practice. As already stated, the theorem does not restrict the complexity of the sets and, therefore, C_1 and C_2 can indeed include signals with multiple constraints. However, as the sets become more complex, so does the calculation of the projection operators. As we shall see, however, in image processing we can usually combine several properties of the signal f and associate them with only one set. As a rule we can combine those properties of the signal, that are easily expressed in the space domain in one set C_1, whose associated projector P_1 can be calculated without too much effort. Similarly, the properties of the signal that are easily expressed in the transform domain can be combined into a second set C_2, and the corresponding projector P_2 can again be calculated without too much effort. An example is given below.

Example 5.4-1 Let C_1 be the set of functions of compact support which are zero for all points $x \notin S_1$ and have a space-domain magnitude which is equal to some prescribed positive function $P(x)$ in a set S_2 with $S_2 \subset S_1$ (i.e., the set S_2 is included in S_1). Thus,

$$C_1 = \{y(x) : y(x) = 0 \text{ for all } x \notin S_1, \text{ and } |y(x)| = P(x) \text{ for all } x \in S_2\}. \tag{5.4-27}$$

It is easily shown that, for any arbitrary $f(x)$,

$$g \overset{\triangle}{=} P_1 f = \begin{cases} 0, & x \notin S_1, \\ f(x), & x \in S_1 \cap S_2^c, \\ P(x), & x \in S_2, f(x) > 0, \\ -P(x), & x \in S_2, f(x) \le 0. \end{cases} \tag{5.4-28}$$

Equation (5.4-28) follows from minimizing the equation

$$\|f - y\|^2 = \int_{x \in S_1^c} |f(x) - y(x)|^2 dx + \int_{x \in S_1 \cap S_2^c} |f(x) - y(x)|^2 dx$$

$$+ \int_{x \in S_2, f(x) > 0} |f(x) - y(x)|^2 dx$$

$$+ \int_{x \in S_2, f(x) \leq 0} |f(x) - y(x)|^2 dx, \tag{5.4-29}$$

where we used the condition $S_2 \subset S_1$. Since minimizing $\|f - y\|^2$ with respect to all functions $y \in C_1$ is equivalent to minimizing each integral on the right-hand side of Eq. (5.4-29) independently of the others, we immediately obtain Eq. (5.4-28) for the function $y \in C_1$ that minimizes $\|f - y\|^2$.   ∎

Nevertheless, there are times when the restriction to $m = 2$ does become a significant problem. For example, in the filter design problem discussed in Chapter 6 (Section 6.4), we seek a solution in the intersection of many non-convex sets. In that case, the theory of *generalized projections in a product space* can be used to advantage. We discuss this approach in Section 5.7.

5.5 TRAPS AND TUNNELS

In Section 5.1 we introduced the idea of a trap. As the reader already knows, a trap represents a point where the SDE has a local minimum. We revisit the idea of a trap here, as well as another undesirable phenomenon known as *tunnel*. We define a trap as a fixed point[†] of the composition operator $T_1 T_2 \cdots T_m$, which is not a fixed point of every individual $T_i = 1, \cdots, m$, i.e., a point which fails to satisfy one or more of the *a priori* constraints yet satisfies

$$f_{n+1} = T_2 T_2 \cdots T_m f_n = f_n. \tag{5.5-1}$$

We say that a point f_n is in a tunnel if Eq. (5.5-1) is almost satisfied, which means that the change in f_n from one iteration to the next is negligible. Traps and tunnels are illustrated in Fig. 5.1-1 and Fig. 5.5-1, respectively. In general, when at least one nonconvex set is involved, traps may exist as demonstrated in Fig. 5.4-1. Since Eq. (5.4-11) can exhibit SDE convergence only when $m = 2$, let us consider this case only. In [3] the following remarks, of practical utility, are demonstrated. We state them here without proof.

(i) The SDE $J(f_n)$ can be used to detect traps. By this we mean that a trap can be detected when we observe no change in $J(f_n) > 0$ from iteration to iteration. To determine the existence of a trap from observations on the SDE, we must show that $J^*(f_{n+1}) = J(f_n)$ implies that $f_{n+1} = f_n$ and vice versa; the asterisk denotes the minimum of $J(f_{n+1})$ with respect to λ_1 and λ_2. In [3, 4] such a

---

[†]Reminder: the fixed point x of an operator L is the point for which $Lx = x$. In engineering, the point x is, typically, a function of n arguments, e.g., a waveform ($n = 1$) or an image ($n = 2$).

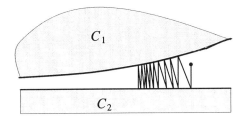

Fig. 5.5-1 A tunnel is a region where set boundaries are nearly parallel and convergence is very slow. Tunnels can occur with convex or non-convex sets.

demonstration is furnished for some restricted cases. One can infer a trap when $J^*(f_{n+1}) > 0$ and $\|f_{n+1} - f_n\| = 0$.[†]

(ii) If P_1 is linear and $P_1T_2f_n = f_n$, then the correct solution f lies in a hyperplane orthogonal to the vector $P_2f_n - f_n$.

Remark (ii) is especially useful. It tells us that when we are in a trap or a tunnel we have to look for a solution along a direction orthogonal to the vector $P_2f_n - f_n$. More research is needed in order to use this fact to improve the algorithm and avoid traps and tunnels. If P_1 is not linear, then the solution does not necessarily lie on an orthogonal direction to $P_2f_n - f_n$, but we can look at this result as approximately true.[‡] Of course, how good this approximation is depends on P_1, P_2, and f_n.

Finally, when dealing with sets that are non-convex or set configurations that have tunnels, the optimization of the relaxation parameters $\lambda_{i,n}$ in Eq. (5.4-12) can have significant impact on the performance of the algorithm. This important point is discussed by Levi in [3, 4]. Examples of generalized projection applications appear in phase retrieval (Chapter 7), neural nets (Chapter 8), and image synthesis (Chapter 9).

5.6 PROOF OF THEOREM 5.4-2

From the definition of the generalized projection operator, we obtain

$$\|P_1T_2f_n - T_2f_n\| \le \|y - T_2f_n\| \qquad \text{for all } y \in C_1, \qquad (5.6\text{-}1)$$

and similarly

$$\|P_2T_2f_n - T_2f_n\| \le \|h - T_2f_n\| \qquad \text{for all } h \in C_2. \qquad (5.6\text{-}2)$$

[†]On a computer, however, it is not always easy to distinguish between traps and tunnels. Finite word length and/or numerical errors may mask the fact that $f_{n+1} \ne f_n$ (a tunnel).
[‡]Unless the morphologies of the non-convex sets are bizarre.

By the definition of $J(T_2 f_n)$ given by Eq. (5.4-16)), we can write

$$J(T_2 f_n) \leq \|y - T_2 f_n\| + \|h - T_2 f_n\| \qquad \text{for all } y \in C_1 \text{ and } h \in C_2. \quad (5.6\text{-}3)$$

In particular, let us choose $y = P_1 f_n$ and $h = P_2 f_n$. Then,

$$J(T_2 f_n) \leq \|P_1 f_n - T_2 f_n\| + \|P_2 f_n - T_2 f_n\|. \qquad (5.6\text{-}4)$$

Now, from the definition of $T_2 \triangleq I + \lambda_2 (P_2 - I)$ and by replacing the first norm by its definition as an inner product, we obtain

$$\begin{aligned} J(T_2 f_n) \quad \leq \quad & [\, \|P_1 f_n - f_n\|^2 - 2\lambda_2 \, \Re\langle P_1 f_n - f_n, P_2 f_n - f_n\rangle \\ & + \lambda_2^2 \|P_2 f_n - f_n\|^2]^{\frac{1}{2}} + |1 - \lambda_2| \, \|P_2 f_n - f_n\|. \quad (5.6\text{-}5) \end{aligned}$$

Let's first assume that $\|P_1 f_n - f_n\| > 0$. Then from the definitions of A_2 and B_2 given by Eqs. (5.4-20) and (5.4-22) and the definition of $J(f_n)$ given by Eq. (5.4-15) we can write

$$\begin{aligned} J(T_2 f_n) - J(f_n) \quad \leq \quad & (\sqrt{1 - 2\lambda_2 B_2 + \lambda_2^2 A_2^2} - 1) \, \|P_1 f_n - f_n\| \\ & + (|1 - 2\lambda_2| - 1) \, \|P_2 f_n - f_n\|. \quad (5.6\text{-}6) \end{aligned}$$

Since we want $J(T_2 f_n) - J(f_n) \leq 0$, we require that the right-hand side of Eq. (5.6-6)) also be non-positive. Therefore, we look for the range of values of λ_2 such that

$$\begin{aligned} & \left(\sqrt{1 - 2\lambda_2 B_2 + \lambda_2^2 A_2^2} - 1 \right) \, \|P_1 f_n - f_n\| \\ & + (|1 - \lambda_2| - 1)\|P_2 f_n - f_n| \leq 0. \quad (5.6\text{-}7) \end{aligned}$$

After some algebraic manipulations, Eq. (5.6-7) becomes

$$\sqrt{1 - 2\lambda_2 B_2 + \lambda_2^2 A_2^2} \leq 1 + A_2 - |1 - \lambda_2| A_2. \qquad (5.6\text{-}8)$$

First, we note that the right-hand side of Eq. (5.6-8) must be non-negative, which can be shown to be equivalent to the requirement

$$-\frac{1}{A_2} \leq \lambda_2 \leq 2 + \frac{1}{A_2}. \qquad (5.6\text{-}9)$$

Let us return to Eq. (5.6-8) and assume Eq. (5.6-9) is satisfied. By squaring both sides of Eq. (5.6-8) and rearranging terms we can write

$$\lambda_2 (A_2^2 - B_2) \leq (A_2^2 + A_2)(1 - |1 - \lambda_2|). \qquad (5.6\text{-}10)$$

From the Schwartz inequality and the definitions of A_2 and B_2, we obtain

$$2A_2 > A_2 + B_2 \geq 0. \tag{5.6-11}$$

By using Eq. (5.6-11) and some algebraic manipulations, the requirement given by Eq. (5.6-10) transforms to

$$0 \leq \lambda_2 \leq \frac{A_2^2 + A_2}{A_2^2 + A_2 - \frac{1}{2}(A_2 + B_2)} \; . \tag{5.6-12}$$

We also note that by using Eq. (5.6-11) we can write

$$\frac{A_2^2 + A_2}{A_2^2 + A_2 - \frac{1}{2}(A_2 + B_2)} \leq \frac{A_2^2 + A_2}{A_2^2} = 1 + \frac{1}{A_2} < 2 + \frac{1}{A_2} \; ;$$

therefore the requirement given by Eq. (5.6-9) is satisfied for every λ_2 that satisfies Eq. (5.6-12).

Next, if $\|P_1 f_n - f_n\| = 0$, then by Eq. (5.6-5)

$$J(T_2 f_n) \leq (|\lambda_2| + |1 - \lambda_2|) \|P_2 f_n - f_n\| = (|\lambda_2| + |1 - \lambda_2|) J(f_n)$$

and the requirement $J(T_2 f_n) - J(f_n) \leq 0$ is satisfied if

$$|\lambda_2| + |1 - \lambda_2| \leq 1 \tag{5.6-13}$$

or

$$0 \leq \lambda_2 \leq 1. \tag{5.6-14}$$

Since by taking $A_2 \to \infty$ in Eq. (5.6-12), we obtain Eq. (5.6-14), we conclude that Eq. (5.6-12) is true for every value of A_2 .

By defining $g_n \triangleq T_2 f_n$ we can repeat the same calculations for $J(T_1 g_n)$ and we can show that the requirement $J(T_1 g_n) - J(g_n) \leq 0$ is satisfied for all λ_1 for which

$$0 \leq \lambda_1 \leq \frac{A_1^2 + A_1}{A_1^2 + A_1 - \frac{1}{2}(A_1 + B_1)} \; , \tag{5.6-15}$$

where A_1 and B_1 are given by Eqs. (5.4-19) and (5.4-21), respectively.

A final remark is that the range of values for λ_1 and λ_2 given by Eqs. (5.6-15) and (5.6-12), respectively, were obtained with a particular choice y and h as described before. Since Eq. (5.6-3) is true for any choice of y, h provided that $y \in C_1$ and $h \in C_2$, it is clear that the set distance reduction property, Eq. (5.4-17), holds for *at least* the range of λ_1 and the range of λ_2 are given by Eqs. (5.6-15) and (5.6-12), respectively.

5.7 GENERALIZED PROJECTION ALGORITHM IN A PRODUCT VECTOR SPACE

From the development in the earlier sections of this chapter it seems rather restrictive that the summed distance error (SDE) reduction property of the generalized projection algorithm is valid when only two sets are involved. In this section, however, we demonstrate that the SDE reduction property can be extended to more general cases when more than two sets are involved. This is made possible by the technique of reformulating the problem in a product space that was introduced earlier in Section 2.9.

For ease of reference we repeat below some of the results developed in Section 2.9. Let $C_1, C_2, \cdots, C_m$ denote m closed, but not necessarily convex, sets in a Hilbert space **H**, and let C_0 denote their intersection set, i.e., $C_0 = \bigcap_{i=1}^{m} C_i$, which is assumed non-empty. Then the problem of finding a point in C_0 has an equivalent formulation in the product space defined by $\mathcal{H} \stackrel{\triangle}{=} \mathbf{H}^m$, i.e., the m-fold Cartesian product of the Hilbert space **H**, whose elements are ordered m-tuples $(\mathbf{x}_1, \mathbf{x}_2, \cdots, \mathbf{x}_m)$, where $\mathbf{x}_i \in \mathbf{H}$ for each $i = 1, 2, \cdots, m$, and will be simply denoted by **X**. The inner product on $\mathcal{H}$ and its induced norm are denoted, respectively, by $\langle\langle \cdot, \cdot \rangle\rangle$ and $||| \cdot |||$, with

$$\langle\langle \mathbf{X}, \mathbf{Y} \rangle\rangle \stackrel{\triangle}{=} \sum_{i=1}^{m} w_i \langle \mathbf{x}_i, \ \mathbf{y}_i \rangle, \tag{5.7-1}$$

and

$$|||\mathbf{X}|||^2 = \langle\langle \mathbf{X}, \mathbf{X} \rangle\rangle = \sum_{i=1}^{m} w_i \|\mathbf{x}_i\|^2, \tag{5.7-2}$$

where $\mathbf{X} = (\mathbf{x}_1, \mathbf{x}_2, \cdots, \mathbf{x}_m) \in \mathcal{H}$, $\mathbf{Y} = (\mathbf{y}_1, \mathbf{y}_2, \cdots, \mathbf{y}_m) \in \mathcal{H}$, and $\langle \cdot, \ \cdot \rangle$ and $\| \cdot \|$ are, respectively, the inner product and the norm defined on **H**.

In the space $\mathcal{H}$ we define the following sets:

$$\mathbf{C} = \{ \ \mathbf{X} : \ \mathbf{X} = (\mathbf{x}_1, \mathbf{x}_2, \cdots, \mathbf{x}_m) \in \mathcal{H}, \ \text{and } \mathbf{x}_i \in C_i \text{ for each } i = 1, 2, \cdots, m \ \}, \tag{5.7-3}$$

and

$$\mathbf{D} = \{ \ \mathbf{X} : \ \mathbf{X} = (\mathbf{x}_1, \mathbf{x}_2, \cdots, \mathbf{x}_m) \in \mathcal{H}, \ \text{and } \mathbf{x}_1 = \mathbf{x}_2 = \cdots = \mathbf{x}_m \ \}. \tag{5.7-4}$$

In words, the set **C** is the set of all ordered m-tuples $(\mathbf{x}_1, \mathbf{x}_2, \cdots, \mathbf{x}_m)$ such that $\mathbf{x}_i, i = 1, 2, \cdots, m$, has membership in C_i. Likewise, the set **D** is the set of all m-tuples $(\mathbf{x}_1, \mathbf{x}_2, \cdots, \mathbf{x}_m)$ such that each component is equal to all others.

Then an m-tuple, say $\mathbf{X}^*$, in the intersection of **C** and **D** will contain a vector in C_0, i.e., a solution to the problem in the original vector space **H**. This is true because, on the one hand, $\mathbf{X}^* \in \mathbf{D}$ so it must be of the form $\mathbf{X}^* = (\mathbf{x}^*, \mathbf{x}^*, \cdots, \mathbf{x}^*)$ for some $\mathbf{x}^* \in \mathbf{H}$; and, on the other hand, $\mathbf{X}^* \in \mathbf{C}$ so $\mathbf{x}^* \in C_i$ for each $i =$

$1, 2, \cdots, m$. Thus, we have $\mathbf{x}^* \in C_0$.

The set $\mathbf{D}$ is convex and closed in $\mathcal{H}$. In fact, $\mathbf{D}$ is a closed subspace. For $\mathbf{X} = (\mathbf{x}_1, \mathbf{x}_2, \cdots, \mathbf{x}_m) \in \mathcal{H}$, its projection onto $\mathbf{D}$ is given in Eq. (2.9-11), i.e.,

$$P_{\mathbf{D}}\mathbf{X} = \left(\sum_{i=1}^{m} w_i \mathbf{x}_i, \sum_{i=1}^{m} w_i \mathbf{x}_i, \cdots, \sum_{i=1}^{m} w_i \mathbf{x}_i \right). \qquad (5.7\text{-}5)$$

The set $\mathbf{C}$ is closed in $\mathcal{H}$. However, it is nonconvex if any of the sets C_i are nonconvex in $\mathbf{H}$. In such a case for $\mathbf{X} = (\mathbf{x}_1, \mathbf{x}_2, \cdots, \mathbf{x}_m) \in \mathcal{H}$, its generalized projection onto $\mathbf{C}$ is given by

$$P_{\mathbf{C}}\mathbf{X} = (P_1 \mathbf{x}_1, P_2 \mathbf{x}_2, \cdots, P_m \mathbf{x}_m), \qquad (5.7\text{-}6)$$

where $P_i, i = 1, 2, \cdots, m$, is the generalized projector onto C_i in $\mathbf{H}$ when C_i is nonconvex and otherwise the projector onto C_i.

To find a point in $\mathbf{C} \cap \mathbf{D}$, the generalized projection algorithm can be used, since there are only *two* sets involved so the Fundamental Theorem of Generalized Projections (Theorem 5.4-2) guarantees the SDE reduction property of the algorithm.

Theorem 5.7-1 (*Extended generalized projections*) *Let*

$$T_{\mathbf{D},n} \overset{\triangle}{=} I + \lambda_{\mathbf{D},n}(P_{\mathbf{D}} - I) \text{ and } T_{\mathbf{C},n} \overset{\triangle}{=} I + \lambda_{\mathbf{C},n}(P_{\mathbf{C}} - I), \qquad (5.7\text{-}7)$$

where I is the identity operator on $\mathcal{H}$. Then the iterates generated by the generalized projection algorithm in its most general form

$$\mathbf{X}_{n+1} = T_{\mathbf{D},n} T_{\mathbf{C},n} \mathbf{X}_n \qquad (5.7\text{-}8)$$

with $\mathbf{X}_0$ arbitrary in $\mathcal{H}$ will satisfy

$$J(\mathbf{X}_{n+1}) \leq J(T_{\mathbf{C},n} \mathbf{X}_n) \leq J(\mathbf{X}_n), \qquad (5.7\text{-}9)$$

where $J(\cdot)$ is the SDE defined by

$$J(\mathbf{X}_n) \overset{\triangle}{=} \|\|P_{\mathbf{C}}\mathbf{X}_n - \mathbf{X}_n\|\| + \|\|P_{\mathbf{D}}\mathbf{X}_n - \mathbf{X}_n\|\|, \qquad (5.7\text{-}10)$$

provided that $\lambda_{\mathbf{D},n}$ and $\lambda_{\mathbf{C},n}$ in Eq. (5.7-7) satisfy

$$0 \leq \lambda_{i,n} \leq \frac{A_{i,n}^2 + A_{i,n}}{A_{i,n}^2 + A_{i,n} - \frac{1}{2}(A_{i,n} + B_{i,n})}, \quad i = \mathbf{C}, \mathbf{D}, \qquad (5.7\text{-}11)$$

and where

$$A_{\mathbf{D},n} \overset{\triangle}{=} \frac{\|\|P_{\mathbf{D}}T_{\mathbf{C},n}\mathbf{X}_n - T_{\mathbf{C},n}\mathbf{X}_n\|\|}{\|\|P_{\mathbf{C}}T_{\mathbf{C},n}\mathbf{X}_n - T_{\mathbf{C},n}\mathbf{X}_n\|\|} \qquad (5.7\text{-}12)$$

$$A_{\mathbf{C},n} \triangleq \frac{|\!|\!| P_{\mathbf{C}} \mathbf{X}_n - \mathbf{X}_n |\!|\!|}{|\!|\!| P_{\mathbf{D}} \mathbf{X}_n - \mathbf{X}_n |\!|\!|} \tag{5.7-13}$$

and

$$B_{\mathbf{D},n} \triangleq \frac{\Re\langle\langle P_{\mathbf{C}} T_{\mathbf{C},n} \mathbf{X}_n - T_{\mathbf{C},n} \mathbf{X}_n, P_{\mathbf{D}} T_{\mathbf{C},n} \mathbf{X}_n - T_{\mathbf{C},n} \mathbf{X}_n \rangle\rangle}{|\!|\!| P_{\mathbf{C}} T_{\mathbf{C},n} \mathbf{X}_n - T_{\mathbf{C},n} \mathbf{X}_n |\!|\!|^2} \tag{5.7-14}$$

$$B_{\mathbf{C},n} \triangleq \frac{\Re\langle\langle P_{\mathbf{D}} \mathbf{X}_n - \mathbf{X}_n, P_{\mathbf{C}} \mathbf{X}_n - \mathbf{X}_n \rangle\rangle}{|\!|\!| P_{\mathbf{D}} \mathbf{X}_n - \mathbf{X}_n |\!|\!|^2} . \tag{5.7-15}$$

In particular, if the pure projectors $P_{\mathbf{C}}$ and $P_{\mathbf{D}}$ are used instead of their relaxed counterparts, then the generalized algorithm in Eq. (5.7-8) simplifies to

$$\mathbf{X}_{n+1} = P_{\mathbf{D}} P_{\mathbf{C}} \mathbf{X}_n. \tag{5.7-16}$$

Clearly, $\mathbf{X}_n \in \mathbf{D}$, so we have

$$J(\mathbf{X}_n) = |\!|\!| P_{\mathbf{C}} \mathbf{X}_n - \mathbf{X}_n |\!|\!|. \tag{5.7-17}$$

From the SDE reduction property in Eq. (5.7-9) we obtain

$$J(\mathbf{X}_{n+1}) \le J(\mathbf{X}_n). \tag{5.7-18}$$

This immediately leads to a *simultaneous* or *parallel* generalized projections algorithm in $\mathbf{H}$, which is summarized in the following:

Corollary 5.7-1 *For every* $\mathbf{x}_0 \in \mathbf{H}$ *and every choice of positive constants* $w_1, w_2,$ $\cdots, w_m$ *such that* $\sum_{i=1}^m w_i = 1$*, the sequence* $\{\,\mathbf{x}_n\,\}$ *generated by*

$$\mathbf{x}_{n+1} = \sum_{i=1}^m w_i P_i \mathbf{x}_n \tag{5.7-19}$$

will satisfy

$$J(\mathbf{x}_{n+1}) \le J(\mathbf{x}_n), \tag{5.7-20}$$

where

$$J(\mathbf{x}_n) \triangleq \left[\sum_{i=1}^m w_i \| P_i \mathbf{x}_n - \mathbf{x}_n \|^2 \right]^{1/2} . \tag{5.7-21}$$

Finally, we present a numerical example to demonstrate the SDE reduction property of the *parallel generalized projections algorithm* (PGPA). For comparison purposes the results from a *sequential generalized projections algorithm* (SGPA) are also presented.

Example 5.7-1 This example is the quantitative version of the one given Fig. 5.4-1. Specifically, in the R^2 space let $C_1, C_2,$ and C_3 be defined as, respectively,

$$C_1 = \{\,\mathbf{y} = (y_1, y_2) : y_1^2 + (y_2 - 3)^2 = 9\,\}, \tag{5.7-22}$$

$$C_2 = \{\ \mathbf{y} = (y_1, y_2) : y_1^2 + (y_2 - 2)^2 = 4\ \}, \qquad (5.7\text{-}23)$$

and

$$C_3 = \{\ \mathbf{y} = (y_1, y_2) : y_2 = 5y_1\ \}. \qquad (5.7\text{-}24)$$

The sets C_1 and C_2 represent two circles on the 2-D plane and C_3 represents a line. Clearly, C_1 and C_2 are both closed but nonconvex, while C_3 is closed and convex. Note that these three sets has only one point in common, that is, $C_1 \cap C_2 \cap C_3 = \{(0,0)\}$.

The generalized projectors P_1 and P_2 onto the set C_1 and C_2, respectively, are given by

$$P_1 \mathbf{x} = \left(\frac{3x_1}{\sqrt{x_1^2 + (x_2 - 3)^2}}\ , \ \frac{3(x_2 - 3)}{\sqrt{x_1^2 + (x_2 - 3)^2}} + 3 \right), \qquad (5.7\text{-}25)$$

and

$$P_2 \mathbf{x} = \left(\frac{2x_1}{\sqrt{x_1^2 + (x_2 - 2)^2}}\ , \ \frac{2(x_2 - 2)}{\sqrt{x_1^2 + (x_2 - 2)^2}} + 2 \right) \qquad (5.7\text{-}26)$$

for $\mathbf{x} = (x_1, x_2) \in R^2$. The projector P_3 onto C_3 is given by

$$P_3 \mathbf{x} = \left(\frac{x_1 + 5x_2}{26}\ , \ \frac{5(x_1 + 5x_2)}{26} \right). \qquad (5.7\text{-}27)$$

First, the parallel generalized projections algorithm (PGPA)

$$\mathbf{x}_{n+1} = \sum_{i=1}^{3} w_i P_i \mathbf{x}_n \qquad (5.7\text{-}28)$$

with $w_1 = w_2 = w_3 = 1/3$ is tested for different starting points $\mathbf{x}_0$. The SDE is now given by

$$J(\mathbf{x}_n) \triangleq \left[\frac{1}{3} \sum_{i=1}^{3} \| P_i \mathbf{x}_n - \mathbf{x}_n \|^2 \right]^{1/2}. \qquad (5.7\text{-}29)$$

Then, a sequential generalized projections algorithm (SGPA)

$$\mathbf{x}_{n+1} = P_3 P_2 P_1 \mathbf{x}_n \qquad (5.7\text{-}30)$$

is tested with the same set of starting points as in the PGPA given above for ease of comparison. The summed distance error now in the R^2 space is given by

$$J'(\mathbf{x}_n) \triangleq \sum_{i=1}^{3} \| P_i \mathbf{x}_n - \mathbf{x}_n \|. \qquad (5.7\text{-}31)$$

By way of comparison, we compute results obtained with both the PGPA and

Table 5.7-1 SDE values of the iterates generated by PGPA and SGPA with starting point $(3, -100)$. Data are given to four decimal places.

Iteration n	PGPA		SGPA	
	$J(\mathbf{f}_n)$	$J'(\mathbf{f}_n)$	$J(\mathbf{f}_n)$	$J'(\mathbf{f}_n)$
1	26.4834	64.9184	0.0152	0.0372
2	8.8295	21.9240	0.0006	0.0014
3	2.9499	7.5145	0.0000	0.0000
4	0.9981	2.6867		
5	0.3589	1.0452		
6	0.1552	0.4634		
7	0.0866	0.2390		
8	0.0560	0.1401		
9	0.0379	0.0892		
10	0.0259	0.0594		
11	0.0178	0.0404		
12	0.0122	0.0278		
20	0.0006	0.0014		

the SGPA using both SDE formulas, Eqs. (5.7-29) and (5.7-31). The reader should recall, however, that Eq. (5.7-29) is, strictly speaking, the SDE appropriate for the PGPA, while Eq. (5.7-31) is appropriate for the SGPA.

Some results are given below. When the starting points are $(3, -100)$ and $(-10, 2)$ both parallel and sequential projections algorithms converge to the correct solution. This is clearly seen in Tables 5.7-1 and 5.7-2. However, when the starting point is $(2, 5)$, several interesting things can be observed from the results in Table 5.7-3: (i) The PGPA shows SDE reduction; (ii) The PGPA fails to converge to the correct solution and falls into a *trap* at the point $(0.7396, 3.7198)$; (iii) The SGPA *fails to exhibit* SDE reduction and falls into a trap at $(1.130, 4.686)$. However, and this may be the critical point, the PGPA *continues to exhibit* SDE reduction in all cases, although, like the SGPA, it is still subject to traps. ■

In summary, from the above numerical results it is clear that the iterates generated by the parallel generalized projections algorithm indeed satisfy the SDE reduction property, while those by the sequential generalized projections algorithm do not. Interestingly, the sequential algorithm can still lead to convergence to the correct solution for some starting points, even when the SDE reduction property fails to hold.

Another observation is that, even though the SDE reduction property holds for the simultaneous generalized projections algorithm, it is still possible for the algorithm to converge to a trap. This is true, since the reformulation of the problem in the product space as a two-set problem does not eliminate the inherent difficulty of converging to a trap when nonconvex sets are involved in the product space.

Table 5.7-2 SDE values of the iterates generated by PGPA and SGPA with starting point $(-10, 2)$.

	PGPA		SGPA	
Iteration n	$J(\mathbf{f}_n)$	$J'(\mathbf{f}_n)$	$J(\mathbf{f}_n)$	$J'(\mathbf{f}_n)$
1	1.2221	3.0255	1.4407	3.5153
2	1.0390	2.8148	0.3934	0.9635
3	0.8329	2.2343	0.0207	0.0508
4	0.6118	1.5750	0.0008	0.0020
5	0.4184	1.0030	0.0000	0.0000
6	0.2758	0.5964		
7	0.1808	0.3444		
8	0.1198	0.2150		
9	0.0804	0.1582		
10	0.0545	0.1134		
11	0.0371	0.0801		
12	0.0254	0.0560		
20	0.0013	0.0060		

Table 5.7-3 SDE values of the iterates generated by PGPA and SGPA with starting point $(2, 5)$.

	PGPA		SGPA	
Iteration n	$J(\mathbf{f}_n)$	$J'(\mathbf{f}_n)$	$J(\mathbf{f}_n)$	$J'(\mathbf{f}_n)$
1	0.8015	2.3118	1.1315	2.0726
2	0.7823	2.2041	1.1350	2.0848
3	0.7799	2.1638	1.1361	2.0889
4	0.7789	2.1418	1.1365	2.0903
5	0.7783	2.1266	1.1367	2.0907
6	0.7779	2.1151	1.1367	2.0909
7	0.7777	2.1060	1.1367	2.0910
8	0.7776	2.0989	1.1367	2.0910
9	0.7775	2.0933	1.1367	2.0910
10	0.7774	2.0889	1.1367	2.0910
11	0.7774	2.0855	1.1367	2.0910
12	0.7774	2.0828		
13	0.7773	2.0807		
14	0.7773	2.0791		
15	0.7773	2.0778		
16	0.7773	2.0768		
17	0.7773	2.0761		
18	0.7773	2.0755		
19	0.7773	2.0750		
20	0.7773	2.0747		

5.8 SUMMARY

When sets are non-convex the extraordinary convergence properties of the method of *projections onto convex sets* (POCS) no longer apply. Nevertheless, because of the large number of practical problems that exist in which constraints are non-convex, it is useful to have at least some theory for this case. In this chapter we reviewed the properties of non-convex constraints and stated a fundamental theorem, which is quite useful in dealing with non-convex sets. This theorem states that, in any problem involving not more than two constraint sets, *summed distance error* (SDE) convergence will always take place, even if non-convex sets are involved. Of course, SDE convergence is not as useful as weak or strong convergence and permits convergence to non-feasible solutions called *traps*.

An alternative approach to dealing with non-convex sets is to adapt the product space concepts in Section 2.9 to the non-convex situation. Product spaces involving non-convex sets allow for an indefinite number of non-convex constraints to be employed in an algorithm that maintains the SDE reduction property.

REFERENCES

1. R. W. Gerchberg and W. O. Saxton, A practical algorithm for the determination of phase from image and diffractions plane pictures, *Optik*, **35**:237–246, 1972.

2. A. Levi, *Image Restoration by the Method of Projections with Applications to the Phase and Magnitude Retrieval Problems,* Ph.D Thesis, Rensselaer Polytechnic Institute, Dept. of ECSE, Troy, NY, Dec. 1983.

3. A. Levi and H. Stark, Restoration from phase and magnitude by generalized projections, chapter 8 in *Image Recovery: Theory and Applications* (H. Stark, ed.), Academic Press, Orlando, FL, 1987.

4. A. Levi and H. Stark, Image restoration by the method of generalized projections with applications to restoration from magnitude, *J. Opt. Soc. Am. A.* **1** : 932–943, 1984.

EXERCISES

5-1. Show that for a closed set C the distance $d(\mathbf{x}, C) = 0$ implies that $\mathbf{x} \in C$, or equivalently, $d(\mathbf{x}, C)$ is always positive for $\mathbf{x} \notin C$.

5-2. In the l^2 space let C be the set of vectors of the following form

$$\mathbf{e}_n = \left(0, 0, \cdots, 0, 1 + \frac{1}{n}, 0, \cdots \right),$$

that is, only the nth component of $\mathbf{e}_n$ is nonzero and has value $1 + \frac{1}{n}$. Show that the set C is closed but nonconvex.

5-3. Show that in the R^n space the set given in Eq. (5.3-5):

$$C_5 = \{(y_1, y_2, \cdots, y_n) : y_i = +1 \text{ or } -1, \quad i = 1, ..., n\}$$

is non-convex and closed. Derive the generalized projection onto C_5.

5-4. Show that in the L^2 space the set given in Eq. (5.3-1)

$$C_1 = \{y(t) : |y(t)| = g(t), \ 0 \le t \le T\},$$

where $g(t)$ is some known non-negative function is nonconvex and closed. Derive the generalized projection onto C_1.

5-5. Show that in the L^2 space the set given in Eq. (5.3-2):

$$C_2 = \{y(t) : y(t) \leftrightarrow Y(\omega), |Y(\omega)| = M(\omega), \omega \in \Omega\},$$

where $M(\omega)$ is some known magnitude function and Ω some known support is nonconvex and closed. Derive the generalized projection onto C_2.

5-6. Show that in the L^2 space the set given in Eq. (5.3-4)

$$C_4 = \{y(t) : y(t) \leftrightarrow Y(\omega), \alpha \le |Y(\omega)| \le \beta, \omega \in \Omega\},$$

where α and β are some known constants and Ω some known support is nonconvex and closed. Derive the generalized projection onto C_4.

5-7. Show that in the $L^2 \times L^2$ space the set given in Eq. (5.3-3)

$$C_3 = \left\{(h, x) : \int_{-\infty}^{\infty} h(\tau)x(t - \tau) = y(t), \ 0 \le t \le T\right\},$$

where $y(t)$ is some known function is nonconvex and closed.

5-8. Consider the following two sets in R^2

$$C_1 = \{\mathbf{x} : \mathbf{x} = (x_1, x_2), x_2 = +1 \text{ or } -1\},$$

and

$$C_2 = \{\mathbf{x} : \mathbf{x} = (x_1, x_2), x_1 = x_2\}.$$

Apply the generalized projection algorithm to find a point common to C_1 and C_2. Examine the effect of initial starting point on the final convergence point of the algorithm.

5-9. In the 2-D $x - y$ plane let C_1 represent the straight line $x = y$ and C_2 represent the unit circle $x^2 + y^2 = 1$. Apply the generalized projection algorithm to find an intersecting point of C_1 and C_2. Examine the effect of initial starting point on the final convergence point of the algorithm.

5-10. In the 2-D $x - y$ plane let C_1 represent the circle of radius 2 centered at $(-1, 0)$ and C_2 represent the circle of radius 2 centered at $(1, 0)$. Apply the generalized projection algorithm to find an intersecting point of C_1 and C_2. Examine the effect of initial starting point on the final convergence point of the algorithm.

5-11. Work out the steps in arriving at the projection in Example 5.4-1.

6

Applications to Communications

6.1 INTRODUCTION

In electrical communication systems, information can be lost due to sampling, quantizing, and bandwidth limitations, as well as external interference, signal compression, and still other factors. In this chapter we explore the application of projection methods to ameliorate the deleterious effects of some of the above phenomena.

Projection methods can also be used in the design of communication electronics. We illustrate this concept by using projection methods to design digital filters, an important component in the processing of information bearing signals.

One of the first well-known applications of *projections onto convex sets* (POCS) was the influential Papoulis algorithm [1], which did for time-signals what the earlier Gerchberg algorithm [2] did for optical fields. The Papoulis algorithm used projections to recover signal components "lost" because of insufficient bandwidth (the so-called *spectrum extrapolation* problem).

Since then, a large number of POCS-type algorithms have been developed for solving problems in communications and signal processing.

6.2 NON-UNIFORM SAMPLING

Any computer processing of continuous time (e.g., speech) or continuous space (e.g., photographic images) signals requires *sampling*. Normally a signal is sampled *uniformly*, which means that the sampling is done on a uniformly spaced, one-dimensional, or two-dimensional lattices (Fig. 6.2-1). As is well known, the

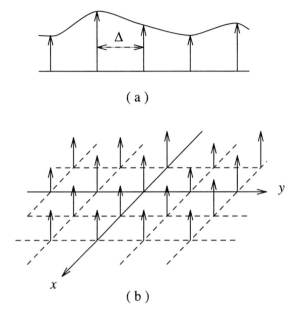

(a)

(b)

Fig. 6.2-1   Sampling on a uniform lattice: (a) one-dimensional case; (b) two-dimensional case.

reconstruction of a continuous, band-limited signal from its uniform samples is achieved through the famous Whittaker-Shannon-Kotelnikov sampling formula[†] [3]. However, there are instances where the sampling of the original signal is not done uniformly, either because it is disadvantageous to do so or because it is impossible to do so. An example of the latter is the estimation of wind velocity from non-uniformly spaced clouds in successive frames of satellite imagery [4]. Other examples of non-uniform sampling are in computer-aided tomography [5, 6] and magnetic resonance imaging [7]. In *adaptive sampling*, one might sample a slowly varying signal infrequently but increase the sampling rate when signal variations occur more rapidly. Once again, this would be a non-uniform sampling situation. Also, non-uniform sampling arises in *image fusion*, where the object is to recover a *high-resolution* image from several undersampled *low-resolution* images that have undergone affine transformations.

The problem of reconstructing a signal from its (non-uniform) samples can easily be solved by projection methods. Before describing how this is done, we first recall that when one is reconstructing a continuous signal from samples, from a practical

[†]Maybe the simplest and best known version of this formula is:

$$f(t) = \sum_n f(nT_s) \frac{\sin 2\pi B(t - nT_s)}{2\pi B(t - nT_s)} ,$$

where $T_s < 1/(2B)$ and B is the bandwidth of the signal.

standpoint, only a *finite number* of samples contribute to the reconstruction of the function, because the interpolating functions invariably decay in amplitude as one moves away from the reconstruction point. We use this fact in what follows.

We are given a sequence of samples $f(t_i), i = 1, 2, \cdots, N$, of an unknown real, band-limited function $f(t)$ and seek to find a real function, say $y(t)$, such that $y(t_i) = f(t_i), i = 1, 2, \cdots, N$, and $Y(\omega) = \mathcal{F}y(t) = 0$ for $|\omega| > 2\pi B$. The known parameter B is the bandwidth of $f(t)$ and the set of all $y(t)$ that satisfy all the constraints is the set of *feasible reconstructions* of $f(t)$. As N becomes large, this set becomes smaller in some sense, eventually becoming a single-point set as $N \to \infty$.[†]

Every problem solved by projection methods has one or more "key" sets, which may not always be obvious at first glance. In this case the key sets are the N sets described by, for $i = 1, 2, \cdots, N$:

$$C_i = \{y(t) : y(t_i) = f(t_i), \text{ and } Y(\omega) = 0 \text{ for } |\omega| \geq 2\pi B\}, \qquad (6.2\text{-}1)$$

where the elements $y \in C_i$ are members of the L^2 space of *real* functions[‡]. In words: C_i is the set of all real functions $\{ y(t) \}$ band-limited to B Hz, whose value at the sampling instant point t_i is the known number $f(t_i)$. The reconstructed function is an element of the set $C_0 \stackrel{\triangle}{=} \cap_{i=1}^{N} C_i$, i.e., the intersection of the N sets given in Eq. (6.2-1). The bandwidth constraint in Eq. (6.2-1) is extremely important: It limits the function $y(t)$ to be super-smooth, i.e., all derivatives of $y(t)$ must exist. It is easy to show that C_i is closed and convex. These and other properties of the set can be determined by the reader.

As usual, the reconstruction algorithm takes the form

$$f_{k+1} = P_N \cdots P_2 P_1 f_k, \qquad (6.2\text{-}2)$$

if all relaxation parameter are set equal to unity. Thus the central task is to compute the projector P_i that projects an arbitrary function $g(t)$ onto C_i. Actually this is done rather straightforwardly in the frequency domain as follows. We start with the Lagrange functional

$$J(y) = \|y - g\|^2 + \lambda [y(t_i) - f(t_i)], \qquad (6.2\text{-}3)$$

and seek a y, say y^*, that minimizes $J(y)$. In shorthand notation, this is restated as

$$y^* = \arg \min_{y \in C_i} J(y). \qquad (6.2\text{-}4)$$

In anticipation of using Parseval's theorem [8], we break up the frequency axis

---

[†]A useful measure of the size of the set is the least upper bound on the distance between two points in the set.

[‡]In keeping with previous notation, capital letters refer to the Fourier transform of the lower case function, e.g., $y(t) \leftrightarrow Y(\omega)$. Also, as usual, $\mathcal{F}$ denotes the Fourier operator.

$\{ -\infty < \omega < +\infty \}$ into two disjoint sets as

$$\{ -\infty < \omega < +\infty \} = \{ \omega : |\omega| \leq 2\pi B \} \cup \{ \omega : |\omega| > 2\pi B \}. \qquad (6.2\text{-}5)$$

We let $\Omega = \{ \omega : |\omega| \leq 2\pi B \}$ and $\Omega^c = \{ \omega : |\omega| > 2\pi B \}$. For $y(t) \in C_i$ the following properties hold:

$$
\begin{aligned}
y(t_i) &= \int_\Omega Y(\omega) e^{j\omega t_i} \frac{d\omega}{2\pi} \\
&= \int_\Omega [Y_R(\omega)\cos\omega t_i - Y_I(\omega)\sin\omega t_i] \frac{d\omega}{2\pi} \\
&\quad + j \int_\Omega [Y_R(\omega)\sin\omega t_i + Y_I(\omega)\cos\omega t_i] \frac{d\omega}{2\pi} \\
&= \int_\Omega [Y_R(\omega)\cos\omega t_i - Y_I(\omega)\sin\omega t_i] \frac{d\omega}{2\pi} , \qquad (6.2\text{-}6)
\end{aligned}
$$

where the subscripts R and I denote *real* and *imaginary*, respectively. Also, since $Y(\omega) = 0$ for $\omega \in \Omega^c$,

$$\|y - g\|^2 = \|Y - G\|^2 = \|Y - G\|_\Omega^2 + \|G\|_{\Omega^c}^2, \qquad (6.2\text{-}7)$$

where

$$\|Y - G\|_\Omega^2 \triangleq \int_\Omega |Y(\omega) - G(\omega)|^2 \frac{d\omega}{2\pi} , \qquad (6.2\text{-}8)$$

and

$$\|G\|_{\Omega^c}^2 \triangleq \int_{\Omega^c} |G(\omega)|^2 \frac{d\omega}{2\pi} . \qquad (6.2\text{-}9)$$

Using these observations and facts enables us to rewrite Eq. (6.2-3) in the frequency domain as

$$
\begin{aligned}
J(y) &= \int_\Omega (Y_R - G_R)^2 \frac{d\omega}{2\pi} + \int_\Omega (Y_I - G_I)^2 \frac{d\omega}{2\pi} + \int_{\Omega^c} |G|^2 \frac{d\omega}{2\pi} \\
&\quad + \lambda \left\{ \int_\Omega [Y_R(\omega)\cos\omega t_i - Y_I(\omega)\sin\omega t_i] \frac{d\omega}{2\pi} - f(t_i) \right\}. \qquad (6.2\text{-}10)
\end{aligned}
$$

To minimize Eq. (6.2-10) we use the variational principle and set

$$\frac{\partial J}{\partial Y_R} = \frac{\partial J}{\partial Y_I} = 0. \qquad (6.2\text{-}11)$$

The result is obtained without much difficulty as [†]

$$y^*(t) = q(t) + [f(t_i) - q(t_i)] \operatorname{sinc}(2B(t - t_i)), \qquad (6.2\text{-}12)$$

[†] The function $\operatorname{sinc}(x)$ is defined as $(\sin\pi x)/(\pi x)$.

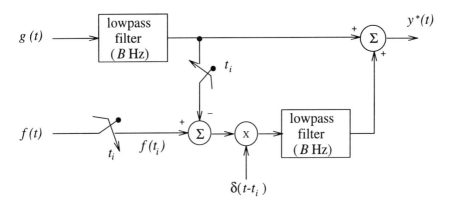

Fig. 6.2-2 Implementation of the projection onto C_i using an analog circuit.

where $q(t) \stackrel{\triangle}{=} g(t) * 2B\mathrm{sinc}(2Bt)$. The notation $*$ is used to mean ordinary convolution. In Fig. 6.2-2 is shown the realization of Eq. (6.2-12) using an analog circuit. We observe that the result is intuitively satisfying: It says that the projection is obtained by first lowpass-filtering $g(t)$, then using the difference between the correct sample and the sampled lowpass-filtered $q(t)$ to modulate a sampling impulse; and finally adding the weighted lowpass impulse responses as a correction term to the original lowpass-filtered $q(t)$.

Of course, $y^*(t)$ in Eq. (6.2-12) is not the complete solution to the problem; the action of P_i on g is merely to create a band-limited function $y^*(t)$, whose value at t_i is $f(t_i)$. To create a band-limited function whose value at t_{i+1} is $f(t_{i+1})$ we must operate on $y^*(t)$ with P_{i+1} and so on. Since all the P_i's, $i = 1, 2, \cdots, N$, have an identical form, there is no need to recompute P_{i+1} once P_i is known. We merely use Eq. (6.2-12) over and over with t_{i+1} replacing t_i until we exhaust the t_i's. This activity comprises a single cycle. Then the process is repeated as in Eq. (6.2-2) until convergence occurs. In [9] it is reported that in typical runs it takes from tens to hundreds of iterations for convergence to occur.

Actually the observation that all the P_i's have the same form prompts the question: Is it possible to project directly onto C_0 and thereby avoid a (possibly) lengthy iteration? The answer is yes and the solution is worked out in [9]. We furnish only the answer here. Define

$$C_0 = \{y(t) : y(t_i) = f(t_i), i = 1, 2, \cdots, N; \text{ and } Y(\omega) = 0 \text{ for } \omega \geq 2\pi B\}.$$
$$(6.2\text{-}13)$$

Then, given an arbitrary function $g(t)$, its *one-step projection* onto C_0 is given by

$$y^*(t) = P_0 g(t) = q(t) + \mathbf{S}(t)^T \mathbf{A}^{-1}(\mathbf{f} - \mathbf{q}),\qquad (6.2\text{-}14)$$

where

$$q(t) \;=\; g(t) * 2B\mathrm{sinc}(2Bt)$$

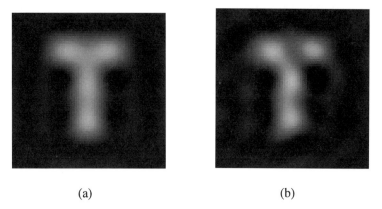

(a)   (b)

Fig. 6.2-3 Reconstruction from non-uniform samples. (a) The letter T band-limited to 4.6 cycles per image dimension; (b) reconstruction from non-unform samples using the one step POCS method.

$$
\begin{aligned}
\mathbf{S}(t) &= [\, 2B\mathrm{sinc}2B(t - t_1), \cdots, 2B\mathrm{sinc}2B(t - t_N)\,]^T \\
\mathbf{A} &= [\, a_{ij}\,]_{N \times N}\,, a_{ij} = 2B\mathrm{sinc}(t_i - t_j),\ i, j = 1, 2, \cdots, N \\
\mathbf{f} &= [\, f(t_1), f(t_2), \cdots, f(t_N)\,]^T \\
\mathbf{q} &= [\, q(t_1), q(t_2), \cdots, q(t_N)\,]^T.
\end{aligned}
$$

Equation (6.2-14) is the one-step projection onto C_0. It furnishes in one step a function $y^*(t)$ that is band-limited and passes through *all* the points $f(t_1), f(t_2), \cdots, f(t_N)$. It is, therefore, a solution to the reconstruction problem. Indeed, as $N \to \infty$ with $t_i = i\Delta, \Delta = [2B]^{-1}$ and $g(t)$ set equal to zero for simplicity, Eq. (6.2-14) reduces to the well-known uniform sampling theorem [10] (Problem 6.4).

At this point, the attentive reader might ask why even bother with the algorithm in Eq. (6.2-2) and risk a possible lengthy iteration if Eq. (6.2-14) gives the solution in one step? The answer is that the inversion of the matrix $\mathbf{A}$, when the number of samples N is large, may pose significant numerical difficulties. Recall that $\mathbf{A}$ is $N \times N$. Also, the one-step algorithm will fail if $\mathbf{A}$ is singular.

One can show that the set C_i in Eq. (6.2-1) is a linear variety. As a result, both the iterative algorithm and the one-step algorithm should yield the same result (Corollary 2.5-2). The reconstruction algorithm can be extended to functions of more than one variable. In Fig. 6.2-3, the 64×64 pixel letter T was band-limited to 4.6 cycles per image dimension (which is 64) and reconstructed from random samples by the one-step methods (Fig. 6.2-3(b)). Essentially an identical result was obtained when the reconstruction was done using the iterative method. For another projection-based approach, see the work by Sauer and Allebach [11].

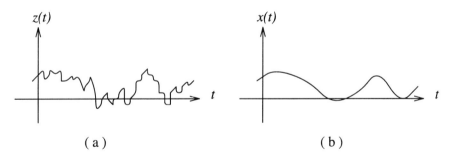

Fig. 6.3-1 (a) An analog function continuous in time and continuous in amplitude (CTCA). (b) A band-limited version of the same function.

6.3 DIGITAL COMMUNICATIONS

Because of the numerous advantages of digital communication systems over their analog counterparts, for example, with respect to noise, reproducibility, and ease of processing, digital systems are replacing analog systems in a host of applications. No doubt this trend will continue as wideband channels become more available. Digital signals are the most egalitarian of waveforms: once a signal has been converted to a sequence of binary symbols it is virtually impossible to (superficially) determine whether the source was noise, music, video, data, or virtually anything else; the playing field has been leveled. This offers the possibility of a common transmission medium for all types of different signals, as well as controlling this information "superhighway" by computer systems. It is for this reason that digital communications and computers are so intertwined. The former could hardly exist without the latter. In its most basic form, an analog-to-digital communication system begins with an analog waveform, say $z(t)$, which is *continuous in time and continuous in amplitude* (CTCA). See Fig. 6.3-1(a) for example. This waveform is first band-limited to B Hz (to reduce noise and prevent information from one channel to leak into another channel, a phenomenon known as *cross-talk*) thereby producing a second, *band-limited*, signal $x(t)$, which is still CTCA (Fig. 6.3-1(b)). Observe that, in going from $z(t)$ to $x(t)$, some information has been thrown away, but this information is often noisy and possibly irrelevant; this particular "lossy" operation is usually not of great concern and we shall assume that $x(t)$ is our source signal. Continuing, the next step is covert $x(t)$ to a *discrete-time* signal. This is done by sampling $x(t)$, usually at a uniform rate of, say, f_s samples per second. The samples are denoted $x(t_1), x(t_2), \cdots$, etc. As discussed in Section 6.2, the *uniform sampling theorem* states that $x(t)$ can be uniquely reconstructed from its samples $\{x(t_i)\}$ if $f_s \geq 2B$.[†] Thus in going from the CTCA signal $x(t)$ to the *discrete time, continuous amplitude* (DTCA) signal $\mathbf{x} \triangleq (\cdots, x(t_1), x(t_2), \cdots)$ there is no

---

[†]Strictly speaking we need $f_s > 2B$ but if the signal has no energy at $2B$ Hz then we write $f_s \geq 2B$.

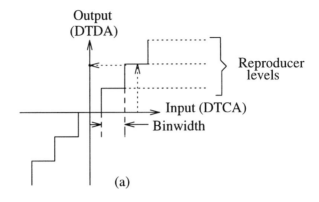

(a)

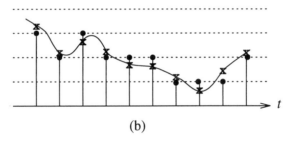

(b)

Fig. 6.3-2   (a) The input-output characteristic of a uniform quantizer. (b) Quantized values are different from sampled values. The symbols "x" denote the sample values and "•" denote the quantized values.

loss of information, according to the sampling theorem, if the sampling frequency is high enough.

At this point we are ready to convert a DTCA signal to a digital signal. This is done by replacing the continuous amplitude range with a *finite* set of values, a process called *quantizing* (Fig. 6.3-2). For example, all samples $\{x(t_i)\}$ that have an amplitude between, say, *two* and *three* are assigned the same value, say 2.5. Thus a whole range of values have been mapped into a single level $q^{(i)}$, which must take its value from the set $\{q_1, q_2, \cdots, q_N\}$. Quantizing is, in effect, a kind of substitution: $x(iT_s)$ is replaced by $q^{(i)} \in \{q_1, q_2, \cdots, q_N\}$, $x((i+1)T_s)$ is replaced by $q^{(i+1)} \in \{q_1, q_2, \cdots, q_N\}$, etc. A common substitution rule is the following: If $q_j - \frac{\delta q}{2} \le x(iT_s) < q_j + \frac{\delta q}{2}$ for some q_j, then $x(iT_s)$ is replaced by q_j. Clearly, given only $q^{(i)}$, it is not possible to know what the precise value of $x(t_i)$ was. This could potentially represent a serious *information loss*. Interestingly this informational loss can be viewed as the effect of wideband noise corrupting the signal. Since $x(t_i)$ can have an infinite set of values while $q^{(i)}$ has to assume one of the N values, the process of converting an analog signal to a digital signal is sometimes described as mapping from an infinite-dimensional space to a finite-dimensional space. Thus the original source signal $x(t)$ has been converted to a

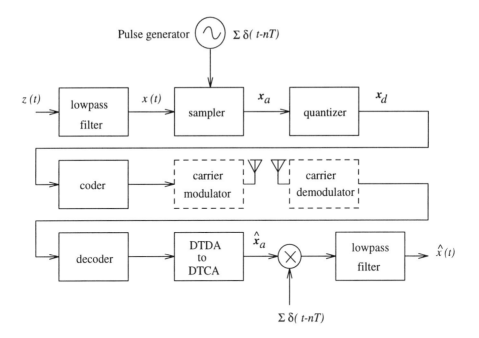

Fig. 6.3-3   A digital communication system.

discrete-time discrete-amplitude (DTDA) sequence.

An important observation, then, is that the elements in the quantized sequence $\{q^{(i)}\}$ generally do not equal the DTCA samples of the original band-limited signal. Thus, while one can reconstruct a band-limited CTCA signal from the DTDA samples, the reconstructed signal will, typically, not be identical with the original CTCA signal $x(t)$.

The set of digital signals $\{q^{(i)}\}$ are usually converted into a binary *code* before transmission. The purpose of the code is to enable easy signaling, e.g., ones and zeros, also to remove redundancy in the original message, i.e., to compress the information (source coding), or to provide protection against noise (channel coding) or, finally, to encipher to provide secrecy. These codes are usually uniquely decipherable and so there is no loss of information involved in coding the digital signal. Hence the receiver can decode the coded waveform unambiguously and furnish $\{q^{(i)}\}$, the DTDA sequence. Now the receiver's task is to generate a best estimate of the source signal $x(t)$. How should it do this? A block diagram of a digital communication system is shown in Fig. 6.3-3; the carrier modulation is shown in the dotted lines since is largely irrelevant to this discussion.

Before proceeding let us make a few observations. If the source signal $x(t)$ had not been quantized, the samples available to the lowpass filter at the receiver would have been $\{x(iT_s)\}$, where $t_{i+1} - t_i = T_s = 1/f_s$ and f_s is the sampling frequency. As stated earlier, the signal is usually *oversampled* so that $f_s > 2B$ or $T_s < (2B)^{-1}$.

The samples $x(t_i)$ can be used to modulate an impulse train $\sum \delta(t - iT_s)$ and the input to the receiver would then be the sequence $\sum x(iT_s)\delta(t - iT_s)$. If the ideal lowpass filter in Fig. 6.3-3 has cut-off frequency B Hz, the output would be $x(t)$, i.e., *the perfectly reconstructed source signal*. Note that at $t = kT_s$ where k is some fixed integer, the value of $x(t)$ is $x(kT_s)$ and the output of the receiver lowpass filter, being a perfect reconstruction of $x(t)$, must equal $x(kT_s)$. Hence, by the uniform sampling theorem, $x(kT_s) = \sum_i x(iT_s)\text{sinc}2BT_s(k - i)$; but since $\text{sinc}(0) = 1$, we get

$$x(kT_s) = x(kT_s) + \sum_{i \neq k} x(iT_s)\text{sinc}2BT_s(k - i). \tag{6.3-1}$$

Thus, for $BT_s < 1/2$,

$$\sum_{i \neq k} x(iT_s)\text{sinc}2BT_s(k - i) = 0, \qquad \text{for } k \text{ an integer.} \tag{6.3-2}$$

This interesting result shows that the samples of the band-limited signals are interdependent. This complex dependency is a result of the organic properties of band-limited functions. Band-limited functions are analytic functions and, among other things, such functions are known forever once they are known over any interval of finite support (see the discussion in Example 2.7-1).

Returning, however, to the case at hand, the input to the receiver lowpass filter is not modulated by the samples $\{x(iT_s)\}$ but by the sequence $\{q^{(i)}\}$. Thus the filter input is $\sum_i q^{(i)}\delta(t - iT_S)$ and the output is the CTCA signal $\hat{x}(t)$ given by

$$\hat{x}(t) = \sum_i q^{(i)}\text{sinc}2B(t - iT_s). \tag{6.3-3}$$

At the instant $t = kT_s$ we obtain

$$\hat{x}(kT_s) = q^{(k)} + \sum_{i \neq k} q^{(i)}\text{sinc}2BT_s(k - i). \tag{6.3-4}$$

Needless to say, $\hat{x}(kT_s)$ is typically not equal to $x(kT_s)$. Why should it be? After all, we threw away information about $x(t)$ in the quantizing process and, hence, expect to make an error! A more subtle question is: If we requantize $\hat{x}(kT_s)$ would we obtain $q^{(k)}$? If the answer is *yes*, then the receiver has done about as well as possible: It has produced an estimate $\hat{x}(t)$, which is *band-limited* and produces the same quantized sequence $\{q^{(i)}\}$ as the original signal did. If the answer is *no*, we have a problem: We have produced an estimate $\hat{x}(t)$ of the original signal $x(t)$ which *does not match* the received data, i.e., the quantized samples.

To ensure that $\hat{x}(kT_s)$ will yield the original quantizing level $q^{(k)}$, we must have, from Eq. (6.3-4), that the second term

$$\sum_{i \neq k} q^{(i)} \operatorname{sinc} 2BT_s(k - i) \tag{6.3-5}$$

is sufficiently small to yield the reproducer level $q^{(k)}$. But, for arbitrary $\{q^{(i)}\}$, this result is rarely true. In other words, we cannot guarantee, in general, that quantizing $\hat{x}(kT_s)$ will yield $q^{(k)}$.

This undesirable state of affairs can be ameliorated at the receiver by post-processing with a convex projection algorithm. What is desired is a reconstructed signal $\hat{x}(t)$ that is band-limited and, upon requantizing, yields the values $\hat{x}(kT_s) = q^{(k)}$ for $k = \cdots, -1, 0, 1, \cdots$. The signal $\hat{x}(t)$ must lie at the intersection of the two sets C_1', C_2', defined by

$$C_1' \triangleq \left\{ y(t) \in L^2 : Y(\omega) = 0 \text{ for } |\omega| > 2\pi B \right\} \tag{6.3-6}$$

and

$$C_2' \triangleq \left\{ y(t) \in L^2 : q^{(k)} - q/2 \leq y(kT_s) < q^{(k)} + q/2, \ \forall \, k \in I \right\}, \tag{6.3-7}$$

where I is the set of real integers. However, in L^2 space, one cannot specify a function by a set of samples values, unless the function possesses a special property such as band-limitedness. Indeed, in L^2 space, two functions can differ at a large number of points and still be regarded as being equal. Hence the set C_2' is poorly specified. One could modify C_2' to C_2'':

$$C_2'' \triangleq \left\{ y(kT) \in l^2 : q^{(k)} - q/2 \leq y(kT_s) < q^{(k)} + q/2, \ \forall \, k \in I \right\}, \tag{6.3-8}$$

i.e., the set of all *sequences* with constrained components. But since C_1' deals with *continuous* functions while C_2'' deals with sequences, this would amount to mixing oil with water and would not solve this problem. Another problem with C_2'' is that it is not *closed*. We can close C_2'' by merely changing the upper inequality from $<$ to $\leq$. However, there is a price to pay for this action, as we shall see shortly in the example.

A more appropriate solution than defining constraints in different spaces is to force $\hat{x}(t)$ to lie in the set[†] C given by

$$C \triangleq \ \{ y(t) \in L^2 : Y(\omega) = 0 \text{ for } |\omega| > 2\pi B \text{ and}$$
$$q^{(k)} - q/2 \leq y(kT_s) \leq q^{(k)} + q/2, \ \forall \, k \in I_N \}, \tag{6.3-9}$$

where I_N is a subset of I containing N elements (samples). In Eq. (6.3-9) q is

---

[†]In the real world we always deal with a finite number of samples. Hence defining the set in terms of N samples reflects reality better and is more convenient mathematically.

the width of the quantizing level and $q^{(k)}$ is the reproducer level. To avoid the computation of the one-step projection onto C, we write C as the intersection of N *larger* sets $C_k, k = 1, 2, \cdots, N$, defined by (from now on all $y(t)$ are assumed in L^2 and we omit further reminder of this assumption in this section)

$$C_k \triangleq \left\{ y(t) : Y(\omega) = 0 \text{ for } |\omega| > 2\pi B \text{ and } q^{(k)} - q/2 \leq y(kT_s) \leq q^{(k)} + q/2 \right\}$$
(6.3-10)

and hence $C = \bigcap_{k=1}^{N} C_k$. We project, in sequence, onto each of the C_k. It is easy to show that C_k in Eq. (6.3-10) is convex. For example, let $y_1, y_2 \in C_k$ and consider $y_3 = \mu y_1 + (1 - \mu) y_2$. Since the sum of band-limited functions is band-limited, y_3 is band-limited. Also, since $\max(y_1(kT_s), y_2(kT_s)) \leq q^{(k)} + q/2$, it follows that $y_3(kT_s)) \leq q^{(k)} + q/2$. Likewise, we have also $y_3(kT_s)) \geq q^{(k)} - q/2$. Thus C_k is convex.

To demonstrate closedness, consider the sequence $y_1, y_2, \cdots$, all in C_k, with limit point y. Then $\lim_{i \to \infty} \|y_i - y\| = 0$. Equivalently, with $Y_i = \mathcal{F}(y_i)$ and $Y = \mathcal{F}(y)$,

$$\lim_{i \to \infty} \|Y_i - Y\| = 0,$$
(6.3-11)

which requires that $Y(\omega)$ has no energy in $|\omega| > 2\pi B$, i.e., $y(t)$ is band-limited. Also with

$$\int_{\Omega} |Y_i(\omega) - Y(\omega)|^2 \frac{d\omega}{2\pi} \triangleq \|Y_i - Y\|_{\Omega}^2,$$
(6.3-12)

$$\int_{\Omega} (Y_i(\omega) - Y(\omega)) e^{j\omega k T_s} \frac{d\omega}{2\pi} \triangleq \left\langle (Y_i - Y), e^{-j\omega k T_s} \right\rangle_{\Omega},$$
(6.3-13)

where $\Omega = \{\omega : -2\pi B \leq \omega \leq 2\pi B\}$, we obtain, by the Schwarz inequality,

$$\left| \left\langle (Y_i - Y), e^{-j\omega k T_s} \right\rangle_{\Omega} \right| \leq \|Y_i - Y\|_{\Omega} \times \|e^{-j\omega k T_s}\|_{\Omega} \to 0 \text{ as } i \to \infty.$$
(6.3-14)

Hence $y_i(kT_s) \to y(kT_s)$. It follows that $q^{(k)} - q/2 \leq y(kT_s) \leq q^{(k)} + q/2$, i.e., $y \in C_k$. What is the projection onto C_k? Actually a comparison of the set in Eq. (6.3-10) with the set defined in Eq. (6.2-1) will show that the projections in the two cases are the same. To demonstrate this, consider an arbitrary function $g(t)$. Then its projection $y^*(t)$ must be band-limited and have a value within the closed interval $[q^{(k)} - q/2, q^{(k)} + q/2]$. Let a_k denote this value. Then the appropriate Lagrange functional is given by

$$\begin{aligned} J(y) &= \|g - y\|^2 + \lambda[y(kT_s) - a_k] \\ &= \|G - Y\|_{\Omega}^2 + \|G - Y\|_{\Omega^c}^2 \\ &\quad + \lambda \left[\int_{\Omega} Y(\omega) e^{j\omega k T_s} \frac{d\omega}{2\pi} - a_k \right], \end{aligned}$$
(6.3-15)

where $\Omega^c = \{\omega : |\omega| > 2\pi B\}$. The solution is given by Eq. (6.2-12) with a_k replacing $f(t_i)$ and kT_s replacing t_i. Define $g_B(t) \triangleq g(t) * 2B\text{sinc}(2Bt)$. Then

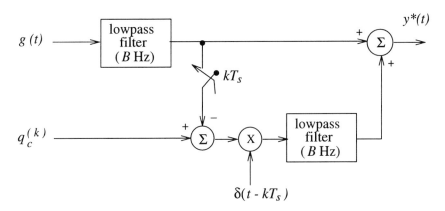

Fig. 6.3-4 Realization of the projector P_k.

the projection $y^*(t) = P_k g(t)$ is given by

$$
y^*(t) = \begin{cases}
g_B(t) - \left\{g_B(kT_s) - (q^{(k)} + q/2)\right\} \operatorname{sinc}[2B(t - kT_s)], \\
\qquad\qquad\qquad\qquad \text{if } g_B(kT_s) > q^{(k)} + q/2; \\
g_B(t) - \left\{g_B(kT_s) - (q^{(k)} - q/2)\right\} \operatorname{sinc}[2B(t - kT_s)], \\
\qquad\qquad\qquad\qquad \text{if } g_B(kT_s) < q^{(k)} - q/2; \\
g_B(t), \text{ otherwise.}
\end{cases}
$$

$$(6.3\text{-}16)$$

Note that when $g(t)$ is already band-limited to B Hz, then $g_B(t)$ is simply $g(t)$.

The overall algorithm for reconstructing a band-limited signal $x(t)$ from quantized samples is

$$x_{l+1}(t) = P_N \cdots P_2 P_1 x_l(t), \quad x_0(t) \text{ arbitrary.} \qquad (6.3\text{-}17)$$

In Eq. (6.3-17) it is convenient to distinguish between *steps* and *cycles*. A step occurs after an operation by a single projector. A cycle occurs when all N projectors operate exactly once on the signal. The following notation is useful:

$$x_{l,i+1}(t) = P_{i+1} x_{l,i}(t); \quad x_{l-1,N} \overset{\triangle}{=} x_l, x_{l,0} \overset{\triangle}{=} x_l. \qquad (6.3\text{-}18)$$

Actually, the order in Eq. (6.3-17) isn't important and we could have written

$$x_{l+1}(t) = P_{p(N)} \cdots P_{p(2)} P_{p(1)} x_l(t), \quad x_0(t) \text{ arbitrary,} \qquad (6.3\text{-}19)$$

where $P_{p(i)} \in \{P_1, P_2, \cdots, P_N\}$ and $P_{p(i)} \neq P_{p(j)}$ for $i \neq j$.

As can be seen from Fig. 6.3-4, where $q_c^{(k)}$ takes on one of the three values $q^{(k)} + q/2, q^{(k)} - q/2$, or $g_B(kT_s)$, the realization of a typical operator P_k is somewhat cumbersome. From Eq. (6.3-16) we see that the projection always yields a band-limited function; hence every projector in Eq. (6.3-17) sees an operand that is band-limited, except for possibly the first one. This suggests that a simplification is possible in the realization of Eq. (6.3-16). Consider the set C_B, defined by

$$C_B = \{y(t) : Y(\omega) = 0 \text{ for } |\omega| > 2\pi B\}. \tag{6.3-20}$$

This set is discussed in Chapter 3 (see Eq. (3.7-8)). The projection of an arbitrary $g(t)$ onto C_B is easily shown to be

$$y^*(t) = P_B g(t) = g(t) * 2B\text{sinc}(2Bt). \tag{6.3-21}$$

If we now begin each cycle with the projector P_B, i.e.,

$$x_{l+1}(t) = P_N \cdots P_2 P_1 P_B x_l(t), \tag{6.3-22}$$

then all of the band-limiting done by the operators $P_i, i = 1, 2, \cdots, N$ becomes superfluous. Then the only step operations that are required are

$$x_{l,i+1}(t) = \begin{cases} x_{l,i}(t) - \left\{x_{l,i}(iT_s) - (q^{(i)} + q/2)\right\} \text{sinc} 2B(t - iT_s), \\ \qquad\qquad\qquad \text{if } x_{l,i}(iT_s) > q^{(i)} + q/2 \\ x_{l,i}(t) + \left\{x_{l,i}(iT_s) - (q^{(i)} - q/2)\right\} \text{sinc} 2B(t - iT_s), \\ \qquad\qquad\qquad \text{if } x_{l,i}(iT_s) < q^{(i)} - q/2 \\ x_{l,i}(t), \quad \text{otherwise.} \end{cases}$$

$$\tag{6.3-23}$$

Equation (6.3-23) can be written in shorthand as

$$x_{l,i+1}(t) = P'_{i+1} x_{l,i}(t), \tag{6.3-24}$$

where P'_i can be interpreted, loosely, as a *conditional projector*, conditioned upon the operand being band-limited. Note that the conditional projector is much simpler to realize than the original projectors P_i.

A numerical implementation of a POCS reconstruction from quantized samples is considered below. In Fig. 6.3-6(a) is shown a band-limited signal consisting of the sum of two shifted sinc signals with $B = 2$ Hz. The sampling rate f_s is 8 samples per second. A uniform quantizer with reproducer levels at $0, \pm 1/2, \pm 1, \pm 3/2, \pm 2, \cdots$, and *binwidth* of one-half is assumed. The samples of the original signal as well as their quantized values are shown in Fig. 6.3-6(b). The conventionally reconstructed signal is shown in Fig. 6.3-6(c); note that the quantized samples of the reconstructed function are not everywhere equal to the original quantized samples (Fig. 6.3-6(d)).

The POCS reconstruction is shown in Fig. 6.3-6(e). We see that there is a failure to reproduce the original quantized samples (Fig. 6.3-6(a)). Why? Recall that by closing the set C''_2 and the subsequent closure of C_k we project points outside the set onto set boundaries that do not necessarily correspond to correct solutions. For example, consider the set $\{1/2 \leq y(kT_s) < 3/2\}$; an arbitrary point $g(kT_s)$ having value, say 2, cannot be assigned a closest value in this set because it is open from above. However, by closing the set as we did, the value of 2 is "corrected" by projecting onto the boundary $3/2$, thereby implying that $y(t)$ has a value in the

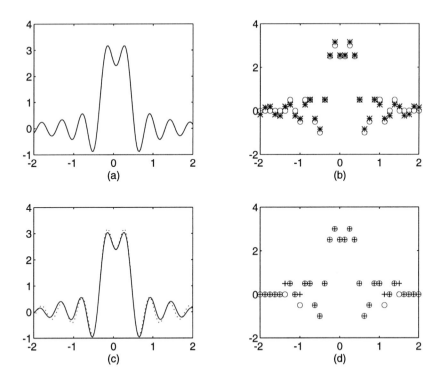

Fig. 6.3-5 (a) A band-limited signal with $B = 2$ Hz. (b) The sampled signal (stars) and the quantized samples (circles). (c) Reconstruction from the quantized samples using a lowpass filter. (d) The quantized samples of the *reconstructed* signal in (c) (crosses) and the quantized sampled values (circles) shown in (b) of the original signal in (a). Note the lack of agreement in a number of places.

interval in the interval $[1/2, 3/2]$. However, this may be in error.

The solution to this problem is to avoid projecting onto the boundary by using *relaxed projectors*. Recall that the relaxed projector $T_i = 1 + \lambda_i(P_i - 1)$, for values of $\lambda_i \neq 1$, does not project onto the boundary of the set. Indeed by choosing λ_i greater than 1, we avoid the problem completely.

When the relaxed projection version of the POCS reconstruction is sampled and quantized, we find exact agreement between the POCS quantized samples and the original quantized samples (Fig. 6.3-6(d) and (f)). Thus POCS did what it set out to do: It reconstructed a 2 Hz band-limited signal whose quantized samples equal those observed at the receiver.

This section has illustrated the great power that POCS can exert on forcing a signal to live up to its constraints. Had the signal been non-negative, limited to a prescribed energy bound, or otherwise restricted, POCS would have handled

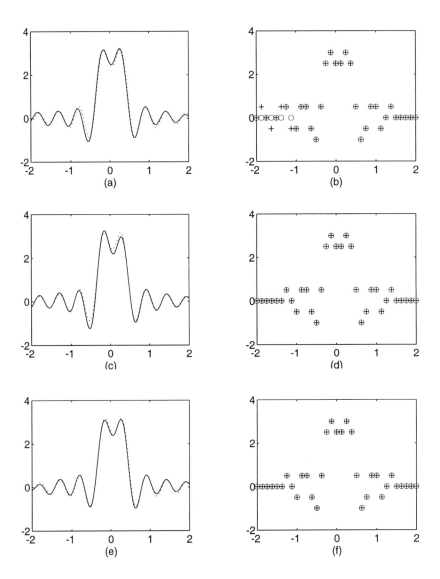

Fig. 6.3-6   (a) Reconstruction using a projection algorithm with $\lambda = 1$. (b) Quantized samples of (a) (crosses) are not equal everywhere to the quantized samples of the original signal (circles). (c) Same as (a) with $\lambda = 1.5$. (d) Quantized samples of (c) (crosses) agree everywhere with quantized samples of the original signal. (e) Same as (a) with $\lambda = 1.95$. (f) Quantized samples of (e) (crosses) agree everywhere with quantized samples of the original signal. Also the reconstruction is closer to the original.

these constraints equally well. Additional information on POCS-type methods for handling quantizing problems is furnished in the work by [12, 13].

6.4 DIGITAL FILTERS

Before the widespread use of computers and *digital signal processing* (DSP) boards, filtering of analog signals was done by analog circuitry involving circuit elements such as resistors, capacitors, inductors, and operational amplifiers. Not only were these circuits sometimes very complex, but they were also designed for a specific task and could not easily be modified to deal with new tasks or new sets of parameters. More modern techniques often use *digital filters* to accomplish the same tasks.

A digital filter is a device that converts one sequence of numbers (called the *input sequence*) into another number sequence (called the *output sequence*) (Fig. 6.4-1). Digital filters are widely used in electrical engineering, from synthesizing human voice, to smoothing noisy waveforms, to enhancing blurred images. In modern digital communication systems, digital filters have replaced analog lowpass and bandpass filters. The input sequence to the digital filter is the sequence of numbers generated from periodic sampling of an analog waveform. Since a DSP board or computer is involved in the actual filtering, finite-precision arithmetic is always involved, i.e., the samples are quantized; however, this is not a serious problem, for the purpose of this discussion.

That linear time-invariant filtering of band-limited analog signals can be done digitally can be demonstrated as follows. Let $x(t), h(t)$, and $y(t)$ denote the analog input, filter impulse response, and output, respectively. Then

$$y(t) = \int_{-\infty}^{\infty} h(\lambda)x(t - \lambda)d\lambda. \qquad (6.4\text{-}1)$$

From the sampling theorem we can write

$$y(t) = \sum_{l} y(l\Delta)\text{sinc } 2B(t - l\Delta) \qquad (6.4\text{-}2)$$

$$x(t) = \sum_{m} x(m\Delta)\text{sinc } 2B(t - m\Delta) \qquad (6.4\text{-}3)$$

$$h(t) = \sum_{n} h(n\Delta)\text{sinc } 2B(t - n\Delta), \qquad (6.4\text{-}4)$$

provided that B is greater than any signal or system bandwidth encountered in the system and $\Delta = 1/2B$. From Eq. (6.4-2) it is clear that the output $y(t)$ is known once the sampled values $\{y(l\Delta)\}$ are known. If we use Eqs. (6.4-2), (6.4-3), and

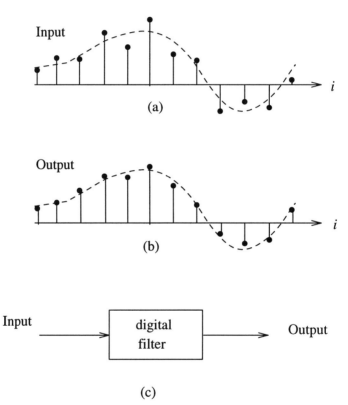

Fig. 6.4-1   A discrete-time waveform (a) is converted into a smoother discrete-time wave-form (b) by the digital filter (c).

(6.4-4) in Eq. (6.4-1) for $t = l\Delta$ we obtain the expression

$$y(l\Delta) = \sum_n \sum_m h(n\Delta)x(m\Delta)I(l, m, n), \qquad (6.4\text{-}5)$$

where

$$I(l, m, n) \triangleq \int_{-\infty}^{\infty} \text{sinc } 2B(\lambda - n\Delta)\text{sinc } 2B(\lambda - (l - m)\Delta)d\lambda. \qquad (6.4\text{-}6)$$

In Eq. (6.4-6) we used the fact that $\text{sinc}(x) = \text{sinc}(-x)$. Finally, we use the result that $I(l, m, n)$ has value 0 for all real integers l, m, n, except when $l - m = n$, whereupon it assumes the value $1/2B$. Hence with $B = 1/2\Delta$ we obtain

$$y(l\Delta) = \sum_n h(n\Delta)x[(l - n)\Delta]\Delta. \qquad (6.4\text{-}7)$$

Quite often the value of Δ is submerged. Then we obtain

$$y(l) = \sum_n h(n)x(l - n), \qquad (6.4\text{-}8)$$

which is the mathematical description of the action of the digital filter. The sequence $\{h(n)\}$ is called the *impulse response of the digital filter* or, less accurately, the *digital filter*. Note that while Eq. (6.4-8) is often called a digital filtering operation it is essentially a *discrete time* operation. The "digital" refers to the operation being done by a DSP board or computer. Since quantizing is involved, there will be some noise in the reconstruction of the analog signal. The amount of noise will depend on how many bits we use to represent the samples: the more bits the less noise.

Design of FIR Filters

Digital filters are often designed as FIR (*finite impulse response*) or IIR (*infinite impulse response*) filters. A big advantage of FIR filters is that they are non-recursive and, hence, are never unstable.

In the design of digital filters we often specify a magnitude characteristic without regard to phase. For example, in the design of lowpass and bandpass filters, we could specify bounds on the magnitude performance of the filter in both the passbands and the stopbands in the frequency domain without specifying phase (Fig. 6.4-2). The phase could subsequently be adjusted or corrected using an *all-pass phase-equalization filter*. However, it is often highly desirable to directly design filters that have *linear phase*. In this way, signals in the passband of the filter are reproduced exactly at the filter output, except for a delay given by the negative of the slope of the phase function. It is well known that if an FIR filter

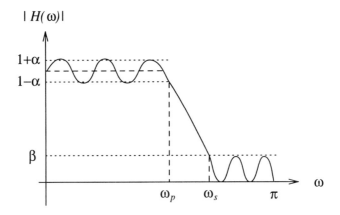

Fig. 6.4-2 A lowpass digital FIR filter can be designed by specifying the allowable ripple in the passband (0 to ω_p) and stopband (ω_s to π).

of length N has impulse response vector[†] $\mathbf{h} = (h(0), h(1), \cdots, h(N-1))^T$, then linear phase is implied by the property

$$h(n) = \begin{cases} h(N-1-n) & \text{for } n = 0, 1, \cdots, N-1 \\ 0 & \text{otherwise.} \end{cases} \tag{6.4-9}$$

The proof of Eq. (6.4-9) is given in many places. See, for example, [14]. The behavior of the FIR filter in the frequency domains is given by its system function (i.e., Z-transform) $\tilde{H}(z)$, defined by

$$\tilde{H}(z) = \sum_{n=0}^{N-1} h(n)z^{-n}, \tag{6.4-10}$$

evaluated at $z = e^{j\omega}$, where ω is a dimensionless normalized frequency. Hereafter we shall write

$$H(\omega) \triangleq \tilde{H}(e^{j\omega}) \triangleq \sum_{n=0}^{N-1} h(n)e^{-j\omega n}, \tag{6.4-11}$$

which is recognized as the Fourier transform of the sequence $\{h(0), h(1), \cdots, h(N-2), h(N-1)\}$. If we use Eq. (6.4-9) in Eq. (6.4-11) and assume N to be even, we obtain

$$H(\omega) = A(\omega)e^{j\phi(\omega)}, \tag{6.4-12}$$

[†] The responses of the filter at discrete time $t = n$ to a unit impulse response at $t = 0$ is $h(n)$ and $h(0), h(1), \cdots, h(N-1)$ are the elements of the impulse response vector.

where

$$A(\omega) \triangleq \sum_{n=0}^{N/2-1} 2h(n) \cos\left[\left(n - \frac{N-1}{2}\right)\omega\right] \qquad (6.4\text{-}13)$$

and

$$\phi(\omega) \triangleq -\frac{N-1}{2}\omega \ . \qquad (6.4\text{-}14)$$

Although N was assumed even in the above derivation, a similar result still holds when N is odd. Consider now the design of an FIR lowpass filter with linear phase that meets the following specifications: in the passband, the magnitude of the filter response must be between $1-\alpha$ and $1+\alpha$ and in the stopband, it cannot exceed β (see Fig. 6.4-2). We put no constraints on the behavior of the filter in the transition band. For the purpose of design we need only require linear phase in the passband: $\Omega_p \triangleq \{\omega : 0 < \omega \le \omega_p\}$. In the stopband: $\Omega_s \triangleq \{\omega : \omega_s < \omega \le \pi\}$, we need only enforce $|H(\omega)| \le \beta$. The solution to this problem by projections proceeds below. As always, the key is to define the appropriate constraint sets. Define C_1, C_2, and C_3 as

$$
\begin{aligned}
C_1 &= \big\{\mathbf{h} \in R^N : h(n) = h(N-1-n), \\
&\quad \text{for } n = 0, 1, \cdots, N-1 \big\} , \qquad (6.4\text{-}15) \\
C_2 &= \big\{\mathbf{h} \in R^N : 1 - \alpha \le A(\omega) \le 1 + \alpha \\
&\quad \text{and } \phi(\omega) = -\omega(N-1)/2 \ \ \text{for } \omega \in \Omega_p \big\} , \quad \text{and} \quad (6.4\text{-}16) \\
C_3 &= \big\{\mathbf{h} \in R^N : \ |H(\omega)| \le \beta \ \ \text{for } \omega \in \Omega_s \big\} . \qquad (6.4\text{-}17)
\end{aligned}
$$

In words, C_1 is the set of all finite-length sequence with appropriate symmetry that imply a Fourier transform with linear phase; C_2 is the set of all finite-length sequence whose Fourier amplitude is appropriately constrained in the passband and whose phase is linear in that band; and C_3 is the set of all finite-length sequences appropriately constrained in the stopband. Note that it might have been tempting to use a Fourier magnitude constraint set, say C_2', given by

$$C_2' = \big\{\mathbf{h} \in R^N : 1 - \alpha \le |H(\omega)| \le 1 + \alpha \ \ \text{for } \omega \in \Omega_p \big\}. \qquad (6.4\text{-}18)$$

However, this set is not convex and, hence, its involvement in a projection algorithm could create *traps* (see Chapter 5).

CONVEXITY OF C_1: Let $h_1(n), h_2(n) \in C_1$ and define $h_3(n) = \mu h_1(n) + (1 - \mu)h_2(n)$ for $0 \le \mu \le 1$. Since $h_1(n) = h_1(N-1-n)$ and $h_2(n) = h_2(N-1-n)$, we have $h_3(n) = \mu h_1(N-1-n) + (1-\mu)h_2(N-1-n) = h_3(N-1-n)$. Hence the set is convex.

CLOSEDNESS OF C_1: Let $\mathbf{h}_k, k = 1, 2, \cdots$ be a sequence of vectors in C_1 with limit point $\mathbf{h}$. Then, by definition, $\sum_{n=0}^{N-1} |h_k(n) - h(n)|^2 \longrightarrow 0$. But $h_k(n) = h_k(N-1-n)$, hence $|h_k(N-1-n) - h(n)|^2 \longrightarrow 0$, which implies that $h_k(N-1-n) \longrightarrow h(n)$. But $h_k(n) \longrightarrow h(n)$; hence $h(n) = h(N-1-n)$. Therefore $\mathbf{h} \in C_1$, and the set C_1 is closed indeed.

PROJECTION ONTO C_1: To simplify matters, assume that all vectors are real. Let **g** be an arbitrary vector, **h** be any vector in C_1, and **h*** the projection of **g** onto C_1. Then

$$\mathbf{h}^* = \arg \min_{\mathbf{h} \in C_1} \sum_{n=0}^{N-1} [g(n) - h(n)]^2. \tag{6.4-19}$$

With $J \triangleq \sum_{n=0}^{N-1} [g(n) - h(n)]^2$, the projection is easily computed once we write (assume N is even) the Lagrange functional as

$$J \triangleq \sum_{n=0}^{N/2-1} \left\{ [g(n) - h(n)]^2 + [g(n + N/2) - h(n + N/2)]^2 \right\} \tag{6.4-20}$$

and use
$$h(n + N/2) = h(N/2 - n - 1). \tag{6.4-21}$$

Then, with
$$\frac{\partial J}{\partial h(l)} = 0 \tag{6.4-22}$$

we obtain
$$h^*(l) = \frac{g(l) + g(N - 1 - l)}{2}. \tag{6.4-23}$$

This is a neat solution since it clearly shows that $h^*(l) = h^*(N - 1 - l)$.

CONVEXITY OF C_2: Let $\mathbf{h_1}$ and $\mathbf{h_2}$ be $\in C_2$. Then $\mathbf{h_3} \triangleq \mu\mathbf{h_1} + (1 - \mu)\mathbf{h_2} \leftrightarrow [\mu A_1(\omega) + (1 - \mu)A_2(\omega)] \exp^{j\phi(\omega)}$. Thus the phase of $\mathcal{F}[\mathbf{h_3}]$ is $\phi(\omega)$ and, since $A_1(\omega)$ and $A_2(\omega)$ are lower and upper bounded by $1 - \delta$ and $1 + \delta$, respectively, so is $A_3(\omega) \triangleq \mu A_1(\omega) + (1 - \mu)A_2(\omega)$ for any $0 \leq \mu \leq 1$.

CLOSEDNESS OF C_2: This is left as an exercise to the reader.

PROJECTION ONTO C_2: The projection of an arbitrary vector $\mathbf{g} \in R^N$ with Fourier transform $G(\omega) =| G(\omega) | \exp^{j\theta_G(\omega)}$ can easily be computed using the Lagrange multiplier method. However, it can also be deduced from Fig. 6.4-3. Either way we obtain

$$\mathbf{h}^* = P_2\mathbf{g} \leftrightarrow H^*(\omega), \tag{6.4-24}$$

where

$$H^*(\omega) = \begin{cases} (1 + \delta) \exp^{j\phi(\omega)} & \text{if cond. A} \\ (1 - \delta) \exp^{j\phi(\omega)} & \text{if cond. B} \\ |G(\omega)| \cos[\theta_G(\omega) - \phi(\omega)] \exp^{j\phi(\omega)} & \text{if cond. C} \\ G(\omega) & \text{if } \omega \in \Omega_p^c, \end{cases} \tag{6.4-25}$$

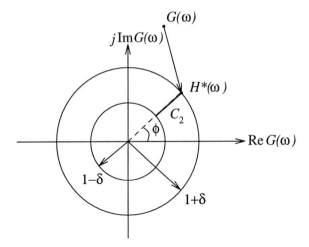

Fig. 6.4-3 The projection onto C_2 is easily diagrammed in the frequency domain.

where conditions A, B, and C apply to all $\omega \in \Omega_p$ and where

$$
\begin{aligned}
\text{cond. A is:} \quad & |G(\omega)| \cos[\theta_G(\omega) - \phi(\omega)] \geq 1 + \delta \\
\text{cond. B is:} \quad & |G(\omega)| \cos[\theta_G(\omega) - \phi(\omega)] \leq 1 - \delta \\
\text{cond. C is:} \quad & 1 - \delta \leq |G(\omega)| \cos[\theta_G(\omega) - \phi(\omega)] \leq 1 + \delta.
\end{aligned}
\tag{6.4-26}
$$

CONVEXITY OF C_3: Let $\mathbf{h_1}$ and $\mathbf{h_2}$ be $\in C_3$. Then $\mathbf{h_3} \overset{\triangle}{=} \mu\mathbf{h_1} + (1 - \mu)\mathbf{h_2} \leftrightarrow [\mu H_1(\omega) + (1 - \mu)H_2(\omega)] \overset{\triangle}{=} H_3(\omega)$ and we must show that that $| H_3(\omega) | \leq \beta$. But for any two complex numbers A and B we have $| A + B | \leq | A | + | B |$. Since $| H_1(\omega) |$ and $| H_2(\omega) |$ are bounded by β, it follows that $| H_3(\omega) | \leq \beta$.

CLOSEDNESS OF C_3: This is left as an exercise to the reader.

PROJECTION ONTO C_3: The projection of an arbitrary $\mathbf{g} \leftrightarrow G(\omega)$ onto C_3 is easily computed with the method of Lagrange multipliers as:

$$
\mathbf{h}^* = P_3\mathbf{g} \leftrightarrow
\begin{cases}
\beta G(\omega)/|G(\omega)| & \text{for } |G(\omega)| > \beta, \omega \in \Omega_s, \\
G(\omega) & \text{for } |G(\omega)| \leq \beta, \omega \in \Omega_s, \\
G(\omega) & \text{elsewhere.}
\end{cases}
\tag{6.4-27}
$$

The FIR filter design algorithm is given by

$$
\mathbf{h}_{k+1} = P_1 P_2 P_3 \mathbf{h}_k, \quad \mathbf{h_0} \text{ arbitrary.}
\tag{6.4-28}
$$

A good choice for $\mathbf{h_0}$ is $\mathbf{h_0} \leftrightarrow H_0(\omega) = H_{ideal}(\omega)$, where $H_{ideal}(\omega)$ is the ideal lowpass response characteristic of the digital filter. Below we furnish numerical results in which the POCS algorithm in Eq. (6.4-28) equals the performance of the famous *Remez Exchange Algorithm* (REA) [15].

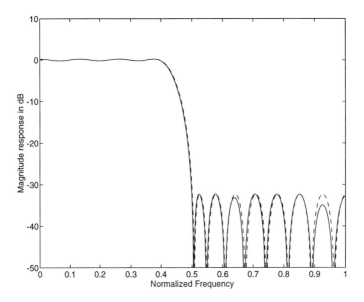

Fig. 6.4-4 Frequency domain characteristics of a lowpass filter designed by Remez Exchange Algorithm (REA) (dashed) and POCS (solid). The Remez result is obtained through the MATLAB routine.

Numerical Examples

We compare POCS with the REA using $N = 31$, $\omega_p = 0.4\pi$, $\omega_s = 0.5\pi$. To implement the REA, we used the MATLAB signal processing module [16] with equal value for the ripple parameters, i.e., $\alpha = \beta$ in the passband and stopband, respectively. Within the limits of the available numerical accuracy, both the REA and POCS algorithms furnished the same results, i.e., $\alpha = \beta = 0.025$. Figure 6.4-4 shows the frequency response characteristic of the POCS and Remez algorithms: as can be seen, they are essentially identical.

The POCS algorithm can be used to design two-dimensional filters with or without circular support. For example, the evolution of a lowpass, zero-phase, four-fold symmetric filter characteristic, designed by POCS, is shown in Fig. 6.4-5. In this case the parameters were $\omega_p = 0.5\pi, \omega_s = 0.6\pi, \alpha = 0.05$, and $\beta = 0.025$. The minimum length of the filter, in terms of equally spaced filter coefficients (taps), is $27 \times 27 = 729$. To design a filter with the same characteristic using the one-dimensional REA algorithm and the transformation

$$T(\omega_1, \omega_2) = -0.5 + 0.5\cos\omega_1 + 0.5\cos\omega_2 + 0.5\cos\omega_1 \cdot 0.5\cos\omega_2 \quad (6.4\text{-}29)$$

requires a minimum filter size of $35 \times 35 = 1225$ taps.

One major advantage of the POCS approach to filter design is the ease with

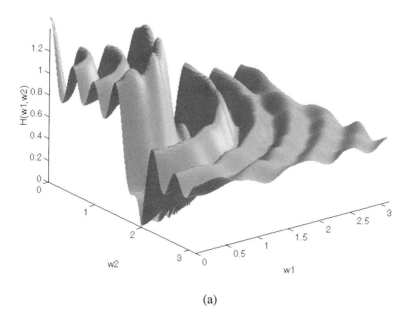

(a)

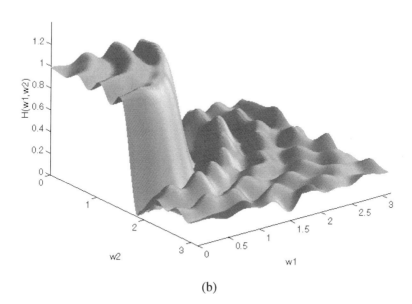

(b)

Fig. 6.4-5 Design of two-dimensional, four-fold symmetric, filter using POCS. Starting point is ideal filter characteristic. The frequency-domain characteristic is shown after 6, 12, 20, and 140 iterations in (a), (b), (c), and (d), respectively.

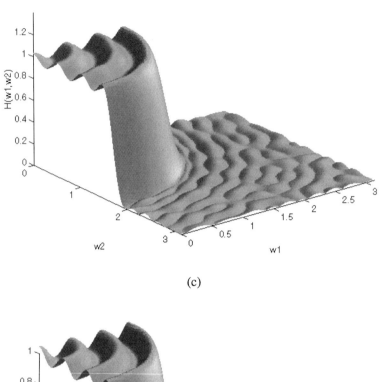

(c)

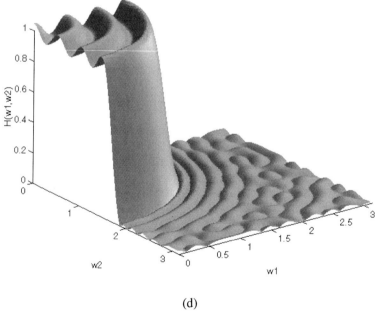

(d)

Fig. 6.4-5 *Continued.*

which other prior knowledge can be incorporated. For example, suppose we wish to design a linear-phase FIR filter whose response to a known input is restricted to lie within certain bounds. Then of use are sets of the form

$$C_4(n) \triangleq \left\{ \mathbf{h} \in R^N : \sigma_n \leq (\mathbf{s} * \mathbf{h})_n \leq \rho_n \right\}, \qquad (6.4\text{-}30)$$

where s is the given input, $*$ denotes convolution, $(\mathbf{s} * \mathbf{h})_n$ denotes the response at time n, and σ_n and ρ_n represent the desired lower and upper bounds, respectively, on the response at time n. To be specific, in addition to designing a linear-phase filter as we did earlier, let us put additional constraints on the filter by bounding its response to a unit step u. Thus we let $\mathbf{s} = \mathbf{u}$ and $\mathbf{a} \triangleq \mathbf{u} * \mathbf{h}$ denotes the *step response*. The nth term of the convolution is given by

$$a(n) = \sum_{i=0}^{N-1} h(i)u(n-i) \qquad \text{for } n = 0, 1, \cdots. \qquad (6.4\text{-}31)$$

The entire system can be written in matrix form as

$$\mathbf{a} = \mathbf{U}\mathbf{h}, \qquad (6.4\text{-}32)$$

where $\mathbf{a} \triangleq (a(0), a(1), \ldots, a(N-1))^T, \mathbf{h} \triangleq (h(0), h(1), \ldots, h(N-1))^T$, and $\mathbf{U}$ is a matrix given by

$$\mathbf{U} \triangleq \begin{bmatrix} 1 & 0 & 0 & \cdots & 0 & \cdots \\ 1 & 1 & 0 & \cdots & 0 & \cdots \\ 1 & 1 & 1 & \cdots & 0 & \cdots \\ & & & \vdots & & \\ 1 & 1 & 1 & \cdots & 1 & \cdots \\ \vdots & \vdots & \vdots & \vdots & \vdots & \vdots \end{bmatrix}. \qquad (6.4\text{-}33)$$

From Eq. (6.4-31) it is easily seen that we can write $a(n)$ as $a(n) = \mathbf{u}_n^T \mathbf{h}$, where $\mathbf{u}_n$ is the vector whose elements are the nth row of $\mathbf{U}$. Then $C_4(n)$ can be written as

$$C_4(n) = \left\{ \mathbf{h} \in R^N : \sigma_n \leq \mathbf{u}_n^T \mathbf{h} \leq \rho_n \right\}. \qquad (6.4\text{-}34)$$

This set is studied in some detail in Section 3.2 of Chapter 3, where it is shown that the projection of an arbitrary vector $\mathbf{g} \in R^N$ onto $C_4(n)$ is given by

$$\mathbf{h}^* = P_{4n}\mathbf{g} = \begin{cases} \mathbf{g} & \text{if } \sigma_n \leq \mathbf{u}_n^T \mathbf{g} \leq \rho_n \\ \mathbf{g} + \frac{\sigma_n - \mathbf{u}_n^T \mathbf{g}}{||\mathbf{u}_n||^2} \mathbf{u}_n & \text{if } \mathbf{u}_n^T \mathbf{h} < \sigma_n \\ \mathbf{g} + \frac{\rho_n - \mathbf{u}_n^T \mathbf{g}}{||\mathbf{u}_n||^2} \mathbf{u}_n & \text{if } \mathbf{u}_n^T \mathbf{g} > \rho_n. \end{cases} \qquad (6.4\text{-}35)$$

Assume that M constraints such as in Eq. (6.4-34) are applied to the response

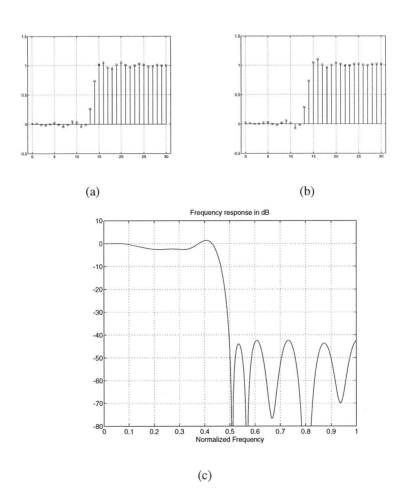

(a)   (b)

(c)

Fig. 6.4-6 (a) The step response of the designed filter using POCS with additional over-shoot constraint on the step response. (b) The step response of the designed filter using POCS without additional overshoot constraint. (c) The magnitude response of the FIR filter designed using POCS with additional constraint on the overshoot of the step response. Although the constraint limits the overshoot in the response, there is some sacrifice in the the passband and stopband performance compared with the frequency response when the overshoot constraint is not applied (see Fig. 6.4-4).

samples. Define $\tilde{P}_4 \triangleq P_{4,0}P_{4,1} \cdots P_{4,M-1}$, i.e., the composition of M projectors onto the sets $C_4(n)$ $n = 0, 1, \cdots, M-1$, respectively. Then the algorithm

$$\mathbf{h}_{k+1} = P_1 P_2 P_3 \tilde{P}_4 \mathbf{h}_k \tag{6.4-36}$$

will yield, assuming all the sets intersect, a filter with the correct step response characteristics. As a numerical example we choose $\sigma_n = \rho_n = 0.05$ for all n and constrain the step response to

$$\sigma_n \leq a_n \leq \rho_n \qquad \text{for } n = 0, 1, \ldots, 13$$
$$1 - \sigma_n \leq a_n \leq 1 + \rho_n \qquad \text{for } n = 18, 19, \ldots, 30 \tag{6.4-37}$$

and leave a_n unconstrained for $n = 14, \ldots, 17$. The step response for this filter is shown in Fig. 6.4-6, where it is compared with the step response of the filter considered earlier. As expected, the constraint has reduced the overshoots and undershoots.

Other constraints can be used. For example if it is desired to limit the total (white) noise power getting through the stopband portion of the filter, one might seek a solution in the set

$$C_5 = \left\{ \mathbf{h} \in R^N : \int_{|\omega| > \omega_s} |H(\omega)|^2 \, d\omega \leq \kappa \right\}. \tag{6.4-38}$$

It is easily shown that this set is convex (Exercise 6-14) and that the projection of an arbitrary vector $\mathbf{g} \in R^N \notin C_5$, is given by

$$\mathbf{h}^* \leftrightarrow H^*(\omega) = \begin{cases} G(\omega) & |\omega| < \omega_s \\ G(\omega) \left[\dfrac{\kappa}{\int_{|\omega| > \omega_s} |G(\omega)|^2 d\omega} \right]^{1/2} & \omega_s < |\omega| < \pi. \end{cases} \tag{6.4-39}$$

Design of IIR All-pass Filters

All-pass filters are used as building blocks in various types of filter structures. One of the chief applications of all-pass filters is in *phase equalization*, i.e., the task of correcting the non-linear phase response of other signal processing units. Recall that a linear phase is required to avoid distorting the signal. Indeed, a linear phase implies that all frequency components of the signal are delayed by the same amount as they travel through the network. Thus at the output of the network, the various frequency components add up in the correct order and, other things being equal, distortion is avoided. Networks in which all frequency components are delayed by the same amount are called *non-dispersive*. The delay or, more precisely, the *group delay* is proportional to derivative of the phase with respect to frequency. Thus if the phase is linear in frequency the group delay is a constant.

All-pass phase-equalization filters typically belong to a class of filters called *infinite impulse response* (IIR) filters. There does not yet exist a single well-established method for their design [17]. In this section we show how POCS might be useful in this regard.

If $h(n), n = 0, 1, \ldots$, denotes the impulse response of an IIR all-pass filter, then it can be shown [17] that its system function (Z-transform) is given by

$$\tilde{H}(z) = \sum_{n=0}^{\infty} h(n)z^{-n} \tag{6.4-40}$$

$$= \frac{a_N + a_{N-1}z^{-1} + \cdots + a_0 z^{-N}}{a_0 + a_1 z^{-1} + \cdots + a_N z^{-N}} \tag{6.4-41}$$

$$\triangleq z^{-N} \frac{D(z^{-1})}{D(z)}, \tag{6.4-42}$$

where N is the *order* of the filter.

The transfer function is computed at $z = e^{j\omega}$. With $H(\omega) \triangleq \tilde{H}(e^{j\omega})$, it is not difficult to show that $\mid H(\omega) \mid = 1$ and the phase, $\theta_H(\omega)$ of $H(\omega)$ is given by

$$\theta_H(\omega) = -N\omega + 2\tan^{-1} \frac{\mathbf{a}^T \mathbf{s}}{\mathbf{a}^T \mathbf{c}}, \tag{6.4-43}$$

where $\mathbf{a}^T \triangleq (a_0, a_1, \cdots, a_N)$ is the all-pass coefficient vector and

$$\mathbf{s} \triangleq (0, \sin\omega, \cdots, \sin N\omega)^T \tag{6.4-44}$$

$$\mathbf{c} \triangleq (0, \cos\omega, \cdots, \sin N\omega)^T. \tag{6.4-45}$$

Now consider a *prescribed* phase, say $\theta(\omega)$, requiring *equalization*, i.e., by an appropriate choice of the coefficient vector $\mathbf{a}$, we require that

$$\theta(\omega) + \theta_H(\omega) = -K\omega, \tag{6.4-46}$$

where $K > 0$ is a restricted constant that is discussed in greater detail below. Equation (6.4-46) may not be physically realizable; hence we modify it to

$$-K\omega - \delta(\omega) \leq \theta(\omega) + \theta_H(\omega) \leq -K\omega + \delta(\omega), \tag{6.4-47}$$

where $\delta(\omega)$ is a tolerance function chosen by the user. For example, if $\delta(\omega) = \delta_o\omega$, the equalized phase must lie within the bounds shown in Fig. 6.4-7, where $K_1 = K - \delta_0$ and $K_2 = K + \delta_0$.

After some algebra, Eq. (6.4-47) can be written as

$$\delta_L(\omega) \leq \tan^{-1} \frac{\mathbf{a}^T \mathbf{s}}{\mathbf{a}^T \mathbf{c}} \leq \delta_U(\omega), \tag{6.4-48}$$

total phase

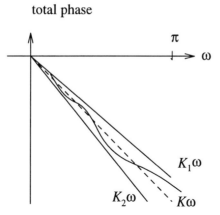

Fig. 6.4-7 Bounds on the deviation of the total phase from linearity.

where

$$\delta_L(\omega) \triangleq \frac{-(K - N)\omega - \delta(\omega) - \theta(\omega)}{2} \qquad (6.4\text{-}49)$$

$$\delta_U(\omega) \triangleq \frac{-(K - N)\omega + \delta(\omega) - \theta(\omega)}{2}. \qquad (6.4\text{-}50)$$

Recall that $\phi = \tan^{-1} x$ is a monotonically increasing function of x as long as we restrict ϕ to a single branch, i.e., a fixed integer n such that $n\pi - \pi/2 < \phi < n\pi + \pi/2$. Hence, in this interval, Eq. (6.4-48) is equivalent to

$$\delta_1(\omega) \le \frac{\mathbf{a}^T \mathbf{s}}{\mathbf{a}^T \mathbf{c}} \le \delta_2(\omega), \qquad (6.4\text{-}51)$$

where

$$\delta_1(\omega) \triangleq \tan \delta_L(\omega) \qquad (6.4\text{-}52)$$

$$\delta_2(\omega) \triangleq \tan \delta_U(\omega). \qquad (6.4\text{-}53)$$

In order to ensure that we are operating in this interval we must restrict K as

$$N - \min_{\omega \in \Omega} \left(\frac{\pi + 2n\pi + \theta(\omega) - \delta(\omega)}{\omega} \right) \le K \le N + \min_{\omega \in \Omega} \left(\frac{\pi - 2n\pi - \theta(\omega) - \delta(\omega)}{\omega} \right)$$

$$(6.4\text{-}54)$$

where Ω is the frequency range of which equalization is to take place. For example, if we choose the principle branch $n = 0$, then for $N = 6, \delta(\omega) = 0.003\omega, \Omega = [0, 0.2\pi]$, and $\theta(\omega) = -\omega^2$, we obtain $4.37 \le K \le 11.62$. Let us assume that K has been properly chosen; then the solution must lie in the intersection of sets of

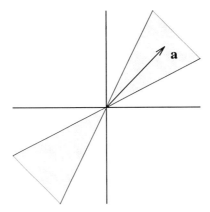

Fig. 6.4-8 The set C_w (shaded) drawn as a double wedge.

the type (one set for each value of ω):

$$C_\omega \stackrel{\triangle}{=} \left\{ \mathbf{a} \in R^{N+1} : \delta_1(\omega) \leq \frac{\mathbf{a}^T \mathbf{s}}{\mathbf{a}^T \mathbf{c}} \leq \delta_2(\omega) \right\}, \qquad (6.4\text{-}55)$$

where $\mathbf{s} = \mathbf{s}(\omega)$, $\mathbf{c} = \mathbf{c}(\omega)$ are defined by Eqs. (6.4-44) and (6.4-45), respectively. A careful examination of C_ω shows that if $\mathbf{a}$ is inside the set so is $\beta\mathbf{a}$, where β is any real number, positive or negative. Hence the set can be drawn as a *double wedge*, i.e., two intersecting hyperplanes (Fig. 6.4-8).

While this set is not convex[†] it can be written as the union of two *convex wedges*, i.e.,

$$C_\omega = C_\omega^+ \cup C_\omega^-, \qquad (6.4\text{-}56)$$

where

$$C_\omega^+ \stackrel{\triangle}{=} \left\{ \mathbf{a} \in R^{N+1} : \mathbf{a}^T \mathbf{c} \geq 0 \text{ and } \delta_1 \mathbf{a}^T \mathbf{c} \leq \mathbf{a}^T \mathbf{s} \leq \delta_2 \mathbf{a}^T \mathbf{c} \right\} \quad (6.4\text{-}57)$$

$$C_\omega^- \stackrel{\triangle}{=} \left\{ \mathbf{a} \in R^{N+1} : \mathbf{a}^T \mathbf{c} \leq 0 \text{ and } \delta_2 \mathbf{a}^T \mathbf{c} \leq \mathbf{a}^T \mathbf{s} \leq \delta_1 \mathbf{a}^T \mathbf{c} \right\}. \quad (6.4\text{-}58)$$

In Eqs. (6.4-57) and (6.4-58) we have omitted some of the ω dependences, for brevity. With an appropriate choice of the function $\delta(\omega)$, neither of the convex wedges will be empty.

For numerical calculations, the variable ω must be discretized to values ω_1, ω_2 $\cdots, \omega_M$; let $C_{\omega_i} \stackrel{\triangle}{=} C_i$. Then the correct coefficient vector must lie in

$$C_0 \stackrel{\triangle}{=} \bigcap_{i=1}^{M} C_i. \qquad (6.4\text{-}59)$$

[†]We leave the demonstration of this fact to the reader.

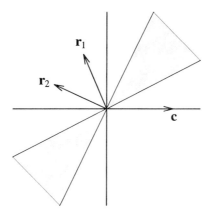

Fig. 6.4-9 When $\mathbf{r_1}$ is to the right of $\mathbf{r_2}$, the set C_w is not empty.

WHEN IS C_i NOT EMPTY? Let $\mathbf{r}_1 \stackrel{\triangle}{=} \mathbf{s} - \delta_1\mathbf{c}$ and $\mathbf{r}_2 \stackrel{\triangle}{=} \mathbf{s} - \delta_2\mathbf{c}$. As long as $\mathbf{r}_1$ is to the right of $\mathbf{r}_2$ the sets C_ω^+ and C_ω^- will not be empty (Fig. 6.4-9). To demonstrate this quantitatively, let $\mathbf{r}_0 \stackrel{\triangle}{=} (\mathbf{r}_1 + \mathbf{r}_2)/2$; then $\mathbf{a}^T\mathbf{r}_0 = 0$ defines a hyperplane going through the origin. We have to show that this hyperplane (and, therefore, all the vectors it contains) satisfies

$$\delta_1 \mathbf{a}^T \mathbf{c} \le \mathbf{a}^T \mathbf{s} \le \delta_2 \mathbf{a}^T \mathbf{c} \tag{6.4-60}$$

for $\mathbf{a}^T\mathbf{c} \ge 0$. But

$$\mathbf{r}_0 = \mathbf{s} - \frac{\delta_1 + \delta_2}{2}\mathbf{c}. \tag{6.4-61}$$

It follows from $\mathbf{a}^T\mathbf{r}_0 = 0$ that

$$\mathbf{a}^T\mathbf{s} = \frac{\delta_1 + \delta_2}{2}\mathbf{a}^T\mathbf{c}. \tag{6.4-62}$$

Hence

$$\delta_1 \mathbf{a}^T \mathbf{c} < \frac{\delta_1 + \delta_2}{2}\mathbf{a}^T\mathbf{c} < \delta_2 \mathbf{a}^T \mathbf{c}, \tag{6.4-63}$$

which, for $\mathbf{a}^T\mathbf{c} > 0$, implies that

$$\delta_1 < \frac{\delta_1 + \delta_2}{2} < \delta_2 \tag{6.4-64}$$

or $\delta_1 < \delta_2$. Thus C_w^+ is not empty. An identical argument shows that C_w^- is not empty for $\delta_1 < \delta_2$. Working back, we find that $\delta_1 < \delta_2$, implies that $\delta_U(\omega) > \delta_L(\omega)$, provided that we limit ourselves to the first branch of the arctan function. But $\delta_U(\omega) > \delta_L(\omega)$ is satisfied for any tolerance function $\delta(\omega) > 0$. Hence, under the conditions stated above, the wedge set will not be empty.

PROJECTION ONTO C_i: The projector P_i is easily deduced from Fig. 6.4-10. As

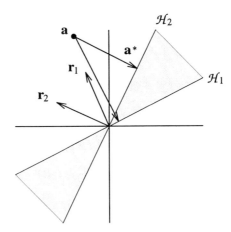

Fig. 6.4-10 The projection of **a** onto the double wedge can be computed by finding the distance to each plane and choosing the smallest.

we can see, the set C_i is bounded by two intersecting hyperplanes, say $\mathcal{H}_1$ and $\mathcal{H}_2$. Plane $\mathcal{H}_j, j = 1, 2$, is defined by its normal vector $\mathbf{r}_j$ or $\mathbf{r}_j/\|\mathbf{r}_j\|$. Given an arbitrary vector $\mathbf{y} \in R^{N+1}$ that lies outside of C_i, its projection is computed by finding the *distance* to $\mathcal{H}_1$ and $\mathcal{H}_2$ and choosing the smallest. Specifically, with d_j denoting the distance from $\mathbf{y}$ to $\mathcal{H}_j$ i.e., $d_j \overset{\triangle}{=} |\mathbf{r}_j^T \mathbf{y}| / \|\mathbf{r}_j\|$, we obtain

$$
\mathbf{a}^* = P_i \mathbf{y} = \begin{cases} \mathbf{y} & \text{if } \mathbf{r}_1^T \mathbf{y} > 0 \text{ and } \mathbf{r}_2^T \mathbf{y} < 0 \\ \mathbf{y} & \text{if } \mathbf{r}_1^T \mathbf{y} < 0 \text{ and } \mathbf{r}_2^T \mathbf{y} > 0 \\ \mathbf{y} - \frac{\mathbf{r}_1^T \mathbf{y}}{\|\mathbf{r}_1\|^2} \mathbf{r}_1 & \text{if } d_1 < d_2 \\ \mathbf{y} - \frac{\mathbf{r}_2^T \mathbf{y}}{\|\mathbf{r}_2\|^2} \mathbf{r}_2 & \text{if } d_1 > d_2. \end{cases} \tag{6.4-65}
$$

At this point we must stop and reflect that the sets $\{C_i\}$ are not convex. As such, there is a possibility of stagnating at a trap. Indeed the topological landscape is quite complex, involving the intersection of many non-convex sets. Thus even *summed distance error convergence* doesn't hold. [†]

Another approach is to work with only a set of single wedges, say the sets $\{C_i^+\}$. While the single wedges are convex, there is the possibility of converging to zero when using an algorithm based on projecting onto such sets. A solution to this problem is to restrict the solution region to coefficient vectors that have a finite length. A straightforward way of doing this is to truncate the cone with an intersecting hyperplane not passing through the origin (see $\mathcal{H}_3$) in Fig. 6.4-11. The result is still convex, but the computation of the projection is more involved.

[†] However, summed distance error reduction property will still hold in product spaces. See Chapter 5 for a discussion of this kind of convergence.

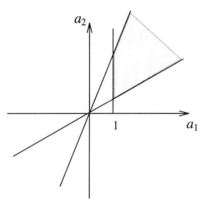

Fig. 6.4-11 A convex set (shaded) that avoids convergence to the zero vector.

Another point is that it may not be possible to equalize the phase over a broad range of frequencies without relaxing the tolerance parameter $\delta(\omega)$. These and other considerations, such as stability, etc., are the subjects of ongoing research [18].

The results of a numerical example are given in Fig. 6.4-12. The aim there was to equalize the phase of a signal whose phase θ varied as $\theta(\omega) = -\omega^2 \text{ sgn } (\omega)$. The details of the example are given in Exercise 6-15. Single-wedge constraints were used without intersecting hyperplanes, to prevent convergence to zero.

6.5 ESTIMATION OF PROBABILITY DENSITY FUNCTIONS

The following paradigm applies to many real-world problems: We make an observation x and must determine whether x is manifestation of an event from either class ω_1 or class ω_2. We give some concrete examples below:

1. An automatic blood-screening test makes measurements on white blood cells and combines them into a vector x. Does x suggest a healthy immune system (class ω_1) or a pathological one (class ω_2)?

2. An automatic X-ray scanner screens for black-lung disease in coal workers. The vector x is a lexicographic ordering of the gray-levels of a coal-worker's chest X-ray. Does x suggests the presence of black-lung disease (class ω_1) or are the coal-worker's lungs healthy (class ω_2)?

3. A strong radar echo suddenly appears at a tactical ground-to-air missile station. Does the echo imply an incoming enemy warhead (class ω_1) or is the echo due to a harmless source (class ω_2), e.g., a flock of geese en route to breeding grounds?

To deal with problems of this type, one must develop an objective rule that uses the information in x and prior knowledge to decide whether the cause of x is class

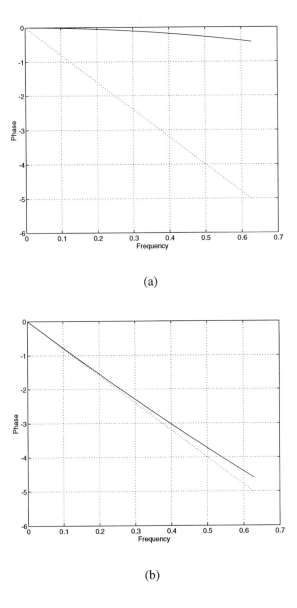

(a)

(b)

Fig. 6.4-12 Phase equalization with an IIR filter designed using projections: (a) the phase response to be equalized; (b) the phase response of the all-pass filter designed using the POCS algorithm; (c) the equalized overall phase response. The dashed curves in each plot represent the desired linear phase to be achieved through equalization. Note that in (c) the equalized phase response completely coincides with the desired linear phase at the scale of the plot. The parameters that are used in this example are $K = 8$, $N = 6$, $\delta = 0.003\omega$, and the frequency range for equalization is $[0, 0.2\pi]$.

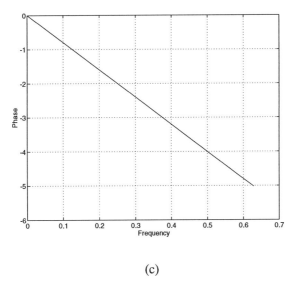

(c)

Fig. 6.4-12   *Continued.*

ω_1 or class ω_2. The objective rule usually follows from the desired goal. If the desired goal is to maximize the probability of a correct decision, then a theoretical calculation (Exercise 6-16) shows that the appropriate rule is:

$$\begin{aligned}
&\text{Assign } \mathbf{x} \text{ to } \omega_1 \quad \text{if } f(\mathbf{x}|\omega_1)P_1 > f(\mathbf{x}|\omega_2)P_2; \\
&\text{Assign } \mathbf{x} \text{ to } \omega_2 \quad \text{if } f(\mathbf{x}|\omega_1)P_1 < f(\mathbf{x}|\omega_2)P_2.
\end{aligned} \tag{6.5-1}$$

In Eq. (6.5-1) $f(\mathbf{x}|\omega_i), i = 1, 2$, is the class-conditional *probability density function* (pdf) and $P_i, i = 1, 2$, is the prior probability of an event from class ω_i. The rule given in Eq. (6.5-1) is known as the *maximum a posteriori* (MAP) decision rule. Many people regard any system that implements the MAP rule as an optimum system. Indeed, the MAP rule is widely used in signal processing, communications, even medical imaging. To implement the MAP rule we need to know the pdf's $f(\mathbf{x}|\omega_i), i = 1, 2$. These are usually estimated from data. The whole purpose of this discussion is to convince the reader that the estimation of pdf's is an important problem. In fact, the estimation of pdf's is important not only in electrical engineering but also in operations research, traffic control, distribution of electric power, etc. In the discussion below we shall try to be as general as possible. We shall use the symbol $f(x)$ for the pdf we wish to estimate.

Our model is that of an underlying random variable (r.v.) X, whose (unknown) pdf is $f(x)$. We make n independent observations on X and call these $X_1, X_2, \cdots, X_n$. The actual numbers we observe are denoted by $x_1, x_2, \cdots, x_n$, and these are called *realizations* of the r.v.'s $X_1, X_2, \cdots, X_n$. The numbers $x_1, x_2,$

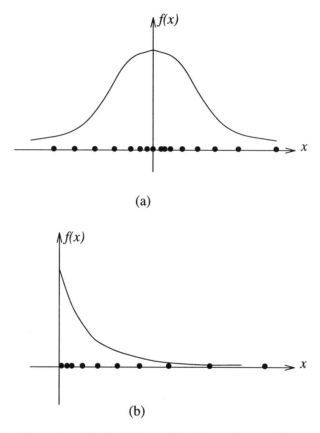

(a)

(b)

Fig. 6.5-1   (a) Observations (black dots) on a Gaussian random variable. (b) Same as (a) for an exponential random variable.

$\cdots, x_n$ are our data, and the $X_i, i = 1, 2, \cdots, n$, are said to be independent, identically distributed (iid) random variables.

For example, let X represent the body weight of a member of a statistically homogeneous population. The individual body weights of the first ten members that we encounter in that group would be denoted as $X_1, X_2, \cdots, X_{10}$. Then the actual recorded weights of these ten members would be denoted by $x_1, x_2, \cdots, x_{10}$.

Figure 6.5-1(a) shows how the data might look if $f(x)$ was a zero-mean Gaussian density; Fig. 6.5-1(b) shows how the data might look if $f(x)$ was the exponential density. Note that if we pick an interval $[a, b]$ and count the number of outcomes, say, k in $[a, b]$, then $P[a \leq X \leq b] \simeq k/n$. The ratio k/n, however, will change if make another n observations on X. Thus k/n is only a rough estimate of $P[a \leq X \leq b]$. Nevertheless this rough estimate is the basis for estimating $f(x)$.

Because of the importance of the problem, many clever ways of estimating $f(x)$ have been derived. Some are listed in [19–22].

Identifying the Membership Sets of $f(x)$

Consider again the event[†] $\{a_i \leq X \leq b_i\}$ and assume that $P[a_i \leq X \leq b_i] \overset{\triangle}{=} p_i$ is known. Then, since

$$P[a_i \leq X \leq b_i] = \int_{a_i}^{b_i} f(x)dx = p_i, \qquad (6.5\text{-}2)$$

the unknown $f(x)$ must have membership in the set $C'(a_i, b_i)$, given by

$$C'(a_i, b_i) = \left\{ y(x) \; : \; \int_{a_i}^{b_i} y(x)dx = p_i \right\}. \qquad (6.5\text{-}3)$$

Since we are free to choose many intervals $[a_i, b_i]$, we can generate a large number of sets $C'(a_i, b_i)$ and the unknown pdf $f(x)$ must have membership in each of the $C'(a_i, b_i), i = 1, 2, \cdots, N$, as well as in the intersection $C'_0 = \cap_{i=1}^{N} C'(a_i, b_i)$. Therefore, an acceptable (sometimes called a *feasible*) solution to the problem is a point in C'_0. If P_i is the projector that projects onto $C'(a_i, b_i)$, then the POCS algorithm

$$f_{k+1} = P_1 P_2 \cdots P_N f_k, \qquad f_0 \text{ arbitrary} \qquad (6.5\text{-}4)$$

will converge to an $f(x) \in C'_0$ that is consistent with the data.

The projection of an arbitrary function $g(x)$ onto $C'(a_i, b_i)$ is worked out in Example 3.2-3 and is given by

$$f(x) = y^*(x) = P_i g(x) = \begin{cases} g(x) & x \notin [a_i, b_i] \\ g(x) + \frac{p_i - \int_{a_i}^{b_i} g(x)dx}{b_i - a_i} & x \in [a_i, b_i]. \end{cases} \qquad (6.5\text{-}5)$$

There are other constraints that we can add. For example, since $f(x)$ is a pdf it follows that it cannot be negative and must have unit area, i.e.,

$$f(x) \geq 0, \text{ and } \int_{\Gamma} f(x)dx = 1, \qquad (6.5\text{-}6)$$

where Γ is the *support* of $f(x)$. Thus $f(x)$ must also have membership in the sets

$$C_+ = \{ y(x) \; : \; y(x) \geq 0 \} \qquad (6.5\text{-}7)$$

and

$$C_{ua} = \{ y(x) \; : \; \int_{\Gamma} f(x)dx = 1 \}. \qquad (6.5\text{-}8)$$

If Γ is unknown it must be estimated. However, this is a technical detail. Thus the

---

[†]We add subscripts to a, b to form a_i, b_i, in order to construct many sets so $i = 1, 2, \cdots$.

POCS algorithm in Eq. (6.5-4) can be modified to

$$f_{k+1} = P_+ P_{ua} P_1 P_2 \cdots P_N f_k, \qquad f_0 \text{ arbitrary.} \qquad (6.5\text{-}9)$$

Actually, the POCS algorithm can be in any order and can incorporate any constraint, as long as these are associated with closed convex sets.

Confidence Intervals for p_i in Eq. (6.5-3)

While all of the previous theory is quite reasonable if p_i is known, generally speaking, p_i is not known. We cannot replace p_i by k_i/n (k_i is the number of outcomes in $[a_i, b_i]$), since p_i is a constant and k_i/n will typically vary from experiment to experiment. The solution is to find a *confidence interval* for p_i. Thus we seek to make the statement that "the probability that p_i lies between the numbers p_i^L and p_i^H is $1 - \alpha$", where α is typically 0.05 or 0.01. If $\alpha = 0.05$ we speak of a 95 percent confidence interval; likewise, if $\alpha = 0.01$ we speak of a 99 percent confidence interval. The confidence interval $[p_i^L, p_i^H]$ depends on α. The smaller α gets, the larger the interval gets. Thus we replace the non-realistic constraint set in Eq. (6.5-3) with [†]

$$C(a_i, b_i) = \left\{ y(x) \; : \; \int_{a_i}^{b_i} y(x)dx \in [p_i^L, p_i^H] \right\}, \qquad (6.5\text{-}10)$$

where the interval $[p_i^L, p_i^H]$ is a $100(1-\alpha)\%$ confidence interval for p_i. At 95 or 99 percent levels of confidence, any function $f(x)$ that belongs to $C(a_i, b_i)$ is regarded as being consistent with the data. The computation of p_i^L, p_i^H is done using results from the theory of statistics. We sketch below how this is done; for further details see [23]. Let $X_j, j = 1, 2, \cdots, n$, be n independent, identically distributed random variables. Let

$$Y_j = \begin{cases} 1, & \text{if } X_j \in [a_i, b_i] \\ 0, & \text{otherwise.} \end{cases} \qquad (6.5\text{-}11)$$

Then each Y_j is a Bernoulli random variable with mean p_i and variance $p_i(1 - p_i)$ and, by the Central Limit Theorem, the r.v. W' (defined below) converges in distribution to a standard normal r.v., i.e.,

$$W' \triangleq \frac{\sqrt{n}(\bar{Y} - p_i)}{\sqrt{p_i(1 - p_i)}} \xrightarrow{d} N(0, 1), \qquad (6.5\text{-}12)$$

where $\bar{Y} = 1/n \sum_{j=1}^{n} Y_j$ (the sample mean). While Eq. (6.5-12) is true, it is not too useful, because the $\sqrt{p_i(1 - p_i)}$ in the denominator involves the unknown p_i. However, according to the weak law of large numbers [23, 24] $\bar{Y} \to p_i$; that being the case, the *mixed convergence theorem* [24] allows us to replace Eq. (6.5-12)

[†]Among some practitioners a "relaxed" set such as $C(a_i, b_i)$ is called a *fuzzy set* as opposed to $C'(a_i, b_i)$ in Eq. (6.5-3) which is called a *crisp set*.

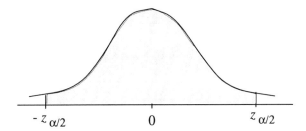

Fig. 6.5-2 Finding the points for a $1 - \alpha$ confidence interval. The area of the shaded region is $1 - \alpha$.

with

$$\tilde{W} \triangleq \frac{\sqrt{n}(\bar{Y} - p_i)}{[\bar{Y}(1 - \bar{Y})]^{1/2}} \xrightarrow{d} W : N(0, 1). \qquad (6.5\text{-}13)$$

Now for any standard normal random variable W there exists numbers $z_{\alpha/2}$ and $-z_{\alpha/2}$ such that (Fig. 6.5-2)

$$P[-z_{\alpha/2} \leq W \leq z_{\alpha/2}] = 1 - \alpha. \qquad (6.5\text{-}14)$$

Using Eq. (6.5-13) in Eq. (6.5-14) leads to

$$P\left[-z_{\alpha/2} \leq \frac{\sqrt{n}(\bar{Y} - p_i)}{[\bar{Y}(1 - \bar{Y})]^{1/2}} \leq z_{\alpha/2}\right] \approx 1 - \alpha, \qquad (6.5\text{-}15)$$

which can also be written as

$$P\left[\bar{Y} - \frac{z_{\alpha/2}\left[\bar{Y}(1 - \bar{Y})\right]^{1/2}}{\sqrt{n}} \leq p_i \leq \bar{Y} + \frac{z_{\alpha/2}\left[\bar{Y}(1 - \bar{Y})\right]^{1/2}}{\sqrt{n}}\right] \approx 1 - \alpha. \qquad (6.5\text{-}16)$$

Let

$$p_i^L = \bar{Y} - \frac{z_{\alpha/2}\left[\bar{Y}(1 - \bar{Y})\right]^{1/2}}{\sqrt{n}} \qquad (6.5\text{-}17)$$

and

$$p_i^H = \bar{Y} + \frac{z_{\alpha/2}\left[\bar{Y}(1 - \bar{Y})\right]^{1/2}}{\sqrt{n}}. \qquad (6.5\text{-}18)$$

Then,

$$P[p_i^L \leq p_i \leq p_i^H] \approx 1 - \alpha. \qquad (6.5\text{-}19)$$

Thus, $[p_i^L, p_i^H]$ is an approximate $100(1 - \alpha)\%$ confidence interval for p_i. Because p_i^L, p_i^H are random variables that depend on the outcome, this interval is also random but contains the unknown p_i $100(1 - \alpha)$ percent of the time.

Projection onto $C(a_i, b_i)$ of Eq. (6.5-10)

To save notation we define $C_i = C(a_i, b_i)$ of Eq. (6.5-10) and use this symbol in what follows. We leave the demonstration of closedness and convexity of C_i as exercise for the reader. To compute the projection of an arbitrary $g(x)$ onto C_i we seek g^* given by

$$g^* = P_i g = \arg \left\{ \min_{y \in C_i} \|g - y\|^2 \right\}. \tag{6.5-20}$$

Stated as a constrained optimization problem, we seek to find

$$\min_{y \in C_i} \|g - y\|^2, \tag{6.5-21}$$

subject to

$$p_i^L \leq \int_{a_i}^{b_i} y(x) dx \leq p_i^H. \tag{6.5-22}$$

Assume first that $q \triangleq \int_{a_i}^{b_i} y(x) dx > p_i^H$. Then an appropriate Lagrange functional is

$$J(y) = \int_\Gamma [g(x) - y(x)]^2 dx + \lambda \left[\int_{a_i}^{b_i} y(x) dx - p_i^H \right]. \tag{6.5-23}$$

There being no constraint on $y(x)$ for $x \notin [a_i, b_i]$, we obtain $g^*(x) = g(x)$ for $x \notin [a_i, b_i]$. Setting $\frac{dJ}{dy} = 0$ yields

$$y(x) = g(x) - \frac{\lambda}{2}, \tag{6.5-24}$$

and from the constraint that

$$\int_{a_i}^{b_i} y(x) dx = \int_{a_i}^{b_i} g(x) dx - \frac{\lambda}{2}(b_i - a_i) = p_i^H, \tag{6.5-25}$$

we obtain, for $q > p_i^H$,

$$g^*(x) = g(x) - \frac{q - p_i^H}{b_i - a_i}. \tag{6.5-26}$$

Repeating the computation for the case $q < p_i^L$ yields

$$g^*(x) = g(x) - \frac{q - p_i^L}{b_i - a_i}. \tag{6.5-27}$$

Finally, if $p_i^L \leq q \leq p_i^H$ then $g(x)$ is already in the set and its projection is itself. Summarizing there result gives

$$
g^* = P_i g = \begin{cases} g(x) & \text{if } x \notin [a_i, b_i], \text{ or } x \in [a_i, b_i] \text{ and } p_i^L \leq q \leq p_i^H \\ g(x) - \frac{q - p_i^H}{b_i - a_i} & \text{if } x \in [a_i, b_i] \text{ and } q > p_i^H \\ g(x) - \frac{q - p_i^L}{b_i - a_i} & \text{if } x \in [a_i, b_i] \text{ and } q < p_i^L. \end{cases}
$$

(6.5-28)

Additional Constraints

In executing the POCS algorithm to estimate pdf's we want to use as much prior knowledge as possible. Clearly the positivity of $f(x)$, its unit area, and the interval probability constraints are important. But there are at least two more constraints that can have significant impact on the quality of the final results: constraint based on the mean, and constraint requiring the reconstructed pdf to be relatively smooth.

In the case of the mean, the constraint set takes the form

$$
C(\mu) = \{ \, y(x) \; : \; \int_\Gamma x \, y(x) \, dx \in [\mu_L, \mu_H] \, \},
$$

(6.5-29)

where μ_L and μ_H are the bounds on an $100(1 - \alpha)\%$ confidence interval estimation for μ. These bounds are computed from

$$
\mu_L = \bar{x} - z_{\alpha/2} \frac{s}{\sqrt{n}} \, , \quad \text{and } \mu_H = \bar{x} + z_{\alpha/2} \frac{s}{\sqrt{n}} \, ,
$$

(6.5-30)

where $\bar{x} = 1/n \sum_{i=1}^n x_i$ (sample mean), $s^2 = 1/(n-1) \sum_{i=1}^n (x_i - \bar{x})^2$ (sample variance). The derivation of Eq. (6.5-30) is given in [25, 26].

Given an arbitrary function $g(x)$, its projection $g^*(x)$ onto $C(\mu)$ is given by

$$
g^* = P_\mu g = \begin{cases} g(x) & \text{if } x \in \Gamma \text{ and } \mu_L \leq \langle x, \, g \rangle \leq \mu_H \\ g(x) - \frac{\langle x, \, g \rangle - \mu_H}{\|x\|_\Gamma^2} x & \text{if } x \in \Gamma \text{ and } \langle x, \, g \rangle > \mu_H \\ g(x) - \frac{\langle x, \, g \rangle - \mu_L}{\|x\|_\Gamma^2} x & \text{if } x \in \Gamma \text{ and } \langle x, \, g \rangle < \mu_L \\ 0 & \text{if } x \notin \Gamma, \end{cases}
$$

(6.5-31)

where $\langle x, \, g \rangle \overset{\triangle}{=} \int_\Gamma x g(x) dx$ and $\|x\|_\Gamma^2 = \langle x, \, x \rangle$.

One would also like to reproduce a *smooth* pdf. There are several constraints that can enforce smoothness. For example, the set

$$
C(\epsilon) = \left\{ \, y(x) \; : \; \left\| \frac{d^2 y}{dx^2} \right\|^2 \leq \epsilon^2 \right\}
$$

(6.5-32)

requires that the energy in the second derivative of its member functions be bounded by ϵ^2. In its numerical implementation the elements of the set $C(\epsilon)$ become vectors and the second derivatives are replaced by finite differences. In this case, the set $C(\epsilon)$ is closed and convex (see Exercise 6-23). The parameter ϵ^2 can be reasonably deduced from the data [26]. The projection onto $C(\epsilon)$ requires some work and is worked out in [25, 26]. With these additional constraints, we can modify the POCS algorithm in Eq. (6.5-9) to

$$f_{k+1} = P_\mu P_\epsilon P_+ P_{ua} P_1 P_2 \cdots P_N f_k, \qquad f_0 \text{ arbitrary.} \qquad (6.5\text{-}33)$$

How effectively does Eq. (6.5-33) reconstruct pdf's from data compared with other methods? We consider this below.

How POCS Compares with Other Methods

Two widely used methods for reconstructing pdf's from data are the method of *penalized likelihood estimation* (PLE) and the method of *windowed kernel estimation* (WKE). They are given and described in the [27]. The reconstruction of a bi-modal pdf involving the weighted sum of two Gaussians is shown in Fig. 6.5-3. The reconstructions furnished by PLE and WKE, based on 100 samples, are so similar that we omit them from the figure.

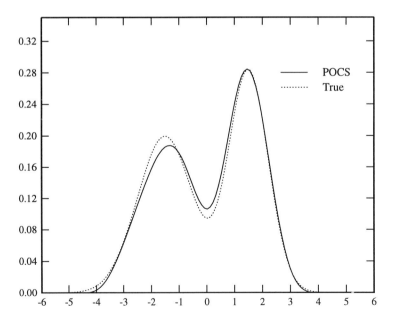

Fig. 6.5-3   POCS reconstruction of a bi-modal pdf from 100 samples.

The reconstructions of the exponential pdf

$$f(x) = e^{-x}u(x) \tag{6.5-34}$$

from 100 samples by POCS, PLE, and WKE are shown in Fig. 6.5-4. The superior results furnished by POCS are due to its ability to enforce the condition $f(x) = 0$ for $x < 0$.

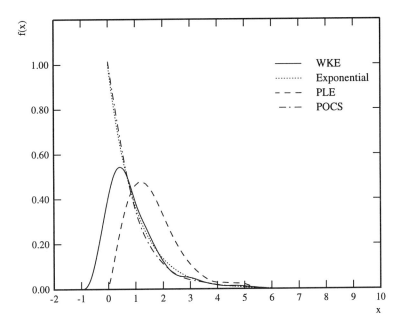

Fig. 6.5-4 Reconstructions of the exponential pdf from 100 samples. Note that POCS correctly reconstructed the zero function for $x < 0$.

6.6 SPREAD SPECTRUM

Spread spectrum (SS) is a modulation technique used in communication systems in which the bandwidth of the transmitted signal is *expanded*, i.e., is made much greater than the signal bandwidth. In this respect it resembles frequency modulation (FM), phase modulation (PM), pulse-code modulation (PCM), and others. However, in the case of SS, the bandwidth of the transmitted signal is determined by some function that is independent of the message signal and is known at the receiver. In this respect it is different from the other modulation schemes above. While bandwidth expansion in FM, PM, and PCM is uniquely related to the message and generates the desirable property of combating independent, additive noise, SS is less effective in doing so. So does SS represent a useless waste of bandwidth? The answer is no. Spread spectrum has certain properties that make it useful in situations requiring the following [28]:

1. Antijam capability—particularly narrow-band jamming;
2. Interference rejection;
3. Multiple access capability;
4. Multipath protection;
5. Covert operations with low probability of intercept; and
6. Secure communications.

Spread spectrum systems were formerly considered primarily for military communications, for reasons that are obvious from its properties. However, SS systems form the basis for a relatively new and powerful civilian communication system called *code division multiple access* (CDMA). The interested reader is referred to [29] for further details.

There are a number of different ways of achieving a spread-spectrum signal. We shall concentrate on *direct-sequence spread-spectrum* (DS-SS) because it has some important advantages over other methods, including best antijam performance and best discrimination against multipath[†]. We note that DS-SS is frequently described as *pseudo-noise spread-spectrum* (PN-SS), which is actually more descriptive of its appearance as a signal.

Let us assume that the original source waveform, carrying the information, is sampled, quantized, and encoded into binary pulses whose values are ± 1. The information thus consists of a message bit sequence $\{a_n = \pm 1\}$, where the subscript n refers to the nth bit in a sequence. The sequence $\{a_n = \pm 1\}$ is transmitted at the rate $1/T_b$ bits per second. Define a unit amplitude pulse $\Pi_T(t)$ by

$$\Pi_T(t) = \begin{cases} 1 & \text{if } 0 < t \leq T \\ 0 & \text{otherwise.} \end{cases} \tag{6.6-1}$$

Then the sequence $\{a_n\}$ can be communicated by the waveform

$$a(t) = \sum_n a_n \Pi_{T_b}(t - nT_b). \tag{6.6-2}$$

Consider now a pseudo-noise (PN) *spreading-code sequence* $\{c(k) = \pm 1\}$, with periodicity K (K an integer). The associated waveform is

$$c(t) = \sum_k c(k) \Pi_{T_c}(t - kT_c), \tag{6.6-3}$$

[†]Multipath refers to a distortion mechanism in which the signal at the receiver is a combination of signals resulting from the original having traveled different paths, thereby experiencing different delays and attenuation

where $T_c \ll T_b$. For simplicity, assume that $KT_c = T_b$, i.e., one whole cycle (consisting of many pulses) of the PN sequence occurs during the time allowed for a single message bit. If $a(t)$ is synchronously[‡] product-modulated by $c(t)$, the result is the spread spectrum waveform

$$p(t) = \sum_k a_{\lfloor k/K \rfloor} c(k) \Pi_{T_c}(t - kT_c), \tag{6.6-4}$$

where $\lfloor x \rfloor$ is the largest integer less than or equal to x. At this point, if desired, the signal $p(t)$ can modulate a high-frequency carrier wave for ease of transmission. At the receiver, by a process called *demodulation*, the DS-SS sequence

$$p(k + nK) \overset{\triangle}{=} a_{\lfloor k/K \rfloor + n} c(nK + k), \quad \forall k \tag{6.6-5}$$

is recovered. The second step at the receiver is the *despreading operation*. Recall that, as a result of the product modulation of $c(t)$ by $a(t)$, the spectrum of the sequence $\{p(k)\}$ is, typically, much broader than the spectrum of the sequence $\{a_n\}$. Indeed this is what gives the spread-spectrum its relative immunity to jamming: A jammer might obliterate some of the band but it cannot, generally, obliterate the entire band. The despreading is done by correlating $\{p(k)\}$ with a synchronized image of $c(k)$ at the receiver. Specifically, for the nth bit, the operation that generates the output $s(n)$ is described by

$$
\begin{aligned}
s(n) &= \sum_{k=0}^{K-1} p(nK + k)c(k) & (6.6\text{-}6) \\
&= \sum_{k=0}^{K-1} a_{\lfloor k/K \rfloor + n} c(nK + k)c(k) & (6.6\text{-}7) \\
&= a_n \sum_{k=0}^{K-1} c(nK + k)c(k) & (6.6\text{-}8) \\
&= a_n \sum_{k=0}^{K-1} c^2(k) & (6.6\text{-}9) \\
&= a_n K. & (6.6\text{-}10)
\end{aligned}
$$

Line 3 follows from $a_{\lfloor k/K \rfloor + n} = a_n$ for $0 \le k \le K - 1$; line 4 follows from the periodicity of the PN spreading sequence $\{c(k)\}$, and Eq. (6.6-10) follows from $c(k) = \pm 1$ and therefore $c^2(k) = 1$ for $k = 0, 1, ..., K - 1$. Since it is the polarity, i.e., the sign of $s(n)$ that contains the information in the message bit a_n, the estimate of a_n, say $\hat{a}_n$, is given by

$$\hat{a}_n = \mathrm{sgn}s(n) \tag{6.6-11}$$

[‡]The start time of the cycle of $c(t)$ is coincident with the start of a message bit.

$$= \text{sgn} K a_n \qquad (6.6\text{-}12)$$

$$= \text{sgn} a_n. \qquad (6.6\text{-}13)$$

Hence in the absence of jammer and thermal noise, perfect recovery is achievable.

In the presence of interference and thermal noise, the recovered discrete-time signal is not $p(k + nK)$ of Eq. (6.6-5), but instead

$$r(k + nK) = \alpha p(k + nK) + j(k + nK) + \nu(k + nK), \qquad (6.6\text{-}14)$$

where $p(k + nK)$ is the transmitted spread-spectrum component, $j(k + nK)$ is the interference signal, $\nu(k + nK)$ is the wideband thermal noise, and α is the channel attenuation. To avoid a significant digression, we temporarily set $\alpha = 1$ and consider the estimation of α later. The output of the discrete correlator is now given by

$$s(n) = \sum_{k=0}^{K-1} r(nK + k)c(k) \qquad (6.6\text{-}15)$$

$$= K a_n + j'(n) + \nu'(n), \qquad (6.6\text{-}16)$$

where

$$j'(n) \triangleq \sum_{k=0}^{K-1} j(nK + k)c(k) \qquad (6.6\text{-}17)$$

and

$$\nu'(n) \triangleq \sum_{k=0}^{K-1} \nu(nK + k)c(k). \qquad (6.6\text{-}18)$$

The estimate of the nth bit a_n from

$$\hat{a}_n = \text{sgn} \sum_{k=0}^{K-1} r(nK + k)c(k) \qquad (6.6\text{-}19)$$

is now subject to error. For this reason some type of noise and interference rejection filter is used before despreading (Fig. 6.6-1).

Before continuing we make an observation: In the absence of jammer and thermal noise it is true that the recovered signal $r(k + nK) = p(k + nK)$ of Eq. (6.6-5); since the spreading sequence $\{c(k)\}$ is known at the receiver, only the *polarity* of $p(k + nK)$ is unknown. Thus $\{p(k)\}$ has all the properties of $\{c(k)\}$, except possibly for a sign inversion. The same cannot be said for the corrupted sequence $\{r(k)\}$ of Eq. (6.6-14). Because of interference and noise it probably will not have the properties of $\{c(k)\}$. Thus it seems reasonable to use prior knowledge to filter $\{r(k)\}$ so that what results has the properties of $\{c(k)\}$. For example, we do a similar thing when a signal, $s(t)$, band-limited to B Hz, is corrupted by independent additive noise $n(t)$. At the receiver, the received signal $r(t) = s(t) + n(t)$ is forced

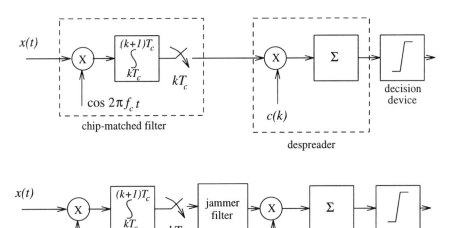

Fig. 6.6-1 Spread-spectrum system: Without interference rejection (top) and with interference rejection (bottom).

to acquire the *a priori* known property of band-limitedness. We do this by projecting $r(t)$ onto the subspace of L^2 functions band-limited to B Hz.

In a recent paper [30], the authors use the method of projections onto convex sets to extract from $\{r(k)\}$ a sequence with the properties of $\{c(k)\}$. We discuss this approach below. Their results suggest that POCS might yield better results than conventional linear, mean-square adaptive filtering. To understand how POCS is applied in this case, we first review the properties of PN sequences.

Properties of PN Sequences

The PN sequence $\{c(k)\}$ is a sequence of plus and minus ones that possess certain properties. Of greatest practical interest are so-called *maximal-length linear shift register sequences* (*m*-sequences). The *m*-sequences have length given by

$$K = 2^m - 1, \quad m = 1, 2, 3, \ldots \tag{6.6-20}$$

Typical values of K for short-length sequences are $K = 7, 15, 31, 63, 127, 255, \ldots$. An important property is the *autocorrelation property* of *m*-sequences described by

$$\rho_c \overset{\triangle}{=} \sum_{k=0}^{K-1} c(k)c(k+n) = \begin{cases} K, & n = 0, \pm K, \pm 2K, \ldots \\ -1, & \text{otherwise.} \end{cases} \tag{6.6-21}$$

In every period of a PN sequence the number of plus ones differs from the number of minus ones by exactly one. Consider the discrete Fourier transform of such a sequence:

$$C(n) \triangleq \text{DFT}[c(k)] = \sum_{k=0}^{K-1} c(k) \exp\left[-j\frac{2\pi nk}{K}\right]. \tag{6.6-22}$$

Therefore

$$|C(0)| = \left|\sum_{k=0}^{K-1} c(k)\right| = 1. \tag{6.6-23}$$

Indeed it can be shown that for K, as in Eq. (6.6-20):

$$|C(n)| = \begin{cases} 1, & n = 0 \\ \sqrt{K+1}, & n = 1, 2, ..., K-1 \end{cases} \tag{6.6-24}$$

Finally note that, for any given PN sequence, the partial sum and, therefore, its magnitude

$$\left|\sum_{k=0}^{l} c(k)\right| = \mu_l \tag{6.6-25}$$

is known for every l.

Constraint Sets for Spread-Spectrum Signals

It is convenient to denote the sequence $y(k), k = 0, 1, ...K - 1$ by the vector $\mathbf{y} = (y(0), y(1), ...y(K-1))^T$ as well as by $\{y(k)\}$. Also we use the notation $\mathbf{y} \leftrightarrow \mathbf{Y}$ or $y(k) \leftrightarrow Y(n)$ to indicate a Fourier transform pair. The constraint sets associated with the spread spectrum problem are the following:

$$C_1(\mathbf{G}) = \left\{\mathbf{y} \in R^K : |Y(n)| \leq G(n) \text{ for all } n \in I_1\right\}, \tag{6.6-26}$$

i.e., $C_1(\mathbf{G})$ denotes the set of all vectors in R^K whose Fourier transform is restricted in magnitude to $|Y(n)| \leq G(n)$ for all n in the (frequency) index set I_1. For simplicity we denote this set by C_1.

APPLICATION TO SPREAD SPECTRUM: Recall from Eq. (6.6-24) that the Fourier transform of $\{c(k)\}$ and, hence, $\{p(k)\}$ must satisfy $|P(n)| = 1$ for $n = 0$ and $|P(n)| = \sqrt{K+1}$ for $n = 1, 2, ..., K-1$.

$$C_2(\mathbf{g}) = \left\{\mathbf{y} \in R^K : |y(k)| \leq g(k) \text{ for all } k \in I_2\right\}, \tag{6.6-27}$$

i.e., C_2 is the of all vectors in R^K whose components $y(k)$ cannot exceed in magnitude the components $g(k)$ for all k in the (time) index set I_2.

APPLICATION TO SPREAD-SPECTRUM: Recall that $p(k)$ satisfies $|p(k)| = 1$ for $k = 0, 1, ..., K-1$. With $g(k) = 1$, Eq. (6.6-27) follows.

$$C_3(\mathbf{x}, d) = \left\{\mathbf{y} \in R^K : |\mathbf{y}^T\mathbf{x}| \leq d, \mathbf{x} \in R^K\right\}, \tag{6.6-28}$$

i.e., C_3 is the set of all vectors in R^K whose inner product with a given real vector x cannot exceed in magnitude a prescribed number d.

APPLICATION TO SPREAD SPECTRUM: There are two cases where C_3 in Eq. (6.6-28) is useful in spread-spectrum:

1. Let $\mathbf{y} = (c(0), c(1), ..., c(K-1))^T$ and $\mathbf{x} = (c(l), c(l+1), ..., c(K-1), c(0), ..., c(l-1))^T$, i.e., a circular shift by l units of the spreading sequence. Then

$$|\mathbf{y}^T\mathbf{x}| = \begin{cases} K & \text{if } l = 0 \\ 1 & \text{if } l = 1, 2, ..., K-1. \end{cases} \qquad (6.6\text{-}29)$$

Replacing $c(k)$ by $p(k)$ in $\mathbf{y}$ or $\mathbf{x}$ yields the same result.

2. Let $\mathbf{y}^T = (c(0), c(1), ..., c(K-1))$ and $\mathbf{x}^T = (\underbrace{11\cdots 1}_{l}\underbrace{00\cdots 0}_{K-l})$. Then

$$|\mathbf{y}^T\mathbf{x}| = |\sum_{k=0}^{l} c(k)| = \mu_l \qquad \text{[Eq. (6.6-25)]}. \qquad (6.6\text{-}30)$$

Since $c(k)$ and $p(k)$ differ, at most, by a sign change, replacing $c(k)$ by $p(k)$ yields the same result. The numbers $\mu_l, l = 0, 1, \cdots$ can be computed *a priori* once the spreading sequence is specified.

The reader will have noticed that we choose to loosen our constrains from equalities to inequalities in defining our sets. For example, if we know that $|c(k)| = |p(k)| = 1$, why does C_2 use the $\leq$ constraint? The answer is that equality constraints in sets C_1 to C_3 would make these sets *non-convex*. We leave it to the reader to demonstrate this fact.

Projections

PROJECTION ONTO C_1: For any fixed n, $G(n)$ is a non-negative number and the set of all complex-valued numbers in the complex plane whose magnitude is bounded by $G(n)$ is represented by a disk of radius $G(n)$ (Fig. 6.6-2). Given a complex number $H(n)$, clearly its nearest neighbor in or on the disk is given by

$$Y^*(n) = \begin{cases} H(n) & \text{if } |H(n)| \leq G(n), n \in I_1 \\ G(n)H(n)/|H(n)| & \text{if } |H(n)| > G(n), n \in I_1 \\ 0 & \text{otherwise.} \end{cases} \qquad (6.6\text{-}31)$$

and the projection of an arbitrary $\mathbf{h}$ onto C_1 is described by

$$y^*(k) = P_1 h(k) \leftrightarrow Y^*(n). \qquad (6.6\text{-}32)$$

ERROR REDUCTION PROPERTY OF P_1: Define $y_o^*(k) = p(k), k = 0, ..., K-1$.

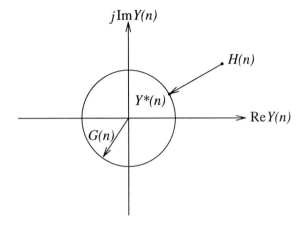

Fig. 6.6-2 Projection onto C_1 as diagrammed in the Fourier domain.

Then, since P_1 is a projector, it immediately follows that

$$||P_1 h - P_1 y_o^*||^2 \le ||h - y_o^*||^2 \qquad (6.6\text{-}33)$$

since P_1 is non-expansive. But $P_1 h = y^*$ and $P_1 y_o^* = y_o^*$. Hence

$$||y^* - y_o^*||^2 \le ||h - y_o^*||^2 \qquad (6.6\text{-}34)$$

and the error never increases.

To show that the error is actually reduced when $h \notin C_1$ form the vectors $\mathbf{Y}^* = [Y_R^*(n) \ \ Y_I^*(n)]^T, \mathbf{H} = [H_R(n) \ \ H_I(n)]^T$, and $\mathbf{Y}_o^* = [Y_{R,o}^*(n) \ \ Y_{I,o}^*(n)]^T$. Assuming that $||\mathbf{H}|| > G$, error reduction occurs for each n if $||\mathbf{Y}^* - \mathbf{Y}_o^*||^2 < ||\mathbf{H} - \mathbf{Y}_o^*||^2$. We note that this relation can be equivalently stated as

$$||\kappa\mathbf{H} - \mathbf{Y}_o^*||^2 < ||\mathbf{H} - \mathbf{Y}_o^*||^2, \qquad (6.6\text{-}35)$$

where $\kappa = G/||\mathbf{H}||$. Expanding both sides of Eq. (6.6-35) and cancelling similar terms yields

$$2(1 - \kappa)\mathbf{H}^T\mathbf{Y}_o^* < (1 - \kappa^2)||\mathbf{H}||^2. \qquad (6.6\text{-}36)$$

Substituting $\mathbf{H}^T\mathbf{Y}_o^* < ||\mathbf{H}|| \ ||\mathbf{Y}_o^*||$ and simplifying Eq. (6.6-36) produces

$$2||\mathbf{Y}_o^*|| < (1 + \kappa)||\mathbf{H}||. \qquad (6.6\text{-}37)$$

We note that $\kappa = ||\mathbf{Y}_o^*||/||\mathbf{H}||$, so that Eq. (6.6-37) reduces to the condition $\kappa < 1$, which is true under the stated assumption that $||\mathbf{H}|| > G$. It is easily verified that a decrease in error for each frequency n leads to an overall decrease in error over the entire signal bandwidth.

PROJECTION ONTO C_2: Given an arbitrary $\mathbf{h} = \mathbf{h}_R + j\mathbf{h}_I$, we wish to find a real

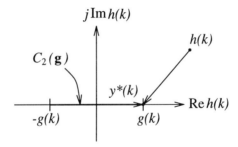

Fig. 6.6-3   The projection $y^*(k)$ of $h(k)$ onto C_2.

vector **y** such that

$$\mathbf{y}^* = \arg \min_{\mathbf{y} \in C_2} \|\mathbf{h} - \mathbf{y}\|. \tag{6.6-38}$$

From Fig. 6.6-3 it is easily seen that the components $y^*(k), k = 0, 1, 2, \cdots$, of the vector $\mathbf{y}^*$ satisfy

$$y^*(k) = P_2 h(k) = \begin{cases} g(k) \ \text{sgn}[h_R(k)] & \text{if } |h_R(k)| \geq g(k), \quad k \in I_2 \\ h_R & \text{if } |h_R(k)| < g(k), \quad k \in I_2 \\ 0 & \text{elsewhere.} \end{cases} \tag{6.6-39}$$

ERROR REDUCTION PROPERTY OF P_2: Since P_2 is non-expansive we immediately obtain

$$\|\mathbf{h}_R - \mathbf{y}_o^*\|^2 \geq \|P_2 \mathbf{h}_R - P_2 \mathbf{y}_o^*\|^2 = \|\mathbf{y}^* - \mathbf{y}_o^*\|^2. \tag{6.6-40}$$

Hence the application of P_2 *never increases* the error.

To show that $\|\mathbf{h}_R - \mathbf{y}_o^*\| > \|\mathbf{y}^* - \mathbf{y}_o^*\|$ for $\mathbf{h}_R \notin C_2$, we proceed as follows. The components of each member vector of C_2 are contained within the interval $[-g(k), g(k)]$ on the real line. Let $h_R(k) > g(k)$ and furthermore let $y^*(k) = y_o^*(k) + n_g(k)$ and $h_R = g(k) + n_h(k)$. We test the inequality $(y^*(k) - y_o^*(k))^2 < (h_R(k) - y_o^*(k))^2$. By noting that $y^*(k) = g(k)$ when $h_R > g(k)$, we get the test

$$n_g^2(k) < (n_g + n_h(k))^2, \tag{6.6-41}$$

which is identical to

$$n_h^2(k) > -2n_g(k)n_h(k). \tag{6.6-42}$$

Since, by assumption, $n_h(k) > 0$ and $n_g(h) \geq 0$ (otherwise $y_o^*(k) \notin C_2$), the inequality is valid. A similar argument holds for the case $h_R(k) < -g(k)$.

PROJECTION ONTO C_3: The projection onto C_3 can easily be computed using the Lagrange multiplier method and we leave this approach as an exercise. We can, however, infer the projection by considering the diagram in Fig. 6.6-4. We choose the direction of the reference vector $\mathbf{x}$ as the x-axis; then $\mathbf{x}/\|\mathbf{x}\|$ is a unit vector along $\mathbf{x}$. The lines L and L' form the boundaries of C_3. Since $\mathbf{y} \in C_3$ is

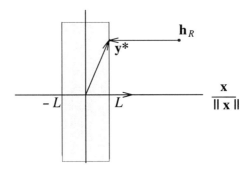

Fig. 6.6-4 The projection $\mathbf{y}^*$ of $\mathbf{h} = \mathbf{h}_R + j\mathbf{h}_I$ onto C_3 involves setting $\mathbf{h}_I = 0$ and operating on $\mathbf{h}_R$ as shown.

real, it can be adjusted to be closest to only the real part $\mathbf{h}_R$ of an arbitrary vector $\mathbf{h} = \mathbf{h}_R + j\mathbf{h}_I$. If $\mathbf{x}^T\mathbf{h}_R \triangleq \tilde{d} > d$, the projection must clearly be on L and the nearest point to $\mathbf{h}_R$ on L is $\mathbf{y}^* = \mathbf{h}_R - \rho\mathbf{x}/||\mathbf{x}||$. Thus

$$\left(\mathbf{h}_R - \rho\frac{\mathbf{x}}{||\mathbf{x}||}\right)^T \mathbf{x} = d, \tag{6.6-43}$$

from which we obtain

$$\rho = \frac{\tilde{d} - d}{||\mathbf{x}||} \tag{6.6-44}$$

and

$$\mathbf{y}^* = \mathbf{h}_R - \frac{\tilde{d} - d}{||\mathbf{x}||}\mathbf{x}. \tag{6.6-45}$$

Thus, the projection involves two steps: (1) remove the imaginary part of $\mathbf{h}$; and (2) add the vector $(d - \tilde{d})\mathbf{x}/||\mathbf{x}||$.

Proceeding in the same fashion for the case where $\tilde{d} < -d$, we obtain

$$\mathbf{y}^* = P_3\mathbf{h} = \begin{cases} \mathbf{h}_R - \frac{\tilde{d}-d}{||\mathbf{x}||}\mathbf{x}, & \text{if } \tilde{d} > d \\ \mathbf{h}_R - \frac{d+\tilde{d}}{||\mathbf{x}||}\mathbf{x}, & \text{if } \tilde{d} < -d \\ \mathbf{h}_R, & \text{if } |\tilde{d}| < d. \end{cases} \tag{6.6-46}$$

ERROR REDUCTION PROPERTY OF C_3: Since P_3 is non-expansive, the application of P_3 never increases the error $||\mathbf{h}_R - \mathbf{y}_o^*||^2$. To show that P_3 *always decreases* the error when $\mathbf{h}_R \notin C_3$, consider the case when $\tilde{d} > d$. Then error reduction is achieved if $||\mathbf{y}^* - \mathbf{y}_o^*||^2 < ||\mathbf{h}_R - \mathbf{y}_o^*||^2$. This condition is equivalent to the condition

$$||(\mathbf{h}_R - \mathbf{y}_o^*) + (d - \tilde{d})\frac{\mathbf{x}}{||\mathbf{x}||}||^2 < ||(\mathbf{h}_R - \mathbf{y}_o^*||^2. \tag{6.6-47}$$

Expanding the left side of Eq. (6.6-47) and canceling terms yields

$$||(d - \tilde{d})\frac{\mathbf{x}}{||\mathbf{x}||^2}||^2 + 2(d - \tilde{d})(\mathbf{h}_R - \mathbf{y}_o^*)^T \frac{\mathbf{x}}{||\mathbf{x}||^2} < 0. \tag{6.6-48}$$

Substituting $\mathbf{x}^T \mathbf{h}_R = \tilde{d}$, $\mathbf{x}^T \mathbf{y}_o^* \le d$, and simplifying[†] Eq. (6.6-48) gives $d + \tilde{d} > 2\mathbf{x}^T \mathbf{y}_o^*$, which is obviously true. A similar result follows from the hypothesis that $\tilde{d} < d$.

Estimation of Channel Attenuation

In our discussion so far we have ignored the fact that the received signal contains an *attenuated* version of the spread-spectrum signal, and not the signal itself. Thus, instead of setting $\alpha = 1$ in Eq. (6.6-14), we consider the estimation of α in

$$r(k + nK) = \alpha p(k + nK) + j(k + nK) + \nu(k + nK), \tag{6.6-49}$$

which is written in vector form as

$$\mathbf{r}_n = \alpha \mathbf{p}_n + \mathbf{j}_n + \boldsymbol{\nu}_n. \tag{6.6-50}$$

To estimate α we first project the received signal $\mathbf{r}_n = (r(nK), \ldots, r(nK + K - 1))^T$ onto a subspace orthogonal to the spreading code vector $\mathbf{c}_n$. This will yield the orthogonal components of the unwanted noise and interference signals. Thus we seek to project onto $C_\perp$, given by

$$C_\perp \overset{\triangle}{=} \{\mathbf{y} \in R^K : \mathbf{y}^T \mathbf{c}_n = 0\}. \tag{6.6-51}$$

The projector onto this set is easily computed to be

$$\mathbf{P}_\perp \overset{\triangle}{=} \mathbf{I} - \frac{\mathbf{c}_n \mathbf{c}_n^T}{||\mathbf{c}_n||^2} = \mathbf{I} - \frac{\mathbf{p}_n \mathbf{p}_n^T}{||\mathbf{p}_n||^2}. \tag{6.6-52}$$

We note that $\mathbf{P}_\perp$ is a matrix with diagonal elements $1 - 1/K$ and off-diagonal elements with magnitude $1/K$. Next we form the vector

$$\begin{aligned} \mathbf{e}_n &= \mathbf{r}_n - \mathbf{P}_\perp \mathbf{r}_n & (6.6\text{-}53) \\ &= \alpha \mathbf{p}_n + (\mathbf{I} - \mathbf{P}_\perp)(\mathbf{j}_n + \boldsymbol{\nu}_n), & (6.6\text{-}54) \end{aligned}$$

which contains the signal and the in-phase components of the noise and interference, and compute $E[\mathbf{e}_n^T \mathbf{e}_n]$. After some algebra and using the *idempotent* property [‡]

[†] Since $d - \tilde{d} < 0$, dividing by $d - \tilde{d}$ changes the direction of the inequality.
[‡] That is, if P is a projector then $P^2 = P$.

of projectors, we finally obtain

$$E[e_n{}^T e_n] = K\alpha^2 + \sigma_j{}^2 + \sigma_\nu{}^2, \qquad (6.6\text{-}55)$$

where $\sigma_j{}^2 \overset{\triangle}{=} E[j^2(k)]$ and $\sigma_\nu{}^2 \overset{\triangle}{=} E[\nu^2(k)]$. Thus for large K i.e., $K\alpha^2 \gg \sigma_j{}^2 + \sigma_\nu{}^2$, a good estimate $\hat{\alpha}$ of α is furnished by

$$\hat{\alpha} = \sqrt{E[e_n{}^T e_n]/K} \qquad (6.6\text{-}56)$$

$$= \sqrt{\alpha^2 + \frac{\sigma_j{}^2 + \sigma_\nu{}^2}{K}} \qquad (6.6\text{-}57)$$

$$\simeq \alpha. \qquad (6.6\text{-}58)$$

In practice, the ensemble average operation is replaced by a time average.

Spread Spectrum Reconstruction Technique

While there are only three constraint sets involved in this problem, the many choices of $\mathbf{x}$ and d in C_3 can lead to an algorithm with considerable complexity if all prior knowledge is used. Following the procedure in [30], we furnish two compositions of projection operators for projecting onto C_3.

THE PROJECTOR $P_{3,\otimes}(l)$: This projector projects onto $C_3(\mathbf{x}_l, d_l)$ when $\mathbf{x}_l = (c(l), \cdots, c(K-1), c(0), c(1), \cdots, c(l-1))^T$ and $d_l = K$ for $l = 0$ and -1 for $l = 1, ..., K - 1$.

THE PROJECTOR $P_{3,\oplus}(l)$: This projector projects onto $C_3(\mathbf{x}_l, d_l)$ when $\mathbf{x}_l = (\underbrace{11\cdots1}_{l}\underbrace{00\cdots0}_{K-l})^T$ and d_l is the magnitude of the partial sums, i.e., μ_l of Eq. (6.6-30). There are many choices for the composition operator $\tilde{P}_3$. Two such choices are:

$$\tilde{P}_3^{(1)} = P_{3,\oplus}(1) P_{3,\otimes}(K-1) \cdots P_{3,\oplus}(K) P_{3,\otimes}(0) \qquad (6.6\text{-}59)$$

and

$$\tilde{P}_3^{(2)} = P_{3,\oplus}(K-2) P_{3,\otimes}(2) P_{3,\oplus}(K-1) P_{3,\otimes}(1) P_{3,\oplus}(K) P_{3,\otimes}(0). \qquad (6.6\text{-}60)$$

While $\tilde{P}_3^{(1)}$ uses $2K$ projectors and all of the information in the partial sums and correlations, $\tilde{P}_3^{(2)}$ uses 6 projectors and only a small portion of the available prior knowledge. Clearly, for a problem of this kind, there is a critical trade-off between processing time and algorithm effectiveness.

Results

In [30] the authors compare the performance of the algorithm $\mathbf{p}_{k+1} = P_2 \tilde{P}_3 P_1 \mathbf{p}_k$ with a linear least-squares filter that uses the *fast transversal filter* (FTF) updating

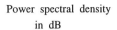

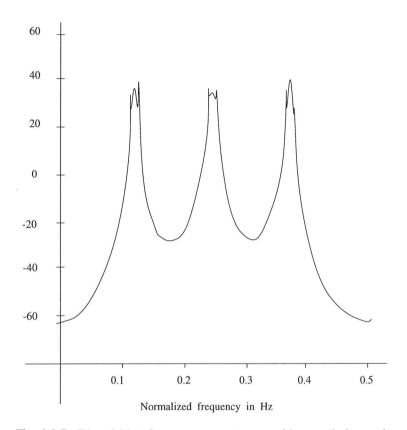

Fig. 6.6-5 Tri-modal interference power spectrum used in numerical example.

algorithm [31]. Here k refers to the iteration index. The full POCS algorithm uses $\tilde{P}_3 = \tilde{P}_3^{(1)}$; the *modified*, i.e., reduced complexity, version uses $\tilde{P}_3 = \tilde{P}_3^{(2)}$. Some parameters of interest are: Number of FTF taps $= 19$; signal-to-noise ratio $= 6$ dB; signal-to-interference ratio $= -20$ dB; number of iterations in either version of the POCS algorithm was 2. The processing gain K had value $K = 27$.

Figure 6.6-5 shows the power spectrum of the tri-modal interference model that was used (SIR$= -20$ dB). The total spectral occupancy of the interference was set at 10 percent of the signal bandwidth.

In Fig. 6.6-6 is shown the performance of POCS, modified POCS, and FTF filtering vis-à-vis *bit error rate* versus *signal-to-noise ratio per bit*. A lower bound on performance (no interference rejection) as well as an upper bound on performance (no interference) are provided for reference. The results show that POCS-based interference-filtering might offer a reasonable alternative to more conventional methods.

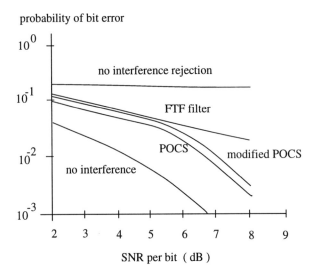

Fig. 6.6-6 The probability of bit error for an SIR of -20 dB and a tri-modal interference covering 10 percent of the signal bandwidth.

6.7 IMAGE COMPRESSION

In order to store an image in a computer it must first be sampled, quantized, and coded. In the process, an enormous number of *bits* are generated. Depending on the type of imagery one is dealing with, even a single image can require in excess of 10^9 bits. Consider the storage requirements for the following types of imagery [32]:

- A low-resolution, TV quality, color video image: 512×512 pixels/color, 8 bits/pixel, and 3 colors $\Longrightarrow \approx 6 \times 10^6$ bits,

- A 24×36-mm (35-mm) negative photograph scanned at 12 μm: 2000×3000 pixels/color, 8 bits/pixel, and 3 colors $\Longrightarrow \approx 144 \times 10^6$ bits,

- A 14×17-inch radiograph scanned at 70μm: 5000×6000 pixels, 12 bits/pixel, $\Longrightarrow \approx 360 \times 10^6$ bits,

- A LANDSAT multispectral image (used in remote sensing): approximately 6000×6000 pixels/spectral band, 8 bits/pixel, and 6 nonthermal spectral bands $\Longrightarrow \approx 1.7 \times 10^9$ bits.

The transmission of images over telephone lines also presents a problem. For example, in [32] Rabbani and Jones consider the transmission of a low-resolution $512 \times 512 \times 8$ bits/pixel $\times 3$ color video image over telephone lines. Using a 9600 baud (bits/s) modem, it would take approximately 11 minutes for just a single image to be transmitted, an unrealistically long time in most applications.

Fortunately, in most cases, it is not necessary to store or send so much data. This is because most images contain a great deal of redundancy. If one could remove this redundancy, storage and transmission problems would be significantly eased. Consider first the redundancy between adjacent frames in a sequence of images such as one sees in TV or cinema. Adjacent frames are often so similar it is difficult to tell them apart. This is called *temporal redundancy*. Next consider the redundancy between different color planes or spectral bands. For example, in the *red-green-blue* (RGB) color system, most of the "red" image will contain most of the "green" image, except for objects that are monochrome red or monochrome "green". This is called *spectral redundancy* or multi-channel redundancy, and is widely exploited in compatible TV. Finally, consider the redundancy in a single monochrome image. Typically, neighboring pixel values are either the same or almost the same. This is due to the fact that most natural images have their energies at DC and low spatial frequencies. This is called *spatial redundancy*. In what follows we shall discuss a scheme that achieves image compression by exploiting the spatial redundancy in an image. The scheme, which is now widely in use, is called the JPEG DCT algorithm.[†] JPEG achieves image compression by normalizing and subsequently quantizing Fourier-type coefficients called DCT coefficients in the frequency domain. The net effect is to set-to-zero the values of many, if not most, of the coefficients in the frequency plane. Hence only a relatively few Fourier coefficients are encoded and compression is achieved.

The standard JPEG compression algorithm falls into a class of compression techniques that are said to be *lossy*, although lossless versions of JPEG exist. In lossy compression, the reconstructed image contains degradations relative to the original image. In contrast, in *lossless* compression the reconstructed image is numerically identical to the original image on a pixel-by pixel basis. The degradations that are manifest in lossy compression reflect the fact that information is discarded. However, such degradations can be ameliorated by the use of prior knowledge. This is a situation where applying POCS is very effective since POCS uses prior information in an efficient way.

Among the undesirable consequences of JPEG compression, a particularly annoying one is a phenomenon known as *blocking artifacts*. Blocking artifacts results when the strong correlations among neighboring blocks of pixels in the image are lost in the compression process. To illustrate how POCS can be used effectively in overcoming degradations in image compression, we focus on the blocking artifact problem. However, for an understanding of how POCS eliminates blocking artifacts, we first need to understand how JPEG compression works.

THE JPEG DCT ALGORITHM: The JPEG DCT encoding is shown in Fig. 6.7-1. The entropy encoding (decoding) process is ignored in this discussion since it is an information lossless process. The following steps are involved:

[†]JPEG are the initials of *Joint Picture Expert Group*. DCT is short for *discrete cosine transform*.

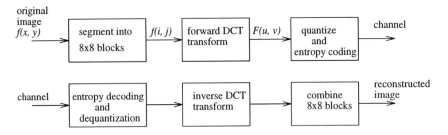

Fig. 6.7-1 The JPEG DCT encoding and decoding process.

1. The image to be encoded is partitioned into contiguous blocks of 8×8 pixel square;

2. Each block is transformed by the DCT given by

$$F(u, v) = \frac{C(u)C(v)}{4} \sum_{j=0}^{7} \sum_{k=0}^{7} f(j, k) \cos \left[\frac{(2j+1)u\pi}{16} \right] \cos \left[\frac{(2k+1)v\pi}{16} \right],$$

(6.7-1)

where $C(u) = 1/\sqrt{2}$ for $u = 0$, and $C(u) = 1$ otherwise and likewise for $C(v)$. Note that the DCT produces an array of 64 coefficients $F(u, v), u = 0, 1, ..., 7; v = 0, 1, ..., 7$;

3. The DCT coefficients are normalized using a user-defined normalization array $Q(u, v)$, which is the same for all blocks. The elements in the normalization array determine the quantization step size; larger values correspond to larger quantization steps. The *human visual system* (HVS) contrast sensitivity function can be used as a guide in determining the elements of $Q(u, v)$. Thus components in $F(u, v)$ that are important from a perceptual point of view would be weighted by the corresponding components in $Q(u, v)$ in a fashion that would enhance that particular component with respect to others;

4. The normalized and quantized DCT value is given by

$$\tilde{F}(u, v) = \text{round} \left[\frac{F(u, v)}{Q(u, v)} \right],$$

(6.7-2)

where *round* is short for rounding to the nearest integer. The rounding step is information *lossy*;

5. The set of DCT coefficients $\{\tilde{F}(u, v)\}$ are further encoded using a *lossless* encoding algorithm such as Huffman encoding [33]. As stated earlier, we shall ignore this step since the Huffman decoding at the receiver identically reproduces the coefficients $\{\tilde{F}(u, v)\}$ of Eq. (6.7-2);

6. At the receiver, after Huffman decoding, the received data is "dequantized" by the operation

$$\hat{F}(u, v) = \tilde{F}(u, v)Q(u, v) \tag{6.7-3}$$

for $0 \leq u \leq 7, 0 \leq v \leq 7$ and for each block;

7. The coefficients $\{f(u, v)\}$ for each block are used to construct a sub-image which is one block of the reconstructed image. The sub-image is obtained via the inverse block DCT (BDCT)

$$\hat{f}(j, k) = \frac{1}{4} \sum_{u=0}^{7} \sum_{v=0}^{7} C(u)C(v) \cos \left[\frac{(2j + 1)u\pi}{16} \right] \cos \left[\frac{(2k + 1)v\pi}{16} \right] \tag{6.7-4}$$

for $0 \leq j, k \leq 7$. The blocks are then reassembled in their respective order and the entire image is thus available for display.

As we can see, in both the encoding and decoding processes, the image is processed on a block-to-block basis. The existing strong correlation among the neighboring blocks is never taken into consideration in either the encoder or the decoder. Hence, the decoded image exhibits discontinuities at the block boundaries, which are the blocking artifacts.

IMAGES AS VECTORS: A digital image is an array of numbers. However, it is often inconvenient to represent an image by a matrix. By scanning the matrix along some well-defined path, the matrix can be converted to a vector which, in turn, can be uniquely reconstructed as the original matrix. In JPEG compression, for example, the DCT array of the original image array is scanned in a zigzag fashion for ease of subsequent encoding. The resulting vector has only a few non-zero entries right at the beginning, followed by a string of zeroes, which are easily encoded.

For simplicity we treat an $N \times N$ image as a $N^2 \times 1$ vector $\mathbf{f}$ by scanning either along rows or columns. This is called *lexicographic ordering* (by row or column) and the resulting vector can be viewed as a vector in the Euclidean space R^{N^2}. For instance, using an example from [32], the digital image $f(j, k)$ can be expressed either as a 8×8 pixel matrix

$$[f(j, k)] = \begin{bmatrix} 139 & 144 & 149 & 153 & 155 & 155 & 155 & 155 \\ 144 & 151 & 153 & 156 & 159 & 156 & 156 & 156 \\ 150 & 155 & 160 & 163 & 158 & 156 & 156 & 156 \\ 159 & 161 & 162 & 160 & 160 & 159 & 159 & 159 \\ 159 & 160 & 161 & 162 & 162 & 157 & 157 & 157 \\ 161 & 161 & 161 & 161 & 160 & 157 & 157 & 157 \\ 162 & 162 & 161 & 163 & 162 & 157 & 157 & 157 \\ 162 & 162 & 161 & 161 & 163 & 158 & 158 & 158 \end{bmatrix} \tag{6.7-5}$$

or as a (row-stacked) 64×1 pixel vector

$$\mathbf{f} = (139, 144, \cdots \cdots, 158, 158)^T. \tag{6.7-6}$$

Likewise, the BDCT of $[f(i,j)]$ can be represent as a matrix

$$[F(u,v)] = \begin{bmatrix} 1260 & -1 & -12 & -5 & 2 & -2 & -3 & 1 \\ -23 & -17 & -6 & -3 & -3 & 0 & 0 & -1 \\ -11 & -9 & -2 & 2 & 0 & -1 & -1 & 0 \\ -7 & -2 & 0 & 1 & 1 & 0 & 0 & 0 \\ -1 & -1 & 1 & 2 & 0 & -1 & 1 & 1 \\ 2 & 0 & 20 & -1 & 0 & 2 & 1 & -1 \\ -1 & 0 & 0 & -1 & 0 & 2 & 1 & -1 \\ -3 & -2 & -4 & -2 & 2 & 1 & -1 & 0 \end{bmatrix} \tag{6.7-7}$$

or as a (row-stacked) 64×1 vector

$$\mathbf{F} = (1260, -1, \cdots\cdots, -1, 0)^T.$$

Indeed, one can find a matrix $\mathbf{B}$, called the *BDCT matrix*, which maps $\mathbf{f}$ to $\mathbf{F}$ as a linear transformation

$$\mathbf{F} = \mathbf{Bf}. \tag{6.7-8}$$

We leave the computation of $\mathbf{B}$ as a challenging exercise for the reader.

Due to the unitary property of the 2-D DCT transform for each block, the BDCT matrix is also unitary and the inverse transform satisfies $\mathbf{B}^{-1} = \mathbf{B}^T$, where the superscript T denotes transpose. Then the inverse DBCT can be written as

$$\mathbf{f} = \mathbf{B}^T \mathbf{F}. \tag{6.7-9}$$

As already stated, in JPEG-DCT coding [32], each coefficient $F(u,v)$ is quantized according to Eq. (6.7-2). This quantization process can be described mathematically by an operator, say Q, from R^{N^2} to R^{N^2} such that

$$\tilde{\mathbf{F}} = Q\mathbf{F}. \tag{6.7-10}$$

Using this notation, the whole process from the original image $\mathbf{f}$ to the received data $\tilde{\mathbf{F}}$ can be conveniently written as

$$\tilde{\mathbf{F}} = Q\mathbf{Bf}. \tag{6.7-11}$$

In the following, we let $\tilde{T}$ denote the concatenation of $\mathbf{B}$ and Q for additional simplicity, i.e., $\tilde{T} \triangleq Q\mathbf{B}$. Then Eq. (6.7-11) can be rewritten as

$$\tilde{\mathbf{F}} = \tilde{T}\mathbf{f}. \tag{6.7-12}$$

With above notation, the image decoding problem can be phrased as: Given the data $\tilde{\mathbf{F}}$, find an $\mathbf{f}$ that satisfies Eq. (6.7-12). A seemingly straightforward approach would be to solve for $\mathbf{f}$ from Eq. (6.7-12) by inverting $\tilde{T}$. Unfortunately, due to the *many-to-one* mapping nature of Q, $\tilde{T}$ is also a *many-to-one* mapping. Therefore Eq. (6.7-12) is not invertible. In general there are many images that satisfy Eq. (6.7-12).

Since finding f by inverting Eq. (6.7-12) is not possible, we consider a projection method solution to the problem. The available constraints are discussed below.

Constraint Set Based on Transform Coefficients

A fundamental constraint set can be defined from Eq. (6.7-12), namely,

$$C_{T_0} = \left\{ \mathbf{y} : (Q\mathbf{B}\mathbf{y}) = \tilde{\mathbf{F}} \right\} \tag{6.7-13}$$

or, equivalently,

$$C_{T_0} = \left\{ \mathbf{y} : (Q\mathbf{B}\mathbf{y})_n = \tilde{F}_n, \forall n \in I \right\}, \tag{6.7-14}$$

where I is the index set of all BDCT coefficients, the subscript T_0 reminds us that the *transform* set is *open*, and Eq. (6.7-14) is merely a restatement of Eq. (6.7-13) on a component-by-component basis. A more convenient way of describing the constraint in Eq. (6.7-14) is suggested from the following example: Assume that for some pair[†] (u_n, v_n) $F_n = 132$ and $Q_n = 70$; then $\tilde{F}_n = 2$ and, after dequantizing, $Q_n \tilde{F}_n = 140$. Since the rounding operation will furnish $\tilde{F}_n = 2$ for *any* $1.5 < F_n/Q_n < 2.5$, we know that the true DCT coefficient F_n must lie between $105 < F(u_n, v_n) < 175$. Note that the numbers 105 and 175 depend on Q_n and the specific quantizing scheme used, e.g., rounding to the nearest integer versus, say, rounding to the smallest integer greater than F_n/Q_n. Hence Eq. (6.7-14) can be converted to

$$C_{T_0} = \left\{ \mathbf{y} : F_n^{\min} < (\mathbf{B}\mathbf{y})_n < F_n^{\max}, \forall n \in I \right\}, \tag{6.7-15}$$

where $F_n^{\min}$ and $F_n^{\max}$ are known *a priori* and are determined by the quantizer; for example, in the example given above $F_n^{\max} = 175$ and $F_n^{\min} = 105$.

The set in Eq. (6.7-15) is convex but not closed. We close the set by adding both end points to the decision interval. Thus the final form of the transform-coefficient-based constraint is

$$C_T \triangleq \bar{C}_{T_0} = \left\{ \mathbf{y} : F_n^{\min} \le (\mathbf{B}\mathbf{y})_n \le F_n^{\max}, \forall n \in I \right\}. \tag{6.7-16}$$

As usual, the projection onto C_T can be computed by the method of Lagrange multipliers. However, in this case, the projection can be obtained by inspection. Consider the two-dimensional case where $\mathbf{B}$ is a 2×2 matrix and $\mathbf{f}$ is an arbitrary 2×1 column vector to be projected onto C_T. With $\mathbf{b}_1$ and $\mathbf{b}_2$ denoting column vectors whose components are the first and second row, respectively, of $\mathbf{B}$, we can write $\mathbf{B}\mathbf{f} = (\mathbf{b}_1^T\mathbf{f}, \mathbf{b}_2^T\mathbf{f})^T$. In Fig. 6.7-2 we show a vector $\mathbf{f}$ for which $\mathbf{B}\mathbf{f}$ is outside the set since $\mathbf{b}_1^T\mathbf{f} > F_1^{\max}$ and $\mathbf{b}_2^T\mathbf{f} < F_2^{\min}$. Clearly, the nearest point in C_T is $P_T\mathbf{f} = (F_1^{\max}, F_2^{\min})^T$. Following this line of reasoning for the general n-dimensional case and denoting the nth component of $\mathbf{B}\mathbf{f}_n$ by $(\mathbf{B}\mathbf{f})_n$ we obtain

---

[†]We denote $F(u_n, v_n)$ by F_n, etc.

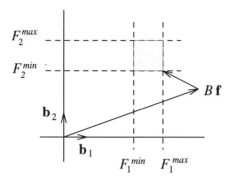

Fig. 6.7-2 The vector $\mathbf{f}$ (not shown) is outside the set C_T since its DCT $\mathbf{Bf}$ is outside the shaded region.

as projection

$$\mathbf{y}^* = P_T\mathbf{f} = \mathbf{B}^T\mathbf{F}, \qquad (6.7\text{-}17)$$

where the components F_n of $\mathbf{F}$ are given by

$$F_n = \begin{cases} F_n^{\min} & \text{if} \quad (\mathbf{Bf})_n < F_n^{\min} \\ F_n^{\max} & \text{if} \quad (\mathbf{Bf})_n > F_n^{max} \\ (\mathbf{Bf})_n & \text{if} \quad F_n^{\min} \leq (\mathbf{Bf})_n \leq F_n^{\max}. \end{cases} \qquad (6.7\text{-}18)$$

Constraints Based on Prior Knowledge

As stated earlier, reconstruction of the image at the decoder, from the transmitted data only, leads to blocking artifacts because the information in the separate blocks does not include the high degree of correlation, i.e., smoothness, across block boundaries. Hence we need to involve the prior knowledge that characterizes this correlation. In additional examples, given in later chapters of this book, we shall see how important the application of, even imprecise, prior knowledge is in getting the desired results. As in all POCS algorithms, the mathematical description of the prior knowledge must be formulated. Clearly, while we recognize that we need to enforce some kind of smoothness constraint across the boundary, the creative part of POCS is to characterize this constraint mathematically. Obviously there are a number of possibilities. One technique is to bound the energy (norm) in the discontinuity of the gray levels across the columns at the block boundaries. For example, in the 32×32 pixel square image shown in Fig. 6.7-3, we could force $\|\mathbf{f}_8 - \mathbf{f}_9\| \leq E$, $\|\mathbf{f}_{16} - \mathbf{f}_{17}\| \leq E$ and $\|\mathbf{f}_{24} - \mathbf{f}_{25}\| \leq E$ and then repeat this process for the row discontinuities. More generally, we could require that $\|\mathbf{f}_j - \mathbf{f}_{j+1}\| \leq E$ for $j = 8, 16, 24, \cdots$. To be specific, let $j = 8i$ for some i and consider the set

$$C_j = \{\mathbf{y} : \|\mathbf{y}_j - \mathbf{y}_{j+1}\| \leq E\}. \qquad (6.7\text{-}19)$$

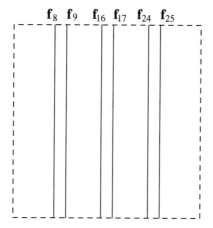

$\mathbf{f}_8$ $\mathbf{f}_9$ $\mathbf{f}_{16}$ $\mathbf{f}_{17}$ $\mathbf{f}_{24}$ $\mathbf{f}_{25}$

Fig. 6.7-3 Reducing blocking artifacts by bounding the energy associated with the differences in gray-level values every eighth column.

Denote an $N \times N$ image $\mathbf{f}$ in its column vector form

$$\mathbf{f} = \{\mathbf{f}_1, \mathbf{f}_2, \cdots, \mathbf{f}_N\}, \tag{6.7-20}$$

where $\mathbf{f}_i$ is the ith column of the image and has N components. To find the projection of an arbitrary $\mathbf{f}$ onto C_j of Eq. (6.7-19), we seek $\mathbf{g}^* = (\mathbf{g}_1, \mathbf{g}_2, \cdots, \mathbf{g}_N)$ such that

$$\mathbf{g}^* = \arg \min_{\mathbf{g} \in C_j} \|\mathbf{f} - \mathbf{g}\|^2. \tag{6.7-21}$$

For notational simplicity, we assume $j = 8$ in the following. Then, to find $\mathbf{g}^*$ we construct the Lagrange functional

$$J(\mathbf{g}) \triangleq \|\mathbf{f} - \mathbf{g}\|^2 + \lambda\{\|\mathbf{g}_8 - \mathbf{g}_9\|^2 - E^2\}. \tag{6.7-22}$$

Written out in component form, the Lagrange functional becomes

$$J(\mathbf{g}) = \sum_{i=1}^{N} \|\mathbf{f}_i - \mathbf{g}_i\|^2 + \lambda\{\|\mathbf{g}_8 - \mathbf{g}_9\|^2 - E^2\}. \tag{6.7-23}$$

From

$$\frac{\partial J}{\partial \mathbf{g}_8} = \frac{\partial J}{\partial \mathbf{g}_9} = \mathbf{0}$$

we obtain

$$\mathbf{g}_8 - \mathbf{f}_8 + \lambda(\mathbf{g}_8 - \mathbf{g}_9) = 0$$

and

$$\mathbf{g}_9 - \mathbf{f}_9 - \lambda(\mathbf{g}_8 - \mathbf{g}_9) = 0,$$

from which we easily obtain, as projection,

$$
\begin{aligned}
\mathbf{g}_8^* &= \alpha\mathbf{f}_8 + (1-\alpha)\mathbf{f}_9 \\
\mathbf{g}_9^* &= (1-\alpha)\mathbf{f}_8 + \alpha\mathbf{f}_9,
\end{aligned}
\tag{6.7-24}
$$

where α is computed from the constraint that $\|\mathbf{g}_8^* - \mathbf{g}_9^*\|^2 = E$. Using Eqs. (6.7-24) in this constraint we obtain for α:

$$
\alpha = \frac{1}{2}\left[\frac{E}{\|\mathbf{f}_8 - \mathbf{f}_9\|} + 1\right].
\tag{6.7-25}
$$

Similarly, we obtain $\mathbf{g}_i^* = \mathbf{f}_i$ for $i \neq 8$ or 9.

Finally, recalling that we set $j = 8$ for simplicity of notation during the computation, we revert back to the more general notation to obtain the projection onto C_j as

$$
\begin{aligned}
\mathbf{g}_j^* &= \alpha\mathbf{f}_j + (1-\alpha)\mathbf{f}_{j+1} & (6.7\text{-}26) \\
\mathbf{g}_{j+1}^* &= (1-\alpha)\mathbf{f}_j + \alpha\mathbf{f}_{j+1} & (6.7\text{-}27)
\end{aligned}
$$

and $\mathbf{g}_i^* = \mathbf{f}_i$ for $i \neq j$ or $j + 1$. In [34], the above reasoning was extended to the whole image. Specifically Yang et al. define the set (subscript c is for *column*)

$$
C_c \triangleq \{\mathbf{f} : \|\mathbf{Qf}\| \leq E\},
\tag{6.7-28}
$$

where E is a scalar upper bound that defines the size of this set; $\mathbf{Q}$ is a linear operator defined as follows: for an $N \times N$ image $\mathbf{f}$ in its column vector form

$$
\mathbf{f} = \{\mathbf{f}_1, \mathbf{f}_2, \cdots, \mathbf{f}_N\},
\tag{6.7-29}
$$

$\mathbf{Qf}$ gives the difference between adjacent columns at the block boundaries of $\mathbf{f}$. For example, for the case of $N = 512$ and 8×8 blocks

$$
\mathbf{Qf} = \begin{bmatrix}
\mathbf{f}_8 & - & \mathbf{f}_9 \\
\mathbf{f}_{16} & - & \mathbf{f}_{17} \\
& \cdot & \\
& \cdot & \\
& \cdot & \\
\mathbf{f}_{504} & - & \mathbf{f}_{505}
\end{bmatrix}.
\tag{6.7-30}
$$

The norm of $\mathbf{Qf}$

$$
\|\mathbf{Qf}\| = \left[\sum_{i=1}^{63}\|\mathbf{f}_{8i-1} - \mathbf{f}_{8i+1}\|^2\right]^{\frac{1}{2}}
\tag{6.7-31}
$$

is a measure of the total intensity variation between the boundary columns of adjacent blocks.

For an image $\mathbf{f} \overset{\triangle}{=} \{\mathbf{f}_1, \mathbf{f}_2, \cdots, \mathbf{f}_N\} \notin C_c$, $\tilde{\mathbf{f}} \overset{\triangle}{=} P_c \mathbf{f} \overset{\triangle}{=} \{\tilde{\mathbf{f}}_1, \tilde{\mathbf{f}}_2, \cdots, \tilde{\mathbf{f}}_N\}$ is given by

$$\tilde{\mathbf{f}}_i = \mathbf{f}_{i+1} + \alpha(\mathbf{f}_i - \mathbf{f}_{i+1}) \text{ and } \tilde{\mathbf{f}}_{i+1} = \mathbf{f}_i - \alpha(\mathbf{f}_i - \mathbf{f}_{i+1})$$

for $i = 8 \cdot k$ and $k = 1, 2, \cdots, 63$; otherwise, $\tilde{\mathbf{f}}_i = \mathbf{f}_i$, $\quad$ (6.7-32)

where $\alpha = \frac{1}{2} \left[\frac{E}{\|\mathbf{Q}\mathbf{f}\|} + 1 \right]$. In Eq. (6.7-32) a 512×512 image and 8×8 blocks are used.

In a similar fashion, a set C_r (subscript r is for *row*) and projector P_r, which captures the intensity variations between the rows of the block boundaries can be defined. Using these convex sets a POCS-based recovery algorithm can be defined in the decoder to reconstruct the compressed images.

Besides the sets previously defined, another useful set is the set that captures the information about the range of the *pixel values* of an image. For example, for an 8-bit $N \times N$ grayscale image, this set is defined by

$$C_a \overset{\triangle}{=} \{\mathbf{y} : 0 \leq y(i, j) \leq 255, 1 \leq i, j \leq N\}, \quad (6.7-33)$$

where the subscript a reminds us that it is a constraint upon the *amplitude* of the image. This type of constraint was encountered in Chapter 3 (Eq. (3.6-22)). For an image $\mathbf{f}$ with components $f(i, j)$, its projection onto C_a is given by

$$\mathbf{f}^* \overset{\triangle}{=} P_a \mathbf{f},$$

where the components of $\mathbf{f}^*$ satisfy

$$f^*(i, j) = \begin{cases} 0, & f(i, j) < 0 \\ 255, & f(i, j) > 255 \\ f(i, j) & \text{otherwise.} \end{cases} \quad (6.7-34)$$

Let us now summarize the various constraint sets encounted above:

- C_T, the constraint set on DCT coefficients, with projector P_T;

- C_c, the constraint set on the energy allowed in *column* gray-level discontinuities, with projector P_c;

- C_r, the constraint set on the energy allowed in *row* gray-level discontinuities, with projector P_r;

- C_a, the constraint set on the allowable range of pixel amplitudes of the image, with projector P_a.

Recovery Algorithm

The perceptive reader might have conjectured that by reducing the discontinuity across the boundary, new discontinuities are introduced in the columns (rows) adjacent to the original boundary columns (rows). Likewise, as these discontinuities are ameliorated by projections, secondary discontinues are produced that "propagate" away from the boundary. The solution to this problem is to apply the column (row) projector to adjacent pairs of columns (row) within the block, as one moves away from the boundary. The projections are repeated until all eight column (rows) discontinuities are ameliorated. We use the symbol $\tilde{P}_c$ ($\tilde{P}_r$) to denote the composition projector that sequentially reduces the blocking artifacts at the boundary columns (rows) and the induced column (row) discontinuities within the block. While the applications of $\tilde{P}_c$ ($\tilde{P}_r$) produces a certain amount of undesirable smoothing, this phenomenon can be controlled by adjusting the control parameter E in accordance with the amount of gray-level fluctuations sensed in a region.

With $\tilde{P}_c$ and $\tilde{P}_r$ defined as above, the POCS recovery algorithm can be described as follows:

- Set $\mathbf{f}_0 = \tilde{\mathbf{f}}$, where $\tilde{\mathbf{f}}$ is the raw JPEG reconstructed image by the inverse BDCT matrix i.e., $\tilde{\mathbf{f}} = \mathbf{B}^T \tilde{\mathbf{F}}$;

- For $k = 1, 2, \cdots$, compute $\mathbf{f}_k$ from

$$\mathbf{f}_k = \tilde{P}_c \tilde{P}_r P_T P_a \mathbf{f}_{k-1};\qquad(6.7\text{-}35)$$

- Continue this iteration until $\|f_k - f_{k-1}\|$ is less than some prescribed bound.

The results of applying the recovery algorithm of Eq. (6.7-35) can be seen from the experimental results shown in Fig. 6.7-4.

In Fig. 6.7-4(a) is shown an image of the famous model "Lena". When the "Lena" image is compressed using the JPEG algorithm at 0.24 bits per pixel (bpp) and JPEG-recovered at the receiver, a very blocky picture results: the blocking artifacts are clearly visible in Fig. 6.7-4(b). The recovered image using the POCS algorithm from Eq. (6.7-35) is shown in Fig. 6.7-4(c).

While other artifacts remain, the blocking artifacts are mostly gone. A refinement of the sets C_c and C_r, which uses adaptive smoothing at the boundaries of blocks and within blocks themselves is described in [34,35]. Using this refinement, many of these artifacts are ameliorated. The reader is referred to the listed references for further details.

6.8 SUMMARY

The flexibility of projection methods enable them to be applied fruitfully to a large number of problems in the area of communication systems engineering. A number

(a)

(b)

Fig. 6.7-4 Example of reducing blocking artifacts: (a) Original image of Lena; (b) Recovered Lena from a JPEG compression algorithm at 0.24 bits per pixel; (c) POCS recovery of Lena image. The POCS algorithm is applied after application of the inverse DCT at the receiver.

(c)

Fig. 6.7-4 *Continued.*

of such applications were furnished in this chapter. For example, we demonstrated how vector-space projections methods can be used to reconstruct a band-limited function from non-uniform samples, either in iterative fashion or by a one-step formula. In a second application of such methods, we showed how post-processing of the data in an oversampled-digital communication system can lead to fewer quantizing errors.

The design of certain types of finite impulse response (FIR) and infinite impulse response (IIR) filters can be accomplished using projections. In the FIR case, POCS-based design yielded results comparable to optimal designs, e.g., the Remez exchange algorithm. In the IIR case we demonstrated how phase-equalization filters can be designed with the appropriate constraints and associated projectors.

Accurate estimation of probability density functions (pdf's) is an important pre-processing operation in the design of optimal systems in communication receivers, radar, and pattern recognizers. We showed that projection methods are very effective in estimating pdf's from sample data, especially if the data are constrained in some *a priori* known way. The application of vector-space methods to spread-spectrum systems for interference rejection yielded results comparable with or superior to some well-known adaptive methods.

Finally, in a practical problem of considerable importance, we showed how projections can be used to reduce blocking artifacts in modern image compression, such as JPEG. In the future we expect to see many applications of projections in the field of communications and allied fields.

REFERENCES

1. A. Papoulis, A new algorithm in spectral analysis and bandlimited extrapolation, *IEEE Trans. Circuits Syst.* **CAS-22**:735–742, Sept. 1975.

2. R.W. Gerchberg, Super-resolution through error energy reduction, *Optica Acta*, **21**(9):709–720, 1974.

3. J. M. Wozencraft and I. M. Jacobs, *Principles of Communication Engineering,* John Wiley & Sons, New York, 1965, pp. 678–683.

4. M. J. Carlotto and V. T. Tom, Practical interpolation of 2-D surfaces using the Gerchberg algorithms, Digest of Topical Meeting on Signal Recovery and Synthesis with Incomplete Information and Partial Constraints, Incline Village, NV, Jan. 12–14, 1983, pp. (WA3-1)–(WA3-4).

5. H. Stark and J. W. Woods, Polar sampling theorems and their applications to computer-aided tomography, *Proc. SPIE*, **231**, Book 1, pp. 230–236, 1980.

6. H. Stark, J. W. Woods, I. Paul, and R. Hingorani, Direct Fourier reconstruction in computer tomography, *IEEE Trans. Acoust. Speech and Sign. Pro.,* **29**:237–245, April 1981.

7. E. Yudilevich and H. Stark, Interpolation from samples on a linear spiral scans, *IEEE Trans. on Medical Imaging,* **TMI-6**:209–219, Sept. 1987.

8. A. Papoulis, *The Fourier Integrals and Its Applications,* McGraw-Hill, New York, 1962.

9. Shu-jen Yeh, Projection method applications to problems in neural networks and signal reconstruction, Ph.D. thesis, Department of Electrical, Computer, and Systems Engineering, Rensselaer Polytechnic Institute, Troy, New York, 1991.

10. Shu-jen Yeh and H. Stark, Iterative and one-step reconstruction from non-uniform samples by convex projections, *J. Opt. Soc. Am., A,* **7**(3), March 1990.

11. K. D. Sauer and J. P. Allebach, Iterative reconstruction of band-limited images from nonuniformly spaced samples, *IEEE Trans. on Circuits and Systems,* **CAS-34**:1497–1506, Dec. 1987.

12. N. T. Thao and M. Vetterli, Oversampled A/D conversion using alternate projections, *Proc. Twenty-Fifth Annu. Conf. Inform. Sci. Syst.,* 1991, pp. 241–248.

13. S. Hein and A. Zakhor, Reconstruction of oversampled band-limited signals from $\Sigma\Delta$ encoded binary sequences, *IEEE Trans. Signal Proc.,* **42**(4):799–811, 1994.

14. A. V. Oppenheim and R. W. Schafer, *Discrete-Time Signal Signal Processing,* Prentice-Hall, Englewood Cliffs, NJ, 1988.

15. J. H. McClellan and T. Parks, A unified approach to the design of optimum FIR linear-phase digital filters, *IEEE Trans. Circuit Theory,* **CT-20**:697–70, Nov. 1973.

16. *The Student Edition of MATLAB: Version 4 User's Guide*, Prentice-Hall, Englewood Cliffs, NJ, 1995.

17. P. A. Regalia et al., The digital all-pass filter: A versatile signal processing building block, *Proc. IEEE*, **76**(1):19–37, 1988.

18. K. Haddad, H. Stark, N. P. Galatsanos, and Y. Yang, FIR and IIR filter designs using projections, Technical Report no.6, Department of Electrical and Computer Engineering, Illinois Institute of Technology, Chicago, 1998.

19. D. W. Scott, *Multivariate Density Estimation: Theory, Practice, and Visualization,* John Wiley & Sons, New York, 1992.

20. E. Parzen, On the estimation of a probability function and the mode, *Ann. Math. Stat.,* **33**:1065–1076, 1962.

21. R. A. Tabia and J. R. Thompson, *Nonparametric Probability Density Estimation,* John Hopkins University Press, Baltimore, MD, 1978.

22. W. E. Jacklin and D. R. Ucci, Locally Optimum Detection, Chapter 9 in *Signal Processing Methods for Audio, Images, and Telecommunications,* (P. M. Clarkson and H. Stark eds.), Academic Press, London, 1995.

23. A. M. Mood and F. A. Graybill, *Introduction to the Theory of Statistics,* 2nd ed., McGraw-Hill, New York, 1963.

24. E. J. Dudewicz and S. N. Mishra, *Modern Mathematical Statistics*, John Wiley & Sons, New York, 1988.

25. Y. Yang and H. Stark, Estimation of probability density function using projections onto convex sets, Chapter 10 in *Signal Processing Methods for Audio, Images, and Telecommunications,* P. M. Clarkson and H. Stark eds., Academic Press, London, 1995.

26. Y. Yang, Projection-based signal processing for visual communication problems, Ph.D. thesis, Department of Electrical and Computer Engineering, Illinois Institute of Technology, Chicago, 1994.

27. STAT/LIBRARY: FORTRAN Subroutines for Statistical Analysis, IMSL, April 1987.

28. R. E. Ziemer and R. L. Peterson, *Digital Communications and Spread Spectrum Systems,* Macmillan, New York, 1995.

29. D. Schilling et al., Spread spectrum for commercial communications, *IEEE Commun. Soc. Mag.,* 66–79, April 1991.

30. J. F. Doherty and H. Stark, Direct-sequence spread spectrum narrowband interference rejection using property restoration, *IEEE Trans. on Communications,* **44**(9):1197–1204, Sept. 1996.

31. J. Cioffi and T. Kailath, Fast recursive least squares transversal filter for adaptive filtering, *IEEE Trans. Acoustics, Speech, Signal Process.,* **ASSP-32**:304–337, April 1984.

32. M. Rabbani and P. W. Jones, *Digital Image Compression Techniques,* SPIE Optical Engineering Press, **TT7**, 1991.

33. D. A. Huffman, A method for the construction of minimum redundancy codes, *Proc. IRE,* **40**:1098–1101, 1952.

34. Y. Yang, N. Galatsanos, and A. Katsaggelos, Regularized reconstruction to reduce blocking artifacts of block discrete cosine transform compressed images, *IEEE Trans on Circuits and Systems for Video Tech.*, **3**(6):421–432, December 1993.

35. Y. Yang, N. Galatsanos, and A. Katsaggelos, Projection-based spatially-adaptive reconstruction of block transform compressed images, *IEEE Trans. on Image Processing,* July 1995.

EXERCISES

6-1. Show that the set C_i in Eq. (6.2-1) is a closed linear manifold.

6-2. Establish the result in Eq. (6.2-6). *Hint*: recall that $y(t_i)$ is assumed real.

6-3. Derive the projection y^* in Eq. (6.2-12) for the non-uniform sampling problem.

6-4. Show that Eq. (6.2-14) reduces to the classic form of the uniform sampling theorem when $N \to \infty, t_i = i\Delta, \Delta = (2B)^{-1}$ and $q(t) = 0$.

6-5. Show that a strictly band-limited function admits to a Taylor series expansion. *Hint*: Show that all derivatives of the function exist.

6-6. Given a quantizer with bin size δ and reproducer level at the center of the bin. Then the quantizing error ϵ between the analog and digital signal can be modeled as a uniformly distributed random variable across the range of $(-\delta/2, \delta/2)$. Assume that the analog sampled signal $x(i\Delta t)$ is a Gaussian random variable with mean zero and variance σ_X^2.
a) Show that the signal-to-noise ratio (SNR) is given by

$$SNR = 12\sigma_X^2/\delta^2.$$

b) If the full quantizer range is $(-4\sigma_X, 4\sigma_X)$ show that the SNR (in dBs) obeys the famous *6-dB rule*, i.e., the SNR increases 6 dB with each additional bit of coding.

6-7. Write a program in MATLAB, BASIC, or C, to yield the results shown in Figs. 6.3-5 and 6.3-6. Assume that $B = 2$, the oversampling rate is 100 percent, and the function $f(x)$ is given by:

$$f(x) = 4(\text{sinc}2\pi B(t + 0.125) + \text{sinc}2\pi B(t - 0.25)).$$

Thus in the range $(-2, +2)$ you should have 32 or 33 points. Use a uniform quantizer with reproducer levels at $0, \pm 1/2, \pm 1$, etc., and binwidth of one-half.

6-8. Show that Eq. (6.4-1) is appropriate for describing the output of a linear, shift-invariant filter to an input $x(t)$. *Hint:* Recall the definition of $h(t)$.

6-9. Show that any FIR filter that satisfies Eq. (6.4-9) is a linear-phase filter.

6-10. Obtain the projection onto the set C_1 of Eq. (6.4-15) by Lagrange multipliers. Check your answer by comparing your result with Eq. (6.4-23).

6-11. Show that the set C_3 in Eq. (6.4-17) is closed and compute its projection using the method of Lagrange multipliers.

6-12. Write a MATLAB computer program for designing an FIR filter with the POCS algorithm described in the text. Use $\alpha = \beta = 0.0245, \omega_p = 0.4\pi, \omega_s = 0.5\pi$, and $N = 31$.

6-13. Write a MATLAB computer program for designing an FIR filter with a POCS algorithm for limiting the size of the step response. Use the following parameters: $\alpha = \beta = 0.12, \omega_p = 0.4\pi, \omega_s = 0.5\pi, N = 31$, step response constraints as given in the text near Eq. (6.4-37).

6-14. Show that the set C_5 in Eq. (6.4-38) is closed, convex, and has as its projector Eq. (6.4-39).

6-15. Write a MATLAB computer program for implementing a projection algorithm for designing an IIR phase-equalizing filter for the phase

$$\phi = -\omega^2 \ \text{sgn}\ (\omega).$$

Use the following parameters: $K = 8, N = 6, \delta = 0.003\omega, \Omega = [0, 0.2\pi]$. A 512 sampling grid is suggested.

6-16. Show that Eq. (6.5-1) corresponds to a minimum error probability test for two classes.

6-17. Show that $C''(a_i, b_i)$ in Eq. (6.5-3) is closed and convex.

6-18. Derive Eq. (6.5-5).

6-19. Show that C_+ and C_{ua} in Eqs. (6.5-7) and (6.5-8), respectively, are convex. Under what condition is C_{ua} closed?

6-20. Show that $C(a_i, b_i)$ in Eq. (6.5-10) is closed and convex for any finite values of a_i, b_i.

6-21. Show that μ_L and μ_H, as given in Eq. (6.5-30), are the interval limits on a $100(1 - \alpha)$ confidence interval for μ.

6-22. Derive Eq. (6.5-31).

6-23. Show that the set $C(\epsilon)$ in Eq. (6.5-32) is convex (this is the easy part). To show that it is closed is more difficult. An easier problem is to show that $C(\epsilon)$ is closed if its members are band-limited functions.

6-24. Show that when the set $C(\epsilon)$ in Eq. (6.5-32) is replaced by

$$C'(\epsilon) = \left\{ \mathbf{y} : \sum_{n=2}^{M} |2y_n - y_{n+1} - y_{n-1}| \le \epsilon^2 \right\},$$

the new set is closed and convex. Note that $C'(\epsilon)$ or another discrete expression is required to implement $C(\epsilon)$ on a computer.

6-25. Show that the projection onto $C_\perp$ of Eq. (6.6-51) is indeed given by Eq. (6.6-52).

7

Applications to Optics

7.1 INTRODUCTION

Even before the appearance of Youla's seminal papers on projection methods [1,2], such methods were already in use in optics. Of course, they weren't perceived as projection methods and the broad underlying theory connecting various projection-based algorithm was not yet appreciated by optical physicists and engineers. For example, the super-resolution algorithms proposed by Gerchberg [3] and Papoulis [4] were examples of relatively simple *convex projection algorithms,* while the Gerchberg-Saxton phase retrieval [5] technique was an application of *generalized projections.*

While many problems in optics can be solved by projections, it is difficult to solve such problems using all-optical methods. A notable exception is Marks' all-optical implementations of the convex projection algorithm for implementing super-resolution [6]. Nevertheless, from X-ray optics in computer-aided tomography, to astronomy, to optical beam forming, to measuring ultra-short pulses [7], projection methods have made a significant impact in solving optical problems of practical importance.

The problems discussed in this chapter have long been of interest to the optics community. They include image sharpening, phase-retrieval, beam forming, color matching, and blind deconvolution. No attempt is made to list all the excellent work in these areas by numerous researchers. The purpose of this chapter is merely to acquaint the reader with some examples of how projection methods can be used in optics.

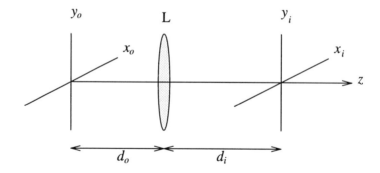

Fig. 7.2-1 Imaging configuration.

7.2 IMAGE SHARPENING

Refer to Fig. 7.2-1. A lens L images an optical field $f_0(x_0, y_0)$ in the object plane (x_0, y_0) at a distance d_0 in front of the lens as a field $f_i(x_i, y_i)$ in the *image plane* (x_i, y_i) at a distance d_i in back of the lens. Elementary optics teaches us that the image comes into focus at a distance d_i when

$$\frac{1}{d_0} + \frac{1}{d_i} = \frac{1}{FL},\tag{7.2-1}$$

where *FL* is the focal length of the lens. Equation (7.2-1) is the famous *lens law* and can be derived using geometrical principles. These same principles predict that the image intensity in the plane at d_i is given by

$$f_i(x_i, y_i) = \frac{1}{M^2} f_0\left(-\frac{x_i}{M}, -\frac{y_i}{M}\right) \triangleq f_g(x_i, y_i),\tag{7.2-2}$$

where *M* is the magnification $(M \triangleq d_i/d_0)$ and $f_g(x_i, y_i)$ is the *ideal geometric image*. Thus, other than undergoing a coordinate inversion and magnification, the ideal geometric image is identical with the object, i.e., no loss of resolution blurring etc. However, when the imaging process is analyzed more carefully, one finds that the image does not have infinite resolution; it is blurred by a phenomenon known as the *finite pupil effect*. Briefly stated, the finite pupil effect refers to the fact that a finite-size lens cannot capture all the rays (wavefronts) coming from the object. Indeed, those rays that are sharply angled with respect to the optical axis are less likely to be intercepted by the lens and it is precisely those rays that carry high-frequency information.

Assuming quasi-monochromatic, incoherent, e.g, non-laser illumination, one can show (for example, see Goodman [8]) that the actual measured image f_m is given by

$$f_m(x_i, y_i) = \int_{-\infty}^{\infty} \int_{-\infty}^{\infty} f_g(\alpha, \beta) h(x_i - \alpha, y_i - \beta) d\alpha d\beta,\tag{7.2-3}$$

where $h(\alpha, \beta)$ is the *intensity point-source* response of the optical system. In the absence of lens aberrations, $h(\alpha, \beta)$ is given by

$$h(\alpha, \beta) = \left| \int_{-\infty}^{\infty} \int_{-\infty}^{\infty} P(\xi, \eta) e^{-j2\pi[\xi\alpha + \eta\beta]} d\xi d\eta \right|^2, \qquad (7.2\text{-}4)$$

where $P(\xi, \eta)$ is the *pupil function* of a diffraction-limited lens. It is given by

$$P(\xi, \eta) = \begin{cases} 1 & \text{for } (\xi, \eta) \in S \\ 0 & \text{otherwise,} \end{cases} \qquad (7.2\text{-}5)$$

and S the set of points *inside* the lens pupil. When there are no other factors present to degrade image quality, the field $f_m(x_i, y_i)$ given in Eq. (7.2-3), with the help of Eq. (7.2-4) and Eq. (7.2-5), is called the *diffraction-limited image* of the object. Now Eq. (7.2-3), being a convolution with what is usually a continuous lowpass kernel, does not contain all the frequency components, in their respective proportion, which exist in the original image. In fact, there may be frequencies in f_g that f_m doesn't contain at all. To see this, rewrite Eq. (7.2-3) in the frequency domain. Let $f_m \leftrightarrow F_m$, $f_g \leftrightarrow F_g$, $h \leftrightarrow H$, where $\leftrightarrow$ indicates a Fourier pair. Then

$$F_m = F_g H, \qquad (7.2\text{-}6)$$

where H is readily computed from Eq. (7.2-4) as

$$H(u, v) = \int_{-\infty}^{\infty} \int_{-\infty}^{\infty} P(\xi, \eta) \bar{P}(\xi + u, \eta + v) d\xi d\eta, \qquad (7.2\text{-}7)$$

i.e., the autocorrelation of the aperture pupil function. Now clearly, if P has finite support so will H; hence we can write that the transfer function $H(u, v)$ must satisfy $H(u, v) = 0$, $|u| > u_c$ and $|v| > v_c$, where u_c, v_c are *horizontal* and *vertical cut-off frequencies,* respectively.[†] Thus f_m contains no horizontal frequencies above u_c and no vertical frequencies above v_c.

The correct re-insertion of some or all of the high-frequency components in the measured image f_m from data and prior knowledge is called *super-resolution.* We show below how projection methods are useful in super-resolution.

Super-resolution by Convex Projection

The aim is to recover the ideal geometric image, f_g, from the measured image f_m and the knowledge that the image is of finite extent (compact support). Three useful

---

[†]These cut-off frequencies can be computed from the dimensions of $P(\xi, \eta)$.

sets are[†]

$$C_1 = \{f(x,y): \ F(u,v)H(u,v) = F_m(u,v) \text{ for all } u,v\}, \quad (7.2\text{-}8)$$
$$C_2 = \{f(x,y): \ f(x,y) = 0 \text{ for all } (x,y) \notin S\}, \quad\quad\quad (7.2\text{-}9)$$
$$C_3 = \{f(x,y): \ f(x,y) \geq 0 \text{ for all } (x,y) \in S\}. \quad\quad\quad (7.2\text{-}10)$$

In words, C_1 is the set of all functions whose weighted spectrum matches the spectrum of the measured image; C_2 is the set of all image functions with compact support S, i.e., the set of all functions of finite extent in both directions; and C_3 is the set of all functions having no negative values. We give below the projections upon C_1, C_2, C_3, without derivation. By now the reader will have sufficient expertise to derive the projection f^* of an arbitrary signal $g(x,y)$ onto these sets.

Onto C_1:

$$f^*(x,y) = P_1 g \leftrightarrow \begin{cases} F_m(u,v)/H(u,v) & \text{for } H(u,v) \neq 0 \\ G(u,v) & \text{otherwise.} \end{cases} \quad (7.2\text{-}11)$$

Onto C_2:

$$f^*(x,y) = P_2 g = \begin{cases} g(x,y) & \text{for } (x,y) \in S \\ 0 & \text{otherwise.} \end{cases} \quad (7.2\text{-}12)$$

Onto C_3:

$$f^*(x,y) = \begin{cases} \Re[g(x,y)] & \text{when } \Re[g(x,y)] \geq 0 \\ 0 & \text{otherwise,} \end{cases} \quad (7.2\text{-}13)$$

where $\Re$ means take the real part. The algorithm

$$f_{n+1} = P_3 P_2 P_1 \ f_n \quad (7.2\text{-}14)$$

with arbitrary starting point $f^{(0)}$ will converge to a point in the intersection of C_1, C_2, and C_3. Because the correct image $f_g(x,y)$ is of finite extent, its Fourier transform is analytic in the frequency domain. Since the spectrum is already known over a compact support, we can, in principle, recover it everywhere. The projection algorithm given in Eq. (7.2-14) is one way to recover the missing spectrum. The positivity constraint represented by set C_3 is not always applicable. For example, in *coherent optical processing*, the image field $f(x,y)$ can be complex. Then, instead of the algorithm in Eq. (7.2-14), we would have $f_{n+1} = P_2 P_1 \ f_n$ or $f_{n+1} = P_1 P_2 \ f_n$. The action of this algorithm is portrayed in one dimension in Fig. 7.2-2. In Fig. 7.2-2(a) is shown the known portion of spectrum; in Fig. 7.2-2(b) is shown the image associated with the spectrum. The action of operator P_2 is shown in (d); likewise, the action of the operator P_1 is shown in the modification of the spectrum from (c) to (e). Note the close parallel to the example furnished

---

[†]It is assumed that the sets consist of elements in $L^2(R^2)$.

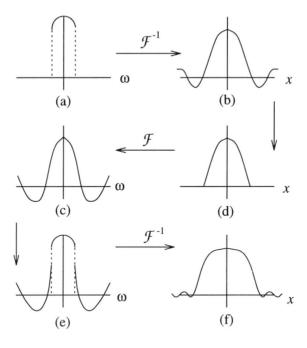

Fig. 7.2-2 Steps involved in super-resolution algorithm: (a) known portion of spectrum; (b) move to space domain; (c) apply space-domain constraint; (d) move to frequency domain; (e) apply frequency-domain constraint; (f) move to space domain etc.

in Fig. 2.7-1 in Chapter 2, where the extrapolation was applied to a time domain signal with a given (known) bandwidth.

There are a number of examples of super-resolution by projections. As stated earlier, Marks [6] realized a projection-recovery algorithm using an all-optical method. Other early examples are furnished by [3,4,9,10]. When the measurement of the spectrum $F_m(u, v)$ is corrupted by noise, the operation in Eq. (7.2-11) tends to enhance the noise when $H(u, v)$ is nearly zero. For this reason, more sophisticated methods are used in practice. See Chapter 9, Section 9.4, for an example.

7.3   THE PHASE RETRIEVAL PROBLEM

The problem of restoring a signal from magnitude information only is of significant practical importance. The problem is alternatively called *restoration from magnitude* or *phase retrieval*. It appears in several fields (see, for example, [11–14] such as astronomy, optics, X-ray crystallography, antenna design and beam forming, and electron microscopy. Typically, the directly available information about an image comes from intensity measurements in the space or spatial-frequency domain (or both) and from some *a priori* knowledge such as the fact that the ideal image is space-limited and the image intensity is non-negative. The phase retrieval problem in astronomy is discussed in some detail in [14].

In discussing the PR problem there are two fundamental issues: (1) Under what conditions does the magnitude alone uniquely specify the signal?; and (2) Can one design an algorithm that recovers the phase from the magnitude measurement and prior knowledge of the signal?

Generally speaking, it is the *Fourier magnitude* one measures and the related uniqueness question refers to whether a function can be uniquely defined by its Fourier magnitude. The problem is often specialized for discrete-time or discrete-space signals since, invariably, the processing is done by computer. Considerable research in the area was done by Hayes [15] for the case of space-limited multi-dimensional *sequences*, i.e., discrete case.

When dealing with uniqueness in the phase retrieval problem it must be noted that all the real functions $f(x,y), -f(x,y), f(x - x_0, y - y_0)$, and $f(-x,-y)$ have the same Fourier transform magnitude function and therefore uniqueness in this case is up to a sign, shift in coordinates, or coordinate reversal. It was found by Bruck and Sodin [16] that a space-limited sequence is uniquely defined by its Fourier transform magnitude if its Z-transform polynomial includes at most one irreducible nonsymmetric factor. Hayes also found conditions for uniqueness when the magnitude of the sequence is not known at all frequencies. Thus, assuming a discrete signal of support length N with only the magnitude of an M-point discrete Fourier transform available, $f(n)$ can still be uniquely determined if its Z-transform polynomial includes at most one irreducible nonsymmetric factor and $M \neq 2N - 1$. It is not an easy task to check a two-dimensional polynomial for irreducibility. However, it is widely believed [17–20] (based partly on experimental results) that two-dimensional images generally satisfy the uniqueness conditions. Moreover, it is known that, within the set of all polynomials in more than one variable, the subset of reducible polynomials is a nondense set of measure zero [15].

Many of these points are discussed in greater detail in [20].

Projection-Based Phase Retrieval

SETS AND PROJECTIONS: In the phase retrieval problem, it is generally assumed that the image $g(\mathbf{x})$ is space-limited and that the magnitude $M(\boldsymbol{\omega})$ of its Fourier transform $G(\boldsymbol{\omega})$ is known or prescribed. The two appropriate sets, then, are:

$$C_1 = \{y(\mathbf{x}) \in L^2 : y(\mathbf{x}) = 0 \text{ for } \mathbf{x} \notin S\} \tag{7.3-1}$$

$$C_2 = \{y(\mathbf{x}) \in L^2 : |Y(\boldsymbol{\omega})| = M(\boldsymbol{\omega}) \text{ for all } \boldsymbol{\omega}\}. \tag{7.3-2}$$

It is often easy to incorporate additional constraints into C_1. For instance:

$$C_{1L} = \left\{ y(\mathbf{x}) \in L^2 : a \le y(\mathbf{x}) \le b \text{ for } \mathbf{x} \in S, \text{ and } y(\mathbf{x}) = 0 \text{ otherwise. } \right\}.$$
(7.3-3)

In C_{1L}, a simple but useful two-level amplitude constraint has been added. It is easy to verify that C_1 or C_{1L} is convex while C_2 is not. The projections of an arbitrary signal $q(\mathbf{x})$ onto C_1, C_{1L}, and C_2 are given by, respectively,

$$P_1 q = \begin{cases} q(\mathbf{x}) & \mathbf{x} \in S \\ 0 & \mathbf{x} \notin S, \end{cases}$$
(7.3-4)

$$P_{1L} q = \begin{cases} q(\mathbf{x}) & \mathbf{x} \in S \text{ and } a \le q(\mathbf{x}) \le b \\ b & \mathbf{x} \in S \text{ and } q > b \\ a & \mathbf{x} \in S \text{ and } q < a \\ 0 & \mathbf{x} \notin S, \end{cases}$$
(7.3-5)

$$P_2 q \leftrightarrow M(\boldsymbol{\omega}) e^{j\phi(\boldsymbol{\omega})},$$
(7.3-6)

where $\phi(\boldsymbol{\omega})$ is the phase of $Q(\boldsymbol{\omega}) \triangleq \mathcal{F}[q(\mathbf{x})]$. Although C_2 is nonconvex, the projection $P_2 q$ is unique. The derivation of these projections is given elsewhere [22,23]. They are not difficult to derive.

ALGORITHM: The general restoration algorithms are given by

$$f_{n+1} = T_1 T_2 f_n$$
(7.3-7)

or

$$f_{n+1} = T_2 T_1 f_n,$$
(7.3-8)

where

$$T_1 \triangleq I + \lambda_1 (P_1 - I)$$
(7.3-9)

and

$$T_2 = I + \lambda_2 (P_2 - I).$$
(7.3-10)

The projector P_{1L} is used to replace P_1 if a curb on amplitude excursions is desired. The I in T_1, T_2 is the identity operator.

The similarity of Eqs. (7.3-7) and (7.3-8) to the Gerchberg-Saxton [5] and Fienup [24] algorithms is discussed in [23]. A good choice of λ_i, $i = 1, 2$, will greatly accelerate convergence. The question is, however, what is a good choice? Levi and Stark [25] show that for the algorithm

$$f_{n+1} = T_1 T_2 f_n$$
(7.3-11)

close to optimum choices of λ_1 and λ_2 (call these $\lambda_{1,0}$ and $\lambda_{2,0}$) are

$$\lambda_{1,0} = 1$$
(7.3-12)

Fig. 7.3-1 The letter "F".

and the $\lambda_{2,0}$ that minimizes

$$J^2(f_{n+1}) = \int_{-\infty}^{\infty} |M(\omega) - F_{n+1}(\omega, \lambda_2)|^2 \frac{d\omega}{(2\pi)^2} \,, \qquad (7.3\text{-}13)$$

where

$$F_{n+1}(\omega, \lambda_2) \triangleq \mathcal{F}\{P_1[f_n + \lambda_2(P_2 f_n - f_n)]\} \,, \qquad (7.3\text{-}14)$$

and $\mathcal{F}$ is the Fourier operator. Finding $\lambda_{2,0}$ requires a search. Typically, as observed from numerical simulations, $\lambda_{2,0}$ has values in the vicinity of 2.

Numerical Results

While research has shown that many functions of engineering interest are uniquely specified by their Fourier transform magnitude, it does not mean, however, that the generalized projection algorithm will always converge to a recognizably correct solution. As explained in Chapter 5, generalized projection (GP) algorithms are subject to traps and tunnels.

Figure 7.3-1(a) shows an image of the letter F and other items. Figure 7.3-2 shows the restoration with incorrect recovery of the image when the solution path goes through a trap. By changing the starting point, the trap is avoided and the algorithm converges to a recognizably correct solution (Fig. 7.3-2(b)).

7.4 BEAM FORMING

The beam forming problem can be stated as follows: What distribution of phase $\phi(\mathbf{x})$ is needed at the source array to generate a prescribed far-field intensity $I(\mathbf{u})$? Here $\mathbf{x}$ is the displacement vector in the input plane and $\mathbf{u}$ is the displacement in the far-field (Fourier) plane. The problem has application in various areas of beam forming, including those associated with radar, antennas, and dynamically recon-

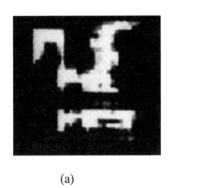

(a)   (b)

Fig. 7.3-2 GP reconstruction of the letter F: (a) Incorrect convergence due to trap; the starting point was the zero image with a support of 40×40 pixel square. (b) Correct convergence results when trap is avoided by selecting different starting point; here the starting point was an image consisting of a small 20×20 pixel square of amplitude 0.72 nested within a large square of size 40×40 with reduced amplitude of 0.36.

figurable optical interconnect [26, 27]. A particular beam forming configuration is shown in Fig. 7.4-1. There, a *spatial light modulator* generates a phase mask by application of an appropriate set of electrical control signals, which produces the desired field in the Fourier plane. The phase profile of the SLM evolves with time as the electrical signals are changed, thereby enabling different points in the far field to be flashed by focused light beams. However, we consider the situation where a phase distribution is sought to produce a given static far-field intensity. The relationship between $I(\mathbf{u})$ and $\phi(\mathbf{x})$ is given by

$$I(\mathbf{u}) = \left| \int_{\Omega} \exp(-j\phi(\mathbf{x}))\exp(-j2\pi \mathbf{u} \cdot \mathbf{x})d\mathbf{x} \right|^{2}, \qquad (7.4\text{-}1)$$

where Ω is the support of the phase function. For Eq. (7.4-1) to model the configuration shown in Fig. 7.4-1 accurately, a number of physical constraints must be satisfied, not the least of which is that the highest frequency in the spectrum of $\phi(\mathbf{x})$ must be less than the inverse wavelength of illumination. The constraint on a real $\phi(\mathbf{x})$ requires that

$$|\exp[j\phi(\mathbf{x})]| = 1. \qquad (7.4\text{-}2)$$

Because the positive square root of $I(\mathbf{u})$ is the Fourier magnitude function $M(\mathbf{u})$, the problem can be restated as follows: For a given $M(\mathbf{u})$, find $\phi(\mathbf{x})$ such that

$$M(\mathbf{u}) = \left| \int_{\Omega} \exp[j\phi(\mathbf{x})] \exp(-j2\pi \mathbf{u} \cdot \mathbf{x})d\mathbf{x} \right|. \qquad (7.4\text{-}3)$$

SETS AND PROJECTIONS: The two sets involved in the phase-mask design problem are C_1, the set of all complex-valued functions having the prescribed Fourier

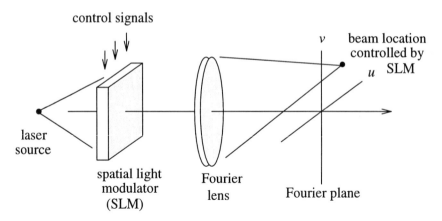

Fig. 7.4-1 Beam forming configuration.

magnitude $M(\mathbf{u})$, and C_2, the set of all complex-valued functions of compact support with unity magnitude. Thus

$$C_1 = \{y(\mathbf{x}) \in L^2 : y(\mathbf{x}) \leftrightarrow Y(\mathbf{u}), \ |Y(\mathbf{u})| = M(\mathbf{u})\} \qquad (7.4\text{-}4)$$

and

$$C_2 = \{y(\mathbf{x}) \in L^2 : y(\mathbf{x}) = 0, \text{ all } \mathbf{x} \in \Omega^c; \ |y(\mathbf{x})| = 1, \text{ all } \mathbf{x} \in \Omega\}. \qquad (7.4\text{-}5)$$

Notice that, except for the change from radians per millimeter (ω) to cycles per millimeter ($\mathbf{u}$), set C_1 is identical to set C_2 in Eq. (7.3-2). Hence its projector is given by Eq. (7.3-6). Recall $\omega = 2\pi\mathbf{u}$. In Eq. (7.4-5) Ω^c is the complement of Ω. It is easily shown that C_2 is nonconvex. The projection of $q(\mathbf{x})$ onto C_2 is given by

$$P_2 q = \begin{cases} 0 & \text{for } \mathbf{x} \in \Omega^c \\ \exp[j\phi_q(\mathbf{x})] & \text{for } \mathbf{x} \in \Omega, \end{cases} \qquad (7.4\text{-}6)$$

where $\phi_q(\mathbf{x})$ is the phase of $q(\mathbf{u})$ being projected onto C_2.

Additional constraints can be imposed as required [28, 29], e.g., binary phase, sub-band energy, etc.

The algorithm for finding $\phi(\mathbf{x})$ or, more appropriately, $f(\mathbf{x}) = \exp[j\phi(\mathbf{x})]$, is given by

$$f_{k+1} = P_2 P_1 f_k, \qquad \text{where } f_0 \text{ is arbitrary,} \qquad (7.4\text{-}7)$$

and a block diagram describing the required set of operations is shown in Fig. 7.4-2. As the reader may have observed, the beam forming problem has elements in common with the phase recovery problem. Note, however, that the present problem is more of an image "discovery" problem than an image "recovery" problem.

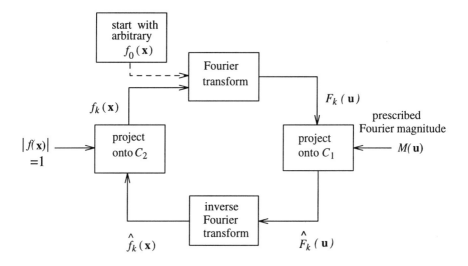

Fig. 7.4-2 Block diagram describing algorithm that attempts to find correct phase distribution for a prescribed far-field intensity.

Numerical Results

Figure 7.4-3(a) shows a prescribed 64-point far-field magnitude $M(u)$. In Fig. 7.4-3(b), the far-field magnitude produced by the synthesized phase computed using Eq. (7.4-7) is shown. Except for a scale change in amplitude and some noise in the high-frequency tails of the recovered $|F(u)|$, the two patterns are nearly identical.

7.5 COLOR MATCHING

Studies have shown that there are three types of color receptors (cones) in the normal human retina, each with a distinct absorption spectra $S_i(\lambda)$ for $i = 1, 2, 3$. Typically, the $S_i(\lambda)$ are somewhat bell-shaped functions of wavelength λ with $S_1(\lambda)$ peaking in the red around 630 nm, $S_2(\lambda)$ peaking in the green around 570 nm, and $S_3(\lambda)$ peaking in the blue around 450 nm. There is significant overlap between $S_1(\lambda)$ and $S_2(\lambda)$ (see Fig. 7.5-1).

According to three-color theory, each color receptor exhibits a color sensation that is proportional to the total intensity absorbed by that color receptor. For instance, if $K(\lambda)$ denotes the spectral intensity distribution of some colored light, then the quantification of the sensation of color can be done through a set of spectral response coefficients $\alpha_1, \alpha_2, \alpha_3$, given by

$$\alpha_i(K) = \int_\Omega S_i(\lambda)K(\lambda)d\lambda, \qquad \text{for } i = 1, 2, 3, \qquad (7.5\text{-}1)$$

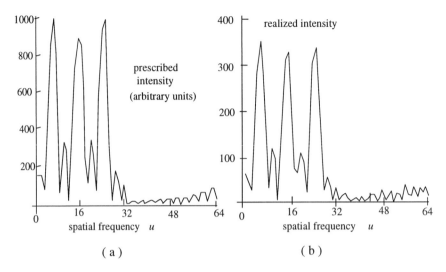

Fig. 7.4-3 Numerical example that shows that computed phase at SLM can reproduce desired far-field intensity: (a) prescribed far-field intensity; (b) achieved far-field intensity by GP-generated phase at SLM.

where Ω is the set of wavelengths in the visible spectrum.

An important result in color theory is that two colors need not be identical in order to be *perceived* as *identical* by the observer. Thus if $K_1(\lambda)$ and $K_2(\lambda)$ are two spectral distributions that produce responses $\alpha_i(K_1)$ and $\alpha_i(K_2)$ for $i = 1, 2, 3$ and if

$$\alpha_i(K_1) = \alpha_i(K_2) \qquad \text{for } i = 1, 2, 3, \qquad (7.5\text{-}2)$$

then the spectra K_1 and K_2 are perceived as being identical and are said to be *matched*. This result is of interest in industries dealing with color, for example, manufacturers of pigments, dyes, paints, and jewelry. It suggests that, possibly, only certain minimal changes need to be made in a spectrum $K(\lambda)$ to match it to another. The previous discussion can be generalized to include color matching by N sensors. Thus, given a set of N color sensors with response functions $S_1, S_2, \cdots, S_N$, we can define a match between two spectra K_1 and K_2 if

$$\alpha_1(K_1) = \alpha_1(K_2) \qquad \text{for } i = 1, \cdots, N. \qquad (7.5\text{-}3)$$

Suppose we are given a color spectrum $K'(\lambda)$. What minimum changes are required for it to match another spectrum $K(\lambda)$? This and related problems can easily be solved using vector-space projections [30]. We demonstrate this below.

SETS AND PROJECTIONS: The appropriate set C_0 is

$$C_0 = \left\{ K(\lambda) \in L^2 : \int_\Omega S_i(\lambda) K(\lambda) \, d\lambda = \alpha_i, \ i = 1, \cdots, N \right\}. \qquad (7.5\text{-}4)$$

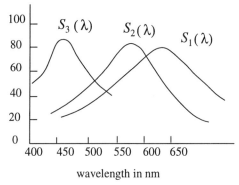

Fig. 7.5-1 Typical absorption spectra of the three color receptors in human retina.

In words, C_0 is the set of all color spectra that are matched, i.e., that appear to be the same to a human ($N=3$) or a machine (arbitrary N). While it is not difficult to find the direct projection onto C_0, in some cases it might be useful to reach C_0 through sequence of iterations. To achieve the latter we note that we can write

$$C_0 = \bigcap_{i=1}^{N} C_i, \tag{7.5-5}$$

where, as usual, $\cap$ denotes set intersection and

$$C_i = \left\{ K(\lambda) \in L^2 : \int_\Omega S_i(\lambda)K(\lambda)\,d\lambda = \alpha_i \right\}, \tag{7.5-6}$$

for $i = 1, 2, \cdots, N$. Note that C_0 belongs to the class of linear constraints given in general form by Eq. (3.2-2). Hence its convexity and closedness follows, since the set in Eq. (3.2-2) was shown to be convex and closed in general. To find the projection of an arbitrary point $K'(\lambda)$ onto C_i we must solve

$$\min_{K \in C_i} \int |K'(\lambda) - K(\lambda)|^2 \, d\lambda. \tag{7.5-7}$$

As usual, this can be done using the method of Lagrange multipliers. The result is (all functions are assumed real):

$$K^*(\lambda) \triangleq P_i K' = K'(\lambda) - \frac{\langle K', S_i \rangle - \alpha_i}{\|S_i\|^2} S_i(\lambda), \tag{7.5-8}$$

where

$$\langle K', S_i \rangle \triangleq \int_\Omega K'(\lambda)S_i(\lambda) \, d\lambda \tag{7.5-9}$$

and .

$$\|S_i\|^2 \triangleq \langle S_i, S_i \rangle. \tag{7.5-10}$$

Equation (7.5-8) is the result we seek: $K^*(\lambda)$ is the point in C_i closest to $K'(\lambda)$. Note that, if $K'(\lambda)$ is already in the set, then $\langle K', S_i \rangle = \alpha_1$ and $P_i K' = K'$. Note also that P_i projects only C_i. To achieve color matching, we must find a point in

$$C_0 = \cap_{i=1}^N C_i. \tag{7.5-11}$$

Any permutation of the projection operators to form a composition of the form $P_{p(1)} P_{p(2)} \cdots P_{p(N)}$, where each $P_{p(i)}$ is distinct from the others and is an element from the set $\{P_1, P_2, \cdots, P_N\}$, will suffice. A pure projection algorithm that will converge to a point in C_0 is then

$$K_{n+1} = P_{p(1)} \cdots P_{p(N)} K_n, \quad n = 0, 1, 2, \cdots \tag{7.5-12}$$

with $K_0 \triangleq K'(\lambda)$, the original spectrum. Although it may not be obvious, the point of convergence in this iterative approach is identical to the direct projection of $K'(\lambda)$ onto C_0. This is not generally true, but it is true in this case because all the constraint sets are linear varieties.[†] The direct projection of an arbitrary $K'(\lambda)$ onto C_0 is easily done using the method of Lagrange multipliers with *multiple constraint*. Thus, to find P_0, the projector onto C_0, we must solve

$$\min_{k \in C_0} \|K'(\lambda) - K(\lambda)\|^2, \tag{7.5-13}$$

where C_0 is given in Eq. (7.5-11). As always we extremize a Lagrange functional, which, in this case, is

$$J(K) \triangleq \int_\Omega |K'(\lambda) - K(\lambda)|^2 \, d\lambda + \sum_{i=1}^N \gamma_i \left[\int_\Omega S_i(\lambda) K(\lambda) \, d\lambda - \alpha_i \right], \tag{7.5-14}$$

where the γ_i, $i = 1, \cdots, N$ are N Lagrange multipliers. [‡] By setting $\partial J / \partial K = 0$ and performing some algebraic operations, we obtain the *one-step color matching projector*. To describe the projector, we make some definitions. Let

$$\mathbf{a} \triangleq (\alpha_1, \alpha_2, \cdots, \alpha_N)^T \quad \text{(a constant vector)}, \tag{7.5-15}$$

$$\mathbf{q} \triangleq (\langle K', S_1 \rangle, \cdots, \langle K', S_N \rangle)^T \quad (\text{ a constant vector}), \tag{7.5-16}$$

[†]This point is discussed in greater detail in Chapter 2, Section 2.5. See *Corollary* 2.5-2 in the same Chapter.
[‡]The use of the symbol λ for wavelength has pre-empted its use for the Lagrange multipliers.

$$\mathbf{S} \triangleq \begin{bmatrix} \langle S_1, S_1 \rangle & \cdots & \langle S_1, S_N \rangle \\ & & \\ & & \\ & & \\ \langle S_N, S_1 \rangle & \cdots & \langle S_N, S_N \rangle \end{bmatrix} \quad \text{(a constant matrix)}, \qquad (7.5\text{-}17)$$

$$\langle S_i, S_j \rangle \triangleq \int_\Omega S_i(\lambda) S_j(\lambda) \, d\lambda, \qquad (7.5\text{-}18)$$

and $\mathbf{s}(\lambda) \triangleq [S_1(\lambda), \cdots, S_N(\lambda)]^T$ (a vector with wavelength dependence).

Using the preceding equations, the one-step projection onto C_0 is

$$\begin{aligned} P_0 K' &= K'(\lambda) - \mathbf{s}^T(\lambda)\mathbf{S}^{-1}(\mathbf{q} - \mathbf{a}) \\ &= K'(\lambda) - (\mathbf{q} - \mathbf{a})^T \mathbf{S}^{-1}\mathbf{s}(\lambda). \end{aligned} \qquad (7.5\text{-}19)$$

Not unexpectedly, when $N = 1$, Eq. (7.5-19) reduces to Eq. (7.5-8). The advantage of the iterative approach of Eqs. (7.5-8) and (7.5-12) over the one-step approach of Eq. (7.5-19) is that in the former we do not have to compute the inverse of the matrix $\mathbf{S}$. On the other hand, the one-step solution does not require iterations and, therefore, is likely to be faster.

Figure 7.5-2 shows the results of a simulation that is given as an exercise in Exercise 7-8. In Fig. 7.5-2 the spectrum to be matched is $K(\lambda)$, i.e., the desired reference spectrum. The initial spectrum, i.e., the mismatched spectrum, is $K'(\lambda)$. The final result is $K^*(\lambda)$ which, while not having the same shape as $K(\lambda)$, matches almost exactly from the point of view of the color sensors.

To the best of our knowledge, Trussell [31] was the first to recognize that many color problems can be solved by projection methods.

7.6   BLIND DECONVOLUTION

The blind deconvolution problem refers to a class of problems of the form

$$g(\mathbf{x}) = \int f(\mathbf{x} - \mathbf{y}) h(\mathbf{y}) d\mathbf{y} + n(\mathbf{x}), \qquad (7.6\text{-}1)$$

where $g(\mathbf{x})$ are the observed data, $f(\mathbf{x})$ is the *unknown* source signal, $h(\mathbf{x})$ is the *unknown* blurring function (impulse response) of the system, and $n(\mathbf{x})$ is noise. Even without noise, recovering both $f(\mathbf{x})$ and $h(\mathbf{x})$ from $g(\mathbf{x})$ alone is a difficult problem.

With or without noise, Eq. (7.6-1) admits to an infinite number of solutions if no other constraints are available for $f(\mathbf{x})$ or $h(\mathbf{x})$ or both. However, once constraints are imposed, the set of feasible solutions to Eq. (7.6-1) can be greatly reduced. A number of algorithms have been proposed for solving the blind-deconvolution problem [32–34]. A commonly used technique is described in [32]. It is an

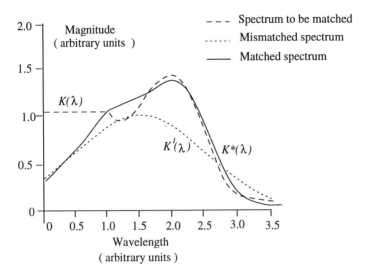

Fig. 7.5-2 Numerical results of color matching problem.

iterative algorithm, whereby the constraints about $f(\mathbf{x})$ and $h(\mathbf{x})$, i.e, the *a priori* knowledge, is enforced. However, this algorithm may not converge. As a result, other algorithms have been proposed to solve the blind-deconvolution problem. As expected, each has some advantages and disadvantages; for a review of some of these methods, see [35].

For the noiseless case, Eq. (7.6-1) becomes

$$g(\mathbf{x}) = \int f(\mathbf{x} - \mathbf{y})h(\mathbf{y})d\mathbf{y}. \tag{7.6-2}$$

As usual, we assume that both $f(\mathbf{x})$ and $h(\mathbf{x})$ in Eq. (7.6-2) are elements of the L^2 space of square integrable functions. The Hilbert space of interest is $\mathcal{H} = \{(u, v) : u \in L^2, v \in L^2\}$, which carries a norm

$$\|(u, v)\| \overset{\triangle}{=} \left[\int |u(\mathbf{x})|^2 d\mathbf{x} + \int |v(\mathbf{x})|^2 d\mathbf{x} \right]^{1/2}. \tag{7.6-3}$$

Define the set

$$C_g \overset{\triangle}{=} \{(u, v) : u * v = g\} \tag{7.6-4}$$

as the set of all solution pairs, i.e., the set of all function pairs $u(\mathbf{x})$ and $v(\mathbf{x})$, such that

$$g(\mathbf{x}) = \int u(\mathbf{x} - \mathbf{y})v(\mathbf{y})d\mathbf{y}. \tag{7.6-5}$$

The symbol $*$ denotes convolution. We shall reserve the symbols f and h for the true source and the true blurring function, respectively. Clearly, the pair $(f, h) \in C_g$.

Even to readers already familiar with the theory of alternating projections, the set defined in Eq. (7.6-4) may seem strange. Unlike the majority of sets used in this book, it is a set whose elements are 2-tuples i.e., function pairs; thus the constraint is simultaneously imposed on two functions. [†] As is to be expected, the projection onto C_g is considerably more difficult to compute than for the vast majority of ordinary constraint sets, i.e., those involving only a single function[‡]. In general, the set C_g contains infinitely many elements, each of which is a feasible solution to the blind-deconvolution problem if no further information is provided about either the signal $f(\mathbf{x})$ or the system $h(\mathbf{x})$ or both. Thus we must use as much prior knowledge about f and h as possible, in order to exclude spurious solutions in the set C_g and obtain a meaningful solution.

PRIOR KNOWLEDGE: Let C_f and C_h denote the constraint sets based on the prior knowledge about the source f and the blurring function h, respectively; i.e.,

$$C_f \overset{\triangle}{=} \{(u,v) : u \text{ satisfies prescribed prior knowledge about } f\},$$

$$C_h \overset{\triangle}{=} \{(u,v) : v \text{ satisfies prescribed prior knowledge about } h\}.$$

Then an element in the set $C_0 \overset{\triangle}{=} C_g \cap C_f \cap C_h$ will be a solution to Eq. (7.6-2), which also satisfies all the available prior knowledge. In general, the set C_0 contains fewer elements than the set C_g, making the solution physically more meaningful. The definitions of the sets C_f and C_h are based on available prior knowledge, and the sets are problem dependent. Usually they are determined by our understanding of the problem. Examples of such sets are

1. *Sets based on the support region of the signal:*

$$C_f \overset{\triangle}{=} \{(u,v) : u(\mathbf{x}) = 0, \ \forall \mathbf{x} \in \Omega_f\}.$$
$$C_h \overset{\triangle}{=} \{(u,v) : v(\mathbf{x}) = 0, \ \forall \mathbf{x} \in \Omega_h\},$$

where Ω_f and Ω_h denote the support of f and h, respectively.

2. *Sets based on the intensity range.* For example,

$$C_f \overset{\triangle}{=} \{(u,v) : u(\mathbf{x}) \in [u_{\min}, u_{\max}]\}, \tag{7.6-6}$$

where $u_{\min}$ and $u_{\max}$ are determined by the specific problem.

3. *Sets based on the physical properties of the system.* For example, if the system is band-limited, then a constraint set can be defined as

$$C_h \overset{\triangle}{=} \{(u,v) : v(\mathbf{x}) \leftrightarrow V(\boldsymbol{\omega}) = 0, \ \forall \boldsymbol{\omega} \notin \Omega_H\}, \tag{7.6-7}$$

[†] In this sense, the set is more like the product-space sets introduced in Chapter 2, Section 2.9.
[‡] We shall encounter 2-tuple sets again in Chapter 8 in our discussion of learning in neural nets.

where Ω_H is the support of $H(\omega)$, the Fourier transform of h.

4. *Sets based on experimental measurement.* For example, if we have some measurement $\bar{h}(\mathbf{x})$ of the system blurring function $h(\mathbf{x})$, then we have

$$C_h \overset{\triangle}{=} \{(u,v) : \|v - \bar{h}\| \le E\}, \tag{7.6-8}$$

where E is some constant that can be determined by the measurement accuracy.

A brief remark is in order: The sets described in examples 1 to 4 have elements that are 2-tuples, but the constraint is applied to only one member of the function pair. In this sense, these sets are more like the ordinary constraint sets used in other examples in this book than like the set in Eq. (7.6-4).

Let P_g, P_f, and P_h, denote the projection operators onto the set C_g, C_f and C_h, respectively. At this point we can afford to be completely general. In any specific problem we would have to specify which *particular constraints* we are applying to f and h. This would define C_f and C_h. The theory of generalized projections guarantees that the recursion

$$(f,h)_{(k+1)} = P_h P_f P_g (f,h)_k, \quad k = 0, 1, 2, \cdots, \tag{7.6-9}$$

where $(f,h)_k$ is the estimated recovery of (f,h) at the kth iteration and $(f,h)_0$ is our initial estimate, will converge to a point in C_0, provided that the algorithm does not stagnate at erroneous solutions called *traps* [see Chapter 5, Section 5.1]. *Note:* In Chapter 5, Section 5.4, it was argued that an algorithm like the one in Eq. (7.6-9), i.e., involving nonconvex sets, has the summed distance error convergence property if the number of constraint sets involved is at most two. However, as the reader can see, Eq. (7.6-9) involves three sets C_h, C_f, and C_g. The statement surrounding Eq. (7.6-9) is still true, because it is readily shown that if $C_h \cap C_f \overset{\triangle}{=} C_{hf}$, then $P_{hf} = P_h P_f$ and thus, despite its appearance, Eq. (7.6-9) can be regarded as projections onto only two sets.

PROJECTORS: The projectors P_f and P_h, corresponding to the sets defined above, are well known (see Chapter 3, Section 3.6). The projector P_g onto set C_g in Eq. (7.6-4) can be derived as follows: consider any element (f', h') in $\mathcal{H}$ but not in C_g; its projection onto set C_g, denoted (f^*, h^*), is solved by minimization of the quantity $\|(u,v) - (f', h')\| \overset{\triangle}{=} \|(u - f', v - h')\|$ under the constraint that $(u,v) \in C_g$. The minimization is done by use of the Lagrange multiplier technique. First we agree to let capital letters denote the Fourier transform of the respective lower-case functions. Thus, $(u, v, f', h', u^*, v^*) \leftrightarrow (U, V, F', H', U^*, V^*)$, respectively. The Lagrange functional can be expressed in the frequency domain by

$$\int \{|U - F'|^2 + |V - H'|^2 d\omega + \lambda \int |UV - G|^2 d\omega \,,$$

where $U = U_R + jU_I, V = V_R + jV_I$, etc. for the other transforms. Upon taking the derivatives with respect to $U_R, U_I, V_R,$ and V_I and setting them to zero, and using the relation that $V = G/U$, we obtain two relations that can be used to obtain the projections (U^*, V^*) in a two-step procedure:

a. Solve for $U^*(\omega) = \mathcal{F}[u^*(\mathbf{x})]$ from

$$|U^*|^4 - |U^*|^2 U^* F' + H' \bar{G} U^* = |G|^2, \qquad (7.6\text{-}10)$$

where the overbar denotes the complex conjugate; and

b. Solve for $V^*(\omega) = \mathcal{F}[v^*(\mathbf{x})]$ from

$$V^*(\omega) = G(\omega)/U^*(\omega). \qquad (7.6\text{-}11)$$

The projection is found implicitly by solving Eq. (7.6-10), which, unfortunately, is non-linear. The projection could be computed numerically. However, there might be more than one solution to Eq. (7.6-10), because the set C_g is nonconvex. Another difficulty in finding the projection is that in the region where $U^*(\omega)$ is small, numerical inaccuracies pose a problem in the computation of $V^*(\omega)$. This also suggests that external noise could be a significant problem.

Because of the numerical difficulties discussed above, as well as the presence of the aforementioned traps, the generalized-projection algorithm presented above is not too practical to implement. Before suggesting an alternative procedure, consider the properties of the functional

$$J_{(v,u)} \overset{\triangle}{=} |g - u * v|^2 = \int |g(\mathbf{x}) - (u * v)(\mathbf{x})|^2 d\mathbf{x} \geq 0. \qquad (7.6\text{-}12)$$

Any element (u, v) in the set C_g is the point at which $J_{(u,v)}$ assumes its global minimum. The algorithm in Eq. (7.6-9) will converge to a feasible solution or stagnate at a trap. At a feasible solution the prior-knowledge constraints defined by the sets C_f and C_h are satisfied and $J_{(u,v)} = 0$, since the solution has membership in C_g. However, to avoid the serious mathematical problems associated with solving Eqs. (7.6-10) and (7.6-11), we propose an alternative, mixed-projection, algorithm that, while still subject to erroneous solutions associated with local minima of J, is significantly easier to implement. The algorithm involves the following steps:

1. Take an educated initial guess of (f, h) and call this (f_0, h_0). Otherwise, if no such educated guess is available, set $f_0 = P_f g$ and $h_0 = 0$.

2. Solve for

$$h_k = \arg\{\min_{h \in C_h} J_{(f_{k-1}, h)}\}. \qquad (7.6\text{-}13)$$

The expression is shorthand for "find the variable h that minimizes $J_{(f_{k-1}, h)}$ subject to h lying in the set C_h and set this value of h equal to h_k."

3. Solve for

$$f_k = \arg\{\min_{f \in C_f} J_{(f, h_k)}\}. \qquad (7.6\text{-}14)$$

4. Set $k = k + 1$; repeat step 2 and step 3 until convergence is achieved.

Some remarks are in order:

1. If $J_{(f_k, h_k)}$ converges to 0, then a global minimum is found, and the resulting (f, h) will also be a stationary point of the earlier generalized-projection-based algorithm in Eq. (7.6-9).

2. If more-accurate information is available about h than about f, we can interchange the role of f with the role of h in the algorithm.

The functional $J_{(u,v)}$ is a quadratic function with respect to u if v is fixed, and vice-versa. Therefore, the minimization in steps 2 and 3 (i.e., Eqs. (7.5-13) and (7.5-14)) can be done by a standard *gradient-projection method* [33, 34]. In step 2, the following iteration will yield the global minimum: for the kth cycle

$$h_k^{l+1} = P_h[h_k^l - \alpha_k \nabla J_{(f_{k-1}, h_k^l)}], \tag{7.6-15}$$

where ∇ is the gradient operator, P_h is the projection operator onto C_h, $h_k^0 = h_{k-1}$, and α_k is a constant chosen to guarantee convergence.

Equation (7.6-15) needs some discussion. First, observe that it is a contraction mapping,[†] since P_h is nonexpansive and, for the proper value of α_k, the operation in brackets is a contraction. Also, the numerical realization of the functional $J_{(f,h)} = \|g - h * f\|^2$, when f is held fixed at, say, the vector value $\mathbf{f}_k$, becomes $J = \|\mathbf{g} - \mathbf{F}_k \mathbf{h}\|^2$, where $\mathbf{g}$ and $\mathbf{h}$ are vectors and $\mathbf{F}_k$ is a *circulant/Toeplitz* matrix of the shifted $\mathbf{f}_k's$. The gradient of J is then easily computed to be

$$\nabla J = 2\mathbf{F}_k^T \mathbf{F}_k \mathbf{h} - 2\mathbf{F}_k^T \mathbf{g}, \tag{7.6-16}$$

where the superscript T denotes the transpose. Let λ_k denote the largest eigenvalue of the matrix $(\mathbf{F_k}^T \mathbf{F_k})$; then it is well known that Eq. (7.6-15) will converge, i.e., be a contraction mapping for $0 < \alpha_k < 1/\lambda_k$ (see Exercise 7-12). For the numerical computation of Eq. (7.6-16) the two-dimensional discrete Fourier transform and the diagonalization properties of circulant matrices by the discrete Fourier transform are used.

Step 3 can be done in a similar fashion:

$$f_k^{l+1} = P_f[f_k^l - \beta_k \nabla J_{(f^l, h_k)}], \tag{7.6-17}$$

where $f_k^0 = f_{k-1}$ and P_f is the projection operator onto C_f.

NUMERICAL EXAMPLE: The original signal $f(\mathbf{x})$ is a 64×64 pixel square image of a triple star, modeled by three disks, as shown in Fig. 7.6-1(a). The blurring

---

[†]A mapping T is said to be a contraction mapping if $\|T\mathbf{x} - T\mathbf{y}\| \leq r\|\mathbf{x} - \mathbf{y}\|$ for some $0 < r < 1$.

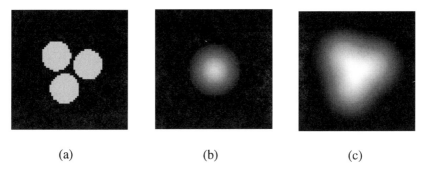

(a)   (b)   (c)

Fig. 7.6-1 A numerical simulation: (a) a triple star; (b) the Gaussian-shaped blurring function; (c) the blurred image of the triple star.

(a)   (b)

Fig. 7.6-2 The correct support of the object and blurring function.

function is modeled by a truncated Gaussian-shaped function

$$h(x_1, x_2) = \begin{cases} 1/(2\pi\sigma^2)\exp[-(x_1^2 + x_2^2)/2\sigma^2] & \text{for } x_1^2 + x_2^2 \leq 4\sigma^2 \\ 0 & \text{otherwise,} \end{cases}$$

$$(7.6\text{-}18)$$

where σ is chosen to be 7 in this example. The blurring function $h(\mathbf{x})$ is shown in Fig. 7.6-11(b). The blurred image $g(\mathbf{x})$ is shown in Fig. 7.6-1(c). The true supports $W_f^{(T)}$ for $f(\mathbf{x})$ and $W_h^{(T)}(\mathbf{x})$ for $h(\mathbf{x})$ are square regions (windows) of sizes $W_f : 33 \times 33$ and $W_h : 29 \times 29$, respectively (Fig. 7.6-2). However in practice the exact support information of the original image and the blurring function is rarely known, and therefore it would be unfair for us to assume such perfect knowledge in evaluating the algorithm. Nevertheless, when both the signal and the blurring function are positive, the sum of the spatial supports of the signal and blurring function is equal to the spatial support of the observed data. The proposed algorithm was tested with the following spatial support constraints:

$$W_f : \quad 37 \times 37 \quad \text{pixel square region} \qquad (7.6\text{-}19)$$

$$W_h : \quad 25 \times 25 \quad \text{pixel square region,} \qquad (7.6\text{-}20)$$

and

$$W_f : \quad 29 \times 29 \quad \text{pixel square region} \tag{7.6-21}$$

$$W_h : \quad 33 \times 33 \quad \text{pixel square region.} \tag{7.6-22}$$

Thus W_f in the first case is *overestimated*, while W_h is *underestimated*. In the second case, W_f is *underestimated*, while W_h is *overestimated*. The initialization image was taken to be

$$f_0(\mathbf{x}) = \left\{ \begin{array}{ll} g(\mathbf{x}) & \text{if } \mathbf{x} \in W_f \\ 0 & \text{otherwise,} \end{array} \right. \tag{7.6-23}$$

and $h_0(\mathbf{x})$ was assigned the value 0. For the assumed supports, the constraint sets are

$$C_f \;=\; \{(u,v): \; u(\mathbf{x}) \geq 0 \text{ for } \mathbf{x} \in W_f \text{ and } u(\mathbf{x}) = 0 \text{ otherwise}\} \tag{7.6-24}$$

$$C_h \;=\; \{(u,v): \; v(\mathbf{x}) = 0 \text{ for } \mathbf{x} \notin W_h\} \tag{7.6-25}$$

and W_f and W_h are as in Eqs. (7.6-19) or (7.6-20). The results are shown in Fig. 7.6-3 and demonstrate that projection-based deconvolution can be effective in recovering the source and blurring functions. Furthermore, the results are significantly better than the raw data, even when the prior knowledge contains errors. A detailed description of this approach, including its performance against competing algorithms such as furnished by Lane [33], is furnished in [35].

7.7 DESIGN OF DIFFRACTIVE OPTICAL ELEMENTS

The beam forming spatial-light modulator discussed in Section 7.3 is an example of a diffractive optical element. In this section we extend the design procedure used in Section 7.3 to a more general geometry.

A diffractive element $t(x, y)$, illuminated by a quasi-monochromatic unit amplitude plane wave, produces a field $g(\zeta, \eta)$ in the (ζ, η) plane, a distance z away, according to the Fresnel diffraction law (see Chapter 3 in [8])

$$g_z(\zeta, \eta) = \frac{1}{\lambda z} \int_{\mathcal{R}} t(x, y) \exp\left\{ j\frac{\pi}{\lambda z}[(x - \zeta)^2 + (y - \eta)^2] \right\} dx\,dy, \tag{7.7-1}$$

where λ is the wavelength of radiation, $\mathcal{R}$ is the set of points inside the aperture, and non-essential constants have been omitted for simplicity. For further simplicity, we shall develop the theory using a one-dimensional configuration such as in Fig. 7.7-1.

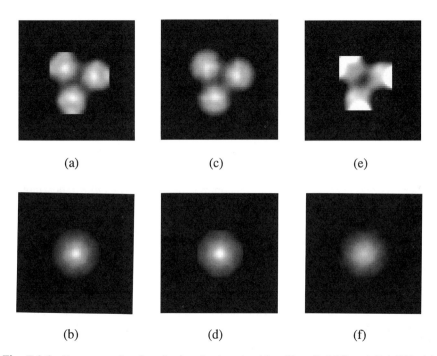

(a)　　　　　　　　(c)　　　　　　　　(e)

(b)　　　　　　　　(d)　　　　　　　　(f)

Fig. 7.6-3 Recovery using the mixed projection algorithm [Eqs. (7.6-15) and (7.6-17)]: (a) recovery when exact knowledge of support is available; (b) recovery with partial knowledge of support ($\hat{W}_f = 37 \times 37, \hat{W}_h = 25 \times 25$); (c) recovery with partial support knowledge ($\hat{W}_f = 29 \times 29, \hat{W}_h = 25 \times 25$).

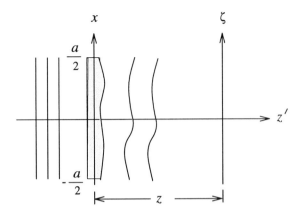

Fig. 7.7-1   Diffraction by an optical element illuminated by a plane wave.

Then, instead of Eq. (7.7-1), we write

$$g_z(\zeta) = \frac{1}{\sqrt{\lambda z}} \int_{-a/2}^{a/2} t(x) \exp\left\{ j\frac{\pi}{\lambda z}(x - \zeta)^2 \right\} dx. \qquad (7.7\text{-}2)$$

Either Eq. (7.7-1) or Eq. (7.7-2) is sometimes called the *Fresnel diffraction integral* and we write $g_z(\zeta) \overset{\triangle}{=} \mathcal{F}_r[t(x)]$, or simply $g_z \overset{\triangle}{=} \mathcal{F}_r[t]$.

In the following we consider a few examples of designing diffractive elements using vector-space projection methods.

Example 7.7-1   Here we consider a Fresnel-field extension of the Fourier-field design problem discussed in Section 7.3. The problem is stated as follows: Find the transmittance $t(x)$ of a phase-only, monochromatic-plane-wave illuminated, diffractive element that produces a given field magnitude $M(\zeta)$ in the Fresnel region a distance z away. In mathematical terms we seek a $t(x)$ such that

$$\left| \frac{1}{\sqrt{\lambda z}} \int_{-a/2}^{a/2} t(x) \exp\left\{ j\frac{\pi}{\lambda z}(x - \zeta)^2 \right\} dx \right| = M(\zeta). \qquad (7.7\text{-}3)$$

The sets of interest in the one-dimensional case are:

$$C_1 = \left\{ y(x) \in L^2 : |\mathcal{F}_r[y]| = M(\zeta) \right\}, \qquad (7.7\text{-}4)$$

and

$$C_2 = \left\{ y(x) \in L^2 : |y(x)| = 1 \text{ for } x \in [-a/2, a/2] \text{ and } 0 \text{ elsewhere} \right\}. \qquad (7.7\text{-}5)$$

Sometimes it is desirable to modify C_1 to C_1' defined as

$$C_1' = \left\{ y(x) \in L^2 : |\mathcal{F}_r[y]| \le c(\zeta) \right\}, \qquad (7.7\text{-}6)$$

where $c(\zeta)$ is a bound on the intensity of the Fresnel diffraction pattern. The set C_2 was encountered in the phase retrieval problem in Section 7.3 and enforces the phase-only condition; the set C_1 is new and requires some discussion. We first note that C_1 is non-convex because of the magnitude constraint. In this sense, it is similar to the Fourier magnitude constraint in Section 7.4. To find the projection onto C_1, we rewrite Eq. (7.7-3) as:

$$\left| \int_{-a/2}^{a/2} \tilde{t}(x, z) \exp\{-j\omega x\}\, dx \right| = M_z(\omega), \qquad (7.7\text{-}7)$$

where

$$\tilde{t}(x, z) \overset{\triangle}{=} t(x)\Omega(x, z) \qquad (7.7\text{-}8)$$

$$\Omega(x, z) \overset{\triangle}{=} \exp\left\{ j\frac{\pi}{\lambda z}x^2 \right\} \qquad (7.7\text{-}9)$$

$$\omega \overset{\triangle}{=} 2\pi\zeta/\lambda z \qquad (7.7\text{-}10)$$

and

$$M_z(\omega) \overset{\triangle}{=} \sqrt{\lambda z} M(\lambda z\omega/2\pi). \qquad (7.7\text{-}11)$$

Let us adopt the notation that when a lower-case function, say $h(x)$, is multiplied by $\Omega(x, z)$, the resulting function is denoted by $\tilde{h}(x, z)$ and its Fourier transform, i.e., $\mathcal{F}[\tilde{h}]$, is denoted by $\tilde{H}(\omega)$.

To find the projection of an arbitrary $h(x)$ upon C_1 we seek to find a $y^*(x)$ such that

$$y^* = \arg\min_{y \in C_1} \|h - y\|. \qquad (7.7\text{-}12)$$

But $\|h - y\| = \|(h\Omega - y\Omega)\bar{\Omega}\| = \|h\Omega - y\Omega\| = \|\tilde{h} - \tilde{y}\|$, where the overbar denotes conjugation. Therefore, $\min\|h - y\| = \min\|\tilde{h} - \tilde{y}\|$. Consider now the set $\tilde{C}_1$ defined by

$$\tilde{C}_1 \overset{\triangle}{=} \left\{ \tilde{y}(x) \in L^2 : \left| \tilde{Y}(\omega) \right| = M_z(\omega) \right\}. \qquad (7.7\text{-}13)$$

The projection of an arbitrary $\tilde{h}(x)$ onto $\tilde{C}_1$ is given by Eq. (7.3-6), i.e.,

$$\tilde{y}^*(x) = \tilde{P}_1\tilde{h} \leftrightarrow M_z(\omega)\exp[j\theta_{\tilde{H}}(\omega)], \qquad (7.7\text{-}14)$$

where $\theta_{\tilde{H}}(\omega)$ is the phase of $\tilde{H} \overset{\triangle}{=} \mathcal{F}[\tilde{h}]$. In Fig. 7.7-2, the point $\tilde{y}^*$ minimizes the distance from $\tilde{h}$ to $\tilde{C}_1$ and, likewise, the point y^* minimizes the distance from h to C_1. Since $\|h - y^*\| = \|\tilde{h} - \tilde{y}^*\|$, the projection of h onto C_1 is given by

$$y^* = \tilde{y}^*\overline{\Omega(x, z)}. \qquad (7.7\text{-}15)$$

Thus the algorithm

$$t_{k+1}(x) = P_1P_2\, t_k(x), \qquad t_0 \text{ arbitrary}, \qquad (7.7\text{-}16)$$

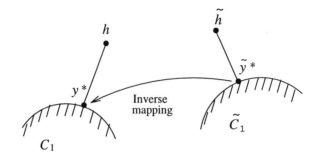

Fig. 7.7-2 Computing the projection onto $\tilde{C}_1$ is equivalent to finding the projection onto C_1.

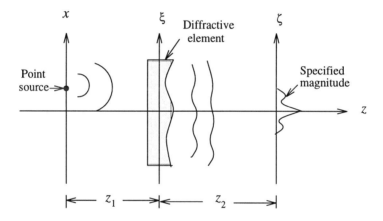

Fig. 7.7-3 A diffractive element that produces a specified optical field magnitude when illuminated by a point source.

where

$$P_1 h = \overline{\Omega} \mathcal{F}^{-1}[M_z(\omega) \exp\{j\theta_{\tilde{H}}(\omega)\}] \tag{7.7-17}$$

$$P_2 h = \begin{cases} 0 & x \notin [-a/2, a/2] \\ \exp\{j\phi_h(x)\} & x \in [-a/2, a/2] \end{cases} \tag{7.7-18}$$

and $\phi_h(x)$ is the phase of $h(x)$, enjoys the summed distance error property. Providing that the sets intersect and the algorithm doesn't settle into a trap (in which case a new starting point should be chosen), Eq. (7.7-16) will converge to a $t(x) \in C_1 \cap C_2$.
∎

Example 7.7-2 Refer to Fig. 7.7-3. It is required that a point source at $x = x_0$ in a plane z_1 units in front of a diffractive element, generates a field with magnitude $M(\zeta)$ in a plane a distance z_2 in back of the element. The analysis is simplified by a twice-repeated application of the Fresnel diffractive integral. Thus the field in the ξ-plane immediately to the left of the element is

$$g_{z_1}(\xi) = \frac{1}{\sqrt{\lambda z_1}} \int_{-\infty}^{\infty} \delta(x - x_o) \exp\left\{ j\frac{\pi}{\lambda z_1}(x - \xi)^2 \right\} dx \qquad (7.7\text{-}19)$$

$$= \frac{1}{\sqrt{\lambda z_1}} \exp\left\{ j\frac{\pi}{\lambda z_1}(x_0 - \xi)^2 \right\}, \qquad (7.7\text{-}20)$$

while the field immediately to the right of the diffractive element is $g_{z_1}(\xi)t(\xi)$. Hence in the ζ-plane we obtain

$$g_{z_2}(\zeta) = \frac{1}{\sqrt{\lambda z_2}} \int_{-a/2}^{a/2} g_{z_1}(\xi)t(\xi) \exp\left\{ j\frac{\pi}{\lambda z_2}(\zeta - \xi)^2 \right\} d\xi. \qquad (7.7\text{-}21)$$

After some algebra, and recalling that it is the magnitude $M(\zeta)$ of $g_{z_2}(\zeta)$ that is of interest, we obtain

$$\left| \int_{-a/2}^{a/2} t(\xi) \exp\left\{ j\frac{\pi}{\lambda z}\xi^2 \right\} \exp\left\{ -j\omega\xi \right\} d\xi \right| = M_z(\omega), \qquad (7.7\text{-}22)$$

where

$$z \triangleq \frac{z_1 z_2}{z_1 + z_2} \qquad (7.7\text{-}23)$$

$$\omega \triangleq \frac{x_0}{\lambda z_1} + \frac{\zeta}{\lambda z_2} \qquad (7.7\text{-}24)$$

and

$$M_z(\omega) \triangleq \lambda\sqrt{z_1 z_2} M(\lambda\omega z_2 - x_0 z_2/z_1). \qquad (7.7\text{-}25)$$

Finally by letting $\tilde{t}(\xi, z) \triangleq t(\xi) \exp\left\{ j\frac{\pi}{\lambda z}\xi^2 \right\}$, Eq. (7.7-22) adopts the same form as Eq. (7.7-7) and, hence, the design of the diffractive element in this example (Fig. 7.7-3) is identical with that of Example 7.7-1.

A numerical example is furnished in what follows. The configuration is that in Fig. 7.7-3, with the point x_0 chosen as $x_0 = 0$ for simplicity. It is desired to find a diffractive element that will image a point such that the magnitude of the image is bounded by the function $c(\zeta)$ from above and attains the value $c(0) = 4$ at the origin. The bounding function $c(\zeta)$ is defined by $c(\zeta) \triangleq 0.85 + 3.15 \, \text{rect}(2\zeta)$. In Fig. 7.7-4 we show the results furnished by a projection algorithm when the initial element is a thin lens of (normalized) radius $a = 1$. After 5 iterations the "lens" has grown to size $a = 2$ and has a phase profile characteristic of a *Fresnel lens* of radius $a = 2$.

The influence of the starting point is obviously important. For example, in Fig. 7.7-5, the initial object from which the appropriate lens evolves, iteration-by-iteration, is a uniform section of glass of radius $a = 1$. After 100 iterations the projection algorithm produces a slightly different Fresnel lens than in the previous case but one that, nevertheless, meets the imaging constraint. Because the starting

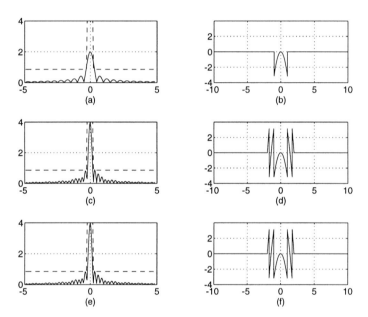

Fig. 7.7-4 Evolution of a lens that meets a design constraint. (a) Image of point source from the initial lens. The image does not meet the design constraint, since the peak is not held at 4 and part of the main lobe is above the lower bound; (b) phase profile of initial lens; (c) image after 5 iterations; (d) corresponding phase profile of evolving lens; (e) image after 15 iterations; (f) final phase profile of lens.

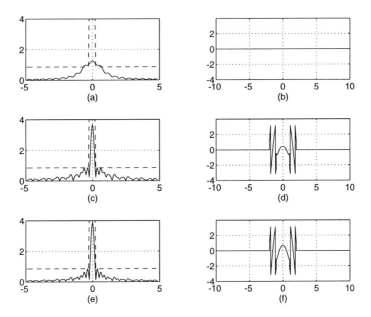

Fig. 7.7-5 Evolution of a lens that meets the design constraint when the starting point is a uniform section of glass. (a) Image of point source from the initial "lens". The image does not meet the design constraint; (b) phase profile of initial lens is a constant; (c) image after 20 iterations. The evolving lens still doesn't meet the design constraint at the origin; (d) corresponding phase profile of evolving lens; (e) image after 100 iterations; (f) final phase profile of lens.

point is further from the solution region, the number of required iterations is much larger. ∎

Example 7.7-3 Piestun and Shamir [36, 37] used the method of convex projections to design diffractive elements that could, among other things, generate so-called non-diffracting beams. Refer to Fig. 7.7-6. For simplicity we modify their original configuration slightly and assume that the diffractive element is illuminated by a unit-amplitude plane wave and the object is to control the spreading of the field in a region of space behind the diffractive element. The way that this problem is solved in [36] is as follows: Given the input field $t(x)$, we find its projection onto the constraint set C_z by calculating the field in a plane at a distance z, projecting this Fresnel-zone function onto the constraints of that plane and performing the inverse Fresnel transform to obtain the corresponding field at the input. If we take n planes we have n constraint sets $C_{z_1}, C_{z_2}, ..., C_{z_n}$. Thus we seek a $t(x) \in C_o \triangleq \cap_{i=1}^{n} C_{z_i}$.

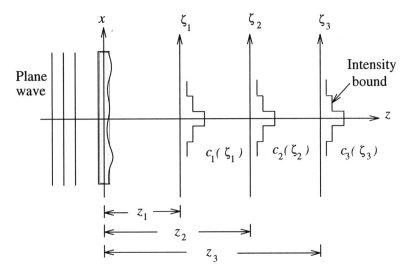

Fig. 7.7-6 Configuration for generating a non-diffracting beam.

Define

$$\check{Y}_i(\zeta_i) \triangleq \frac{1}{\sqrt{\lambda z_2}} \int_{-\infty}^{\infty} y(x) \exp\left\{ j\frac{\pi}{\lambda z_i}(x - \zeta_i)^2 \right\} dx \qquad (7.7\text{-}26)$$

$$= \mathcal{F}_r(y). \qquad (7.7\text{-}27)$$

Then a constraint set useful for this problem is

$$C_{z_i} = \left\{ y(x) \in L^2 : |\check{Y}_i(\zeta_i)| \le c_i(\zeta_i) \right\}, \qquad (7.7\text{-}28)$$

where $c_i(\zeta_i) \ge 0$ is a prescribed bound on the field magnitude and need not depend on i. For example, the one-dimensional equivalent of what is used in [36] is

$$c(\zeta_i) = 1 + 3 \operatorname{rect}\left(\frac{\zeta_i}{7}\right) + 3 \operatorname{rect}\left(\frac{\zeta_i}{3}\right) \qquad . \qquad (7.7\text{-}29)$$

We note in passing, that if $\check{F}(\zeta) \triangleq \mathcal{F}_r[f]$ and $\check{G}(\zeta) \triangleq \mathcal{F}_r[g]$, then

$$\|\check{F} - \check{G}\| = \|f - g\| \qquad (7.7\text{-}30)$$

and, hence, the minimization can be done in the Fresnel domain. The proof of Eq. (7.7-30) can be obtained in direct fashion by starting with the definition of $\mathcal{F}_r[\cdot]$. However, a simpler method is to recognize that, except for a constant factor,

$$\mathcal{F}_r[f] = \mathcal{F}[\tilde{f}] = \tilde{F}(\omega), \qquad (7.7\text{-}31)$$

where $\tilde{f}(x,z) = f(x)\exp\left\{j\frac{\pi}{\lambda z}x^2\right\}$. Then

$$\|f - g\| = \|\tilde{f} - \tilde{g}\| \tag{7.7-32}$$
$$= \|\tilde{F} - \tilde{G}\| \tag{7.7-33}$$
$$= \|\check{F} - \check{G}\|. \tag{7.7-34}$$

Line 1 follows from the argument advanced in Example 1; line 2 follows from Parseval's theorem; and line 3 merely echoes the fact that the Fourier transform of any $\tilde{f}(x,z)$ is the Fresnel transform of $f(x)$. Thus Parseval's theorem applies to Fresnel transforms.

The projection of an arbitrary $h(x)$, with Fresnel transform $\check{H}(\zeta_i)$, onto C_{z_i} of Eq. (7.7-28) is easily computed to be

$$y^*(x) = P_{z_i}h(x) \overset{\mathcal{F}_z}{\leftrightarrow} \check{Y}^*(\zeta_i), \tag{7.7-35}$$

where $\check{Y}^*(\zeta_i)$ is given by

$$\check{Y}^*(\zeta_i) = \min\left\{|\check{H}(\zeta_i)|, c_i(\zeta_i\right\} exp[j\theta_{\check{H}}(\zeta_i)], \tag{7.7-36}$$

and where $\theta_{\check{H}}(\zeta_i)$ is the phase of $\check{H}(\zeta_i)$. Thus the "projection", i.e., correction to fit the constraint, is done in the Fresnel plane, which is permissible because of Parseval's theorem.

Actually, in [36], a slightly modified constraint set from that of Eq. (7.7-28) is used. The one-dimensional equivalent of what is used there is

$$C_{z_i} = \left\{y(x) \in L^2 : |\check{Y}_i(\zeta_i| \le c_i \text{ for } \zeta_i \ne 0 \text{ and } \check{Y}_i(0) = M\right\}, \tag{7.7-37}$$

where M is a prescribed constant. In L^2 space, constraints on isolated samples carry no weight unless the underlying function is additionally restricted, e.g., band-limited. However, when the projection onto C_{z_i} is realized on a computer, there is an implicit switch to the discrete, i.e., vector case, and the constraint at $\zeta_i = 0$ carries weight. Hence, while the authors used the projection

$$y^*(x) \overset{\mathcal{F}}{\leftrightarrow} Y^*(\zeta_i) = \begin{cases} M & \text{if } \zeta_i = 0 \\ \min\{|H(\zeta_i)|, c_i(\zeta)\} exp[j\theta_H(\zeta_i)] & \text{otherwise,} \end{cases} \tag{7.7-38}$$

they managed to generate a good approximation of a non-diffracting beam over a designated region of four meters (Fig. 7.7-7).

■

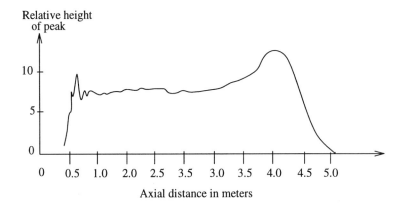

Relative height
of peak

0   0.5   1.0   2.0   2.5   3.5   3.0   3.5   4.0   4.5   5.0

Axial distance in meters

Fig. 7.7-7 Results obtained by Piestun and Shamir using POCS. Maximum beam height remains relatively constant for nearly four meters. Over this distance beam divergence remains small.

7.8 SUMMARY

In this chapter we discussed some of the many applications of projection methods to significant problems in optics. Even before the rigorous theory of convex and generalized projections percolated to the optical engineering and scientific communities, researchers were already using projection-type methods for solving important problems in optics. In particular, the super-resolution problem, as tackled by Gerchberg [3] and the phase-retrieval problem, as attacked by Gerchberg and Saxton [5] were early examples of projection methods, before these were fully understood.

This being an *applications* chapter, we furnished a number of examples where projection methods could fruitfully be applied to optics problems. These included super-resolution, phase recovery, beam forming, color matching, blind deconvolution, and design of diffractive elements. Space limitations prevent us from describing other applications and, for that matter, going into great detail regarding the applications we do describe.

No doubt, as time goes on, projection methods will find new applications. For example, not long ago, researchers used projection methods [7] to measure ultrashort laser pulses by imposing constraints based on the polarization-gate geometry of their optical system and the experimentally obtained data. The authors write: "In experimental measurements, the generalized-projection algorithm achieves lower errors than previous algorithms."

REFERENCES

1. D. C. Youla, Generalized image restoration by the method of alternating projec-

tions, *IEEE Trans. Circuits Syst.* **CAS-25**(9):694–702, 1978.

2. D. C. Youla and H. Webb, Image restoration by the method of convex projections: Part 1–theory, *IEEE Trans. Med. Imaging* **MI-1**:81–94, Oct. 1982.

3. R. W. Gerchberg, Super-resolution through error energy reduction, *Optica Acta*, 2(1):709–720, 1974.

4. A. Papoulis, A new algorithm in spectral analysis and bandlimited extrapolation, *IEEE Trans. Circuits Syst.* **CAS-22**:735–742, 1975.

5. R. W. Gerchberg and W. O. Saxton, A practical algorithm for the determination of phase from image and diffraction plane pictures, *Optik*, **35**:237–246, 1972.

6. R. J. Marks II, Coherent optical extrapolation of two-dimensional signals: Processor theory, *Applied Optics* **19**:1670–1672, 1980.

7. K. W. Delong, D. N. Fittinghoff, and R. Trebino, Pulse retrieval in frequency-resolved optical gating using the method of generalized projections. *Opt. Lett.*, **19**:2152, Dec. 15, 1994.

8. J. W. Goodman, *Introduction to Fourier Optics* (See Chapter 6), McGraw-Hill, San Francisco, 1968.

9. A. Lent and H. K. Tuy, An interactive method for the extrapolation of band-limited functions, *J. Math Anal. Applic.*, **83**:553–565, 1981.

10. M. I. Sezan and H. Stark, Image restoration by the method of convex projections: Part 2–applications and numerical results, *IEEE Trans. Med. Imaging*, **MI-1**:95–101, 1982.

11. G. N. Ramachandran and R. Srinivasan, *Fourier Methods in Crystallography*, Wiley-Interscience, New York, 1970.

12. W. O. Saxton, *Computer Techniques for Image Processing Electron Microscopy*, Academic Press, New York, 1978.

13. H. A. Ferwerda, The phase reconstruction problem for wave amplitudes and coherence functions, in *Inverse Source Problems in Optics* (Chapter 2), (H. P. Baltes, ed.), Springer-Verlag, Berlin, 1978.

14. L. S. Taylor, The phase retrieval problem, *IEEE Trans. Ant. and Prop.* **AP-29**(2):386–391, 1981.

15. J. C. Dainty and J. R. Fienup, Phase retrieval and image reconstruction for astronomy, chapter 7 in *Image Recovery: Theory and Application*, (H. Stark, ed.), Academic Press, Orlando, FL, 1987.

16. M. H. Hayes, The reconstruction of a multidimensional sequence from the phase or magnitude of its Fourier transform, *IEEE Trans. Acoust., Speech, Sig. Proc.* **ASSP-30**:140–154, 1982.

17. Y. M. Bruck and L. G. Sodin, On the ambiguity of the image reconstruction problem, *Opt. Commun.* **30**:304–308, 1979.

18. H. A. Ferwerda, The phase problem in object reconstruction and interferometry, *Technical Digest of the Opt. Soc. Am.* **THAI-1**, 1983.

19. R. H. T. Bates, Fourier phase problems are uniquely solvable in more than one dimension. I: Underlying theory, *Optik* **61**:247–262, 1982.

20. J. R. Fienup, Reconstruction of an object from the modulus of its Fourier transform. *Opt. Letters* **3**:27–29, 1978.

21. M. H. Hayes, The unique reconstruction of multi-dimensional sequences from Fourier transform magnitude or phase, chapter 6 in *Image Recovery: Theory and Application*, (H. Stark, ed.), Academic Press, Orlando, FL, 1987.

22. A. Levi, *Image Restoration by the Method of Projections with Applications to the Phase and Magnitude Retrieval Problems,* Ph.D Thesis, Rensselaer Polytechnic Institute, Dept. of ECSE, Troy, NY, Dec. 1983.

23. A. Levi and H. Stark, Image restoration by the method of generalized projections with applications to restoration from magnitude, *J. Opt. Soc. Am. A* **1**:932–943, 1984.

24. J. R. Fienup, Phase retrieval algorithms: a comparison, *Appl. Opt.*, **21**:2758–2769, 1982.

25. A. Levi and H. Stark, Restoration from phase and magnitude by generalized projections, Chapter 8 in *Image Recovery: Theory and Application*, (H. Stark, ed.), Academic Press, Orlando, FL, 1987.

26. D. L. Flannery and J. L. Horner, Fourier optical signal processors, *Proc. IEEE*, **77**:1511–1527, 1989.

27. C. Warde and A. D. Fisher, Spatial light modulators: Applications and functional capabilities, in *Optical Signal Processing* (J. Horner, ed.), Academic Press, San Diego, 1988, pp.478–523.

28. H. Stark, W. C. Catino, and J. L. LoCicero, Design of phase gratings by generalized projections, *J. Opt. Soc. Am. A,* **8**:566–571, 1991.

29. W. C. Catino, J. L. LoCicero, and H. Stark, Design of continuous and quantized phase holograms by generalized projections, *J. Opt. Soc. Am. A,* **14**:2715–2725, 1997.

30. Y. Yang and H. Stark, Solutions of several color-matching problems using projection theory, *J. Opt. Soc. Am. A* **11**:1–8, 1994.

31. H. J. Trussell, Applications of set theoretic methods to color systems, *Color Res. Appl.*, **16**:31–41, 1991.

32. G. Ayers and J. C. Dainty, Iterative blind deconvolution method and its applications, *Opt. Lett.* **13**:547–549, 1988.

33. R. G. Lane, Blind deconvolution of speckle images, *J. Opt. Soc. Am.* A, **9**:1508–1514, Sept. 1992.

34. S. L. S. Jacoby, J. S. Kowalik, and J. T. Pizzo, *Iterative Methods for Nonlinear Optimization Problems*, Prentice-Hall, Englewood Cliffs, NJ, 1972.

35. Y. Yang, N. P. Galatsanos, and H. Stark, A projection-based approach to the blind deconvolution problem, *J. Opt. Soc. Am.* A, **11**:89–96, 1994.

36. R. Piestun and J. Shamir, Control of wave-front propagation with diffractive elements, *Opt. Lett.*, **19**:771–773, June 1, 1994.

37. R. Piestun, B. Spektor, and J. Shamir, Diffractive optics for unconventional light distribution, in SPIE Vol. 2404: *Optoelectronic and Micro-optical devices*, SPIE Publishers, San Jose, CA, 1995.

EXERCISES

7-1. Derive Eqs. (7.2-1) and (7.2-3). *Hints:* In Fig. 7.2-1 assume that a ray emerging from the (x_o, y_o) plane that goes through the *front* focal point emerges as a paraxial ray after passage through the lens. Also, a paraxial ray entering the lens emerges as a skewed ray passing through the *rear* focal point i.e., the focal point on the right of the lens. Finally, any ray passing through the center of the lens emerges undeflected, i.e., as if the lens wasn't there.

7-2. Show that Eq. (7.2-7) follows from Eq. (7.2-4).

7-3. Verify that the projection onto C_1 of Eq. (7.2-8) is given by Eq. (7.2-11).

7-4. Show that $f(x,y), -f(x,y), f(x-x_o, y-y_o)$, and $f(-x,-y)$ all have the same Fourier magnitude.

7-5. Show that C_1 in Eq. (7.3-1) is convex but that C_2 in Eq. (7.3-2) is not.

7-6. Let $g(x) = 4\exp[-x(1+j)]\, u(x)$ and let $|M(u)| = \exp[-|u|]$. Compute the projection of $g(x)$ onto C_1 of Eq. (7.4-4) and C_2 of Eq. (7.4-5) if Ω is the region $1 \le x \le 2$.

7-7. In the color matching problem in Section 7.5 we are given $S_i(\lambda) = \exp[-0.5(\lambda-5)^2]$ for $0 \le \lambda \le 10$ and 0 otherwise. Also, we are given that $\alpha_i = 6$ and $K'(\lambda) = \lambda$ for $0 \le \lambda \le 10$ and $K'(\lambda) = 10\ \exp[-(\lambda-10)^2]$ for $\lambda > 10$; compute the projection onto C_i of Eq. (7.5-6).

7-8. We are given four sensors with the following spectral sensitivity functions:

$$S_1(\lambda) \;=\; \exp[-4(\lambda-1)^2]\, I_\Lambda$$

$$S_2(\lambda) = S_1(\lambda - 1)$$
$$S_3(\lambda) = S_2(\lambda - 1)$$
$$S_4(\lambda) = \begin{cases} \lambda - 1.5, & \text{if } 1.5 \le \lambda \le 2.5 \\ 3.5 - \lambda, & \text{if } 2.5 \le \lambda \le 3.5 \\ 0, & \text{elsewhere.} \end{cases}$$
$$\Lambda = \{\lambda : \lambda \in [0, 3.5]\}$$
$$I_\Lambda = \begin{cases} 1, & \text{if } \lambda \in \Lambda \\ 0, & \text{if } \lambda \notin \Lambda . \end{cases}$$

Also:

$$K(\lambda) = \begin{cases} 1, & \text{if } 0 \le \lambda \le 1 \\ \exp(1 - \lambda) + 0.5[1 + \cos[\pi(\lambda - 2)]], & \text{if } 1 \le \lambda \le 3 \\ \exp(1 - \lambda), & \text{if } 3 \le \lambda \le 3.5 \\ 0, & \text{elsewhere.} \end{cases}$$

a) Show that the response vector to $K(\lambda)$ is

$$\alpha^* = (0.884552, 1.02886, 0.225009, 0.730532)^T.$$

b) Assume that we have a color source given by $K'(\lambda) = \exp[-0.5(\lambda - 1.5)^2]\, I_\Lambda$. Show that the response to this source is

$$\alpha = (0.747151, 0.747677, 0.300817, 0.603613)^T.$$

c) Write and implement a POCS-based program to match $K(\lambda)$.
Hint: See Fig. 7.5-2, which shows a POCS correction of $K'(\lambda)$ to match $K(\lambda)$.

7-9. Assume that we have available six color sources as basis colors. Their spectral densities are defined as

$$R_j(\lambda) = tri(\lambda - j/2), \quad j = 1, \ldots, 6,$$

where

$$tri(x) = \begin{cases} 2x + 1, & \text{if } -0.5 \le x \le 0 \\ 1 - 2x, & \text{if } 0 \le x \le 0.5 \\ 0, & \text{otherwise.} \end{cases}$$

Find the vector β^* such that $K^*(\lambda) = \sum_j \beta_j^* R_j(\lambda)$ will match the spectrum $K(\lambda)$ of Exercise 7-8.

7-10. For the set C_g in Eq. (7.6-4), show that a reasonable *inner product* is
$\langle (u_1, v_1)(u_2, v_2) \rangle = \langle u_1, u_2 \rangle + \langle v_1, v_2 \rangle$.

7-11. In the blind deconvolution problem in Section 7.6, the gradient-projection algorithm involves a projection at *every* cycle. Show that if the projector is applied only *once*, say at the end of the cycle, there is the possibility of failing to find the true minimum. *Hint*: Consider finding the minimum of the function $f(x,y) = 0.5(x^2 + y^2)$ over the line set $x + y = 1$.

7-12. Show that Eq. (7.6-15) is a contraction mapping for $0 < \alpha_k < 1/\lambda_k$ where λ_k is the largest eigenvalue of the matrix $(\mathbf{F_k}^T\mathbf{F_k})$.

7-13. Show that Eq. (7.7-3) can be rewritten as Eq. (7.7-7) using the definitions given in Eqs. (7.7-8) to (7.7-11).

7-14. Write a MATLAB program that uses projection methods to duplicate the results given in Fig. 7.7-4.

8

Applications to
Neural Nets

8.1 INTRODUCTION

Projection methods can fruitfully be applied to neural nets in a number of ways. For example, the dynamics of the Hopfield net can be modeled as a sequence of alternating projections which, in turn, is useful in explaining why convergence to one of the stored *library vectors* doesn't always occur. Also projection methods can improve the performance of certain types of nets when additional prior knowledge is available. Finally, projection methods can be used to derive learning rules for multi-layer *feed-forward nets*, which can increase the learning rate of such nets. In this chapter we discuss these topics after reviewing the operation of simple neural nets. The chapter is relatively self-contained, and little prior knowledge is required on the reader's part.

8.2 THE BRAIN AS A NEURAL NETWORK

According to the New York Times [1]:

> Neuroscientists are pretty sure that inside each of our heads are 100 billion to a trillion neurons, constantly forging new connections and unraveling old ones in response to signals from the senses. Detecting regularities amid the confusion, the brain connects neuron to neuron forming circuitry that somehow corresponds to patterns in the outside world. And then it finds patterns among the patterns.
>
> Once it has cobbled together one circuit representing your dog and another circuit re-senting you neighbor's dog, the brain can notice that they are similar and abstract the

concept "dog." Now this idea can be recorded by snapping together another constellation of neurons. And so on up the scale. The structures that stand for dog, cat, raccoon and bear can be abstracted into the concept "quadruped."

Order in the Cortex

If this is what we mean by thinking, then intelligence is a measure of how well we do it. But does it take a hundred neurons to represent a dog, or a thousand, or a million? No one knows. Even the mechanism by which a new cerebral connection is forged is still a matter of intense dispute.

Details aside, it seems perfectly plausible that some people have just the right wiring or the right balance of enzymes and other molecules to make connections more rapidly than others, or to detect more subtle regularities. But without understanding how brains record and create ideas, we cannot really say how much of the machinery is a genetic gift and how much is acquired through the hard labor of learning.

The biggest intelligence test of all is the brain's ability to understand itself. By that measure we are in remedial education.

Artificial network structures modeled after the human brain that emulate human recall or can learn by training are called *neural networks.*

A *neural network is a massively parallel distributed processor that has a natural propensity for storing experimental knowledge. It resembles the brain in two respects*:

1. *Knowledge is acquired by the network through a learning process.*

2. *Interneuron connection strength sometimes called synaptic weights are used to store the knowledge.* [Adapted from Haykin [2]]

There are many different types of neural networks. The two types we shall discuss in this chapter are the *Hopfield net* and the *two-layer feed-forward net.* All neural nets use neurons as a fundamental unit. A neuron is shown in Fig. 8.2-1.

In Fig. 8.2-1(a), the quantity θ_i is called the *threshold* or *bias*; it is frequently set to zero. In Fig. 8.2-1(b) the bias is shown as an interconnection weight operating on an input $x_0 = -1$. The action of the ith neuron can be described mathematically by

$$y_i = \eta(z_i), \tag{8.2-1}$$

$$z_i = \sum_{j=1}^{L} w_{ij}x_j - \theta_i \qquad i = 1, ..., N, \tag{8.2-2}$$

where L is the number of inputs to a neuron, N is the number of neurons, and $\eta(\cdot)$ is a non-linear zero-memory transmittance sometimes called the neuron *response function.* Frequently used forms for $\eta(\cdot)$ are

$$\eta(x) = \text{sgn}(x) = \begin{cases} 1 & x \geq 0 \\ -1 & x < 0 \end{cases} \tag{8.2-3}$$

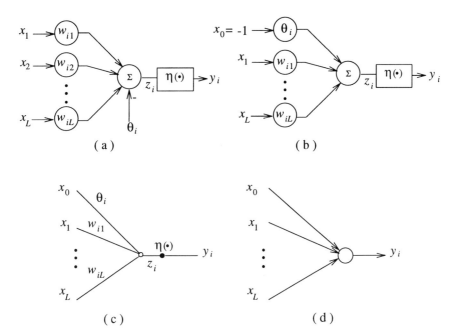

Fig. 8.2-1 Various representations of an (artificial) neuron: (a) the input $\{x_i\}$ are amplified, summed, and passed through a non-linear device; (b) homogeneous realization in which activation threshold is taken as part of the input; (c) signal-flow representation of a neuron in which links represent operations and nodes are summing points; (d) architectural representation of a neuron.

and

$$\eta(x) = \frac{1}{1 + e^{-\alpha x}} \ . \tag{8.2-4}$$

The non-linearity in Eq. (8.2-3) is often used in the Hopfield net; the one in Eq. (8.2-4) is often used in the feed-forward net. Several neuron response functions are shown in Fig. 8.2-2.

8.3 THE HOPFIELD NET

The Hopfield net functions as an *associative content-addressable memory* (ACAM). In an ACAM the memory is accessed not by a numerical address as, say, in a digital computer, but rather by a possibly degraded (e.g., partially specified or noisy) version of the stored memory pattern. [†]

[†] The pattern is usually a known numerical vector that contains encoded data, alphanumeric information, even pictures and other graphics. The word *associative* refers to the fact that the network *associates* an output vector with a user supplied input vector.

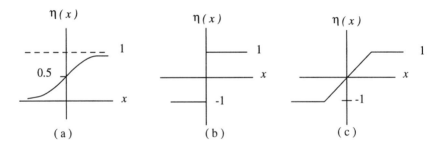

Fig. 8.2-2 Non-linear response functions commonly used in modeling neurons: (a) sigmoidal with offset of 0.5; (b) bi-level hard limiter; (c) saturated linear. Levels are normalized to peak values.

In his 1982 paper, Hopfield furnished the following vivid example of how an ideal ACAM might perform [3]:

> Suppose that an item stored in memory is "H.A. Kramers & G.H. Wannier *Physi Rev.* *60*, 252 (1941)." A general content-addressable memory would be capable of retrieving this entire memory item on the basis of sufficient partial information. The input "& Wannier (1941)" might suffice. An ideal memory could deal with errors and retrieve this reference even from the input "Wannier, (1941)."

In Fig. 8.3-1 we show an architectural graph of a Hopfield net for $L = 3$ neurons. Each open circle represents a neuron and incorporates both the linear (summing) and non-linear (two-sided limiting) operations. A functionally equivalent diagram is shown in Fig. 8.3-2. There the linear and non-linear operations are shown separately. The items stored in memory are sometimes called *library vectors*; we use this nomenclature below.

Principles of Operation of the Hopfield Net

In the following, we assume that the ACAM stores N, L-dimensional, library vectors: $\mathbf{x}_n \overset{\triangle}{=} (x_{n1}, x_{n2}, ..., x_{nL})^T, n = 1, 2, ..., N$. Thus, the network has L nodes $(L \geq N)$, whose states preferentially converge to the library vector to be recalled when they are initialized by a *degraded* (e.g., partially specified and/or noisy) version of the library vector. The library vectors can be binary, e.g., $x_{ni} = \pm 1$ (the bipolar case) or continuous-valued, e.g., $-1 \leq x_{ni} \leq 1$, $i = 1, 2, ..., L$.

In the discrete-time, binary Hopfield ACAM [3], the *state* of the network at time k is described by vector $\mathbf{v}^{(k)}$. The state component associated with the ith node, $v_i^{(k)}$, at time k, changes to its next value, $v_i^{(k+1)}$, at time $k + 1$, according to

$$v_i^{(k+1)} = \eta \left(\sum_{j=1}^{L} w_{ij} v_j^{(k)} \right), \quad i = 1, 2, ..., N, \quad (8.3\text{-}1)$$

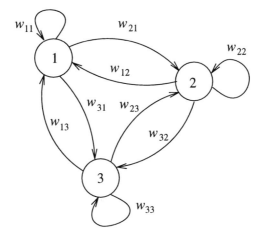

Fig. 8.3-1 A graph of a three-neuron Hopfield net. In the original configuration the autocorrections had weights $w_{ii} = 0$. Gindi et al. [5] showed that if $w_{ii} > 0$, the *energy decreasing property* of the Hopfield net was still valid.

where w_{ij} corresponds to the interconnection weight between the ith and the jth node and $\eta(\cdot)$ defines the node nonlinearity. (Note that the bias input to the ith node is assumed to be zero.) For bipolar vectors, the nonlinearity $\eta(\cdot)$ is defined by

$$\eta(z) = \text{sgn}(z) \overset{\triangle}{=} \begin{cases} 1 & z \geq 0 \\ -1 & z < 0 \end{cases} \tag{8.3-2}$$

for $z \in R$ (the real line).

The interconnection weights $\{w_{ij}, i, j = 1, 2, ..., L\}$ form the $L \times L$ symmetric interconnection matrix $\mathbf{W}_1 \overset{\triangle}{=} [w_{ij}]$, which is defined via the *outer-product learning rule* [3]:

$$w_{ij} \overset{\triangle}{=} \begin{cases} \sum_{n=1}^{N} x_{ni} x_{nj}, & i \neq j \\ 0, & i = j. \end{cases} \tag{8.3-3}$$

That is, $\mathbf{W}_1 = \mathbf{X}\mathbf{X}^{\mathbf{T}} - N\mathbf{I}$, where $\mathbf{X} \overset{\triangle}{=} [\mathbf{x}_1\mathbf{x}_2...\mathbf{x}_N]$ is the matrix whose columns are the library vectors and $\mathbf{I}$ denotes the $L \times L$ identity matrix. Note that in Hopfield's original net the *autoconnections* on the nodes are excluded and, hence, $w_{ii} = 0$. [The nonlinear block in Fig. 8.3-2 implements some nonlinearity such as the one given in Eq. (8.3-2).] *Hopfield* [3] showed that the *energy functional* $E(k)$ defined by

$$E(k) \overset{\triangle}{=} -\sum_{i=1}^{L}\sum_{j=1}^{L} w_{ij} v_i^{(k)} v_j^{(k)} \tag{8.3-4}$$

decreases monotonically upon the *update rule* given in Eq. (8.3-1) and that the network converges to a stable state corresponding to a local minimum of E. This is the famous *energy minimizing property* of the Hopfield network that can be

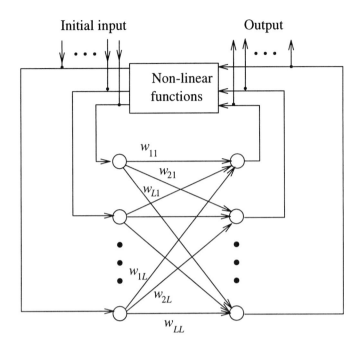

Fig. 8.3-2 Non-linear "filter" description of an L-neuron Hopfield net.

exploited when solving optimization problems. For example, see Yeh et al. [4].

Gindi et al. [5] considered a modified binary Hopfield network, in which nonzero autoconnections had value $w_{ii} = N$, and showed that the stable states of the network remain the same by allowing nonzero diagonal terms in Eq. (8.3-4). Further, the nonzero-diagonal network was shown to outperform the original network.

It should be emphasized that in both the zero-diagonal and the nonzero-diagonal cases, the library vectors do not necessarily form stable states. That is, the network states may not converge to themselves when they are initialized by a library vector [3,5–7]. An empirical result [3,8] is that convergence to the library vector occurs with high probability if $N < 0.15L$. It is pointed out in [8] that it is relatively easy to select a set of N library vectors that satisfy the condition $N < 0.15L$, and yet do not exhibit *self-convergence*. However, these vectors usually have many elements in common. The library vectors correspond to stable states of the network if the learning rule defined in Eq. (8.3-1) is modified, such that the interconnection matrix is given by [9]:

$$\mathbf{W}_2 \stackrel{\triangle}{=} \mathbf{X}(\mathbf{X}^T\mathbf{X})^{-1}\mathbf{X}^T. \tag{8.3-5}$$

This result will be obvious in the case of the generalized projections (GP) formulation discussed below.

The interconnection matrix $\mathbf{W}_2$ was used in [9–12] and it is identical to the interconnection matrix associated with the *orthogonalization learning rule* discussed in [9]. Two things to note here are: (i) in the case of linearly independent vectors,

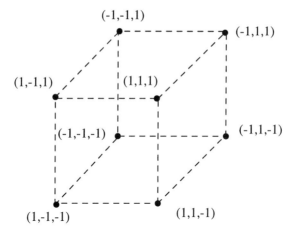

Fig. 8.3-3 The elements of the set C_{sgn} for $L = 3$.

the inverse $(\mathbf{X}^T \mathbf{X})^{-1}$ always exists; and (ii) in the case of orthonormal library vectors, $\mathbf{W}_2$ reduces to $\mathbf{W}_1 + N\mathbf{I}$, and therefore, library vectors do correspond to stable states if they are orthonormal.

Projection Formulation of the Binary Hopfield ACAM

As stated often throughout this book, the key to solving problems via projection methods is to define the appropriate sets. The key to defining the appropriate constraint sets is to be found in the dynamic behavior of the Hopfield net. The Hopfield dynamics are described by

$$\mathbf{v}^{(k+1)} = \text{sgn}(\mathbf{W}\mathbf{v}^{(k)}), \tag{8.3-6}$$

where $\mathbf{W} = \mathbf{W}^T$ is a general interconnection matrix, e.g. $\mathbf{W} = \mathbf{X}\mathbf{X}^T - N\mathbf{I} \triangleq \mathbf{W}_1$ if the outer-product learning rule is involved. The vector $\mathbf{v}^{(k)} \triangleq \left[v_1^{(k)}, v_2^{(k)}, \cdots, v_N^{(k)} \right]$ denotes the *state vector* and $\text{sgn}(\cdot)$ operates on each component of a vector as in Eq. (8.3-2). Recall that the columns of $\mathbf{X}$ are the binary vectors $\mathbf{x}_i, i = 1, 2, \cdots, N$.

We first note that all state vectors must have components $v_i^{(k)} = \pm 1, i = 1, ..., L$. Therefore, we seek a solution in the set (Fig. 8.3-3)

$$C_{\text{sgn}} \triangleq \left\{ \mathbf{y} \in R^L : y_i = \pm 1, \ i = 1, 2, ..., L \right\}. \tag{8.3-7}$$

We also want all library vectors, not necessarily orthogonal ones, to be stable states, i.e., if the input $\mathbf{v}^{(0)}$ at time zero is the library vector $\mathbf{x}_j$, e.g., $\mathbf{v}^{(0)} = \mathbf{x}_j$, then $\mathbf{v}^{(1)} = \mathbf{v}^{(2)} = ... = \mathbf{x}_j$, for $j = 1, 2, ..., N$. One way to *ensure* this property is to project upon the set

$$C_s \triangleq \left\{ \mathbf{y} \in R^L : \mathbf{y} = \sum_{n=1}^{N} \alpha_n \mathbf{x}_n, \quad \alpha_n \in R \right\}. \tag{8.3-8}$$

That is, C_s is the linear subspace spanned by all the N library vectors, i.e., $C_s = \text{span}\{\mathbf{x}_1, \mathbf{x}_2, \cdots, \mathbf{x}_N\}$. For example, the projection of any of the library vectors $\mathbf{x}_j$ upon C_s will yield

$$\mathbf{y}^* = P_s \mathbf{x}_j = \mathbf{x}_j. \tag{8.3-9}$$

How can we make this assertion without even having computed P_s? The answer is that $\mathbf{x}_j \in C_s$ since we can write that $\mathbf{x}_j = \sum_{n=1}^{N} \alpha_n \mathbf{x}_n$, with $\alpha_n = 0$ for all $n \neq j$, and $\alpha_j = 1$.

The projection onto C_{sgn} can be determined by inspection. Given an arbitrary vector $\mathbf{q} = (q_1 \ q_2 \ ... \ q_L)^T \in R^L$, consider its projection onto C_{sgn}. To minimize

$$\|\mathbf{q} - \mathbf{y}\|^2 \triangleq \sum_{i=1}^{L} (q_i - y_i)^2, \tag{8.3-10}$$

subject to $y_i = \pm 1$, we set

$$y_i^* = \begin{cases} 1 & q_i \geq 0 \\ -1 & q_i < 0 \end{cases} \tag{8.3-11}$$

or

$$\mathbf{y}^* = P_{\text{sgn}}\mathbf{q} = (\text{sgn}(q_1), \text{sgn}(q_2), ..., \text{sgn}(q_L))^T. \tag{8.3-12}$$

The computation of the projection upon C_s is slightly more involved. Given an arbitrary $\mathbf{q}$, we wish to minimize

$$\left\| \mathbf{q} - \sum_{n=1}^{N} \alpha_n \mathbf{x}_n \right\|^2, \tag{8.3-13}$$

which can be rewritten as

$$\min_{\boldsymbol{\alpha}} \|\mathbf{q} - \mathbf{X}\boldsymbol{\alpha}\|^2, \tag{8.3-14}$$

where $\boldsymbol{\alpha} = (\alpha_1, \alpha_2, ..., \alpha_N)^T$ and $\mathbf{X}$ is the $L \times N$ matrix $(L \geq N)$ of library vectors. But the solution to Eq. (8.3-14) is the *least-squares solution* of the equation $\mathbf{X}\boldsymbol{\alpha} = \mathbf{q}$ and is, therefore, given by [†]

$$\boldsymbol{\alpha}^* = (\mathbf{X}^T\mathbf{X})^{-1}\mathbf{X}^T\mathbf{q}. \tag{8.3-15}$$

[†]It is assumed that all the N library vectors are linearly independent. Therefore the matrix $\mathbf{X}^T\mathbf{X}$ is non-singular.

(We leave the details as an exercise for the reader.) Hence the vector $\mathbf{y}^*$ that minimizes $\|\mathbf{q} - \mathbf{X}\boldsymbol{\alpha}\|^2$ is

$$
\begin{aligned}
\mathbf{y}^* &= \mathbf{X}\boldsymbol{\alpha}^* \\
&= \mathbf{X}(\mathbf{X}^T\mathbf{X})^{-1}\mathbf{X}^T\mathbf{q} \qquad &(8.3\text{-}16) \\
&\stackrel{\triangle}{=} P_s\mathbf{q}. &(8.3\text{-}17)
\end{aligned}
$$

Suppose $\mathbf{q} = \mathbf{x}_1$, i.e., the first library vector. Then, since $\mathbf{x}_1 = \mathbf{X}\boldsymbol{\alpha}_1$ when $\boldsymbol{\alpha}_1 = (1, 0, ..., 0)^T$, we obtain

$$
\begin{aligned}
\mathbf{y}^* &= \mathbf{X}(\mathbf{X}^T\mathbf{X})^{-1}\mathbf{X}^T\mathbf{X}\boldsymbol{\alpha}_1 \\
&= \mathbf{x}_1. &(8.3\text{-}18)
\end{aligned}
$$

Furthermore, since $\mathbf{x}_1 \in C_{\text{sgn}}$, we have $P_{\text{sgn}}P_s\mathbf{x}_1 = P_{\text{sgn}}\mathbf{x}_1 = \mathbf{x}_1$, etc., for $\mathbf{x}_2, \mathbf{x}_3, ..., \mathbf{x}_N$. Thus, library vectors are fixed points of the individual operators P_{sgn}, P_s and their composition $P_{\text{sgn}}P_s$. Consider now the algorithm

$$
\begin{aligned}
\mathbf{v}^{(k+1)} &= P_{\text{sgn}}P_s\mathbf{v}^{(k)} &(8.3\text{-}19) \\
&= \text{sgn}(\mathbf{X}(\mathbf{X}^T\mathbf{X})^{-1}\mathbf{X}^T\mathbf{v}^{(k)}). &(8.3\text{-}20)
\end{aligned}
$$

But $\mathbf{W}_2 \stackrel{\triangle}{=} \mathbf{X}(\mathbf{X}^T\mathbf{X})^{-1}\mathbf{X}^T$ is the (superior) orthogonalization learning rule that ensures that library vectors are stable states. Hence the general projection algorithm given in Eq. (8.3-19) can be written as

$$
\mathbf{v}^{(k+1)} = \text{sgn}(\mathbf{W}_2\mathbf{v}^{(k)}), \qquad (8.3\text{-}21)
$$

which is *identical* with the Hopfield dynamics given in Eq. (8.3-6), with the interconnection matrix $\mathbf{W}$ replaced by $\mathbf{W}_2$ of Eq. (8.3-5). Thus the operation of the Hopfield net can be described as two successive projections: one onto convex set, another onto a non-convex set. Furthermore, by having selected the set C_s in a reasonable manner, we have automatically derived the orthogonalization learning rule, which allows library vectors to be stable states of the Hopfield net.

Projections and Net Dynamics

In addition to obtaining the desirable $\mathbf{W}_2$ interconnections matrix, the projection approach yields interesting clues about the behavior of the Hopfield net. Since the set C_{sgn} is not convex, Levi's generalized projection theorem applies.[†] Recall that Levi [13] showed that, for a problem involving two constraint sets, if at least one of the sets is non-convex, then convergence is not guaranteed, but the *set-distance error reduction* property still holds. For the case being considered, i.e.,

---

[†] This theorem is discussed in Chapter 5.

$\mathbf{v}^{(k+1)} = P_{\text{sgn}} P_s \mathbf{v}^{(k)}$, the SDE reduction property states that

$$Z(\mathbf{v}^{(k+1)}) \leq Z(P_s \mathbf{v}^{(k)}) \leq Z(\mathbf{v}^{(k)}),$$

where

$$Z(\mathbf{y}) \triangleq \|P_s \mathbf{y} - \mathbf{y}\| + \|P_{\text{sgn}} \mathbf{y} - \mathbf{y}\|. \tag{8.3-22}$$

If we think of a stable state, say $\hat{\mathbf{v}}$, as a point where no further change is observed in the network, then the fact that C_{sgn} is not convex, allows for the existence of undesirable (stable) *trap-states*.

For a trap-state, $Z(\hat{\mathbf{v}}) > 0$ (Fig. 5.1-1). Must trap-states always exist when one or more sets are non-convex? Not necessarily, but at least one is alerted about their possible existence. Knowing about these states can reduce the time spent looking for a "bug" in the program.

The projection formulation also implies the existence of non-trap states that are *not library vectors*. These states are characterized by $Z(\hat{\mathbf{v}}) = 0$, so that $\hat{\mathbf{v}} \in C_0 \triangleq C_{\text{sgn}} \cap C_s$, but $\hat{\mathbf{v}} \neq \mathbf{x}_i$ $i = 1, ..., N$. These stable states are *linear combinations of library vectors*, as predicted by the set C_s, i.e., C_s contains not only library vectors but also linear combinations of library vectors. Such spurious-stable states have been observed in practice.

A typical scenario is illustrated in Fig. 8.3-4 in the case of $N = 2, L = 3$. The set C_{sgn} is the set of vertices of the cube and the set C_s is a two-dimensional subspace (a plane) in R^3. Thus, $C_0 = C_{\text{sgn}} \cap C_s$ is the set of vertices that lie on the plane. Observe that only two out of the four vertices in C_0 are library vectors, i.e., $\mathbf{x}_1$ and $\mathbf{x}_2$. In Fig. 8.3-4, two case are illustrated:

(i) starting with $\mathbf{v}_0$, we reach the stable state vector $\mathbf{v}_2$ which, corresponds to the library vector $\mathbf{x}_1$; and

(ii) starting with $\hat{\mathbf{v}}_0$, we reach the stable state vector $\hat{\mathbf{v}}_2$, which is not a library vector ('no match' condition). The best strategy to handle the 'no-match' condition is a future research topic. A possible strategy, which may not be the best, is to compute the Hamming distance between the non-matching stable state and the library vectors for all patterns and select the pattern with the minimum Hamming distance. This strategy can be implemented using the MAXNET network discussed in [8,9].

8.4 POCS AND THE CONTINUOUS HOPFIELD ACAM

A direct extension of the binary Hopfield ACAM is when the library vectors are *continuously valued*. We shall find the projection formulation in this case is particularly useful, since it predicts the non-existence of trap states. We assume that the library vector components satisfy $-1 \leq x_{ni} \leq 1$, $n = 1, 2, ..., N$; $i = 1, 2, ..., L$, where N is the number of library vectors and L is the number of components per vector.

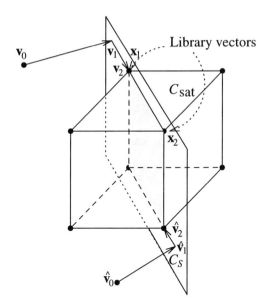

Fig. 8.3-4 Convergence to library vectors depends on starting point. If $\mathbf{v}_0$ is the starting point, then convergence occurs when $\mathbf{v} = \mathbf{v}_2 = \mathbf{x}_1$, which is a library vector. When $\hat{\mathbf{v}}_0$ is the starting point, then convergence occurs when $\hat{\mathbf{v}} = \hat{\mathbf{v}}_2$ which is not a library vector.

A direct continuous-valued extension of the network described in Eqs. (8.3-6) and (8.3-5) can be defined by

$$\mathbf{v}^{(k+1)} = \text{sat}(\mathbf{W}_2\mathbf{v}^{(k)}), \qquad (8.4\text{-}1)$$

and

$$\mathbf{W}_2 = \mathbf{X}(\mathbf{X}^T\mathbf{X})^{-1}\mathbf{X}^T, \qquad (8.4\text{-}2)$$

where sat$(\cdot)$ is the unity-slope saturation nonlinearity, whose action on the vector $\mathbf{y}^{(k)} \triangleq \mathbf{W}_2\mathbf{v}^{(k)}$ is defined by

$$\text{sat}(\mathbf{y}^{(k)}) \triangleq \left[\text{sat}(y_1^{(k)}),\ \text{sat}(y_2^{(k)}),\ \cdots,\ \text{sat}(y_L^{(k)})\right]^T \qquad (8.4\text{-}3)$$

and

$$\text{sat}(y_i^{(k)}) \triangleq \begin{cases} -1 & y_i^{(k)} < -1 \\ y_i^{(k)} & -1 \leq y_i^{(k)} \leq 1 \\ 1 & y_i^{(k)} > 1. \end{cases} \qquad (8.4\text{-}4)$$

The sat$(\cdot)$ nonlinearity is shown in Fig. 8.2-2 (c).

We consider the following sets:

$$C_{\text{sat}} \triangleq \left\{\mathbf{y} \in R^L : -1 \leq y_i \leq 1,\ \ i = 1, 2, ...L\right\}, \qquad (8.4\text{-}5)$$

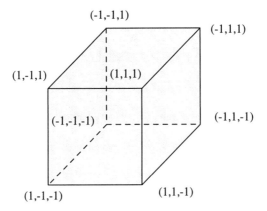

Fig. 8.4-1   The set C_{sat} for $L = 3$ is a cube.

$$C_s \triangleq \left\{ \mathbf{y} \in R^L : \mathbf{y} = \sum_{n=1}^{N} \alpha_n \mathbf{x}_n, \alpha_n \in R \right\},$$ (8.4-6)

where $\mathbf{x}_n$, $n = 1, 2, \cdots, N$, are the library vectors. The set C_{sat} is a closed convex set and defines an L-dimensional hypercube (Fig. 8.4-1).

The projection of an arbitrary $\mathbf{q} \in R^L$ is given by [14]

$$P_{\text{sat}} \mathbf{q} = \text{sat}(\mathbf{q}).$$ (8.4-7)

(See Exercise 8-4.) The set C_s is the same set defined in the binary-valued case, because the learning rule is the same in both the binary and continuous-valued cases. Thus, from Eqs. (8.3-16) and (8.4-7), the state update rule given in Eq. (8.3-1) can be expressed as

$$\mathbf{v}^{(k+1)} = P_{\text{sat}} P_s \mathbf{v}^{(k)}, \qquad k = 0, 1, \ldots ,$$ (8.4-8)

which is indeed in the form of a POCS algorithm.

Therefore, for an arbitrary initialization, the network state converges to a stable state vector in $C_0 \triangleq C_{\text{sat}} \cap C_s$. That is, the set of stable states corresponds to the set of points common to both the hypercube and the subspace spanned by the library vectors. The set C_0 is illustrated in Fig. 8.4-2 for the case of $N = 2$ and $L = 3$. As in the binary case, the network state may converge to a vector in C_0, which is not a library vector. In that case the closest library vector (in the R^L-norm sense) might be taken as the solution. Here too, the best strategy to handle the "no-match" condition is a further research topic. Finally, note that *trap-states cannot occur in the continuous-valued Hopfield ACAM, since the sets are convex.*

The convergence properties of the continuous-valued Hopfield ACAM are similar to Hopfield's continuous-valued, continuous-time model, for which he showed that the energy-minimizing stable states lie within the L-dimensional hypercube in the

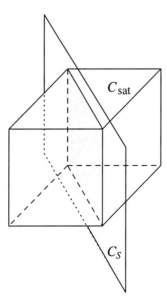

Fig. 8.4-2 The solution set (shaded region) for the continuous Hopfield ACAM.

case of sigmoidal nonlinearity [6].

8.5 A HOPFIELD-NET BASED CLASSIFIER

A Hopfield ACAM can be modified to function as a classifier[†] by following the
Hopfield ACAM by a *perceptron* (Fig. 8.5-1). When used as an ACAM, the
Hopfield network takes the degraded input at time $k = 0$ and furnishes, ideally,
upon convergence, one of the stored library vectors. For the Hopfield model to work
as a classifier it is necessary to add a mechanism that determines which of the, say,
K classes the output belongs to. This has to be done because the Hopfield network
by itself is less a classifier than a memory. The classification itself can be done
by following the Hopfield network with a perceptron network. A Hopfield neural
network classifier for K classes is shown in Fig. 8.5-2; there the library vectors are
shown to be L-dimensional vectors. Library vectors and, for that matter, any other
vectors used to train a pattern classifie, are sometimes called *exemplars.*

   The basic perceptron element is a single neuron with a prescribed non-linearity
(Fig. 8.5-1). The design of a perceptron[‡] depends on several factors, including the
choice of cost functions to be optimized. In Fig. 8.5-2 the input to the perceptron is
the vector $\mathbf{x} = [x_1, x_2, ..., x_L]^T$ and the output is the vector $\mathbf{y} = [y_1, y_2, ..., y_K]^T$.

[†]A classifier is a machine that classifies patterns according to which class or set they belong to.
[‡]Also called a *single layer perceptron.*

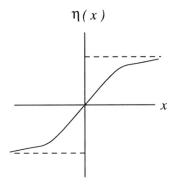

Fig. 8.4-3   Sigmoidal nonlinearity in Hopfield content-addressable memory.

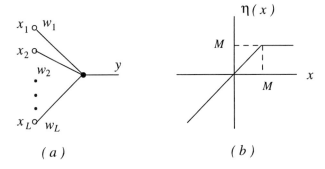

(a)   (b)

Fig. 8.5-1   (a) Signal flow diagram of a single perceptron element. A perceptron element is essentially a neuron with a prescribed nonlinearity; (b) perceptron-element nonlinearity.

Typically the perceptron action is described by

$$y_i = \eta(c_i + \sum_j w_{ij}x_j), \qquad i = 1, 2, ...K, \tag{8.5-1}$$

where $\eta(z)$ is some nonlinear function, e.g., $\eta(z) = z \cdot u(z)$, $u(z)$ being the unit step function, c_i is the bias associated with the ith output neuron, and w_{ij} are real-valued weights associated with the links. For example, suppose the components of $\mathbf{x}'$ are ± 1's and the cost function to be minimized is the Hamming distance. Assume that we use the above-mentioned $\eta(z)$. If it is known *a priori* that the variability in the $\mathbf{x}$ vectors is caused by noise in the binary symmetric channel, then with

$$c_i = \Delta - L/2 \tag{8.5-2}$$

and

$$w_{ij} = x_j^i/2, \tag{8.5-3}$$

where x_j^i is the jth component of an exemplar of class i, only output nodes corre-

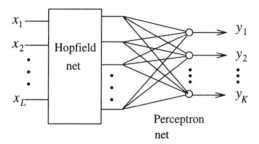

Fig. 8.5-2 A Hopfield net followed by a perceptron for pattern classification.

sponding to prototypes with a Hamming distance less than Δ from the input will have positive outputs [9]. Moreover, the best match will yield the largest output.

If the mechanism for intra-class variability is not known, then the perceptron must learn the correct weights from training samples. The algorithm for teaching the perceptron what the correct weights are depends on the cost function used, whether the training samples form separable classes, and the means used to extremize the cost function. In what follows, we discuss a *perceptron learning algorithm* (PLA) for the linearly separable, two-class case.

Suppose that we have a set of N training samples, $\mathbf{x}_n$, $n = 1, 2, ..., N$, represented by L-dimensional vectors in the Euclidean space R^L equipped with the usual inner product and norm. We assume that N_1 samples belong to class, say A, and N_2 samples belong to class, say B, such that $N = N_1 + N_2$. A PLA determines the weight vector $\mathbf{w}$, which defines a linear discriminant functional g as[†].

$$g(\mathbf{y}) \stackrel{\triangle}{=} \langle \mathbf{w}, \mathbf{y} \rangle, \qquad (8.5\text{-}4)$$

such that a sample $\mathbf{x}_n$ is classified correctly if either of the two conditions is satisfied:

$$\langle \mathbf{w}, \mathbf{x}_n \rangle > 0 \qquad \text{and} \qquad \mathbf{x}_n \in A,$$

$$\langle \mathbf{w}, \mathbf{x}_n \rangle < 0 \qquad \text{and} \qquad \mathbf{x}_n \in B.$$

Clearly, if the samples that belong to B are *replaced by their negatives*, the weight vector satisfies

$$\langle \mathbf{w}, \mathbf{x}_n \rangle > 0, \quad n = 1, 2, ..., N, \quad \text{i.e., for } all \ \mathbf{x}_n \text{ either from } A \text{ or } B. \qquad (8.5\text{-}5)$$

If such a weight vector exists, the classes A and B are said to be *linearly separable*. The equation $\langle \mathbf{w}, \mathbf{x} \rangle = 0, \mathbf{x} \in R^L$ defines a separating hyperplane in R^L.

The separating weight vector is, in general, not unique. The set of admissible weight vectors can be constrained by introducing a *margin* M. In that case,

---

[†]The inner product of two real vectors $\langle \mathbf{w}, \mathbf{y} \rangle$ is computed as $\mathbf{w}^T \mathbf{y} = \mathbf{y}^T \mathbf{w}$.

Eq. (8.5-5) is replaced by

$$\langle \mathbf{w}, \mathbf{x}_n \rangle > M, \quad n = 1, 2, ..., N. \tag{8.5-6}$$

If a weight vector satisfying Eq. (8.5-6) exists, the classes are said to be linearly separable with a margin M. Here too, the separating vector is not unique, but any solution satisfying the margin M is clearly preferred to one that offers none.

A Classical Perceptron Learning Algorithm (PLA)

A well-known perceptron learning algorithm (PLA) for the two-class, linearly separable problem has been derived in [14] as a result of solving the following minimization problem:

$$\min_{\mathbf{w}} \left(\sum_{n \in \mathcal{N}} \frac{(\langle \mathbf{w}, \mathbf{x}_n \rangle - M)^2}{\|\mathbf{x}_n\|^2} \right), \tag{8.5-7}$$

where $\mathcal{N}$ is the index set corresponding to the indices of misclassified sample, i.e., $\langle \mathbf{w}, \mathbf{x}_n \rangle < M$ for $n \in \mathcal{N}$. A basic iterative descent procedure on Eq. (8.5-7) yields

$$\mathbf{w}^{(k+1)} = \mathbf{w}^{(k)} + \lambda_{l(k)} \frac{M - v_{l(k)}}{\|\mathbf{x}_{l(k)}\|^2} \mathbf{x}_{l(k)}, \tag{8.5-8}$$

where k is the iteration index, i.e., $k = 0, 1, 2, \cdots$, and

$$v_{l(k)} \overset{\triangle}{=} \eta(\langle \mathbf{w}, \mathbf{x}_{l(k)} \rangle), \tag{8.5-9}$$

$$l(k) \overset{\triangle}{=} (k \bmod N) + 1 \tag{8.5-10}$$

and η is a nonlinear function defined by (Fig. 8.5-1(b))

$$\eta(z) \overset{\triangle}{=} \begin{cases} M, & z \geq M \\ z, & z < M. \end{cases} \tag{8.5-11}$$

In Eq. (8.5-10) $k \bmod N$ is read as k modulo N and denotes the remainder obtained on division of k by N and is one of the integers $0, 1, ..., N - 1$. The parameter $\lambda_{l(k)}$ is the relaxation parameter.

Thus, learning is achieved by updating the weight vector according to Eq. (8.5-8) whenever a sample is misclassified; note that for a properly classified sample $\mathbf{w}^{(k+1)} = \mathbf{w}^{(k)}$. If, after a number of iterations, the value of the weight vector, say $\hat{\mathbf{w}}$, is unaffected by subsequent samples, then the PLA is said to converge to $\hat{\mathbf{w}}$. In the case of separable classes, the iteration given in Eq. (8.5-8) is convergent, provided that $\lambda_{l(k)} \in (0, 2)$ [15]. The connection weight vector $\mathbf{w} = [w_1, w_2, \cdots, w_L]^T$ can be determined using Eq. (8.5-8).

A Projection-Based Learning Algorithm

A projection-based formulation can be obtained by defining the family of constraint sets C_n, $n = 1, 2, \cdots, N$, as:

$$C_n \overset{\triangle}{=} \left\{ \mathbf{y} \in R^L : \langle \mathbf{y}, \mathbf{x}_n \rangle \geq M \right\}, \tag{8.5-12}$$

where $\mathbf{x}_n$ is the nth training sample and M is the predetermined margin. The set C_n is closed and convex and the corresponding projection of an arbitrary vector $\mathbf{q}$ is computed as [15]

$$P_n \mathbf{q} \overset{\triangle}{=} \mathbf{q} + \frac{M - v_n}{\|\mathbf{x}_n\|^2} \mathbf{x}_n, \tag{8.5-13}$$

where

$$v_n \overset{\triangle}{=} \eta(\langle \mathbf{q}, \mathbf{x}_n \rangle), \tag{8.5-14}$$

$\eta(\cdot)$ is as in Eq. (8.5-11), and $\mathbf{q}$ is an arbitrary vector in R^L. The relaxed projection operator T_n is given by

$$T_n \overset{\triangle}{=} (1 - \lambda_n)I + \lambda_n P_n$$

so that

$$
\begin{aligned}
T_n \mathbf{q} &= \mathbf{q} - \lambda_n \mathbf{q} + \lambda_n \mathbf{q} + \lambda_n \frac{M - v_n}{\|\mathbf{x}_n\|^2} \mathbf{x}_n \\
&= \mathbf{q} + \lambda_n \frac{M - v_n}{\|\mathbf{x}_n\|^2} \mathbf{x}_n.
\end{aligned} \tag{8.5-15}
$$

In particular, with $\mathbf{q} \overset{\triangle}{=} \mathbf{w}^k$, $T_n \mathbf{q} \overset{\triangle}{=} \mathbf{w}^{k+1}$ and $n \overset{\triangle}{=} l(k)$, Eq. (8.5-15) becomes identical with Eq. (8.5-8). Thus the action of the relaxed projector is identical with the updating rule of the classical PLA.

Now, if we assume that the classes are separable, a separating weight vector exists and belongs to the intersection $C_0 = \cap_{n=1}^{N} C_n$. Therefore, the sequence $\left\{ \mathbf{w}^{(k)} \right\}$ generated by the projection algorithm

$$\mathbf{w}^{(k+1)} = T_N \, T_{N-1} \, \ldots \, T_1 \, \mathbf{w}^{(k)}, \qquad k = 0, 1, \ldots \tag{8.5-16}$$

converges to a separating vector $\mathbf{w}^* \in C_0$. In particular, one iteration cycle of the projection algorithm (N relaxed projections) corresponds to N iterations of the PLA given in Eq. (8.5-8).

ADVANTAGES OF THE PROJECTION FORMULATION: The POCS formulation has two important consequences:

(i) The vector-space projections theory allows for additional constraints to further constrain the admissible set of weight vectors,

(ii) As long as the additional constraints are themselves convex, there is no need to prove convergence of the algorithm incorporating the additional constraints, since

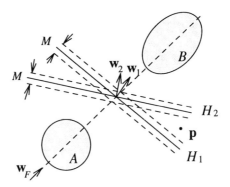

Fig. 8.5-3 POCS-based learning allows for additional constraints such as an angle constraint that improves classification.

the latter is guaranteed by the fundamental theorem of POCS. (See Theorem 2.5-1 in Chapter 2.)

If a desirable property of the weight vector corresponds to a convex-type constraint, it can be incorporated into the POCS-based learning algorithm. For instance, the orientation of the separating hyperplanes can be constrained, such that the angle between the weight vector and the *Fisher Discriminant direction*[†] $\mathbf{w}_F$ [14] does not exceed a predetermined value. The closed convex constraint set restricting the *cosine* of the angle between $\mathbf{w}$ and $\mathbf{w}_F$ to be at least β is defined as

$$C_\beta \triangleq \left\{ \mathbf{y} \in R^L : \frac{|\langle \mathbf{y}, \mathbf{w}_F \rangle|}{\|\mathbf{w}_F\|} \geq \beta \|\mathbf{y}\| \right\}. \tag{8.5-17}$$

The projection operator projecting onto C_β is derived in [16]. Its properties are discussed in some detail in Section 3.5 of Chapter 3.

The rationale of the angle constraint is illustrated in the case of a hypothetical situation shown in Fig. 8.5-3. In Fig. 8.5-3 both H_1 and H_2 are valid separating hyperplanes, defined by the normal weight vectors $\mathbf{w}_1$ and $\mathbf{w}_2$, respectively. The angle between $\mathbf{w}_1$ and $\mathbf{w}_F$ is smaller than the angle between $\mathbf{w}_2$ and $\mathbf{w}_F$. Note that the point $\mathbf{p}$, which, intuitively, is more than likely to belong to class B, is correctly classified by the H_1 separating hyperplane, but incorrectly classified by H_2. Without the additional angle constraint on the weight vector, H_2 would have been as likely to have been chosen as H_1 but, with the angle constraint enforced, H_2 would not have been a feasible solution.

An optimum lower bound for the relaxation parameters has been derived in [17].

[†]The Fisher Discriminant direction $\mathbf{w}_F$ is given by $\mathbf{w}_F = c\mathbf{K}^{-1}(\boldsymbol{\mu}_A - \boldsymbol{\mu}_B)$, where c is a normalizing constant, $\mathbf{K}$ is the covariance matrix of the two classes, and $\boldsymbol{\mu}_A$ and $\boldsymbol{\mu}_B$ are the means of classes A and B, respectively.

The so-called "per-step" optimization procedure results in

$$\lambda_n^{\text{opt}} \geq 1, \qquad n = 1, 2, ..., N. \tag{8.5-18}$$

Thus, the allowable $(0, 2)$ range for the relaxation parameters can be constrained to the $[1, 2)$ range. Kumaradjaja [15] reported significant improvement in the convergence rate of the PLA when $\lambda_n \in [1, 2)$, as opposed to $\lambda_n \in (0, 1)$.

8.6 LEARNING IN MULTI-LAYER NETS

The multi-layer feed-forward net is an extension of the single-layer perceptron discussed in Section 8.5. As is well known, the single-layer perceptron can yield good results as a pattern classifier when the classes are separable by hyperplanes, but fails when this condition is not met. The *exclusive-or* problem (Fig. 8.6-1(a)) cannot be solved by a single-layer neural net nor, for example, can the classification of samples in the distance-from-the-origin problem (Fig. 8.6-1(b)). On the other hand, many problems of practical interest can be solved by two-layer neural nets, including the ones cited above. There exist a number of learning algorithms for training multi-layer neural nets, of which the *back-propagation learning rule* (BPLR) is the most popular. The BPLR is an iterative gradient search algorithm, designed to minimize the mean-square error between the actual output of a multi-layer feed-forward net and the desired output. It requires that the neuron excitation functions be continuous and differentiable. As pointed out by Lippman [8], the BPLR often gives good performance, despite the fact that it is a gradient search technique and therefore could converge to a local minimum instead of a global one. Techniques for avoiding local minima include using a small *learning rate parameter*, using multiple sets of random starting weights and others. The BPLR exhibits slow convergence in many cases. Variations of the BPLR that increase the speed of convergence have been proposed. For example, the incorporation of a 'momentum' term of adjustment of weights was proposed in [8].

One way to view the slow convergence of the BPLR is to recall how it controls learning in the network. In Fig. 8.6-2 is shown, for the sake of illustration, a two-layer neural-net, with $\mathbf{x}_i$ denoting the actual input, $\mathbf{y}_i^* = (d_1, d_2, \cdots, d_{N_0})^T$ the desired output, and $\mathbf{y}_i$ the actual output. For each i, that is, for each specific *training* sample, the BPLR seeks to find the adjustments in the weight matrices $\mathbf{W}$ and $\mathbf{T}$ that will minimize the error between $\mathbf{y}_i^*$ and $\mathbf{y}_i$. The trouble is that the adjustments made in case i may not be suitable for $i - 1$ or $i + 1$, etc. If the degree-of-freedom of the net (that is, the number of nodes and interconnects) is large enough, then hopefully there will result at least one pair of $\mathbf{W}$ and $\mathbf{T}$ matrices that will classify correctly, if not all, most of the training patterns, if the training runs are numerous. Success in correctly classifying *training* patterns generally suggests that the *test* patterns, assuming stationary statistics, will be mostly correctly classified as well. However, by sequentially optimizing the net for each training sample without

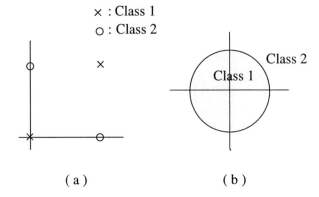

$\times$: Class 1
$\circ$: Class 2

(a)   (b)

Fig. 8.6-1 Example of problems not solvable by a single-layer neural net; (a) the exclusive-or problem; (b) the distance from the origin problem.

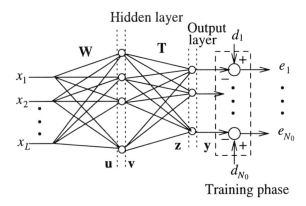

Training phase

Fig. 8.6-2 A two-layer feed-forward neural net. The structure in the dashed region is applied during the training phase.

consideration of the others, the BPLR fails to develop an efficient strategy for the training of network.

A *projection-method learning rule* (PMLR) seeks to impose the multiple con-straints imposed by the collection of training samples in *every iteration*. While this seems like a reasonable thing to do, the exact projection-based rule is difficult to compute. Indeed, the PMLR that is finally implemented is an *approximation* to the exact set-theoretic learning rule derived from projections. Before deriving the PMLR, we review how the BPLR works. We do this to contrast the two learning rules and compare their performance.

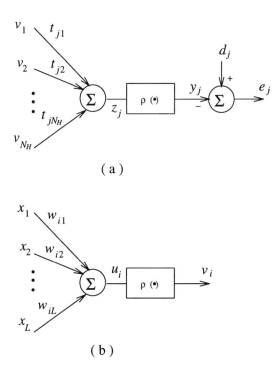

$$(a)$$

$$(b)$$

Fig. 8.7-1 (a) The interconnections for the jth output neuron; (b) the interconnections for the ith hidden-layer neuron.

8.7   BACK-PROPAGATION RULE FOR THE TWO-LAYER NET

Figure 8.7-1 is an enlarged view of Fig. 8.6-2 centered about the jth output neuron. The basic idea behind the BPLR is very simple: We apply an N-dimensional training vector $\mathbf{x}$ at the input (the extreme left in Fig. 8.6-2) and compare the output y_j of the jth output neuron with the *desired output* d_j. Then we make a *correction* in the interconnection proportional to both the size of the error and the strength of the input. This is the so-called *delta rule*. From Fig. 8.7-1 we see that

$$z_j = \sum_{i=1}^{N_H} t_{ji} v_i \tag{8.7-1}$$

and

$$y_j = \rho(z_j), \tag{8.7-2}$$

where N_H is the number of hidden neurons and $\rho(\cdot)$ is a differentiable and invertible non-linearity. The error e_j is given by

$$e_j \overset{\triangle}{=} d_j - y_j \tag{8.7-3}$$

and the (scaled) sum of squared errors is given by

$$\epsilon \triangleq \frac{1}{2} \sum_{j=1}^{N_0} e_j^2, \qquad (8.7\text{-}4)$$

where N_0 is the number of output neurons. The delta rule is derived from the gradient algorithm that seeks to minimize ϵ with respect to t_{ji}. The gradient algorithm is

$$t_{ji}^* = t_{ji} - \eta \frac{\partial \epsilon}{\partial t_{ji}} \qquad (8.7\text{-}5)$$

and the change in interconnection weight is

$$\Delta t_{ji} = -\eta \frac{\partial \epsilon}{\partial t_{ji}} \quad . \qquad (8.7\text{-}6)$$

To compute $\dfrac{\partial \epsilon}{\partial t_{ji}}$ we merely use the chain-rule of calculus:

$$\frac{\partial \epsilon}{\partial t_{ji}} = \frac{\partial \epsilon}{\partial e_j} \frac{\partial e_j}{\partial y_j} \frac{\partial y_j}{\partial z_j} \frac{\partial z_j}{\partial t_{ji}} \qquad (8.7\text{-}7)$$

$$= e_j(-1)\rho'(z_j)v_i. \qquad (8.7\text{-}8)$$

Equation (8.7-8) follows from Eqs. (8.7-1) to (8.7-4), respectively. Thus

$$\Delta t_{ji} = \eta e_j \rho'(z_j) v_i \triangleq \eta \delta_j v_i, \qquad (8.7\text{-}9)$$

where δ_j is the localized gradient at the jth output node and is given by

$$\delta_j = e_j \rho'(z_j) = (d_j - y_j)\rho'(z_j). \qquad (8.7\text{-}10)$$

To compute the correction for the **W** matrix is slightly more involved. Again, we write

$$w_{ji}^* = w_{ji} - \eta \frac{\partial \epsilon}{\partial w_{ji}} \qquad (8.7\text{-}11)$$

as the basic gradient algorithm. Using the chain rule, we obtain

$$\frac{\partial \epsilon}{\partial w_{ji}} = \frac{\partial \epsilon}{\partial v_j} \frac{\partial v_j}{\partial u_j} \frac{\partial u_j}{\partial w_{ji}} \quad . \qquad (8.7\text{-}12)$$

From Fig. 8.7-1 we observe that

$$u_j = \sum_{i=1}^{L} w_{ji} x_i \qquad (8.7\text{-}13)$$

and

$$v_j = \rho(u_j). \tag{8.7-14}$$

It follows, therefore, that

$$\frac{\partial u_j}{\partial w_{ji}} = x_i \tag{8.7-15}$$

and

$$\frac{\partial v_j}{\partial u_j} = \rho'(u_j). \tag{8.7-16}$$

That leaves only the computation of $\frac{\partial \epsilon}{\partial v_j}$. The problem is that the sum of squared errors ϵ is essentially "removed" one layer form the hidden layer output v_j. So we must proceed by *back-propagating* the error signals to the left. We write:

$$
\begin{aligned}
\frac{\partial \epsilon}{\partial v_j} &= \frac{\partial}{\partial v_j} \left[\frac{1}{2} \sum_{k=1}^{N_0} e_k^2 \right] \\
&= \sum_{k=1}^{N_0} e_k \frac{\partial e_k}{\partial v_j} \\
&= \sum_{k=1}^{N_0} e_k \frac{\partial e_k}{\partial z_k} \frac{\partial z_k}{\partial v_j} .
\end{aligned}
\tag{8.7-17}
$$

Since $e_k = d_k - \rho(z_k)$ and $z_k = \sum_{i=1}^{N_H} t_{ki} v_i$, it follows that

$$\frac{\partial \epsilon}{\partial v_j} = -\sum_{k=1}^{N_0} e_k \rho'(z_k) t_{kj}. \tag{8.7-18}$$

Hence

$$
\begin{aligned}
\frac{\partial \epsilon}{\partial w_{ji}} &= -\rho'(u_j) \sum_{k=1}^{N_0} e_k \rho'(z_k) t_{kj} x_i \\
&= -\rho'(u_j) \sum_{k=1}^{N_0} \delta_k t_{kj} x_i
\end{aligned}
\tag{8.7-19}
$$

and

$$
\begin{aligned}
\Delta w_{ji} &= -\eta \frac{\partial \epsilon}{\partial w_{ji}} \\
&= \eta \sum_{k=1}^{N_0} \delta_k t_{kj} \rho'(u_j) x_i \\
&= \eta \tilde{\delta}_j x_i,
\end{aligned}
\tag{8.7-20}
$$

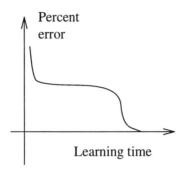

Fig. 8.7-2 Typical learning behavior of a neural net using the *back-propagation learning rule* (BPLR).

where $\tilde{\delta}_j$, the local gradient for a hidden neuron, is given by

$$\tilde{\delta}_j = \sum_{k=1}^{N_0} \delta_k t_{kj} \rho'(u_j). \tag{8.7-21}$$

While the above computation is for a two-layer network, it can be used for arbitrary multi-layer networks. Thus, for the *second* hidden layer (if there were one) to the left, the error signals would be back-propagated *twice*, etc., for additional layers.

The non-linearity is often chosen to be the so called sigmoidal non-linearity

$$\rho(x) = \frac{1}{1 + e^{-\alpha x}}, \quad -\infty < x < \infty. \tag{8.7-22}$$

The parameter $\alpha > 0$ controls the amount of non-linearity; the larger α is, the more non-linear $\rho(\cdot)$ becomes.

An important point is that the variables that appear in Eqs. (8.7-1) to (8.7-21) depend on the training sample index, i.e., $n = 1, ..., N_s$, where N_s is the total number of training samples. Thus $d_j = d_j(n)$, $y_j = y_j(n)$, $z_j = z_j(n)$, $e_j = e_j(n)$, $\epsilon = \epsilon(n)$ even $\mathbf{T} = \mathbf{T}(n)$ and $\mathbf{W} = \mathbf{W}(n)$. After each training sample is presented to the network, the interconnections are recomputed. This recomputing is done without efficiently "remembering" what the optimum interconnection was for the previous sample. As a result, the BPLR requires many cycles of training before it finally learns the correct responses for all of them. The personality of the BPLR tends to be of the "Eureka! I finally get it!" type. Its training performance shows some initial learning skills followed by a long period of bafflement, followed by a sudden show of brilliance when all training samples yield the desired output (i.e., 100 percent correct response). A typical learning behavior pattern is shown in Fig. 8.7-2.

8.8 PROJECTION METHOD LEARNING RULE

We adopt the notation used in Fig. 8.6-2 and furnish the derivation for the two-layer (i.e., one hidden layer) network; the method is extendable to networks with more layers. As before, we let L be the number of input nodes, N_H be the number of hidden-layer nodes, and N_0 be the number of output nodes. The notation $(\mathbf{W}, \mathbf{T})$ refers to a two-tuple matrix pair that specifies the connection weights between layers of nodes and bias levels of neurons; that is, the adjustable bias level of a neuron is viewed as the connection weight from an additional node that has a fixed output value of one. The mathematical working space is the collection of all such two-tuple matrix pairs with proper dimensions, with addition and scalar multiplication operations defined as

$$(\mathbf{W}_1, \mathbf{T}_1) + (\mathbf{W}_2, \mathbf{T}_2) \triangleq (\mathbf{W}_1 + \mathbf{W}_2, \mathbf{T}_1 + \mathbf{T}_2), \qquad (8.8\text{-}1)$$

$$\alpha(\mathbf{W}, \mathbf{T}) \triangleq (\alpha\mathbf{W}, \alpha\mathbf{T}), \qquad (8.8\text{-}2)$$

and inner product defined as

$$\langle (\mathbf{W}_1, \mathbf{T}_1), (\mathbf{W}_2, \mathbf{T}_2) \rangle \triangleq \sum_{i,j} w_{ij}^{(1)} w_{ij}^{(2)} + \sum_{k,l} t_{kl}^{(1)} t_{kl}^{(2)}, \qquad (8.8\text{-}3)$$

and with an inner-product derived norm[†]. This space is isomorphic to the $N_H \times (L + 1) + N_0 \times N_H$ dimensional Euclidean space, and it represents all possible variations of a two-layer network with a specific topology (i.e., a given number of input nodes, hidden nodes, and output nodes), and specific neuron activation functions.

The set $C_i \triangleq \{(\mathbf{W}^*, \mathbf{T}^*) : \rho(\mathbf{T}^* \rho(\mathbf{W}^* \mathbf{x}_i)) = \mathbf{y}_i^*\}$ is the set of all matrix pairs $(\mathbf{W}^*, \mathbf{T}^*)$ that produces the desired output $\mathbf{y}_i^*$ when the training sample $\mathbf{x}_i$ is applied. The subscript $i, (i = 1, ..., N_s)$ indicates the ith training sample or the constraint set imposed by that sample. Note that we encountered sets involving elements of two-tuples in Chapter 7, Section 7.6 on *blind deconvolution*.

We leave it to the reader to show that C_i is not convex (Exercise 8-11). However, experience with other problems involving nonconvex sets has shown that the undesirable trap case does not always present itself. (See, for example, Chapter 5, Fig. 5.1-2.)

The function $\rho(\cdot)$ is the point-wise non-linearity that characterizes the action of the neurons. Strictly speaking, the neuron activation functions for the two layers need not be the same, although for simplicity, we assume they are.

In finding the projector upon C_i, we drop the subscript i to simplify the notation.

[†]The reader familiar with matrix theory will recognize that $\|(\mathbf{W}, \mathbf{T})\|^2 = \sum_{i,j} w_{ij}^2 + \sum_{i,j} t_{i,j}^2$ i.e., the sum of the Frobenius norm-squared of $\mathbf{W}$ and $\mathbf{T}$.

For any $(\mathbf{W}^*, \mathbf{T}^*) \in C$, we have

$$\mathbf{y}^* = \rho(\mathbf{T}^* \rho(\mathbf{W}^* \mathbf{x})), \tag{8.8-4}$$

where $\mathbf{x}$ is an L-dimensional training vector $(x_1, ..., x_L)^T$ and $\mathbf{y}^*$ is an N_0-dimensional vector $(y_1, ..., y_{N_0})^T$. We assume that $\rho(\cdot)$ is invertible and differentiable. Then with $\mathbf{z} \stackrel{\triangle}{=} \rho^{-1}(\mathbf{y})$ and $\mathbf{z}^* \stackrel{\triangle}{=} \rho^{-1}(\mathbf{y}^*)$, the projector P is such that $P((\mathbf{W}, \mathbf{T})) = (\mathbf{W}^*, \mathbf{T}^*)$, subject to

$$\mathbf{z}^* = \mathbf{T}^* \rho(\mathbf{W}^* \mathbf{x}). \tag{8.8-5}$$

Thus the problem of finding the projection $(\mathbf{W}^*, \mathbf{T}^*)$ is a constrained minimization problem with the following Lagrange functional:

$$L = \sum_{i,j}(w_{ij}^* - w_{ij})^2 + \sum_{k,i}(t_{ki}^* - t_{ki})^2 + \sum_k \lambda_k \left[\sum_i t_{ki}^* \rho \left(\sum_j w_{ij}^* x_j \right) - z_k^* \right].$$
$$\tag{8.8-6}$$

By setting

$$\frac{\partial L}{\partial w_{ij}^*} = \frac{\partial L}{\partial t_{ki}^*} = 0 \tag{8.8-7}$$

and using the constraint in Eq. (8.8-5), we obtain the following learning rule:

$$t_{ki}^* = t_{ki} + \tilde{e}_k v_i^*, \tag{8.8-8}$$

$$w_{ij}^* = w_{ij} + \sum_k \tilde{e}_k t_{ki}^* \rho'(u_i^*) x_j, \tag{8.8-9}$$

where

$$\tilde{e}_k \stackrel{\triangle}{=} \frac{z_k^* - \tilde{z}_k}{\|\mathbf{v}^*\|^2}, \tag{8.8-10}$$

$$v_i^* \stackrel{\triangle}{=} \rho \left(\sum_j w_{ij}^* x_j \right), \tag{8.8-11}$$

$$u_i^* \stackrel{\triangle}{=} \sum_j w_{ij}^* x_j \, \rho^{-1}(v_i^*), \tag{8.8-12}$$

$$\tilde{z}_k \stackrel{\triangle}{=} \sum_i t_{ki} v_i^*, \tag{8.8-13}$$

$$\|\mathbf{v}^*\|^2 \stackrel{\triangle}{=} \sum_i (v_i^*)^2, \tag{8.8-14}$$

$$\rho'(\alpha) \stackrel{\triangle}{=} \frac{\partial \rho}{\partial \alpha} \, . \tag{8.8-15}$$

The reader will observe that the learning rule given by Eqs. (8.8-8) and (8.8-9) is quite complex and difficult to implement, at least for software implementation. In Eq. (8.8-8) $\{t_{ki}^*\}$ depends in a non-linear way on $\{w_{ij}^*\}$, and so does $\{w_{ij}^*\}$ in Eq. (8.8-9) on $\{t_{ki}^*\}$.

An obvious simplification is to decouple $\{t_{ki}^*\}$ from $\{w_{ij}^*\}$ in Eq. (8.8-8), and linearize Eq. (8.8-9) in $\{w_{ij}^*\}$. Then what results is

$$t_{ki}^* = t_{ki} + e_k v_i, \tag{8.8-16}$$

$$w_{ij}^* = w_{ij} + \sum_k e_k t_{ki}^* \rho'(u_i) x_j, \tag{8.8-17}$$

where

$$e_k \overset{\triangle}{=} \frac{z_k^* - z_k}{\|\mathbf{v}\|^2}, \tag{8.8-18}$$

$$v_i \overset{\triangle}{=} \rho\left(\sum_j w_{ij} x_j\right), \tag{8.8-19}$$

$$\|\mathbf{v}\|^2 \overset{\triangle}{=} \sum_i v_i^2, \tag{8.8-20}$$

$$u_i \overset{\triangle}{=} \sum_j w_{ij} x_j \rho^{-1}(v_i), \tag{8.8-21}$$

$$z_k \overset{\triangle}{=} \sum_i t_{ki} v_i. \tag{8.8-22}$$

Recall that $\mathbf{z}^* \overset{\triangle}{=} \rho^{-1}(\mathbf{y}^*)$ is specified once the desired output $\mathbf{y}^*$ is specified. Thus, while the solution given in Eqs. (8.8-16) and (8.8-17) are only approximation to the projection solution, they are in closed form and do not require solutions of coupled non-linear equations.

Finally, by introducing a relaxation parameter μ, Eqs. (8.8-16) and (8.8-17) can be rewritten as,

$$\Delta t_{ki}^* \overset{\triangle}{=} t_{ki}^* - t_{ki} = \mu e_k v_i, \tag{8.8-23}$$

$$\Delta w_{ij}^* \overset{\triangle}{=} w_{ij}^* - w_{ij} = \mu \sum_k e_k t_{ki}^* \rho'(u_i) x_j. \tag{8.8-24}$$

We call the above two equations the *projection-method learning rule* (PMLR).

For the sake of comparison, [†] the back-propagation learning rule is listed here:

$$\Delta t_{ki}^{(BP)} = \eta \delta_k^{(BP)} v_i, \tag{8.8-25}$$

[†] We change the indices for ease of comparison.

$$\Delta w_{ij}^{(BP)} = \eta \sum_k \delta_k^{(BP)} t_{ki} \rho'(u_i) x_j, \tag{8.8-26}$$

where η is a parameter like μ, i.e., the *learning rate parameter*, and $\delta_k^{(BP)}$ is defined as

$$\delta_k^{(BP)} \triangleq \rho'(z_k) \left[y_k^* - \rho(z_k) \right]. \tag{8.8-27}$$

It is interesting to note that the PMLR and the BPLR resemble each other in form, despite the fact that they are based on quite different principles. Both are in the form of the *generalized delta rule*. However, as we shall see in the simulation, the characteristics of their behavior are quite different, and the difference becomes more drastic as the neuron activation function gets more nonlinear.

Let us summarize in words the essential difference, as we understand it, between the PMLR and the BPLR. In the BPLR, an attempt is made to adjust the interconnections for each training sample so as to minimize the error between the actual and desired outputs. The underlying *conjecture* is that the minimization for each pattern will result in interconnections that are effective for the ensemble.

In the PMLR, the interconnections effective for the ensemble of N_s training samples $\mathbf{x}_i$ are represented by the intersection of N_s sets. The goal is to reach a point in this intersection by an iterative algorithm that always works when the N_s sets are convex but may not work when one or more sets is not convex.

The *minimal disturbance principle* [18] may provide an auxiliary view of the PMLR. That is, the *projection* onto a constraint set associated with a training sample is the set of updated connection weights for the net closest to the previous weights while correctly classifying the new training sample.[‡]

It is interesting to note that the formula for the PMLR at the output layer is similar in form to the Adaline learning rule of Widrow and Hoff [19].

8.9 EXPERIMENTS AND RESULTS

The previous discussions on the BPLR and PMLR make for rather serious reading and contain many complex algebraic relations, which do not reveal the power of the projection approach. To give a tangible demonstration of the power of PMLR, we furnish the results of some numerical experiments. We simulate the PMLR and the BPLR on the two-layer network and test them on the XOR problem and a classification problem. The simulation procedure, including the training and the monitoring portions, is shown as a flowchart in Fig. 8.9-1.

[‡]Strictly speaking, the statement is true only for exact projection operations. With the approximation we made in the PMLR, this statement is only approximately true.

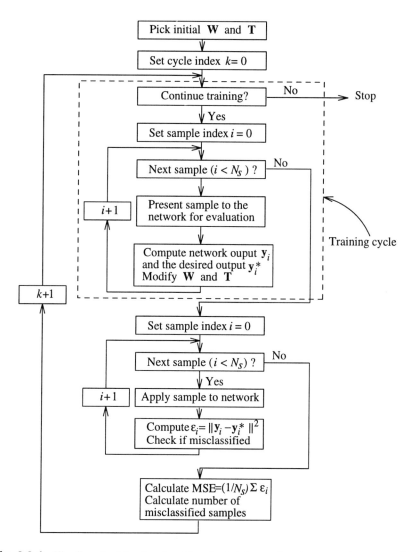

Fig. 8.9-1 The flowchart for the algorithm for training and testing a two-layer neural net.

Table 8.9-1   Average number of learning cycles required to achieve a mean square error of 10^{-3} for the XOR problem. The numbers in the table are average results over 10 to 20 runs of experiments with different initial weightings and biases.

Number of hidden nodes	Back-Propagation Learning Parameter η			Projection Method Learning Parameter μ			
	0.25	0.50	0.75	0.10	0.15	0.20	0.25
3	6200	2935	1776	846	739	651	948
8	2791	1349	894	117	80	61	49
12	2584	1281	850	99	66	50	41

XOR Problem

We start with a set of randomly assigned initial weights and biases, and apply the PMLR or BPLR using the four patterns repeatedly, i.e., each cycle of presentation of the learning samples consists of the successive presentation of the four patterns 00, 01, 10, 11. The network has a certain number of hidden neurons, whose biases are adjustable. The two output nodes are not allowed free bias. The nonlinear activation function of the hidden and output neurons is the function $1/[1 + \exp(-x)]$. The desired output values to be learned are set to 0.99 and 0.01, for representing logic 1 and 0, respectively. Thus when a class 1 pattern appears, i.e., the sequence 00 or 11, the output nodes should signal (0.99,0.01). If a class 2 pattern appears, i.e., the sequence 01 or 10, the output nodes should signal (0.01, 0.99).

The results show that the PMLR learns faster than the BPLR for this problem under every condition tested. The number of cycles of presentation needed for the mean square error to reach and stay less than 10^{-3}, averaged over 10 to 20 experimental trials, are listed in Table 8.9-1. The mean square error, whose computation is shown in Fig. 8.9-1, is computed after each training cycle to monitor the progress of the network's learning.

A parameter that affects the speed of learning is the number of hidden neurons. It was reported that, for the BPLR, the number of presentations needed decreases in proportion to the log of the number of hidden neurons [20]. We tested three different numbers of hidden nodes and observed that the PMLR also learned faster as the number of hidden nodes increased. The PMLR outperforms the BPLR in all three cases.

Distance-from-Origin Classification

In this example we are given two classes whose samples are presented by 2-tuples (x_1, x_2). For class 1 the samples satisfy $x_1^2 + x_2^2 > 1$. Like the previous problem, this case is not amenable to solution by a linear discriminant function. The problem appears as an example in [21]. The training set consists of 100 randomly generated

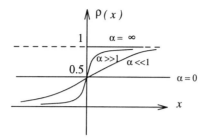

Fig. 8.9-2 The effect of the parameter α on the behavior of the activation function.

samples with approximately 50 samples per class. The samples are presented to the network one-by-one for learning and cycled iteratively, as in the manner shown in Fig. 8.9-1. The network uses 2 input ports, 8 hidden neurons with adjustable biases, and 2 output neurons. The activation function is $1/[1 + exp(-\alpha x)]$; the parameter α can be used as a measure of the nonlinearity of the function, as the plot in Fig. 8.9-2 shows: the larger the α, the more nonlinear the function. We use the same sets of initial weights and biases for comparing the performance of the BPLR and the PMLR. We chose the desired output vector to be (0.01,0.99) for class 1 and (0.99,0.01) for class 2. For neuron activation functions that exhibit strong nonlinear features, e.g., $\alpha > 1$, significantly different learning patterns for the PMLR and BPLR are observed (Figs. 8.9-3 and 8.9-4).

As described earlier, back-propagation learning is characterized by three stages: in the first few iterations it learns quickly; then it reaches a stagnation level where additional learning cycles have little effect on the misclassification rate, and the improvement on mean square error is extremely slow. This stagnation level can last for hundreds of cycles. Finally, a critical point is reached, beyond which the network learns rapidly again and corrects the errors in a few tens of cycles. In Fig. 8.9-4 is shown the misclassification performance for different values of η for $\alpha = 1$. It can be seen that the learning improves with increasing η. Eventually, however, the learning becomes unstable and performs poorly when η is too large.

The PMLR exhibits a very different learning pattern. The learning rate of the PMLR is, up to a point, much faster than in the BPLR, and is much less dependent on the learning parameter. The behavior of the PMLR when $\alpha = 1$ is shown in Fig. 8.9-5. However, despite the fact that the PMLR achieves a high correct classification rate after relatively few cycles of training, it continues to misclassify a few of the training samples (typically two or three out of 100), regardless of additional training, and the final error is larger than that obtained by the BPLR. It seems that the PMLR learns what makes the classes different much faster than the BPLR, but finally is not as accurate as the BPLR. However, we should recall that these classification rates apply to training samples, and this does not necessarily imply that the network trained by the BPLR will outperform the network trained by the PMLR when a larger set of test data is applied.

The number of learning cycles required for the BPLR and PMLR to achieve

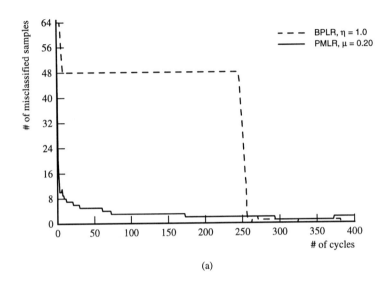

(a)

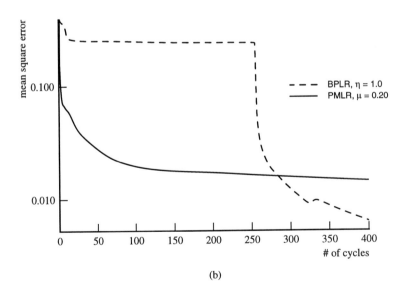

(b)

Fig. 8.9-3 The learning behaviors exhibited by the BPLR and PMLR: (a) misclassification history; (b) mean-square error history.

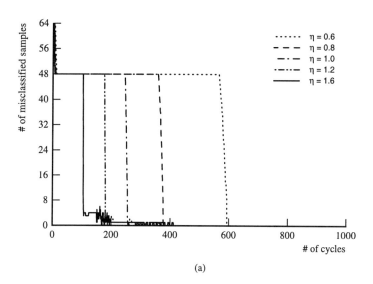

(a)

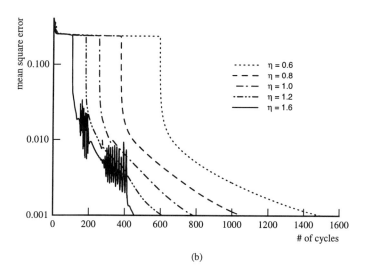

(b)

Fig. 8.9-4 Learning behavior of the BPLR for different values of the learning parameter η: (a) misclassification history; (b) mean-square error history.

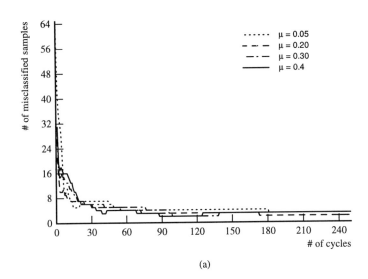

(a)

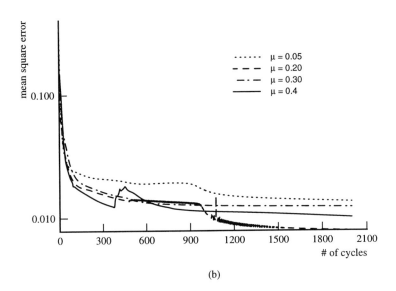

(b)

Fig. 8.9-5 Learning behavior of the PMLR for different values of the learning parameter μ: (a) misclassification history; (b) mean-square error history.

Table 8.9-2 Number of learning cycles required to achieve and remain 96% correct classification of training samples in the distance-from-origin classification problem

	Back-Propagation Learning Parameter η						Projection Method Learning Parameter μ				
α	0.6	0.8	1.0	1.2	1.4	1.6	0.05	0.10	0.15	0.20	0.25
0.5	72	55	44	37	31	27	20	58	232	43	133
1.0	594	377	256	181	139	107	55	120	71	61	58
2.0			> 4000				30	70	249	68	28

and retain at least 96 percent correct classification of the training samples is listed in Table 8.9-2 for several values of the nonlinearity parameter α and learning parameters η and μ. These number show the relative ease (or difficulty) with which the network can be trained using the different rules. For $\alpha = 1$, the PMLR never required more than 120 cycles of learning ($\mu = 0.1$) and as little as 55 cycles of learning ($\mu = 0.05$). The learning rate of the BPLR is critically dependent on the learning parameter η. For $\eta = 0.6$, it took nearly 600 cycles of learning; fastest learning occurred with $\eta = 1.6$ (107 cycles); increasing η beyond this value caused the BPLR to become unstable.

The real benefit of using the PMLR over the BPLR is apparent when α is large, i.e., the neurons operate mostly in the nonlinear region. The learning history for $\alpha = 2$ is shown in Figs. 8.9-7 and 8.9-6 for the two learning rules.

In Fig. 8.9-6 we see that, with the PMLR, 96 percent correct classification of the training samples takes place in 100 or so iterations, which is similar to the case when $\alpha = 1$. On the other hand, the BPLR, for every value of the learning parameter η we tried (from 0.1 to 3.0), is untrainable, despite thousands of cycles of training (Fig. 8.9-7).

In Table 8.9-3 and Table 8.9-4 are shown how well the PMLR and BPLR classify a set of 1000 test samples as a function of the training time (the number of training cycles). As before, the training involves only 100 training samples, which are *not* a subset of the test samples. If network adjustment on a single sample requires X seconds, then a training cycle is bounded by $100X$ seconds, and N training cycles will require at most $N \cdot 100X$ seconds; X is roughly the same for the PMLR and the BPLR in our simulation. The results in Tables 8.9-3 and 8.9-4 give the percentage of correct classification as a function of N for various values of the learning parameters μ (for the PMLR) and η (for the BPLR). In Table 8.9-3, the correct classification rates are shown for $\alpha = 1$. As before, the PMLR gives 90 percent or better classification rates after only 20 cycles of training, while the BPLR becomes very sensitive to the learning parameter η. Best results are obtained for $\eta = 1.6$, in which case 90 percent or better classification occurs around 120 cycles. Thus, under optimum conditions for the BPLR, the speed advantage of the PMLR is a factor of six. When $\eta = 1$, the PMLR learns 10 times faster than the BPLR.

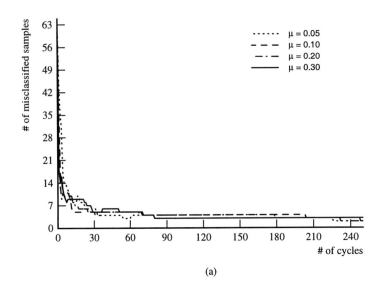

(a)

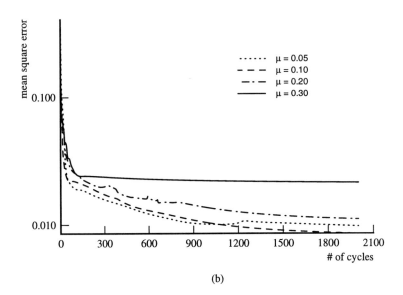

(b)

Fig. 8.9-6 Learning behavior of the PMLR for highly non-linear neurons ($\alpha = 2$): (a) misclassification history; (b) mean-square error history.

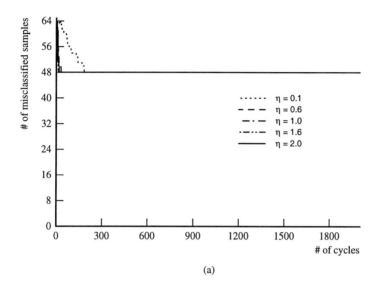

(a)

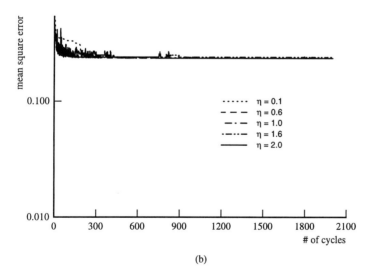

(b)

Fig. 8.9-7 Learning behavior of the BPLR for highly non-linear neurons ($\alpha = 2$): (a) misclassification history; (b) mean-square error history.

Table 8.9-3 Correct classification rate, in percent, out of 1000 test samples, as the function of the number of training cycles. The nonlinear activation function has $\alpha = 1.0$.

Projection Method Learning, $\alpha = 1.0$								
# cycles	10	20	30	60	90	120	150	180	210
$\mu = 0.05$	85	94	96	96	96	95	95	95	95
$\mu = 0.10$	92	92	91	93	93	94	94	94	94
$\mu = 0.15$	91	91	92	93	93	94	94	94	94
$\mu = 0.20$	90	91	92	93	93	93	93	93	93
$\mu = 0.25$	88	91	92	93	93	93	93	93	93
$\mu = 0.30$	86	91	92	93	93	93	93	93	93

Back-Propagation Learning, $\alpha = 1.0$								
# cycles	10	20	30	60	90	120	150	180	210
$\eta = 0.6$	45	51	51	51	51	51	51	51	51
$\eta = 0.8$	51	51	51	51	51	51	51	51	51
$\eta = 1.0$	51	51	51	51	51	51	51	51	51
$\eta = 1.2$	51	51	51	51	51	51	51	84	95
$\eta = 1.4$	51	51	51	51	51	51	96	94	95
$\eta = 1.6$	51	51	51	51	51	95	95	94	94

Table 8.9-4 Correct classification rate, in percent, out of 1000 test samples, as the function of the number of training cycles. The nonlinear activation function has $\alpha = 2.0$.

Projection Method Learning, $\alpha = 2.0$								
# cycles	10	20	30	60	90	120	150	180	210
$\mu = 0.05$	81	86	91	94	94	95	96	96	96
$\mu = 0.10$	87	92	92	95	95	96	96	96	96
$\mu = 0.15$	90	90	92	94	94	94	94	94	95
$\mu = 0.20$	86	90	91	95	95	95	96	96	96
$\mu = 0.25$	85	89	92	94	94	94	94	94	94
$\mu = 0.30$	88	90	92	94	94	94	93	93	93

Back-Propagation Learning, $\alpha = 2.0$								
# cycles	10	20	30	60	90	120	150	180	210
$\eta = 0.6$	44	48	51	51	51	51	51	51	51
$\eta = 0.8$	44	51	51	51	51	51	51	51	51
$\eta = 1.0$	40	51	51	51	51	51	51	51	51
$\eta = 1.2$	43	40	40	42	51	51	51	51	51
$\eta = 1.4$	45	51	51	51	51	51	51	51	51
$\eta = 1.6$	46	51	51	51	51	51	51	51	51

Table 8.9-4 shows the correct classification rate for $\alpha = 2$. Using a 90 percent or better classification rate as the criterion, the PMLR requires about 30 training cycles and this result, as usual, is essentially independent of μ (in the range [0.05,0.3]). The BPLR fails to meet the 90 percent criterion for any value of η, even at 210 cycles. Further runs (not shown in the table) show this to be true, even after 10,000 training cycles.

Three conclusions can be drawn for this particular example from the results shown in three tables: 1) The PMLR typically learns faster than the BPLR, a fact that becomes more evident as α (the nonlinearity parameter) increases; 2) learning by the PMLR is relatively insensitive to α, a result convincingly not true for the BPLR; and 3) learning by the PMLR is relatively independent of the value of the learning parameter in the working range, while, for the BPLR, it can be critically dependent.

In Fig. 8.9-8 is a scatter diagram of the 100 random training samples that were used to train both the BPLR and the PMLR in obtaining the above result. The reader will notice that there is a dearth of training samples at the 4 o'clock and 8 o'clock directions. Figure 8.9-9 shows the distribution of the test samples and the classification result by the network with $\alpha = 2$ trained by the PMLR. The correct classification rate in this case is 94 percent. Most of the errors appear at 4 o'clock and 8 o'clock. The interesting observation to be made is that errors are made where there is little or no training data, e.g., at 4 and 8 o'clock. Thus one may conclude that the PMLR works only as well as it is trained. This conclusion is true for pattern recognition algorithms in general, as well as in humans.

8.10 SUMMARY

In this chapter we demonstrated how projection methods can be usefully applied in the analysis, modeling, and training of neural nets. In the first part of the chapter we showed how the dynamics of the discrete-time, discrete-state, Hopfield net can be viewed as successive projections onto two constraint sets. This point-of-view quickly led to the *orthogonalization learning rule* and predicted the existence of undesirable stable states, including so-called *traps* and stable states created by linear combinations of library vectors. In the discrete-time, continuous-state, Hopfield net, the trap states were shown not to exist.

We also derived an efficient learning algorithm, for the multi-layer feed-forward neural network. This learning algorithm has its roots in vector-space projections, but involves non-convex sets and, hence, is subject to the restrictions of generalized projections. The global strategy of moving toward the intersection of all sets in the projection method helps in explaining the greater efficiency of this learning rule over the back-propagation learning rule. We formulated the PMLR for two-layer nets (i.e., with one hidden layer) and demonstrated, by simulation on the XOR problem and the distance-from-origin classification problem, that it outperforms the well-known BPLR. The PMLR learns much faster, but finally gives somewhat inferior

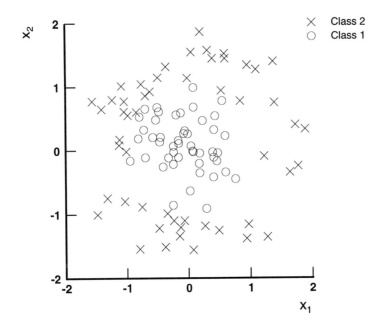

Fig. 8.9-8 The 100 samples for training a neural net to determine whether a sample is inside or outside the unit circle.

training classification results than the BPLR. Strong nonlinearity affects the BPLR severely, while the PMLR is insensitive to it; this is the significant advantage other than the speed of learning.

In applications where precision of the output values is required, it may be a good idea to use PMLR to get a fast estimate, followed by the BPLR to "fine tune" the network weights.

The extension for the PMLR to more complicated networks, for example, networks with more layers, lateral connections, cross-layer connections, or feedback connections, is subject to further research.

REFERENCES

1. G. Johnson in the *New York Times: Current Events*, p.45, October 23, 1994.

2. S. Haykin, *Neural Networks*, Macmillan, New York, 1994.

3. J. J. Hopfield, Neural networks and physical systems with emergent collective

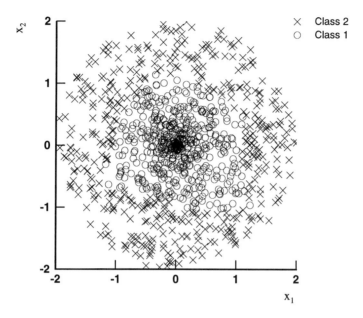

Fig. 8.9-9 Classification of 1000 test samples using a two-layer neural net trained by the PMLR.

computational abilities, *Proc. Natl. Acad. Sci. USA*, **79**:2554–2558, 1982.

4. S -J. Yeh, H. Stark, and M. I. Sezan, Hopfield-type neural networks, in *Digital Image Restoration*, (A. K. Katsaggelos,ed.) Springer-Verlag, Heidelberg, Germany, 1991, Chapter 3, pp.57–88.

5. G. R. Gindi, A. F. Gmitro, and K. Parthasarathy, Hopfield model associative memory with nonzero-diagonal terms in memory matrix, *Appl. Opt.*, **27**:129–134, 1988.

6. J. J. Hopfield, Neurons with graded response have collective computational properties like those of two-state neurons, *Proc. Natl. Acad. Sci. USA*, **81**:3088–3092, 1984.

7. J. J. Hopfield and D. W. Tank, Neural computation of decisions in optimization problems, *Bio. Cyber.*, **52**:141–152, 1985.

8. R. P. Lippman, An introduction to computing with neural nets, *IEEE ASSP Magazine*,**4**:4–22, April 1987.

9. R. P. Lippman, B. Gold, and M. L. Malpass, A comparison of Hamming and Hopfield neural nets for pattern classification, *Technical Report 769*, MIT Lincoln Laboratory, Cambridge, MA, May 1987.

10. R. J. Marks II, Class of continuous level associative memory neural nets, *Appl. Opt.*, **26**:2005–2010, 1987.

11. R. J. Marks II, S. Oh, and L. E. Atlas, Alternating projection neural networks, *IEEE Trans, Circuits Syst.*, **CAS-36**:846–857, 1989.

12. P. M. Grant and J. P Sage, A comparison of neural network and matched filter processing for detecting lines in images, In *Proc. of the Neural Networks for Computing Conference*, (J. Denker, ed.) American Institute of Physics, Snowbird, UT, 1986.

13. A. Levi and H., Stark, Image restoration by the method of generalized projections with application to restoration from magnitude, *J. Opt. Soc. Am. A*, **1**:932–943, 1984.

14. R. O. Duda and P. E. Hart, *Pattern Classification and Scene Analysis*, John Wiley & Sons, New York, 1973.

15. R. Kumaradjaja, Application of projection onto convex sets in pattern recognition, Master's Thesis, Department of Electrical, Computer, and System Engineering, Rensselaer Polytechnic Institute, Troy, NY, 1986.

16. H. Peng and H. Stark, Signal recovery with similarity constraint, *J. Opt. Sci. Am. A*, **6**:844–851, 1989.

17. A. Levi and H. Stark, Signal restoration from phase by projections onto convex sets, *J. Opt. Soc. Am.*, **73**:810–822, 1983.

18. B. Widrow, R. G. Winter, and R. A. Baxter, Layered neural nets for pattern recognition, *IEEE Trans. ASSP*, **36**:1109–1117, 1988.

19. B. Widrow and M. E. Hoff Jr., Adaptive switching circuits, *IRE WESCON Conv. Re.*, 96–104, 1960, IRE Press, New York.

20. D. E. Rumelhart, G. E. Hinton, and R. J. Williams, Learning internal representations, *Parallel Distributed Processing*, Vol.1 (D. E. Rumelhart and J. L. McClelland eds.), MIT Press, Cambridge, MA, 1986, pp.318–362.

21. S. J. Yeh and H. Stark, A fast learning algorithm for multilayer neural networks based on projection methods, in *Neural Networks: Theory and Applications* (R. J. Mammone and Y. Zeevi, eds.), Academic Press, Boston, 1991, pp.323–345.

EXERCISES

8-1. Show that the energy E as given in Eq. (8.3-4) either decreases or stays the same for a change of state in the Hopfield net.

8-2. Show that the solution to the *least-squares problem* in Eq. (8.3-14) is furnished by Eq. (8.3-15).

8-3. Show that the sets C_{sat} and C_s are convex.

8-4. Show that the projection onto C_{sat} is given by Eq. (8.4-7).

8-5. Consider the vectors $\mathbf{x}^{(1)} = (1,1,1)^T$ and $\mathbf{x}^{(2)} = (-1,-1,-1)^T$. These vectors represent the two prototypes of class 1 and class 2, respectively. Thus a vector whose Hamming distance to the class 1 prototype is smaller than its Hamming distance to the class 2 prototype and meets a maximum distance criterion Δ gets associated with class 1, etc., for class 2. A vector $\mathbf{y} = (-0.7, -0.8, 0.95)^T$ is processed by a Hopfield net to yield $\mathbf{x} = (-1.0, -1.0, 1.0)^T$; this is applied to a perceptron net. Show that for $\Delta = 1$ there is no positive output at either perceptron output port. However, for $\Delta = 2$, the class 1 output port is high while the class 2 port remains at 0. *Hint*: Use $K = 2, L = 3$ and the correction weights given in Eq. (8.5-3).

8-6. Prove that when a a gradient algorithm is applied to Eq. (8.5-7), the result is Eq. (8.5-8).

8-7. Consider a perceptron net that attempts to classify 4-vectors into two classes. Class 1 consists of all binary-component vectors closest to $\mathbf{x}^{(1)} = (-1, -1, 1, 1)^T$; Class 2 vectors consist of all vectors closest to $\mathbf{x}^{(2)} = (1, 1, -1, -1)^T$. Closeness is measured by the Hamming distance.
   a) Compute the levels at the two perceptron outputs for an input $\mathbf{y} = (1, 1, 1, -1)^T$.
   b) Follow the perceptron with a MAXNET to determine which class the vector $\mathbf{y}$ belongs to.

8-8. Show that the projection onto C_n of Eq. (8.5-12) is given by Eq. (8.5-13).

8-9. Determine whether the set C_β in Eq. (8.5-17) is convex and compute its associated projector.

8-10. Demonstrate the consistency of Eqs. (8.8-1) to (8.8-3). Show that the inner product as defined leads to a norm that satisfies the requirements for a distance.

8-11. Show that the set C_i introduced in the beginning of Section 8.8 is non-convex.

9

Applications to
Image Processing

9.1 INTRODUCTION

Image processing [1] refers to a broad class of activities in which the central element is the manipulation of data to create a desirable image. The image itself is most often a likeness or representation of a physical object or process † and manifests itself in a number of ways such as the light reflected or transmitted by silver grains in photograph film, or as light distribution on a screen or retina. In a computer, images are *stored* as arrays of numbers; on photographic film storage is in the form of exposed silver halide grains. Image processing is most often done on a digital computer, although certain operations can be done optically (e.g., Fourier-plane spatial filtering), chemically (e.g., dark-room bleaching of the emulsion), or electronically (e.g., converting color-image signals to a black/white image in compatible TV).

One of the most dramatic applications of computer image processing has been in *medical imaging*. Medical imaging techniques, such as *computer-aided tomography* (CAT), *magnetic resonance imaging* (MRI), and *positron-emission tomography* (PET), would be virtually impossible to realize without a computer-based technique called *image reconstruction.*

Projection methods, both convex and generalized, have seen extensive use in image processing, especially medical imaging [2, 3]. However, medical imaging is a highly specialized activity, which requires considerable background explanation.

---

†Although images of non-physical objects can be made in many ways, e.g., by drawing, by computer or computer-generated holograms, by fractals etc..

Such a detour would not serve our purposes and would significantly add to the length of this book. Hence the reader is referred to such specialized journals as the *IEEE Transactions on Medical Imaging* and the *Journal of the Optical Society of America*, among others, for applications of projection methods in this area.

In this chapter we furnish a few examples of projections in image processing. These include image noise-removal, image synthesis for photolithography, image deblurring, high-resolution image realization from low-resolution sequences, and, finally, restoration of quantum-limited images.

9.2 NOISE SMOOTHING

There are important classes of images that tend to be noisy, that is, significant features tend to be masked or distorted by the presence of unwanted (usually) high-frequency fluctuations in luminance or gray level. *Emission tomography* produces very noisy images, a fact that reduces its usefulness as a diagnostic tool. Astronomical imaging of distant stars, night vision devices, and holography using coherent laser illumination, all produce noisy images. The mechanisms may be different, but in each case, the noise is unwelcome and subtracts from the information content of the image.

There are many noise-smoothing techniques. Some are "dumb" in the sense that they smooth everything, including edges, which convey information and therefore shouldn't be smoothed. Others are "smart", in the sense that they try to avoid smoothing in the vicinity of edges.[†] In this section we discuss how "dumb" smoothing can be implemented by the method of vector-space projections. To implement smart smoothing using POCS requires more complex sets. In Chapter 6, Section 6.7, we gave an example of smart smoothing, whereby edge artifacts (so-called blocking artifacts) are smoothed over, while reducing the effect of the smoother on the rest of the image.[‡]

As always, the key to applying POCS effectively is to define the appropriate sets, compute the projection onto these sets, and incorporate the projectors into an image-processing algorithm designed to meet some criteria implied by the constraints.

Sets and Projections

We begin by considering a bound on a derivative norm in the one-dimensional case as follows.

The set C of real-valued functions with a uniform bound on first-derivative

---

[†]An example of such smoothing is where a gradient is used to compute the location of edges and care is taken not to smooth, i.e., blur across the edges.
[‡]The JPEG example furnished in Chapter 6, Section 6.7, could just as well have been discussed in this chapter.

norms is

$$C(\mu) = \left\{ g(x) \in L^2(R) : g'(x) \in L^2(R) \text{ and } \|g'(x)\| \le \mu \right\}. \tag{9.2-1}$$

The derivative g' means the derivative of g, that is, a function that obeys the integration-by-parts formula. The norm in Eq. (9.2-1) is the $L^2(R)$ norm. The projection is easily computed as follows. To find the projection of an arbitrary signal q onto $C(\mu)$ we must do the operation

$$\min_{g} \|q - g\|^2,$$

subject to

$$\int |g'(x)|^2 dx \le \mu^2.$$

To solve this problem, we use Parseval's theorem and the Lagrange multiplier's method. The Lagrange functional to be extremized is

$$J(g, \gamma) = \|Q(\omega) - G(\omega)\|^2 + \gamma \left[\int_{-\infty}^{\infty} \omega^2 |G(\omega)|^2 \frac{d\omega}{2\pi} - \mu^2 \right], \tag{9.2-2}$$

where we use $g \leftrightarrow G$, $q \leftrightarrow Q$ to denote Fourier pairs. We let $G(\omega) \stackrel{\triangle}{=} G_R(\omega) + jG_I(\omega)$, likewise for $Q(\omega)$, and G_R and G_I, denote the real and imaginary parts of $G(\omega)$, respectively. To find the minimum, we set $\partial J/\partial G_R = \partial J/\partial G_I = 0$ and solve the resulting equations. The result is [$g^*(x)$ denotes the projection]

$$g^*(x) = Pq(x) \leftrightarrow G^*(\omega) = \begin{cases} Q(\omega), & \text{if } \|q'(x)\| \le \mu \\ \frac{Q(\omega)}{1 + \gamma \omega^2}, & \text{otherwise.} \end{cases} \tag{9.2-3}$$

In Eq. (9.2-3) $\gamma > 0$ is a constant that depends on μ, according to

$$\int_{-\infty}^{\infty} \frac{\omega^2}{(1 + \gamma \omega^2)^2} |Q(\omega)|^2 \frac{d\omega}{2\pi} = \mu^2. \tag{9.2-4}$$

Define

$$I(\gamma) \stackrel{\triangle}{=} \int_{-\infty}^{\infty} \frac{\omega^2}{(1 + \gamma \omega^2)^2} |Q(\omega)|^2 \frac{d\omega}{2\pi} \; ; \tag{9.2-5}$$

then, since $I'(\gamma) < 0$ for $\gamma > 0$ [except in the trivial case of $Q(\omega) = 0$], $I(\gamma)$ is a monotonically decreasing function of γ, and the value of γ that satisfies Eq. (9.2-4) can be found by the Newton-Raphson method or a similar type algorithm. The extension of the projection in Eq. (9.2-3) to the two-dimensional case is readily done. Consider a function $g(x, y)$ with partial derivatives g_x and $g_y \in L^2(R^2)$. Here

$$g_x \stackrel{\triangle}{=} \frac{\partial g}{\partial x} \quad \text{and} \quad g_y \stackrel{\triangle}{=} \frac{\partial g}{\partial y} \; . \tag{9.2-6}$$

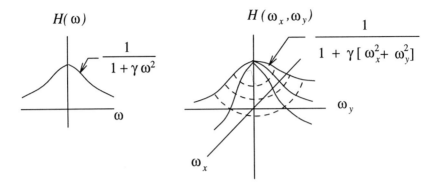

Fig. 9.2-1 Noise smoothing based on POCS: (a) A filter that furnishes a uniform bound on first-derivative norms; (b) A filter that furnishes a uniform bound on the gradient norm.

The gradient of $g(x, y)$ is the vector

$$\nabla g \triangleq g_x \hat{i} + g_y \hat{j}, \tag{9.2-7}$$

where $\hat{i}$ and $\hat{j}$ are unit vectors in the x and y directions, respectively. Then, by Parseval's theorem,

$$\|\nabla g\|^2 \triangleq \int_{-\infty}^{\infty} \int_{-\infty}^{\infty} |\nabla g|^2 dx dy = \frac{1}{4\pi^2} \int_{-\infty}^{\infty} \int_{-\infty}^{\infty} (\omega_x^2 + \omega_y^2)|G(\omega_x, \omega_y)|^2 d\omega_x d\omega_y. \tag{9.2-8}$$

Thus the condition $\|\nabla g\|^2 \leq \mu^2$ can be restated as

$$\frac{1}{4\pi^2} \int_{-\infty}^{\infty} \int_{-\infty}^{\infty} (\omega_x^2 + \omega_y^2)|G(\omega_x, \omega_y)|^2 d\omega_x d\omega_y \leq \mu^2. \tag{9.2-9}$$

Define the set

$$C_2(\mu) = \left\{ g(x, y) \in L^2(R^2) : g_x \text{ and } g_y \in L^2(R^2) \text{ and } \|\nabla g\| \leq \mu \right\}. \tag{9.2-10}$$

Then, using the technique in the previous case, the projection onto C_2 is found to be

$$g^*(x, y) = P_2 q(x, y) \leftrightarrow G^*(\omega_x, \omega_y) = \begin{cases} Q(\omega_x, \omega_y), & \text{if } \|\nabla f\| \leq \mu \\ \frac{Q(\omega_x, \omega_y)}{1+\gamma[\omega_x^2 + \omega_y^2]}, & \text{otherwise,} \end{cases} \tag{9.2-11}$$

where, as before, γ is found by the Newton-Raphson method (Fig. 9.2-1).

As a third example, define the set $C_3(\mu)$ of real-valued functions with known bounds on the Laplacian L^2 norms:

$$C_3(\mu) = \left\{ g(x,y) \in L^2(R^2) : g_{xx} \text{ and } g_{yy} \in L^2(R^2) \text{ and } \|\nabla^2 g\| \leq \mu \right\}.$$
$$(9.2\text{-}12)$$

Constraining the Laplacian

$$\nabla^2 g \overset{\triangle}{=} g_{xx} + g_{yy} = \frac{\partial^2 g}{\partial x^2} + \frac{\partial^2 g}{\partial y^2} \qquad (9.2\text{-}13)$$

of an image function $g(x,y)$ is a technique often used in image-restoration problems to smooth the image. After extremizing the Lagrange functional

$$
\begin{aligned}
J(g,\gamma) \;=\;& \|Q(\omega_x,\omega_y) - G(\omega_x,\omega_y)\|^2 \qquad\qquad (9.2\text{-}14) \\
+\;& \gamma \left[\int_{-\infty}^{\infty} \int_{-\infty}^{\infty} (\omega_x^2 + \omega_y^2)^2 |G(\omega_x,\omega_y)|^2 d\omega_x d\omega_y - \mu^2 \right],
\end{aligned}
$$

we obtain

$$g^*(x,y) = P_3 q(x,y) \leftrightarrow G^*(\omega_x,\omega_y) = \begin{cases} Q(\omega_x,\omega_y), & \text{if } \|\nabla^2 f\| \leq \mu \\ \dfrac{Q(\omega_x,\omega_y)}{1 + \gamma(\omega_x^2 + \omega_y^2)^2}, & \text{otherwise.} \end{cases}$$
$$(9.2\text{-}15)$$

As usual, γ can be found by solving

$$\frac{1}{4\pi^2} \int_{-\infty}^{\infty} \int_{-\infty}^{\infty} \frac{\omega_x^2 + \omega_y^2}{1 + \gamma(\omega_x^2 + \omega_y^2)^2} |Q(\omega_x,\omega_y)|^2 d\omega_x d\omega_y = \mu^2 \qquad (9.2\text{-}16)$$

by iteration, say, by the Newton-Raphson method.

Examples of Image-Restoration Algorithms

The previous constraint sets and their projectors can be combined with other well-known constraints to implement novel restoration algorithms. In particular, four useful constraint sets are:

1. Space-limitedness:

$$C_{SL}(X) = \left\{ g(\mathbf{x}) \in L^2(R^2) : g(\mathbf{x}) = 0 \text{ for } \mathbf{x} \notin X \right\}. \qquad (9.2\text{-}17)$$

2. Band-limitedness:

$$C_{BL}(B) = \left\{ g(\mathbf{x}) \in L^2(R^2) : g(\mathbf{x}) \leftrightarrow G(\boldsymbol{\omega}) = 0, \boldsymbol{\omega} \notin B \right\}. \qquad (9.2\text{-}18)$$

3. Amplitude-limitedness:

$$C_{AL}(a,b) = \left\{ g(\mathbf{x}) \in L^2(R^2) : a \leq g(\mathbf{x}) \leq b \right\}. \qquad (9.2\text{-}19)$$

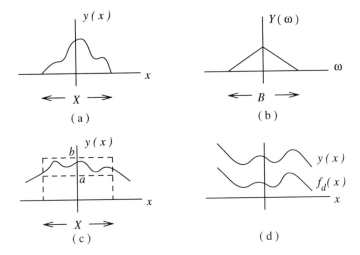

Fig. 9.2-2 Element of the four sets defined in Eqs. (9.2-17) to (9.2-20), respectively.

4. Similarity:

$$C_S(f_d, \alpha) = \left\{ g(\mathbf{x}) \in L^2(R) : \left\langle g, \frac{f_d}{\|f_d\|} \right\rangle \geq \alpha \|g\| \right\}, \qquad (9.2\text{-}20)$$

where $\langle \cdot, \cdot \rangle$, as usual, denotes the inner product operation, $0 \leq \alpha \leq 1$, and f_d is a non-zero known vector.

Typical elements of these sets are shown in one-dimension (for simplicity) in Fig. 9.2-2. The derivation of the projectors P_{SL}, P_{BL}, P_{AL} are left as exercises. If help is needed, they are derived in Chapter 3 of this book and also in Chapters 2 and 4 of [3]. The similarity constraint is useful when we expect that there will be only a small change in shape between the present (unknown) image and a known prior image f_d, e.g., two adjacent frames in a video sequence.

Some typical restoration algorithms are [4]:

1. Restoration of a space-limited, amplitude-limited image with known bound on the Laplacian norm:

$$f_{k+1} = P_{SL} P_{AL} P_3 f_k; \qquad (9.2\text{-}21)$$

2. Restoration of a band-limited, amplitude-limited image, similar to a reference image f_d with degree of similarity α and with a known bound on the gradient norm:

$$f_{k+1} = P_{BL} P_{AL} P_S P_2 f_k; \qquad (9.2\text{-}22)$$

3. Restoration of a space-limited, amplitude-limited image with known bounds on the Laplacian and gradient norms:

$$f_{k+1} = P_{SL} P_{AL} P_3 P_2 f_k. \qquad (9.2\text{-}23)$$

Fig. 9.3-1 Predistorted "4" that yields desired "4" in microphotography (after Saleh [5]). Note the indents and serifs added at corners to compensate for photographic distortions.

Other restoration algorithms can be constructed. The order of the operators, while possibly having some effect on the *rate* of convergence, will not affect the ultimate convergence of the algorithms to a solution consistent with prior constraints.

"Dumb" noise-smoothing algorithms, while useful, are not often used by themselves. They are usually combined with image sharpening procedures. Section 9.4 discusses a smarter procedure, which combines noise-smoothing with image sharpening using vector-space projections.

9.3 IMAGE SYNTHESIS

The section closely follows portions of the work done by B. E. A. Saleh, as reported in [5].

The goal of image synthesis is to find a function or pre-image which, after experiencing distortion by an optical system, yields the desired result. For example, in Fig. 9.3-1 is shown a predistorted number *four*, which will yield a desired *four* after very high levels of photographic reduction through an aberrating optical microphotographic camera [5, 6]. Note the indents at interior corners and the pointed extensions (serifs) at exterior corners that compensate for the distortions introduced by the photographic process.

Most imaging systems are spatially band-limited. That is, they can image periodic objects such as sine-waves, line-pairs, gratings, etc., so long as the fundamental frequency is less than some number B. The units of B are typically cycles (or line pairs) per millimeter. Frequencies above B cannot be imaged with enough contrast to be visible. Also, many systems produce binary images, i.e., images that are either white or black. For example, *half-tone images* are essentially binary as are pho-

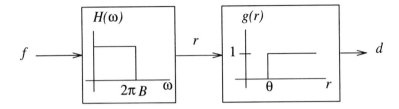

Fig. 9.3-2   Representation of diffraction-limited imaging system, followed by a hard limiter.

toresists in microlithography, or images recorded on very high-gamma photographic films.

In a first example of image synthesis, we consider the system shown in Fig. 9.3-2. This system represents a diffraction-limited imaging system with transfer function $H(\omega)$, followed by a band-limiting point nonlinearity that models a binary recoding process such as photoresist or high gamma film. The point nonlinearity is described by

$$g(r) = \begin{cases} 1, & r \geq \theta \\ 0, & r < \theta, \end{cases} \tag{9.3-1}$$

where θ is the threshold constant and is assumed independent of position. Since the desired image $d(\mathbf{x})$ is binary (e.g., it can only have values 0 and 1) and defined *a priori*, we can define regions $\mathcal{D}_+$ and $\mathcal{D}_-$ such that

$$d(\mathbf{x}) = \begin{cases} 1, & \mathbf{x} \in \mathcal{D}_+ \\ 0, & \mathbf{x} \in \mathcal{D}_-. \end{cases} \tag{9.3-2}$$

Equations (9.3-1) and (9.3-2) suggest that any function $r(\mathbf{x})$ that is band-limited[†] to B and satisfies

$$\begin{aligned} r(\mathbf{x}) \geq \theta \text{ for } \mathbf{x} \in \mathcal{D}_+ \\ r(\mathbf{x}) < \theta \text{ for } \mathbf{x} \in \mathcal{D}_- \end{aligned} \tag{9.3-3}$$

is a solution to the problem. There then exists an infinite number of input functions that will yield $r(\mathbf{x})$, for example, any function $f(\mathbf{x}) = r(\mathbf{x}) + s(\mathbf{x})$, where[‡] $s(\mathbf{x})$ is any function whose Fourier transform $S(\omega) = 0$ for $\omega \in$ supp $R(\omega)$. It is unnecessary to assume that $H(\omega)$ is constant over the band. As long as $H(\omega) \notin 0$ for $\omega \in$ supp $R(\omega)$, one can easily compute $f(\mathbf{x})$ once $r(\mathbf{x})$ is specified. We leave the details as an exercise for the reader.

The convex projection solution to finding an appropriate $r(\mathbf{x})$ is most easily illustrated in one dimension. For the sake of being specific, let the threshold θ have value $\theta = 0$. Then we seek a function $r(x)$ such that

---

[†] In two dimensions the bandwidth of a lowpass signal can be taken as the radius of the smallest circle, centered at the origin of the frequency domain, which encloses the support of its Fourier transform.
[‡] As usual, capital letters represent the Fourier transform of the lower case functions, unless otherwise stated. In the above $R(\omega)$ is the Fourier transform of $r(\mathbf{x})$ and "supp" is short for "support of."

1. $R(\omega) \triangleq \mathcal{F}[r(x)] = 0, \qquad |\omega| > 2\pi B;$

2. $r(x) \geq 0$ for $x \in \mathcal{D}_+$ and $r(x) < 0$ for $x \in \mathcal{D}_-$.

The appropriate sets are

$$C_1 = \{y(x) \in L^2 : Y(\omega) = 0, |\omega| > 2\pi B\}, \qquad (9.3\text{-}4)$$

and

$$\tilde{C}_2 = \{y(x) \ \in L^2 : y(x) \geq 0 \text{ for } x \in \mathcal{D}_+ \text{ and } \ y(x) < 0 \text{ for } x \in \mathcal{D}_-\}. \qquad (9.3\text{-}5)$$

The set $\tilde{C}_2$ is, unfortunately, not closed. For example, suppose we are given a sequence $y_1(x), y_2(x), \cdots, y_n(x), \cdots$ such that $y_i(x) < 0$ for $x \in \mathcal{D}_-$, and $y_n(x) \to 0$; clearly the sequence is in $\tilde{C}$, but its limit $y(x) = 0$ is not in the set. We could close the set by defining

$$\tilde{C}'_2 = \{y(x) \in L^2 : y(x) \geq 0 \quad \text{for } x \in \mathcal{D}_+ \text{ and } y(x) \leq 0 \text{ for } x \in \mathcal{D}_-\}. \quad (9.3\text{-}6)$$

However, this creates an ambiguity at those values of x where $y(x) = 0$; fortunately, since $y(x)$ must ultimately be band-limited, the set of points $\{x : y(x) = 0\}$ is very thin (in mathematical parlance one would say *not dense*) and creates little in the way of practical problems. An appealing way of defining the set C_2 is

$$C_2 = \{y(x) \in L^2 : \text{ for } x \in \mathcal{D}_+, y(x) \geq 0 \text{ and } \text{ for } x \in \mathcal{D}_-, y(x) \leq 0\}. \quad (9.3\text{-}7)$$

It is easily shown that the sets C_1 and C_2 are convex. The convexity of C_1 is demonstrated in Section 3.7. For demonstrating the convexity of C_2 let $y_1, y_2 \in C_2$. Then clearly $y_1 \geq 0, y_2 \geq 0$ for $x \in \mathcal{D}_+$ and for any $0 \leq \mu \leq 1$, $\mu y_1 + (1 - \mu)y_2 \geq 0$ for $x \in \mathcal{D}_+$. Likewise, for all $x \in \mathcal{D}_-$ and $y_1, y_2 \in C_2$, $\mu y_1 + (1 - \mu)y_2 \leq 0$ for all $x \in \mathcal{D}_-$. The projection upon C_1 is discussed in Chapter 3 and is given here without proof. Let $q(x)$ be an arbitrary signal and let $P_1 q$ be its projection onto C_1. Then

$$y^*(x) = P_1 q(x) \leftrightarrow Y^*(\omega) = \begin{cases} Q(\omega), & \text{for } |\omega| \leq 2\pi B \\ 0, & \text{for } |\omega| > 2\pi B. \end{cases} \qquad (9.3\text{-}8)$$

The projection upon C_2 requires some discussion. Given an arbitrary $q(x)$, define $\mathcal{Q}_+$ and $\mathcal{Q}_-$ as the point sets: $\mathcal{Q}_+ \triangleq \{x : q(x) \geq 0\}$, $\mathcal{Q}_- \triangleq \{x : q(x) < 0\}$. The projection y^* is defined by

$$y^* = \arg \min_{\text{all } y \in C_2} \|q - y\|^2. \qquad (9.3\text{-}9)$$

Hence we wish to minimize

$$\int_{-\infty}^{\infty} |q(x) - y(x)|^2 dx. \qquad (9.3\text{-}10)$$

On reflection it is readily seen that the projection y^* is obtained as follows:

$$y^*(x) = P_2 q(x) = \begin{cases} q(x), & \text{when } x \in \mathcal{D}_+ \text{ and } q(x) \geq 0; \\ q(x), & \text{when } x \in \mathcal{D}_- \text{ and } q(x) < 0; \\ 0, & \text{otherwise .} \end{cases} \qquad (9.3\text{-}11)$$

The algorithm

$$r_{n+1}(x) = P_1 P_2 r_n(x), \qquad r_0 \text{ arbitrary} \qquad (9.3\text{-}12)$$

converges to a point $r_\infty(x) \in C_1 \cap C_2$. An application of the algorithm is shown in Fig. 9.3-3.

Binary Imaging Using Coherent Systems

In the previous example the signal $r(x)$ was assumed to be real. Indeed, if we restrict $r(x)$ to be non-negative, it could model the (incoherent) field *intensity* prior to detection by the point non-linearity. It has, however, been demonstrated that the use of coherent, i.e., complex illumination can improve the resolution in optical lithography [7, 8]. In the words of Saleh [5]: "The idea is that using complex optical fields to coherently generate the desired image offers another degree of freedom (the phase) that can be utilized to generate the output images of higher resolution."

A model for coherent imaging using high-contrast detection is shown in Fig. 9.3-4. Let us again assume the one-dimensional situation. The appropriate sets are

$$C_1 = \left\{ y(x) \in L^2 : Y(\omega) = 0, \ |\omega| > 2\pi B \right\}, \qquad (9.3\text{-}13)$$

and

$$C_2 = \left\{ y(x) \in L^2 : |y(x)| \geq \sqrt{\theta} \text{ for } x \in \mathcal{D}_+, \text{ and } |y(x)| \leq \sqrt{\theta} \text{ for } x \in \mathcal{D}_- \right\}. \qquad (9.3\text{-}14)$$

The set C_2 is non-convex. The *non-convexity* of C_2 is demonstrated using the following argument: Let $y_1(x) = -y_2(x) = \sqrt{\theta}$ and $\mu = 0.5$, then $y_3(x) = 0$ and thus is not in the set.

Although C_2 is not convex, we can still expect *summed distance error* (SDE) convergence[†] since the number of constraint sets is two. The projection onto C_1 is given by Eq. (9.3-8). The projection onto C_2 will be given without proof; we leave its development as an exercise for the reader. To describe the projection, we let $q(x)$ be an arbitrary complex signal with magnitude $|q(x)|$ and phase $arg\,[q(x)] \triangleq \psi(x)$.

[†] See Chapter 5, Eq. (5.4-3).

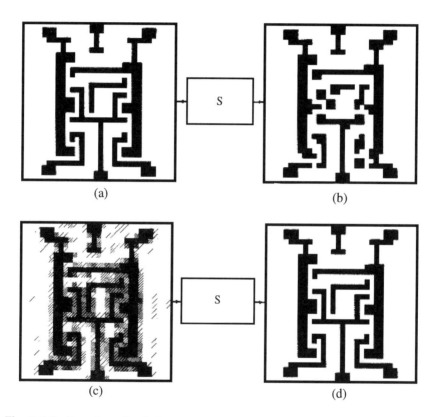

Fig. 9.3-3 Two-dimensional signal synthesis using POCS: (a) desired photolithographic mask for microcircuit design; (b) result of applying desired mask to imaging system. Output mask suffers from distortions; (c) synthesized real-valued mask by POCS produces the mask (d), which is identical to the desired mask in (a).

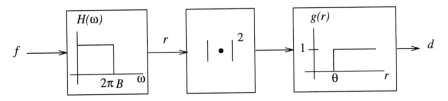

Fig. 9.3-4 Model of a coherent imaging system using high-contrast detection.

Then the projection of $q(x)$ onto C_2 is given by

$$y^*(x) = P_2 q(x) = \begin{cases} q(x) & |q(x)|^2 \geq \theta \text{ and } x \in \mathcal{D}_+ \\ \sqrt{\theta} e^{j\psi(x)} & |q(x)|^2 \geq \theta \text{ and } x \in \mathcal{D}_- \\ \sqrt{\theta} e^{j\psi(x)} & |q(x)|^2 < \theta \text{ and } x \in \mathcal{D}_+ \\ q(x) & |q(x)|^2 < \theta \text{ and } x \in \mathcal{D}_-. \end{cases} \quad (9.3\text{-}15)$$

The generalized projection algorithm takes the form

$$r_{n+1}(x) = P_1 P_2 r_n(x), \quad (9.3\text{-}16)$$

which, while superficially resembling Eq. (9.3-12), is quite different in its action, since the projection P_2 is quite different from its counterpart in Eq. (9.3-11).

Applications

We quote from Saleh [5]:

> The resolving power of an imaging system is determined by its response to two points or two thin lines. The minimum distance between the two points or lines for which the system's response has two discernible peaks is the resolution distance. According to Sparrow's criterion [9], the resolving limit is that for which the two peaks in the response barely merge into one peak with maximum flatness (vanishing second derivative). If the line response function $H(\omega)$ is rectangular with a cutoff angular frequency $2\pi B$ (ideal band-limited system), then Sparrow's two line resolution distance is $d = 0.66/B$).

The performance of the generalized projection algorithm in Eq. (9.3-16) was tested by Saleh and Nashold [10] on a one-dimensional binary signal. They took the first failure of the synthesis technique (i.e., the failure to correctly generate pulses of a given separation width D for a given systems bandwidth B) as an indication of its limit of performance. This first failure defines the resolution limit for the image. The results of their research is best described in their own words:

1. When the desired pattern was itself used as a mask (with zero phase), the resolution limit was approximately 1.75/B.

2. Improvement was obtained by assigning to every other pulse of the pattern a $180°$ phase shift while keeping the amplitudes unchanged. This is the method used in

[7, 8]. Resolution was then improved to approximately $1.09/B^\dagger$.

3. If the iterative (i.e., projection) algorithm was used to find a better mask, but the mask was constrained to be real-valued (this is implemented by using a real-valued initial guess in the algorithm), resolution was brought to about $0.87/B$.

4. When the iterative (i.e., projection) algorithm was used in its most general form (allowing the mask to be complex-valued), resolution improved to about $0.44/B$.

Some results are shown in Fig. 9.3-5. The success of the method of vector space projections in this signal synthesis problem suggest that there may be other applications, especially in communications, e.g., signal design for distorting channels.

9.4 RESTORATION OF BLURRED AND NOISY IMAGES

In the third example of applications of projections to image processing, we consider in the following the restoration of a blurred and noisy image of a "cameraman" shown in Fig. 9.4-1. A clever approach to solving this problem using vector space projections was devised by Sezan and Trussell [11], and it is their work from which the material in this section is adapted.

To understand how blurred images can be made sharp again, it is instructive to review the so-called *non-causal* Wiener filter, named in honor of the great 20th century mathematician Norbert Wiener [12], who derived the optimum causal[‡] smoothing and prediction filters for minimizing the mean-square error between the actual output and the desired output. There are several versions of the Wiener *least mean-square* (LMS) filter, but one of the most commonly used by image processing types is given by

$$\Phi(u,v) = \frac{\bar{H}(u,v)W_s(u,v)}{|H(u,v)|^2 W_s(u,v) + W_\nu(u,v)} \quad , \tag{9.4-1}$$

where $\Phi(u,v)$ is the transfer function of the filter, $H(u,v)$ is the transfer function of the linear space-invariant optical blurring system, $W_s(u,v)$ is the power spectrum of the signal, $W_\nu(u,v)$ is the power spectrum of the noise, u,v are horizontal and vertical spatial frequencies in the Fourier domain, and the overbar means complex conjugation. We discuss the behavior of the Wiener filter below.

[†]Recall that improved resolution implies a finer pulse separation so that as the resolution *increases*, the pulse separation *decreases*.

[‡]Designing a causal filter, i.e., one that does not anticipate the input, is more difficult. Causality is a bigger issue in time-dependent signals than in space-dependent ones. The "causal" direction in space can be any direction, while the causal direction on the time axis is to the right, i.e., the cause precedes the effect.

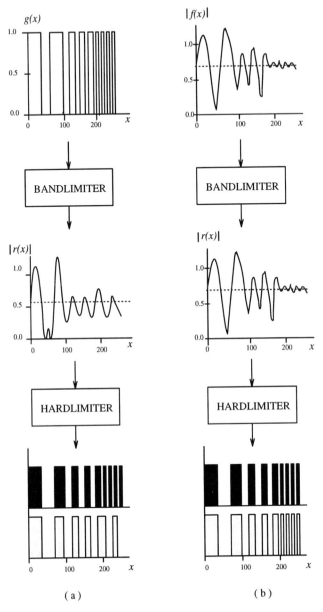

Fig. 9.3-5 Synthesis of a one-dimensional binary image through a coherent system: (a) the pattern shown at bottom left results when the input $f(\mathbf{x}) = g(\mathbf{x})$, i.e., the desired pattern itself is used as a mask. Blackened pattern is the desired pattern, shown for comparison; Notice the lack of agreement; (b) magnitude of the projection-designed complex input $f(\mathbf{x})$ and the corresponding output pattern, shown in bottom right. The desired pattern is shown (black) for comparison. Note that they are identical. (Adapted from Saleh and Nashold [5, 10].)

Wiener Filter

To keep the notation simple, we shall restrict the discussion to the one-dimensional case, except when it is necessary to go to two dimensions. Consider the problem of finding the best estimate of a signal $s(x)$, which has been blurred and distorted by noise $\nu(x)$, so that the observed data $d(x)$ has the form:

$$d(x) = s(x) * h(x) + \nu(x), \qquad (9.4\text{-}2)$$

where $*$ denotes convolution, $h(x)$ is the impulse response of the imaging system causing the blur, $H(\omega)$ is the Fourier transform of $h(x)$, and $\nu(x)$ is independent additive noise. We now ask: What linear space invariant operation can be performed on $d(x)$ so that the result of the operation, say $\hat{s}(x)$, is closest to the original signal $s(x)$ in the mean-square sense? Stated mathematically, this problem becomes: Find $\phi(x)$ or its Fourier transform $\Phi(\omega)$, so that

$$E[d(x) * \phi(x) - s(x)]^2 \rightarrow \ \min. \qquad (9.4\text{-}3)$$

The operator $E[\cdot]$ is the *expectation* or *averaging* operator.

The solution to this problem is Eq. (9.4-1), rewritten in one-dimension as

$$\Phi(\omega) = \frac{\bar{H}(\omega)W_s(\omega)}{|H(\omega)|^2 W_s(\omega) + W_\nu(\omega)} \ , \qquad (9.4\text{-}4)$$

where, as before, the overbar means complex conjugation. The filter is called *non-causal* because $\phi(x)$ is generally not zero for $x < 0$. Consider the form of $\Phi(\omega)$ when $H(\omega) = 1$, i.e., there is no blurring. Then

$$\Phi(\omega) \overset{\triangle}{=} \Phi_{sf}(\omega) = \frac{W_s(\omega)}{W_s(\omega) + W_\nu(\omega)} \ , \qquad (9.4\text{-}5)$$

which is the famous Wiener non-causal *smoothing filter* (hence the subscript *sf*) for reducing noise. When $W_s(\omega) >> W_\nu(\omega)$, $\Phi(\omega) \approx 1$ and when $W_\nu(\omega) >> W_s(\omega)$, $\Phi(\omega) = 0$, which is quite reasonable, i.e., when the signal power spectrum is high the filter should not disturb the input signal, and when the signal power spectrum is low, the filter adjusts the output close to zero. To see how the Wiener filter sharpens an image when $H(\omega) \neq 1$, consider the case when $W_\nu(\omega) = 0$, i.e., no noise. Then

$$\Phi(\omega) \overset{\triangle}{=} \Phi_{if}(\omega) = \frac{1}{H(\omega)} \ , \qquad (9.4\text{-}6)$$

which is the well-known *inverse filter*. Typically, the transfer function $H(\omega)$ causes blurring by attenuating high frequencies. The inverse filter restores the attenuated high frequencies in the blurred images, which results in *image sharpening*. One can see, therefore, that $\Phi(\omega)$ in Eq. (9.4-4) does both sharpening and smoothing simultaneously. It tends to sharpen the signal when the signal is strong and reduces the noise when the signal is weak. Indeed, with $W_s'(\omega) = |H(\omega)|^2 W_s(\omega)$ one could write

$$\Phi(\omega) = \frac{1}{H(\omega)} \left[\frac{W'_s(\omega)}{W'_s(\omega) + W_\nu(\omega)} \right] , \qquad (9.4\text{-}7)$$

which separates the Wiener filter into the product of two filters: one that sharpens and one that primarily reduces noise.

Before discussing the processing of two-dimensional images, we make one final observation. By choosing an upper bound function $\delta(\omega) \geq 0$ we can implement the Wiener solution to any degree of accuracy, using POCS, by projecting upon the set

$$C_w \stackrel{\triangle}{=} \left\{ y(x) \in L^2 : |\Phi(\omega)D(\omega) - Y(\omega)|^2 \leq \delta(\omega) \text{ for } \omega \in \Omega \right\}, \qquad (9.4\text{-}8)$$

where $D(\omega)$ and $Y(\omega)$ are the Fourier transforms of the data $d(x)$ and set-element $y(x)$, respectively, and Ω is the set of frequency points of interest.

Restoration by Convex Projections

In the two-dimensional digital signal processing case, the image is viewed as a sequence of numbers that represent the gray levels in discrete picture units called *pixels* for *picture elements*. A typical medium resolution image might contain 65,000 pixels, while high and low-resolution image might contain around 10^6 and 4000 pixels, respectively. In the two-dimensional discrete case, Eq. (9.4-2) is modified to

$$d(m, n) = \sum_{k=0}^{N-1} \sum_{l=0}^{N-1} s(k, l)h(m - k, n - l) + \nu(m, n), \qquad (9.4\text{-}9)$$

where $m, n = 0, 1, ..., N - 1$. The discrete Fourier transform of Eq. (9.4-9) yields

$$D(u, v) = S(u, v)H(u, v) + N(u, v), \qquad (9.4\text{-}10)$$

where $u, v = 0, 1, ..., N-1$. The one-dimensional non-causal Wiener filter of Eq. (9.4-4) is modified, in two-dimensions, to

$$\Phi(u, v) = \frac{\bar{H}(u, v)W_s(u, v)}{|H(u, v)|^2 W_s(u, v) + W_\nu(u, v)} , \qquad (9.4\text{-}11)$$

which has the appearance of Eq. (9.4-1), but the variables (k, l) and (u, v) are now discrete. Finally, the set C_w of Eq. (9.4-8) at a particular frequency pair (u, v) is modified to

$$C_w(u, v) \stackrel{\triangle}{=} \left\{ \{y(k, l)\} : |\Phi(u, v)D(u, v) - Y(u, v)|^2 \leq \delta(u, v) \right\}. \qquad (9.4\text{-}12)$$

Generally speaking, for the sake of convenience of computations, images [†] such as $\{s(k, l)\}, \{d(k, l)\}, \{\nu(k, l)\}$ and $\{y(k, l)\}$ are often reduced to vectors by a process

---

[†] $\{y(k, l)\}$ is the set of image values for pixels at $k, l = 0, 1, \cdots, N - 1$.

called *lexicographic ordering*. Lexicographic ordering has already been encountered in Chapter 6, Section 6.7. For example, the image $s(k, l)$ $k, l = 0, 1, ..., N - 1$ can be organized into a matrix $\mathbf{M}_s$ given by

$$
\mathbf{M}_s = \begin{bmatrix}
s(0,0) & \cdot \ \cdot \ \cdot & s(0,l) & \cdot \ \cdot \ \cdot & s(0, N-1) \\
\cdot & & & & \cdot \\
\cdot & & & & \cdot \\
\cdot & & & & \cdot \\
s(k,0) & \cdot \ \cdot \ \cdot & s(k,l) & \cdot \ \cdot \ \cdot & s(k, N-1) \\
\cdot & & & & \cdot \\
\cdot & & & & \cdot \\
\cdot & & & & \cdot \\
s(N-1,0) & \cdot \ \cdot \ \cdot & s(N-1,l) & \cdot \ \cdot \ \cdot & s(N-1, N-1)
\end{bmatrix}
\tag{9.4-13}
$$

or into a vector **s** given by

$$
\mathbf{s} = (s(0,0), \cdots, s(0, N-1), \cdots\cdots, s(N-1,0), \cdots, s(N-1, N-1))^T. \tag{9.4-14}
$$

Note that the element $s(k, l)$ maps to the vector component $s(kN + l)$. This is called lexicographic *row ordering*. The first row of the image matrix becomes the top N elements of the vector, the next immediately below, etc. One can compute the Fourier transform vector **S** of the image vector **s** by merely multiplying by a two-dimensional Fourier transform matrix **F**, which has dimensionality $N^2 \times N^2$. Thus $\mathbf{S} = \mathbf{Fs}$ is a frequency domain vector of length N^2. These details are important because it means that the sequence $y(k, l)$, $k, l = 0, 1, ..., N - 1$ or its Fourier transform $Y(u, v)$, $u, v = 0, 1, ..., N - 1$, in $C_w(u, v)$ can be represented by vectors **y** or **Y**, respectively, through lexicographic ordering. The distance between **y** and a vector, say, **q** is given by

$$
\|\mathbf{y} - \mathbf{q}\| = \left\{ \sum_{i=0}^{N^2-1} |y(i) - q(i)|^2 \right\}^{1/2}
\tag{9.4-15}
$$

$$
= \left\{ \sum_{i=1}^{N^2-1} |Y(i) - Q(i)|^2 \right\}^{1/2} = \|\mathbf{Y} - \mathbf{Q}\|.
\tag{9.4-16}
$$

The fact that $\|\mathbf{y} - \mathbf{q}\| = \|\mathbf{Y} - \mathbf{Q}\|$ is merely another form of Parseval's theorem[†]. These and other details are covered by any of several excellent textbooks on digital processing, for example, see [13].

Using the *triangle inequality* $|A \pm B| \le |A| + |B|$, it is rather straightforward to show that $C_w(u, v)$ is convex. We leave this and the computation of the projection

---

[†] Slight variations of this result occur in the literature, depending on how the discrete Fourier transform is defined. For example, one might have $\|\mathbf{y} - \mathbf{q}\| = \gamma \|\mathbf{Y} - \mathbf{Q}\|$, where γ is some constant.

onto $C_w(u, v)$ as an exercise. To show closedness we must show that given a sequence $\mathbf{y}_1, \mathbf{y}_2, \cdots, \mathbf{y}_m, \cdots$ in $C_w(u, v)$ with limit point $\mathbf{y}$, that $\mathbf{y}$ is in $C_w(u, v)$. If the array $\{y_m(k, l)\}$ is in $C(u, v)$, then its Fourier transform $Y_m(u, v)$ at frequencies u, v must satisfy $|\Phi D - Y_m| \leq \delta^{1/2}$. But

$$
\begin{aligned}
|\Phi D - Y| &= |\Phi D - Y_m + Y_m - Y| \\
&\leq |\Phi D - Y_m| + |Y_m - Y|.
\end{aligned}
\tag{9.4-17}
$$

By letting $m \to \infty$, we obtain $|\Phi D - Y| \leq \delta^{1/2}$.

The projection of an arbitrary vector $\mathbf{q}$ onto C_w is derived in [14] and is given by

$$
\mathbf{y}^* = P_{uv}\mathbf{q} \leftrightarrow Y^*(u, v) =
\begin{cases}
\Phi(u, v)D(u, v) - \sqrt{\delta(u, v)}\frac{\Delta(u,v)}{|\Delta(u,v)|}, \\
\qquad\qquad \text{if } |\Delta(u, v)|^2 > \delta(u, v) \\
Q(u, v), \quad \text{otherwise},
\end{cases}
\tag{9.4-18}
$$

where $\Delta(u, v) \overset{\Delta}{=} \Phi(u, v)D(u, v) - Q(u, v)$ and $Q(u, v)$ is the Fourier transform of $\mathbf{q}$. The set $C_w(u, v)$ applies the Wiener constraint at only a single pair of frequencies (u, v) and leaves spectral components at all other frequencies alone. Typically we would like to apply this constraint at all frequencies of significance in the signal power spectrum. We thus construct constraint sets $C_w(u_1, v_1), C_w(u_2, v_2), \cdots$, with projectors $P_{u_1, v_1}, P_{u_2, v_2}, \cdots$. Following Sezan and Trussell in [11], we denote the composition of these projectors by

$$
\tilde{P} \overset{\Delta}{=} P_{u_1 v_1} P_{u_2 v_2} \cdots P_{u_k v_k}.
\tag{9.4-19}
$$

We mention in passing that the composition $\tilde{P}$ is equivalent to the projector onto the set (Exercise 9-11)

$$
C_n \overset{\Delta}{=} \{ \{y(k, l)\} : |\Phi(u, v)D(u, v) - Y(u, v)|^2 \leq \delta(u, v) \text{ for all } u, v \}.
\tag{9.4-20}
$$

Two other sets are needed to restore an image such as "cameraman." The first is the non-negativity constraint on the restored image:

$$
C_{nn} \overset{\Delta}{=} \{ \mathbf{y} : y(k, l) \geq 0 \text{ for } k, l = 0, 1, ..., N - 1 \}.
\tag{9.4-21}
$$

The projection of an arbitrary element $\mathbf{q}$ onto C_{nn} is given by

$$
y^*(k, l) =
\begin{cases}
q(k, l) & \text{if } q(k, l) \geq 0 \\
0 & \text{otherwise.}
\end{cases}
\tag{9.4-22}
$$

The properties of C_{nn} are those of the set C in Eq. (3.6-1), which is discussed in Chapter 3, Section 3.6. The second set imposes a local smoothing constraint on the restored image. Specifically, it limits how much the restored image can deviate from a smooth *prototype image* $\{z(k,l)\}$ at the point (k,l). The prototype image is obtained by *smoothing* the corrupted image $\{d(k,l)\}$ with an averaging kernel $\sigma(k,l)$. Typically $\{\sigma(k,l)\}$ might be a 5×5 pixel-square *uniform window*. Thus the prototype image is described by

$$z(k,l) = \sigma(k,l) * d(k,l) \qquad (9.4\text{-}23)$$

and the smoothing constraint set C_s takes the form

$$C_s(k,l) = \left\{ \mathbf{y} \in R^{N^2} : |z(k,l) - y(k,l)|^2 \leq \eta^2(k,l) \right\}. \qquad (9.4\text{-}24)$$

The demonstration of convexity and closure of $C_s(k,l)$ is left as an exercise for the reader. Likewise, it takes a little work, but it is not difficult to show that the projection of an arbitrary lexicographic vector $\mathbf{q}$ onto $C_s(k,l)$ is given by

$$\mathbf{y}^* = P_{kl}\mathbf{q} = \begin{cases} z(k,l) + \eta(k,l) & q(k,l) > z(k,l) + \eta(k,l) \\ z(k,l) - \eta(k,l) & q(k,l) < z(k,l) - \eta(k,l) \\ q(k,l) & \text{otherwise.} \end{cases} \qquad (9.4\text{-}25)$$

The set $C_s(k,l)$, like the set $C_w(u,v)$ in Eq. (9.4-12), is a *point constraint* set. We wish to apply this type of constraint to all points in the restored image. Hence our solution must lie in the intersection of the sets, i.e, $\cap_{l,k} C_s(k,l)$. To find a point in the intersection, we use the composition

$$\tilde{P}_s \overset{\triangle}{=} P_{k_1,l_1} P_{k_2,l_2} \cdots, \qquad (9.4\text{-}26)$$

where $P_{k_i l_i}$ is the projector associated with set $C_s(k_i, l_i)$. The composition $\tilde{P}_s$ is equivalent to the projector onto the set (Exercise 9-14)

$$C_s = \left\{ \mathbf{y} \in R^{N^2} : |z(k,l) - y(k,l)|^2 \leq \eta^2(k,l) \text{ for all } k,l \right\}. \qquad (9.4\text{-}27)$$

The technical problem of how to determine reasonable values of the parameters $\delta(u,v)$ in Eq. (9.4-12) and $\eta(k,l)$ in Eq. (9.4-24) is discussed in greater detail in other places (for example, [11]). The POCS algorithm incorporating the above constraints is given by

$$\mathbf{s}_{n+1} = P_{nn}\tilde{P}\tilde{P}_s\mathbf{s}_n \qquad \mathbf{s}_0 = \mathbf{d}, \qquad (9.4\text{-}28)$$

where $\mathbf{s}_{n+1}$ is the updated estimate of the true image $\mathbf{s}$ and the starting point $\mathbf{s}_0$ is taken as the blurred and noisy unprocessed image $\mathbf{d}$.

The reconstruction of the blurred "cameraman" by Wiener filtering and POCS is shown Fig. 9.4-1. While the image is much sharper with Wiener filtering, there are undesirable artifacts that decrease the quality of the image (Fig. 9.4-1 (lower-right)). The reconstruction using POCS is shown in Fig. 9.4-1 (lower-left): The sharpness is maintained but many of the artifacts are gone. The parameters of the simulation and other details can be found in the original works [11, 14].

9.5 HIGH-RESOLUTION IMAGE RESTORATION

The problem we consider here is a variant of the problem first considered in Section 4.8 of Chapter 4. We are given a series of low-resolution images taken by, say, a camera as it passes over a scene and the problem is to combine these low-resolution images into a single high-resolution image. The initial convex projection solution to this problem was given in [15]. Tekalp, Ozkan, and Sezan [16] extended the problem by: 1) allowing for noise; and 2) using a motion estimation algorithm to predict the relative motion between a sequence of low-resolution images (sometimes called *frames*). It is reasonably assumed that, as the camera moves, there will be motion (so-called *interframe motion*) between each successive frame and, without prior knowledge of this motion, it must be estimated.

Following the procedure discussed in [15, 16], we assume that we have available J low-resolution frames, each of size $M \times M$ pixels, and that the goal is to reconstruct a single $N \times N$ high-resolution image, where $M < N$. The low-resolution frames are denoted by $\{d_j(m,n), m,n = 0, 1, ..., M-1$ and $j = 1, 2, ..., J\}$. The (unknown) high-resolution image is given by $f(k,l)$, $k,l = 0, ..., N-1$. There are, therefore, N^2 unknowns (the N^2 gray level values at the N^2 locations) and JM^2 equations. In the absence of noise, each low-resolution frame is assumed to be related to the high-resolution image, according to

$$d_j(m,n) = \sum_{k=0}^{N-1} \sum_{l=0}^{N-1} f(k,l) h_j(m,n;k,l), \qquad j = 1, ..., J, \qquad (9.5\text{-}1)$$

where $h_j(m,n;k,l)$ is the (possibly space-variant) *point-spread function* (PSF) of the low-resolution detector/optics configuration. From Eq. (9.5-1) we see that if $JM^2 = N^2$ then the N^2 high-resolution gray levels $f(k,l)$ can be exactly determined. [†] In the presence of noise, there generally will be more unknowns than equations and some other type of solution is solicited, e.g., least-squares. When noise, $\nu(m,n)$ is included, then Eq. (9.5-1) is modified to

$$d_j(m,n) = \sum_{k=0}^{N-1} \sum_{l=0}^{N-1} f(k,l) h_j(m,n;k,l) + \nu_j(m,n). \qquad (9.5\text{-}2)$$

The subscript j on h_j and ν_j is the frame number and suggests that the PSF and noise may change from frame-to-frame.

In order to compute $h_j(m,n;k,l)$ we reason as follows: suppose we *had available* the high-resolution (HR) image $f(k,l)$; by what process is the low-resolution

---

[†]We assume all the equations are linearly independent and there is no noise.

Fig. 9.4-1 Upper-left: the original image "cameraman"; upper-right: blurred and noisy test image "cameraman". The image is degraded by an out-of-focus blur simulated by an eight-pixel diameter uniform circular point spread function and additive Gaussian noise at 40 dB SNR. Lower-right: Wiener restoration of "cameraman"; lower-left: POCS restoration. Note the fewer number of artifacts in the POCS restoration.

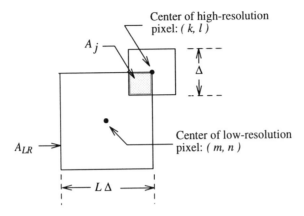

Fig. 9.5-1 Geometry describing low-resolution pixel superimposed upon high-resolution pixels.

(LR) image $d_j(m,n)$ generated? From Eq. (9.5-2) we see that $d_j(m,n)$ is built-up from a weighted superposition of the HR pixels $f(k,l), k,l = 0,1,\cdots,N-1$. Assume that the HR pixels are square with side Δ and that an LR pixel has area $L^2\Delta^2$. We observe, upon reflection, that to prevent a *bias* in brightness, i.e., to keep the average brightness level the same in both high-resolution and low-resolution images, the energy accumulated by the LR pixel must be normalized by $L^2\Delta^2$, i.e., its area. We observe, also from Fig. 9.5-1, that of the N^2 pixels in the high-resolution image, most will lie outside the the larger LR footprint; some will lie wholly inside; and some will straddle the boundary. Pixels outside the footprint make no contributions to the output $d_j(m,n)$; those wholly inside contribute their brightness in their entirety; and those partially inside contribute in proportion to the amount of HR pixel area inside the LR pixel. Now we assume that all pixels, and all portions of the HR pixels, within the LR pixel footprint contribute uniformly, i.e., without any directional or other bias. By introducing the "pixel-contribution" parameter[‡] $r_j(m,n;k,l)$ such that

$$0 \le r_j(m,n;k,l) \le 1 \ , \tag{9.5-3}$$

we can compute an equivalent point-spread function (PSF) $h_j(m,n;k,l)$ as

$$h_j(m,n;k,l) \;=\; \frac{r_j(m,n;k,l)\Delta^2}{L^2\Delta^2} \tag{9.5-4}$$

$$=\; \frac{A_j(m,n;k,l)}{A_L} \ , \tag{9.5-5}$$

where $A_j(m,n;k,l)$ is the area of overlap between the LR pixel centered at (m,n) and a HR pixel centered at (k,l), and A_L is the area of the LR pixel (Fig. 9.5-1).

[‡]$r_j(m,n;k,l)$ is the fraction of overlap between the pixel at (k,l) and the detector at (m,n).

Clearly, $h_j(m, n; k, l)$ is completely specified once $r_j(m, n; k, l)$ is known. In turn, $r_j(m, n; k, l)$ is specified once it is known what displacement took place between the jth frame and, say, the first (reference) frame. Since the motion of the camera with respect to the fixed high-resolution image may not be known, it has to be estimated. Tekalp et al. [16] used the *Fogel motion estimation* algorithm [17] to estimate the relative displacement between the reference and successive frames of the low-resolution sequence. We shall omit a discussion of how motion estimation is done, as this would take us too far afield. The point is that, if the displacement can be well estimated from frame-to-frame, $h_j(m, n; k, l)$ is well specified for all values of its argument and convex projection methods can be used to reconstruct the high-resolution image.

Reconstruction by POCS

Consider again Eq. (9.5-1), rewritten as

$$0 = d_j(m, n) - \sum_{k=0}^{N-1} \sum_{l=0}^{N-1} f(k, l) h_j(m, n; k, l), \qquad (9.5\text{-}6)$$

where $d_j(m, n)$ is the data (known) and $h_j(m, n; k, l)$ is the PSF (known). If we do not know $f(k, l)$ we might try to solve this equation by trying different functions $y(k, l)$ in place of the unknown $f(k, l)$, i.e.,

$$\epsilon'_j(m, n; \mathbf{y}) \overset{\triangle}{=} d_j(m, n) - \sum_{k=0}^{N-1} \sum_{l=0}^{N-1} y(k, l) h_j(m, n; k, l). \qquad (9.5\text{-}7)$$

Thus a good choice of $\mathbf{y}$ is one that makes the error or *residual* ϵ'_j small; a poor choice would make ϵ'_j large. A fortuitous choice of $y(k, l) = f(k, l)$ would make the residual *zero*. In the presence of noise, the total residual is

$$\epsilon_j(m, n; \mathbf{y}) \overset{\triangle}{=} \epsilon'_j(m, n; \mathbf{y}) + \nu_j(m, n) \qquad (9.5\text{-}8)$$

and we might seek a solution $y(k, l)$ that forces this residual to be magnitude bounded[†]. Thus a reasonable constraint set is

$$C_j(m, n) \overset{\triangle}{=} \{ \mathbf{y} : |\epsilon_j(m, n; \mathbf{y})| \leq \eta \}, \qquad (9.5\text{-}9)$$

where η is some suitable chosen constant, e.g., $\eta = 3\sigma$, where σ is the standard deviation of the noise $\nu(m, n)$; this is sometimes called the *three-sigma* bound.

We leave the demonstration of convexity and closedness of $C_j(m, n)$ as an exercise. To compute the projection of an arbitrary image vector $\mathbf{q}^{‡}$ upon $C_j(m, n)$,

---

[†]Other bounds are possible, e.g. bounded *norms*.
[‡]Recall that $q(k, l)$, $k, l = 0, 1, ..., K - 1$ can be converted to a vector by lexicographic ordering.

it is easiest to consider the three cases: (i) $\epsilon_j(m, n; \mathbf{y}) > \eta$; (ii) $\epsilon_j(m, n; \mathbf{y}) < -\eta$; and (iii) $-\eta \leq \epsilon_j(m, n; \mathbf{y}) \leq \eta$. For example, to find the projection when case (i) applies, write the Lagrange functional

$$J(\mathbf{y}) = \sum_{k=0}^{N-1} \sum_{l=0}^{N-1} |q(k, l) - y(k, l)|^2 + \lambda \left[\epsilon_j(m, n; \mathbf{y}) - \eta \right], \tag{9.5-10}$$

where $\epsilon_j(m, n; \mathbf{y})$ is given by Eq. (9.5-8), differentiate with respect to $y(k, l)$ and set equal to zero. The projection so calculated for case (i) is

$$y^*(k, l) = q(k, l) + \frac{\epsilon_j(m, n; \mathbf{q}) - \eta}{\|h_j\|_F^2} h_j(m, n; k, l). \tag{9.5-11}$$

Similarly, for case (ii)

$$y^*(k, l) = q(k, l) + \frac{\epsilon_j(m, n; \mathbf{q}) + \eta}{\|h_j\|_F^2} h_j(m, n; k, l); \tag{9.5-12}$$

and for case (iii) $y^*(k, l) = q(k, l)$. In the above $\|h_j\|_F^2$ is the *Frobenius norm* of the matrix $[h_j(m, n; k, l)]_{N \times N}$ and is given by

$$\|h_j\|_F^2 = \sum_{k=0}^{N-1} \sum_{l=0}^{N-1} h_j^2(m, n; k, l) \tag{9.5-13}$$

and $\epsilon_j(m, n; \mathbf{q})$ is the residual when $y(k, l)$ is replaced by $q(k, l)$ for $k, l = 0, 1, \cdots, N - 1$, i.e.,

$$\epsilon_j(m, n; \mathbf{q}) = d_j(m, n) - \sum_{k=0}^{N-1} \sum_{l=0}^{N-1} q(k, l) h_j(m, n; k, l). \tag{9.5-14}$$

Note that $C_j(m, n)$ is a *pointwise* constraint set. We desire that, ultimately, all the constraints imposed by all J low-resolution frames at each of the M^2 points be applied. Hence we seek a solution, imposed by the global constraint, which is the intersection of JM^2 sets, i.e., $\cap_{m,n,j} C_j(m, n)$. We denote $\tilde{P}$ or, more generally, $\tilde{T}$ (the composition of relaxed projectors) as the composition of the JM^2 pointwise projections.

The version of the high-resolution reconstruction algorithm used by Tekalp et al. [16] is given by

$$f_{n+1} = P_{nn}\tilde{T}f_n, \tag{9.5-15}$$

where P_{nn} is the projector onto the *non-negativity* constraint set C_{nn} of Eq. (9.4-21). In the experiment described below, all the T's in the composition $\tilde{T}$ used a relaxation parameter of $\lambda = 0.1$.

Fig. 9.5-2 A sequence of 16 low-resolution images, each of size 64 × 64.

Numerical Example

In [16], authors Tekalp et al. carefully describes a clever numerical experiment to test out the efficacy of the POCS reconstruction. We use their words to describe the experiment:

> An example of the performance of the proposed POCS algorithm is shown in Fig. 9.5-2 and Fig. 9.5-3 where the POCS reconstruction is compared with the low-resolution image frames and the result of direct spatial interpolation. The 16 low-resolution images of an aerial scene, each of which is 64 × 64 pixels, are shown in Fig. 9.5-2. These images are simulated as follows. The 256 × 256 aerial image is first convolved with a 4 × 4 uniform rectangular kernel to simulate the blur due to the imaging sensor, and then contaminated by white Gaussian noise at 30-dB SNR. The 16 low-resolution frames are obtained by subsampling the degraded 256 × 256 image by 4 in each direction, 16 times, each time starting from a different pixel within the first 4 × 4 block. Thus, each frame is shifted 1/2 pixel (high-resolution pixels) with respect to the previous one. We have selected the upper left image in Fig. 9.5-2 as the reference frame and considered the problem of increasing its resolution by a factor of 4 in each dimension. We have used the Fogel motion estimation algorithm [17] to estimate the relative displacement vector field between the reference and the other frames and used

(a)

(b)

Fig. 9.5-3 Results of high-resolution reconstruction using a POCS algorithm: (a) bicubic interpolation of the first frame in the sequence (i.e., the reference frame) by a factor of 4 in each direction. The POCS algorithm is initialized by this image; (b) the 256×256 high-resolution reconstruction, obtained using the POCS-based algorithm and the hierarchical block matching for estimating the interframe motion.

the estimated values in our experiments rather than the actual values. In Fig. 9.5-3(a), the result of direct bicubic spatial interpolation of the reference frame by a factor of 4 in each direction is shown. The proposed POCS algorithm is initialized by this image. The result obtained after 10 iterations is shown in Fig. 9.5-3(b). In this experiment, the noise variance defining the bound of the residual constraint set has been estimated from the degraded image and the confidence parameter has been set as $c = 1$. The high-quality of the result obtained by the POCS algorithm is rather obvious in Fig. 9.5-3(c). The superiority of the POCS results over direct spatial interpolation is indeed due to the ability of the POCS algorithm to make use of the new information contained in the neighboring frames.

9.6 RESTORATION OF QUANTUM-LIMITED IMAGES

From a mathematical viewpoint, we may think of an image[†] as a function $f(\cdot)$, which assigns to each point (x, y), in the plane a value $f(x, y)$ which denotes an image property such as transmittance, brightness, reflectance, or optical density. An image of a person (the object) bears a likeness of that person. Indeed, when confronted with a really good image, it may be difficult for the viewer to distinguish between the image and the original object. Images are *recorded* when photons strike light sensitive detectors such as silver halide emulsions, photosensitive selenium [‡] compounds such as in TV cameras, or crystal detectors such as found in tomography machines. When the number of photons reaching the detectors is reduced, the image starts to develop a noisy, grainy quality. This noise is called *quantum* or *photon noise* and increases as the light level is reduced. Eventually, as the light is reduced still further, instead of a locally homogeneous gray levels, the image looks like a "thousand points of light" against a dark background. Each point of light represents a single photon recorded at the detector. At such low light levels the observer may not be able to recognize the object. Figure 9.6-1 shows images of the U.S. Capitol at varying light levels measured in total photon counts: Without prior knowledge the images in the upper left and right cannot be recognized as the U.S. Capitol.

A quantum or photon-limited image is one in which the principle source of noise is due to the statistical variability in one or more of the following imaging processes: photon production, photon interaction with matter, and photon detection.

Photon-limited images occur in many settings [18]. They occur in night-vision systems, where natural lighting is at a minimum. They also occur in astronomical and stellar imaging [19, 20] and *computerized tomography* (CT), especially *positron-emission tomography* (PET) [21]. The problem in PET is that the amount of radioactive substance given to a patient must be sharply curtailed to avoid radiation damage to living tissue. In CT one naturally wants to keep the amount of X-ray radiation passing through the body to a minimum; hence the number of photons gathered at the detectors is limited.

[†]Our discussion assumes monochrome, analog, and static images.
[‡]Useful in such devices because its electrical conductivity varies with the intensity of light.

Fig. 9.6-1 Photon-limited images at various bright levels. At very low light levels it becomes difficult to recognize the U.S. Capitol. Lowest photon count image is in upper-left; next lowest is in upper-right; highest is on bottom-right.

The light that comes from the objects that eventually produces the images consists of photons randomly emitted in time and space. Bright regions in a luminous object emit many photons per unit time, while darker regions produce correspondingly fewer ones. The strength of emission of photons is often characterized by a photon rate parameter $\lambda(\mathbf{x}; t)$: the average number of photons emitted per second, per unit area centered at $\mathbf{x} = (x, y)$. During the time interval from 0 to T, the average number of emitted photons per unit area at $\mathbf{x}$ is $\int_0^T \lambda(\mathbf{x}, t) dt$. If the emission rate is *independent* of time t, then

$$\int_0^T \lambda(x, y) dt = T\lambda(x, y). \tag{9.6-1}$$

By setting $T = 1$, we can eliminate the time-dependence without any loss of generality. In a classic paper written early in the twentieth century [22], the *Poisson* nature of photon absorption in photographic film was used to derive the famous sigmoidal *density-log exposure* ($D \log E$) curve of silver-halide photographic film. In considering the restoration of quantum-limited images, we shall assume that the Poisson probability law reigns. While we have not yet defined precisely what we mean by "quantum-limited image restoration," we expect that, after processing, the resulting image will have a smoother appearance without significant loss of important detail or edge information.

Poisson Law

Let $\lambda(x, y)$ be the *Poisson rate parameter*, i.e., the average number of counts[†] per unit area at location (x, y). Then the average number of photons M emitted from a source with surface $\mathcal{A}(x, y)$ centered at (x, y) is

$$\bar{M} = \int \int_{\mathcal{A}(x,y)} \lambda(\alpha, \beta) d\alpha d\beta. \tag{9.6-2}$$

If the surface is small enough and its area is Δ, then [‡]

$$\bar{M} \approx \lambda(x, y)\Delta. \tag{9.6-3}$$

Equations (9.6-2) and (9.6-3) are consequences of the Poisson law, which states the following: With X denoting the photon counts, say, from a surface $\mathcal{A}(x, y)$ centered about (x, y), with area Δ, the *probability* of observing exactly m counts, is given by

$$P[X = m] = \frac{e^{-\bar{M}} (\bar{M})^m}{m!} \quad , \quad m = 0, 1, \cdots \tag{9.6-4}$$

where $\bar{M}$ is as in Eq. (9.6-2). If $\lambda(x, y)$ is essentially constant across $\mathcal{A}(x, y)$, say at λ_{xy}, then

$$P[X = m] = \frac{e^{-\lambda_{xy}\Delta} [\lambda_{xy}\Delta]^m}{m!} . \tag{9.6-5}$$

From Eq. (9.6-4) it is easily demonstrated that

$$E[X] = Var[X] = \bar{M}, \tag{9.6-6}$$

where $E[X]$ and $Var[X]$ are the expected value and variance of X, respectively. A measure that is partially useful in indicating image quality is the signal-to-noise ratio (SNR), given by

$$SNR = 20 \log_{10} \frac{E[X]}{\sqrt{Var(X)}} = 10 \log_{10} \bar{M} , \tag{9.6-7}$$

which depends only on the average count. For $\bar{M} = 10^6$ the SNR is 40 dB higher that for $\bar{M} = 10^2$. Note that the noise actually increases when $\bar{M}$ increases but the signal strength increases faster. For very high counts the noise is submerged in

---

[†] These are counts of emitted, incident, or absorbed photons
[‡] The overbar here refers to the *average* of M, not its complex conjugate.

the dominant signal. Also note that the SNR is a local measure: It depends on the count and therefore λ_{xy} in a region around (x, y). In bright zones of the image the SNR will be larger than in dark zones.

Projection Method Reconstruction

In the beginning of this section we stated that, mathematically speaking, an image was a function $f(\cdot)$ of the two position variables (x, y) that denotes some physical property such as transmittance, etc. In the *quantum limited image restoration problem*, it is convenient to replace the position-dependent transmittance, reflectance, brightness, or optical density with a count variable, the count being absorbed photons. Also, since the restoration will be done by computer, we segment the image into a mosaic of $L \times L$ pixels. Let $X_i, i = 1, 2, ..., L^2$, be the number of counts in pixel i. Then, assuming the Poisson Law:

$$P\left[X_i = m\right] = \frac{e^{-\lambda_i \Delta} (\lambda_i \Delta)^m}{m!} \quad , \qquad (9.6\text{-}8)$$

where λ_i is the Poisson rate parameter, i.e., the average number of counts per unit area at location i, and Δ is the area of a pixel, assumed constant for all pixels in the image. Without loss of generality, we assume a scaling in which $\Delta = 1$. It then follows that $E\left[X_i\right] = \lambda_i$ and $Var\left[X_i\right] = \lambda_i \stackrel{\Delta}{=} \sigma_i^2$. The quantum-limited restoration problem refers to obtaining estimates of λ_i for all i in the image, given only the counts and any prior information.

Consider a region $\mathcal{R}$ within the image field, consisting of N pixels and define $Y \stackrel{\Delta}{=} \sum_{i \in \mathcal{R}} X_i$. Then Y is the total count in the region of N pixels and is known to be Poisson with mean and variance given by [23, 24]

$$E[Y] = \sum_{i \in \mathcal{R}} \lambda_i, \qquad Var[Y] = \sum_{i \in \mathcal{R}} \lambda_i. \qquad (9.6\text{-}9)$$

Since the $X_i, i \in \mathcal{R}$, are independent (adjacent pixels do not overlap), by invoking the Central Limit Theorem we can argue that the random variable

$$Z \stackrel{\Delta}{=} \frac{Y - \sum_{i \in \mathcal{R}} \lambda_i}{\left[\sum_{i \in \mathcal{R}} \lambda_i\right]^{1/2}} \stackrel{d}{\rightarrow} N(0, 1) \ , \qquad (9.6\text{-}10)$$

i.e., tends to a standard normal in distribution. To find a constraint on $\sum_{i \in \mathcal{R}} \lambda_i$ we proceed as before. From tables of the *standard normal distribution* we find numbers $z_{\alpha/2}$ and $-z_{\alpha/2}$, so that the probability that $Z \in [-z_{\alpha/2}, z_{\alpha/2}]$ is given by

$$P\left[-z_{\alpha/2} \le Z \le z_{\alpha/2}\right] = 1 - \alpha. \qquad (9.6\text{-}11)$$

Using Eq. (9.6-10), in Eq. (9.6-11), with $x \stackrel{\Delta}{=} \sum_{i \in \mathcal{R}} \lambda_i$, we write the equivalent expression

$$P\left[Y - z_{\alpha/2}x^{1/2} \leq x \leq Y + z_{\alpha/2}x^{1/2}\right] = 1 - \alpha. \qquad (9.6\text{-}12)$$

For a particular outcome, say $Y = M$, i.e., M photons in the region $\mathcal{R}$, Eq. (9.6-12) becomes

$$P\left[M - z_{\alpha/2}x^{1/2} \leq x \leq M + z_{\alpha/2}x^{1/2}\right] = 1 - \alpha. \qquad (9.6\text{-}13)$$

A $100(1 - \alpha)$ percent confidence interval $[x_L, x_H]$ on x is obtained by solving

$$x_L = M - z_{\alpha/2}x_L^{1/2}$$

$$x_H = M + z_{\alpha/2}x_H^{1/2}$$

and recalling that $x \geq 0$. The result of such a calculation yields

$$x_H = M + z_{\alpha/2}\left[\sqrt{\left(z_{\alpha/2}\right)^2 + M} + \frac{z_{\alpha/2}}{2}\right] \qquad (9.6\text{-}14)$$

and

$$x_L = M - z_{\alpha/2}\left[\sqrt{\left(z_{\alpha/2}\right)^2 + M} - \frac{z_{\alpha/2}}{2}\right]. \qquad (9.6\text{-}15)$$

Thus the appropriate soft constraint set is

$$C = \left\{\boldsymbol{\lambda} : \sum_{i \in \mathcal{R}} \lambda_i \in [x_L, x_H]\right\}, \qquad (9.6\text{-}16)$$

where $\boldsymbol{\lambda}$ is a vector whose components consist of the rate parameters associated with every pixel (i, j) in the image. Of course, in the quantum-limited image restoration problem, we shall need many sets such as in Eq. (9.6-16). Thus a more appropriate nomenclature is

$$C_j = \left\{\boldsymbol{\lambda} : \sum_{i \in \mathcal{R}_j} \lambda_i \in \left[x_L^{(j)}, x_H^{(j)}\right]\right\}, \qquad (9.6\text{-}17)$$

where $\mathcal{R}_j$ refers to the jth region and $j = 1, ..., n$. A reasonable name for C_j is the *rate-constraint* set for region $\mathcal{R}_j$. For a "linear-filtering model" of the restoration problem it is illuminating to rewrite C_j above as

$$C_{kl} = \left\{\boldsymbol{\lambda} : \sum_{i=0}^{L-1}\sum_{j=0}^{L-1} \lambda(i,j)a(k-i, l-j) \in \left[x_L^{(k,l)}, x_H^{(k,l)}\right]\right\}, \qquad (9.6\text{-}18)$$

where $a(k, l)$ are the elements of a mask and are either *zero* or *one* (zero if the pixel is outside the mask support, one if inside), $\lambda(i, j)$ is the count in pixel (i, j) and the image is assumed square with L pixels on a side. The confidence interval bounds clearly depend on the mask position, since these bounds depend on the total included count.

The reader may recall that the set C_{kl} in Eq. (9.6-18) is of the soft linear type constraint discussed in some detail in Section 3.3. Therefore, it is closed and convex and its projector can be inferred from its general form in Eq. (3.3-10). Indeed, for an arbitrary vector γ, its projection onto C_{kl} is given by

$$
\boldsymbol{\lambda}^* \triangleq P\boldsymbol{\gamma} = \begin{cases} \boldsymbol{\gamma} + (\frac{x_L - \sum_{i \in \mathcal{R}} \gamma_i}{N})\mathbf{a} & \text{if } \sum_{i \in \mathcal{R}} \gamma_i < x_L \\ \boldsymbol{\gamma} & \text{if } x_L \leq \sum_{i \in \mathcal{R}} \gamma_i \leq x_H \\ \boldsymbol{\gamma} + (\frac{x_H - \sum_{i \in \mathcal{R}} \gamma_i}{N})\mathbf{a} & \text{if } \sum_{i \in \mathcal{R}} \gamma_i > x_H, \end{cases} \tag{9.6-19}
$$

where $\mathbf{a}$ is the vector of the elements $a(k,l)$ in Eq. (9.6-18). The projectors P_j onto C_j or P_{kl} onto C_{kl} have the same generic form as Eq. (9.5-11). The notation merely gets more cumbersome, since all parameters require indexing, e.g., either with the index j or indices (k, l).

The overall algorithm is as follows

$$
\boldsymbol{\lambda}^{k+1} = P_{p(1)} P_{p(2)} ... P_{p(n)} \boldsymbol{\lambda}^k, \tag{9.6-20}
$$

where $P_{p(j)}$ are projectors onto sets such as C_j in Eq. (9.6-17). In Eq. (9.6-20), the projectors can be in any order, hence the permutation subscripts, $\boldsymbol{\lambda}^{k+1}$ represents the new (vector) estimate of the rates and $\boldsymbol{\lambda}^k$ the previous estimate. Note that each of the projectors projects onto a different constraint set.

Experimental Results

We demonstrate the efficacy of the convex projection algorithm, Eq. (9.6-20), in restoring the quantum-limited image of a famous standard: The so-called *Shepp-Logan phantom* (SLP) of a head section[25] (Fig. 9.6-2). The image consists of 128×128 pixels and includes objects of different sizes and gray tones. The computer simulation uses the SLP as the "patient" in a tomographic configuration and reconstructs a facsimile of it from raysums at peripheral detectors using the *filtered-convolution back-projection* (FCBP) algorithm. To simulate an 8 million count quantum-limited image, the detector readings were replaced by a Poisson process, whose mean was set equal to the original raysum value. The sum of all raysum counts were then normalized to 8 million.

The basic FCBP reconstruction of the SLP using a *ramp filter* is shown in Fig. 9.6-3(a). Better results are obtained when a *Hamming filter* is used; the SLP reconstructed by the FCBP algorithm, using a Hamming filter is shown in Fig. 9.6-3(b). Can the POCS method improve on this results?

[†] A standard, well-designed phantom is useful as a test object for comparing the performance of different algorithms. The SLP was already used in an example in Section 4.8.

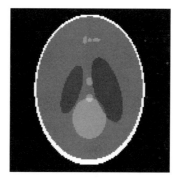

Fig. 9.6-2 Shepp-Logan phantom of a head section.

The POCS restoration proceeds as follows: (1) Choose an initial *uniform* image of count value equal to the average of the noisy image to be restored; (2) Choose a value of $z_{\alpha/2}$ in Eq. (9.6-11) that, from tests, is large enough to significantly improve the outcome, but not so large as to force the POCS algorithm to stagnate at the initialization ($z_{\alpha/2} = 13$ was used in the experiment); (3) Carefully choose the size of the region in which to take the count M. The region should be small enough to preserve detail but large enough to include a reasonable value of M. Another option is to use an adaptive window, which always includes an *a priori* determined count ($M = 5000$ was found to be reasonable in the experiment).

The POCS restoration using the parameters cited above is shown in Fig. 9.6-3(c). The resulting image shows a significant improvement over that obtained by the Hamming filter in Fig. 9.6-3(b). Many more details about this experiment can be found in the thesis by J. Wurster [26] and in [27].

9.7   SUMMARY

In this chapter we demonstrated some applications of projection methods to image processing. We began by showing how vector-space projection methods could be used to design a class of linear filters for smoothing noisy images. Next we showed how the projections could be used to design signals that, after passing thorough the distorting media, yield the desired results. The digital restoration of blurred and noisy image by projections using constraints based on the Wiener filter was discussed next. This was followed by a demonstration of how projections could be used to extract a high-resolution image from a sequence of low-resolution images in which there is interframe displacement due to motion. Finally, we addressed the problem of restoring quantum-limited images using convex projections and demonstrated a projection-based restoration of the Shepp-Logan phantom of a head section.

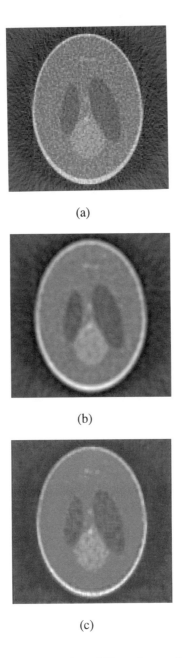

(a)

(b)

(c)

Fig. 9.6-3 Quantum-limited reconstruction of Shepp-Logan phantom: (a) FCBP reconstruction using a ramp filter; (b) FCBP reconstruction using a Hamming filter; (c) POCS
reconstruction using adaptive filtering.

REFERENCES

1. R. C. Gonzalez and R. E. Woods, *Digital Image Processing*, Addison-Wesley, Reading, MA, 1992.

2. G. T. Herman, *Image Reconstruction from Projections*, Springer-Verlag, New York, 1979.

3. H. Stark, ed., *Image Recovery: Theory and Application*, Academic Press, Orlando, FL, 1987.

4. H. Stark and E. Olsen, Projection-based image restoration, *J. Opt. Soc. Am.* A, **9**(11): 1914–1919, 1992.

5. B. E. A. Saleh, Image synthesis: discovery instead of recovery, Chapter 12 in *Image Recovery: Theory and Application*, (H. Stark, ed.), Academic Press, Orlando. FL, 1987, pp.448–463.

6. G. W. W. Stevens, *Microphotography*, Butler and Tanner, London, 1968.

7. M. D. Levenson, N. S. Viswanathan, and R. A. Simpson, Improving resolution in photolithography with a phase-shift mask, *IEEE Trans., Electron Devices*, **ED-29**:1828, 1982.

8. D. S. Goodman, M. D. Levenson, H. Santin, and V. Viswanathan, Improved photolithographic resolution with a phase-shifting mask, In *Proceedings, Microcircuit, 1982, the 1982 Grenoble Photolithography Conference.*

9. E. Hecht and A. Zajac, *Optics*, Addison-Wesley, Reading, MA, 1974.

10. B. E. A. Saleh and K. M. Nashold, Image reconstruction: Optimum amplitude and phase masks in photolithography, *Applied Optics*, **24**:1432–1437.

11. M. I. Sezan and H. J. Trussell, Prototype image constraints for set-theoretic image restoration, *IEEE Trans. Signal Process.*, **39**:2275–2285, 1991.

12. N. Wiener, *The Extrapolation, Interpolation and Smoothing of Stationary Times Series*, John Wiley & Sons, New York, 1949.

13. A. K. Jain, *Fundamentals of Digital Image Processing*, Prentice-Hall, Englewood Cliffs, NJ, 1989.

14. H. Stark and M. I Sezan, Image processing using projection methods, in *Real-Time Optical Information Processing*(J. Horner and B. Javidi, eds.), Academic Press, Orlando, FL, 1994, pp.185–232.

15. H. Stark and P. Oskoui, High-resolution image recovery from image-plane arrays, using convex projections, *J. Opt. Soc. Am. A.* **6**:1715–1726, 1989.

16. A. M. Tekalp, M. K. Ozkan, and M. I. Sezan, High-resolution image reconstruction from lower-resolution image sequences and space-varying image restoration,

in *IEEE Int. Conf. Acoust., Speech, Signal Process.*, San Francisco, CA, March 23-26, 1992, pp. III-169–172.

17. S. V. Fogel, Estimation of velocity vector field from time-varying image sequences, *Comput. Vision, Image Proc., Image Understanding*, **53**:253–287, 1991.

18. M. N. Wernick and G. M. Morris, Image classification at low light levels, *J. Opt. Soc. Am.* A **3**:2179–2187, 1986.

19. X. Shi and R. K. Ward, Reconstruction of photon-limited stellar images by generalized projections with use of the self-cross spectrum," *J. Opt. Soc. Am.* A **11**:1589–1598, 1994.

20. M. J. Northcott, G. R. Ayers, and J. C. Dainty, Algorithms for image reconstruction from photon-limited data using the triple correlation, *J. Opt. Soc. Am* A **5**:986–992, 1988.

21. C.-T. Chen, V. E. Johnson, W. H. Wong, X. Hu, C. E. Metz, Bayesian image reconstruction in positron emission tomography, *IEEE Trans. Nucl. Sci.* **NS-38**:687–692, 1990.

22. L. Silberstein, Quantum theory of photographic exposure, *Phil. Mag.*, **44**, 1922.

23. A. Papoulis, *Probability, Random Variables and Stochastic Processes*, McGraw-Hill, New York, 1965.

24. H. Stark and J. W. Woods, *Probability, Random Processes, and Estimation Theory for Engineers*, 2nd ed., Prentice-Hall, Englewood Cliffs, NJ, 1994.

25. L. A. Shepp and B. F. Logan, The Fourier reconstruction of a head section, *IEEE Trans. Nucl. Sci.*, **NS-21**(3):21–43, 1974.

26. J. L. Wurster, Enhancement and pattern recognition of photon-limited images, Ph.D. Thesis, Department of Electrical and Computer Engineering, Illinois Institute of Technology, May 1995.

27. H. Stark, J. L. Wurster, and Y. Yang, Restoration of quantum-limited images by convex projections, *J. Opt. Soc. Am. A*, **12**:2586–2592, 1995.

EXERCISES

9-1. Work out the details in going from Eq. (9.2-2) to Eq. (9.2-3), i.e., show that the projection onto $C(\mu)$ is indeed given by Eq. (9.2-3).

9-2. Show that $I(\gamma)$ in Eq. (9.2-5) has the property that $I'(\gamma) < 0$ for $\gamma > 0$. The prime indicates the derivative with respect to γ.

9-3. Explain why $I'(\gamma) < 0$ makes it easy to use the Newton-Raphson method to find the appropriate value of γ.

9-4. Prove that the projection onto $C_2(\mu)$ of Eq. (9.2-10) is given by Eq. (9.2-11).

9-5. Show that the projection onto $C_3(\mu)$, i.e., the set of constrained Laplacian norms, is given by Eq. (9.2-15).

9-6. Show that the sets $C_{SL}(X), C_{BL}(B), C_{AL}(a,b)$, and $C_S(f_d, \alpha)$, defined in Section 9.2, are convex and give their associated projectors.

9-7. Show that for the set C_2, defined in Eq. (9.3-14), the appropriate projection is given by Eq. (9.3-15).

9-8. It is desired to create a grating by imaging another grating $f(x) = rect[x/0.05] \star \sum_n \delta(x - 0.1n)$, where $\star$ denotes convolution, through a diffraction-limited coherent optical system with bandwidth $B = 10$ cycles per millimeter (cpmm). The grating to be designed is given by $g(x) = rect[x/0.02] \star \sum_n \delta(x - 0.1n)$. The ideal coherent optical system has transfer function $H(\nu) = rect[\nu/11]$, where ν is spatial frequency measured in cpmm. In the image plane there is available a point nonlinearity described by $u(r - \theta)$, where $r = r(x)$ is the output of the ideal coherent optical system. Compute $r(x)$ and determine the appropriate bias θ so that $g(x)$ can be realized.

9-9. Show that the solution to the problem stated in Eq. (9.4-3) is the *Wiener filter* of Eq. (9.4-4).

9-10. Prove that the projection onto C_w of Eq. (9.4-12) is correctly given by Eq. (9.4-18).

9-11. Show that the composition $\tilde{P}$ in Eq. (9.4-19) is equivalent to the projector onto the set C_n in Eq. (9.4-20).
Hint: use Parseval's theorem.

9-12. Show that the set C_{nn} in Eq. (9.4-21) is closed and convex and compute its associated projector.

9-13. Repeat Exercise 9-12 for the set $C_s(k, l)$ defined in Eq. (9.4-24). Show that the associated projector is given by Eq. (9.4-25).

9-14. Show that the composition $\tilde{P}_s$ in Eq. (9.4-26) is equivalent to the projector onto the set C_s in Eq. (9.4-27).

9-15. Prove convexity and closure of the set $C_j(m, n)$ in Eq. (9.5-9). Also demonstrate that Eq. (9.5-11) is the correct expression for the projection.

9-16. Derive the $100(1 - \alpha)$ percent confidence interval given in Eq. (9.6-13).

9-17. Prove that Eq. (9.6-19) is the projection onto the *rate-constraint* set in Eq. (9.6-17).

9-18. The one-dimensional equivalent of the set C_{kl} in Eq. (9.6-18) is $C_k = \{\boldsymbol{\lambda} : \sum_i \lambda(i)a(k - i) \in [b_1, b_2]\}$. Compute the projection onto C_k for arbitrary real $a(i)$.

Index

Absolute displacements, 10
Absorption spectra, 291, 293
Adaline learning rule, 346
Adaptive sampling, 206
Aliasing artifacts, 170
All-optical methods, 281
All-pass coefficient vector, 234
All-pass filter, 233
All-pass phase equalization filter, 223
Amplitude, 271
Amplitude-limitedness, 367
Analytic continuation, 71
Analytic functions, 214
Angle between two vectors, 13
Angle constraint, 336
Antenna design, 285
Antennas, 288
Appropriate constraint sets, 325
Arbitrary initialization, 330
Associative content-addressable memory, 321
Astronomy, 285
Asymptotically regular operator, 49, 51, 54
Autoconnections, 323
Autocorrelation property, 253
Averaging operator, 377
Back-propagating, 341
Back-propagation learning, 337, 342, 349, 353
Band-limited function, 70, 120, 207, 217, 274
Band-limited signal, 211, 218
Band-limitedness, 215, 367
Bandpass filters, 221

Bandwidth constraint, 120, 207
    Closedness, 121
    Convexity, 120
    Projection, 121
Bandwidth expansion, 249
Bandwidth, 70, 120
Basis, 5
BDCT matrix, 266
Beam forming, 281, 285, 288, 302, 312
Binary code, 213
Binary imaging, 372
Binary symmetric channel, 332
Bit error rate, 261
Black-lung disease, 239
Blind deconvolution, 23, 184, 281, 295–296, 312
Blocking artifacts, 263, 265, 268, 272, 274
Blurred images, 221
Blurring function, 301
Boundary point, 38
Bounded sequences, 11
Bounded value constraint, 118
BPLR, 349, 353, 357
Carrier modulation, 213
Cauchy sequence, 8
Central Limit Theorem, 244, 392
Channel coding, 213
Chessboard distance, 11
Circulant matrix, 300
City-block distance, 11
Classical Fourier series, 26
Classifier, 331

Closed and convex set, 60
Closed convex set, 42, 53
Closeness to convergence, 66
Closure, 41, 53
Code division multiple access, 250
Coherent optical processing, 284
Color matching by N sensors, 292
Color matching, 281, 291, 294, 312
Column projector, 272
Communication electronics, 205
Communications, 205
Compact support, 283
Complement of a set, 2
Composition projector, 272
Composition, 49, 327
Compositions of projection operators, 260
Computed tomography, 162
Computed tomography
   fan-beam, 163
   parallel-beam, 163
Computer-aided tomography, 150, 363
Computerized tomography, 389
Concatenation of operators, 49
Conditional projector, 218
Cone, 113, 291
Confidence intervals, 244
Constrained optimization problem, 246
Continuous functions, 215
Continuous-valued Hopfield ACAM, 330
Contraction mapping, 300
Convergence
   in norm, 26
   weak, 26
Convex hull, 88
Convex projection algorithm, 57
Convex set, 34
Convex wedges, 236
Convexity, 34
Convolution, 209
Correlation, 17
   normalized, 17
Cost functions, 331
Countable orthonormal basis, 24
Cramer's rule, 134
Crisp set, 244
Cross-talk, 211
Crystal detectors, 389
Cycles, 217
Data sets, 68
DCT, 263
Delta rule, 339–340
Demodulation, 251
Density-log exposure curve, 390
Derivative norm, 364
Design of diffractive elements, 312
Design of FIR filters, 223
Design of IIR All-pass Filters, 233–234

Design problems, 69
Despreading operation, 251
Detector footprint, 170
Detector response function, 171
Diffraction-limited image, 283
Diffraction-limited imaging system, 370
Diffractive element, 302, 304, 306, 309
Diffractive optical element, 302
Diffractive optical elements
   design of, 302
Digital communication systems, 221
Digital communications, 211
Digital filter, 221
   finite impulse response, 223
   impulse response, 223
   infinite impulse response, 223
   linear phase, 223
   non-dispersive, 233
   non-recursive, 223
   step response, 231
Digital image, 265
Digital signal processing, 221
Digital signals, 211
Direct sequence spread-spectrum, 250
Discrete cosine transform
   in JPEG, 263
Discrete Fourier transform, 378
Discrete time, 223
   continuous amplitude, 211
Discrete-time discrete-amplitude, 213
Discrete-time signal, 211
Distance of a point to a set, 185
Distance type constraint, 102
Distance type constraint
   Closedness, 103
   Convexity, 102
   Projection, 103
Distance, 7
Distance-from-origin classification problem, 353,
   357
Distance-from-origin classification, 348
Distance-from-the-origin problem, 337
Distributed processor, 320
Dot product, 13, 135
Double wedge, 236, 238
Dumb vs. smart noise smoothing, 364
Electron microscopy, 285
Emission tomography, 364
Energy functional, 323
Energy preservation property, 24
Energy type constraint, 105
Error-type constraint, 102
Estimation of Channel Attenuation, 259
Estimation of pdf, 241
Euclidean distance, 135
Euclidean inner product, 14, 21
Euclidean norm, 9, 135

p-norm, 9
Euclidean space, 21, 25, 57
Examples of vector spaces
   R^n, 4
   real polynomials, 4
   real-valued functions, 5
Exclusive-or problem, 337
Exemplars, 331
Expectation, 377
Extrapolated parallel projection method, 82
Extrapolation, 285
FCBP, 394
Feasible reconstructions, 207
Feasible solution, 299, 336
Feed-forward net, 319, 321
Filtered-convolution back-projection, 394
Finite pupil effect, 282
Finite-precision arithmetic, 221
FIR filter design, 227
Fisher Discriminant direction, 336
Fixed point of an operator, 39
Fixed point, 46, 50
Fogel motion estimation algorithm, 387
Fogel motion estimation, 385
Four-fold symmetric filter, 228
Fourier expansion formula, 24
Fourier expansion, 26
Fourier magnitude constraint, 184, 305
Fourier magnitude, 286, 289
Fourier transform pair, 119–120
Fourier transform vector, 379
Fourier transform, 119, 298, 305
   forward, 119
   inverse, 119
   magnitude, 120
   phase, 120
Fourier-field, 304
Frequency modulation, 249
Frequency-domain constraints, 119
Frequency-domain, 285
Fresnel diffraction integral, 304
Fresnel diffraction law, 302
Fresnel diffraction pattern, 305
Fresnel domain, 310
Fresnel lens, 307
Fresnel region, 304
Frobenius norm, 343, 386
Function pairs, 20
Fundamental theorem of POCS, 49
Fundamental Theorem of POCS, 86
Fundamental theorem of POCS, 336
Fuzzy set, 244
Gaussian density, 242
General content-addressable memory, 322
General interconnection matrix, 325
General restoration algorithms, 287
Generalization of the Euclidean norm, 22

Generalized delta rule, 346
Generalized projection algorithm, 196
Generalized projections, 181, 187, 281, 298, 324
   extension of, 197
   Fundamental theorem of, 189
   in a product space, 192, 196
Generalized Projections
   performance measure, 188
Generalized projector, 187
Generalized-projection algorithm, 299–300
Geometric image, 283
Geometrical principles, 282
Gerchberg-Papoulis algorithm, 67, 72
Gerchberg-Saxton algorithm, 188, 281
Global minimum, 299–300
Gradient-projection method, 300
Gram-Schmidt orthonormalization, 16
Gram-Schmidt process, 31, 135, 148, 151–152
Grayscale image, 271
Greedy path, 83
Ground-to-air missile, 239
Group delay, 233
Half-tone images, 369
Hamming distance, 332, 361
Hard linear constraint, 98
Hertz, 70
Hidden neuron, 342
Hidden-layer neuron, 339
High gamma film, 370
High-contrast detection, 372
High-resolution image realization, 364
High-resolution image, 378
Hilbert space, 21–23, 25–26, 92
Hilbert spaces, 21
Hopfield ACAM, 331
Hopfield dynamics, 325, 327
Hopfield model, 331
Hopfield net, 185, 319–323
   energy minimizing property, 323
   library vectors, 322
   stable states of, 327
Hopfield neural network classifier, 331
Huffman encoding, 264
Human visual system, 264
Human voice, 221
Hypercube, 330
   L-dimensional, 330
Hyperplane, 94, 237
Ideal geometric image, 282
Idempotent property, 259
Identity operator, 46, 287
Image compression, 262
Image deblurring, 364
Image decoding, 266
Image discovery, 290
Image fusion, 206
Image processing, 363

Image reconstruction, 363
Image restoration, 68, 367
Image sharpening, 281, 377
Image synthesis, 369–370
Impulse response vector, 224
In-phase components, 259
Independent
   identically distributed random variables, 244
Induced Euclidean norm, 22
Induced norm, 15
Induction, 52
Informational loss, 212
Inner product spaces, 13, 29
Inner product, 13, 25
Inner-product convergence, 185
Integrable function, 12
Intensity point-source, 283
Intensity range, 297
Interference rejection, 253, 274
Interference, 252
Interframe displacement, 395
Interframe motion, 382
Interpolating functions, 207
Intersecting hyperplanes, 238
Intersection set, 2
Irreducible nonsymmetric factor, 286
Iterative algorithm, 210
Iterative gradient search algorithm, 337
Jammer noise, 252
Jordan-von Neumann's Theorem, 15
JPEG compression, 263, 265
Key sets, 207
Known phase constraint, 122
   Closedness, 123
   Convexity, 122
   Projection, 124
Known segment constraint, 117
   Closedness, 117
   Convexity, 117
   Projection, 118
Known support constraint, 115
   Closedness, 116
   Convexity, 116
Lagrange functional, 40, 106, 141, 207, 216, 226,
   246, 269, 294, 344, 365, 386
Lagrange multiplier's method, 106, 141, 365
Lagrange multipliers, 278, 293
Laplacian L^2 norms, 367
Laplacian, 367
Large-scale sparse matrix, 150
Largest lower bound, 11
Learning process, 320
Learning rate parameter, 337, 346
Least upper bound, 11
Least-squares solution, 134, 157, 326
Lens law, 282
Level of similarity, 110

Levi's generalized projection theorem, 327
Lexicographic ordering, 167, 265
Library vector, 325
Library vectors, 319, 328
Limit inferior, 74
Limit points, 41
Limit superior, 74
Linear combinations of library vectors, 328
Linear constraint set
   in L^2, 95, 97
   in R^n, 94
Linear discriminant function, 348
Linear discriminant functional, 333
Linear equality, 98
Linear inequality, 98
Linear phase, 223, 225
Linear type constraint, 91
   Closedness, 92
   Convexity, 92
   Projection, 93
Linear variety, 55–56, 93, 118, 121, 159, 210, 294
Linear vector space, 2
Linear-phase FIR filter, 231
Linearly separable classes, 333
Linearly separable with a margin M, 334
Local minimum, 337
Low-resolution data, 170
Low-resolution image, 378
Lowpass filters, 221
Magnetic resonance imaging, 150, 363
Magnification, 282
Margin, 333
Matched colors, 292
Maximal-length linear shift register sequences,
   253
Maximum a posteriori decision rule, 241
MAXNET, 328
Medical imaging, 363
Medium-resolution image, 378
Microlithography, 370
Minimal disturbance principle, 346
Minimum distance property, 156
Minkowski's inequality, 9, 12
Mixed convergence theorem, 244
Mixed projection algorithm, 299, 303
Monochromatic-plane-wave, 304
Monochrome reflectivity, 171
Multi-channel redundancy, 263
Multi-layer feed-forward net, 337
Multi-layer feed-forward neural network, 357
Multiple constraint, 294
Neural nets, 319
Neural network, 320
Neuron activation function, 346
Neuron, 320
   bias, 320
   threshold, 320

Neurons, 320
Newton's method, 107, 127
Newton-Raphson method, 127, 365, 367
No-match condition, 330
Nodes, 322
Non-causal filter, 377
Non-causal Wiener filter, 378
Non-closed set, 41
Non-convex set, 182
Non-diffracting beam, 309, 311
Non-expansive operator, 48, 50, 257
Non-negativity constraint, 114
Non-negativity constraint, 380, 386
Non-negativity constraint
   Closedness, 114
   Convexity, 114
   Projection, 115
Non-uniform samples, 274
Non-uniform sampling, 205
Nonlinear activation function, 348
Norm, 7
   L^2, 12
   as a measure of energy, 12
Normed spaces, 29
Normed vector space, 7
Norms
   l^p, 11
   induced, 14
Objective rule, 239
One-step algorithm, 210
One-step color matching projector, 294
One-step projection, 209
Open set, 267
Operator
   asymptotically regular, 49
   fixed point of, 39
   range of, 53
Opial's theorem, 72
   proof of, 72
Optical interconnect, 289
Optics, 285
Order of the filter, 234
Orthogonal complement, 45
Orthogonal components, 259
Orthogonalization learning rule, 324, 327, 357
Orthonormal bases, 25
Orthonormal basis, 24
   countable, 24
   maximal, 24
Orthonormal set, 17
Outer-product learning rule, 323
Overshoot constraint, 232
Parallelogram law of norm, 15
Parallelogram law of vectors, 3
Parseval's formula, 24, 26, 28
Parseval's identity, 123
Parseval's relation, 120

Parseval's theorem, 207, 311, 366, 399
Passband, 225
Penalized likelihood estimation, 248
Per-step optimization, 337
Perceptron element, 331
Perceptron learning algorithm, 333–334
Perceptron, 331, 333
Phase equalization, 233, 240
Phase modulation, 249
Phase recovery, 68, 312
Phase retrieval, 281, 285
Phase-retrieval problem, 312
Photographic film, 363
Photolithography, 364
Photon noise, 389
Photoresist, 370
Pixel, 166, 378
PMLR, 349, 353, 357
POCS algorithm, 57
   pure, 57
   relaxed, 57
POCS, 33
   non-intersecting sets, 83, 85
   parallel projections, 78, 81–82, 85
   fundamental theorem of, 57
Point constraint, 381
Point spread function, 382
Poisson law, 391–392
Poisson rate parameter, 391–392
Positron-emission tomography, 150, 363, 389
Post-processing, 215
Power spectrum, 375
Pre-Hilbert space, 21
Predetermined margin, 335
Prior knowledge, 69, 268
Probability density function, 241
Product spaces, 33
Projection algorithm
   parallel or simultaneous, 81,
   sequential
      serial, 81
Projection formulation, 328
Projection onto a closed subspace, 44
Projection operator, 38
Projection, 38
Projection
   existence, 35
   uniqueness, 35
Projection-based deconvolution, 302
Projection-based learning, 338, 345, 353
Projections onto convex sets, 33, 86, 181
Projector, 38
Projector
   linear, 53
Property sets, 69
Prototype image, 381
Pseudo-noise spread-spectrum, 250

Pulse-code modulation, 249
Pupil function, 283
Pure projection algorithm, 60
Quantization, 266
Quantized sequence, 214
Quantizing errors, 274
Quantizing, 205, 212
Quantum noise, 389
Quantum-limited image restoration problem, 392–393
Quantum-limited images, 364, 389, 395
Quasi-monochromatic plane wave, 302
Radar, 288
Radon transform, 163
Random variable
    Bernoulli, 244
    exponential, 242
    Gaussian, 242
Rate of convergence, 369
Rate-constraint set, 393, 399
Raysum, 163
Real functions, 207
Reasonable wanderer, 53
Recovery problem, 68, 290
Recovery process, 69
Rectification, 115
Reference vector, 110
Relaxation parameter, 47, 58, 334
Relaxation parameters
    choice of, 64
Relaxed projection algorithm, 58, 60, 65
Relaxed projector, 46–47, 65, 87
    over-project, 47
    under-project, 47
Relaxed projectors, 219
Remez exchange algorithm, 227–228
Reproducer level, 216
Resolution enhancement, 170
Restoration by convex projections, 378
Restoration from magnitude, 187, 285
Row projector, 272
Sampling grid, 164
Sampling lattice, 164
Sampling, 205
Scaling of a set, 89
Scanning, 174, 265
Schwarz inequality, 14, 189
SDE reduction property, 328
Selenium, 389
Self-convergence, 324
Separating hyperplane, 333
Separating weight vector, 333
Sequence
    bounded, 11
    convergence, 8
    limit, 8
Set boundaries, 218

Set in a vector space, 2
Set
    intersection, 2
    union, 2
    boundary point, 8
    bounded, 83
    closed, 8
    closure, 8
    complement of, 2
    elements, 2
    interior point, 8
    limit point, 8
    open, 8
    orthogonal, 16
    orthonormal, 16
Set-theoretic learning rule, 338
Shepp-Logan phantom, 168, 173
Sigmoidal non-linearity, 331, 342
Signal oversampling, 213
Signal processing, 205
Signal recovery, 70
Signal synthesis problem, 375
Signal-to-interference ratio, 261
Signal-to-noise ratio, 261, 391
Similarity constraint, 109
Similarity Constraint
    Closedness, 110
    Convexity, 110
    Projection, 111
Similarity, 368
Simply-connected region, 183
Single layer perceptron, 331
Singleton set, 58
SNR, 391
Soft constraint set, 393
Soft linear constraint, 98
    in L^2, 100
    in R^n, 100
Soft Linear Type Constraint, 97
    Closedness, 99
    Convexity, 98
    Projection, 99
Solution set, 138
Solutions of Linear Equations, 133
Source coding, 213
Source signal, 211
Space domain, 285–286
Space-limitedness, 367
Span of a set, 5
Span, 16–17, 326
Sparrow's criterion, 374
Spatial light modulator, 289
Spatial redundancy, 263
Spatial support, 301
Spatial-frequency domain, 286
Spatial-light modulator, 302
Spatially band-limited, 369

Spectral intensity, 291
Spectral redundancy, 263
Spectrum extrapolation, 205
Spectrum-spectrum signals, 254
Spread spectrum, 249
Spread-spectrum systems, 253, 274
Spreading-code sequence, 250
Spurious-stable states, 328
Stagnation level, 349
Standard basis of R^n, 6
Standard normal distribution, 392
Stationary point, 300
Step response, 231
Steps, 217
Stopband, 225
Stored library vectors, 331
Straight-line distance, 10
Strong convergence, 27–28
Subspace, 2
Summed distance error convergence property, 298
Summed distance error convergence, 238, 372
Summed distance error property, 306
Summed distance error, 185, 202
Super-resolution, 72, 281, 283, 285, 312
Support, 116, 243, 297, 370
Synaptic weights, 320
Synthesis problems, 70
System function, 224
System of linear equations, 133
   consistent, 134
   critically determined, 133
   decomposition into subsystems, 159
   inconsistent, 134
   modified, 149
   over-determined, 133
   solution, 133
   under-determined, 133
Tangent plane, 43
   normal to, 43
   of a set, 43
Taylor series expansion, 70
Temporal redundancy, 263
The JPEG DCT Algorithm, 263
Thermal noise, 252
Three-color theory, 291
Threshold constant, 370
Time-domain constraints, 114
Toeplitz matrix, 300
Transfer function, 234
Transition band, 225
Translation of a set, 88
Trap, 181–182, 192, 299, 306
   attractive, 181
   repulsive, 181
Trap-states, 328, 330
Traps, 225, 298, 357
Triangle inequality, 102, 379

of distance, 7
of norm, 7
Triple star, 300
Tunnel, 145, 192
Two line resolution distance, 374
Two-class
   linearly separable problem, 334
Two-layer feed-forward net, 320
Two-layer network, 346
Two-layer neural net, 347
Uniform sampling theorem, 211
Unilateral constraint, 100
Union set, 2
Uniquely decipherable code, 213
Unit-sphere, 10
   for $p = 1$, 10
   for $p = 2$, 10
   for $p = \infty$, 10
Unitary property, 266
Update rule, 323
Vector space, 1
   complex, 2
   real, 2
   addition, 1
   complete, 8
   dimension of, 6
   finite-dimensional, 6
   infinite-dimensional, 6
   scalar multiplication, 1
   scale multiple of a vector, 1
   sum of vectors, 1
   zero vector, 2
Vector spaces
   L^2 function pairs, 20, 23
   L^2, 19
   l^2, 19, 22
   L^2, 23
   l^2, 25, 28
   l^p, 11
   L^p, 11
   R^n, 9
Vectors
   linear combination of, 5
   linear dependent, 5
   linearly independent, 5
   orthogonal, 13, 16
   perpendicular, 13
View angle, 163
Wavelength, 302
Weak convergence, 26, 28, 66, 185
Weak limit, 26
Weight vector, 333, 336
Weighted error, 105
Weighted Euclidean inner product, 30
Weighted Euclidean norm, 9
   weighting factors, 9
Weighted least-squares, 86

Weighted-error type constraint, 105, 107
White Gaussian noise, 387
Whittaker-Shannon-Kotelnikov sampling formula, 206
Wiener filter, 399
  inverse filter, 377
  LMS, 375

  non-causal, 375
  smoothing, 377
Windowed kernel estimation, 248
X-ray crystallography, 285
XOR problem, 357
Zorn's lemma, 24